ABYSMAL

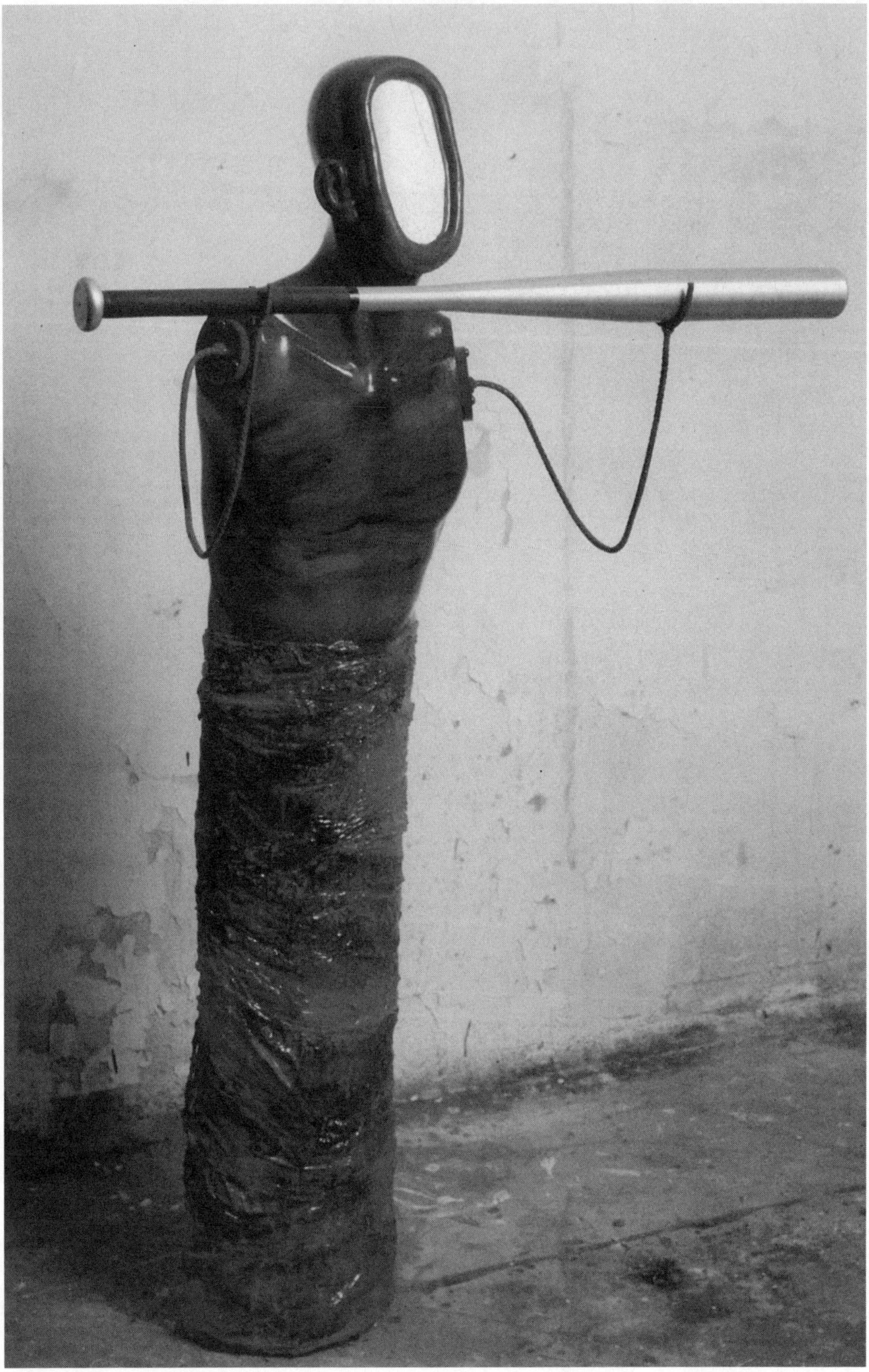

Kent Karlsson, *Absolut kunskap/Frihetens väsen.* 1992. Sculpture, mixed media, h. 127 cm. Bob Kelly Collection, Los Angeles. 

# ABYSMAL

## A CRITIQUE OF CARTOGRAPHIC REASON

*Gunnar Olsson*

*The University of Chicago Press* Chicago & London

GUNNAR OLSSON is professor emeritus of economic geography at Uppsala University.

The University of Chicago Press, Chicago 60637
The University of Chicago Press, Ltd., London

Printed in the United States of America

16 15 14 13 12 11 10 09 08 07 5 4 3 2 1

The publication of this book was supported by a grant from Riksbankens Jubileumsfond, Stockholm.

ISBN-13 (cloth): 978-0-226-62930-8
ISBN-10 (cloth): 0-226-62930-9

Library of Congress Cataloging-in-Publication Data

Olsson, Gunnar, 1935–
Abysmal : a critique of cartographic reason / Gunnar Olsson.
p. cm.
Includes bibliographical references and index.
ISBN 0-226-62930-9 (cloth : alk. paper)
1. Philosophy, Modern—20th century. 2. Philosophy, Modern—21st century. 3. Cartography. 4. Rhetoric. 5. Reason. I. Title.
B804.047 2007
190—dc22
2006016776

♾ The paper used in this publication meets the minimum requirements of the American National Standard for Information Sciences—Permanence of Paper for Printed Library Materials, ANSI Z39.48-1992.

This book is printed on acid-free paper.

# CONTENTS

# CONFESSION

# DESIRES NON-SUPPRESSED

Stand in awe and sin not. *Psalm 4:4*

**Living forwards and understanding backwards**—a crucial condition of what it means to be human. Therefore it was with mixed feelings that on the first day of the ninth month of the third millennium (AD) I found myself promoted to the rank of emeritus, a lifelong sabbatical inching towards the ultimate. And as so many others in a similar situation, I too felt the urge of staging some kind of retrospective, in my case an exhibition of scattered works from the previous decade. Not merely a retrospective, though, but a self-referential catalogue raisonné to go with it. Such was the original idea of cut-and-paste, not so bad and seemingly not so cumbersome.

But ideas are ideas and life is life and what initially was conceived as a backward-looking *retro*spective gradually turned into a forward-looking *pro*spective. For not only did the 9/11 of 2000 come with the conventional birthday gifts but it coincided with the first public show of *Mappa Mundi Universalis,* an abstract sculpture by Gunnael Jensson, close acquaintance. Now, in hindsight, I think the present volume may be read as a record of the silent conversations I have subsequently had with this material expression of desires non-suppressed.

And where has all that got me? To yet another game of American football and to yet another chiasm of thought-and-action![1] A kick between the goal posts.

The beginning of that story reaches far back, at least to the early 1960s, when as a curious research student I had just been recruited to the Regiment of Quantitative Geography, lured into the ranks less by the siren songs of power, more by the beauty of the Jean Arp-like models. Politics, however, was never my cup of tea, perhaps because I felt an urge to do it my way, perhaps because the zealots seemed never to laugh. Rather than joining the self-congratulating crowd I gradually came to see Swedish politics not as the profitable farce it may well be but as an updated version of classical tragedy—everything beautifully right at the beginning; everything horribly wrong at the end; no one to blame in between.

To understand what eventually followed, the present book included, one should know that even though the tragic developments often made me furious, they never got me depressed. On the contrary, for in my youthful excitement they presented the social sciences with a tremendous challenge, easier to state than to do anything about. To be precise: if human action is structured like a tragedy, how can we then rely on the principle of truth preservation for tying our premises and conclusions together? Surely the most common purpose of human action is to topple truth, not to preserve it, to falsify rather than preserve what is now the case. Less a matter of formal logic more an instance of creative imagination.

If, as it seems, the realities of power cannot be unambiguously rendered in the syntax of conventional reason, does that mean that if one wants to accumulate knowledge about human action, then one must find a language better suited to the task? The answer is yes, except that answer does itself create problems of another and equally serious kind. The reason is that it is not enough to render the world as it is; I must also be believed when I do it. It is exactly this issue of logic and rhetoric that has been at the core of my entire life, the activities of teaching and research inseparably united. The freshman and the emeritus are in it together, the reflections in the mirror not many but one.

This is not the occasion to repeat how the space cadet eventually turned into a cartographer of the taken-for-granted.[2] Suffice it to say that out of the original models of spatial form rose the question of how the highly predictable patterns of human interaction should be interpreted. Which *are* the relations between the picture of a point pattern and the story of how those particular locations have been generated? Is it at all possible to draw valid inferences about human behavior from a map of spatial distributions, to reason from form to process?

The short answer is that such inferences are not valid, a conclusion which in 1968 hit the disciplinary ship of Geography below the waterline. The trustworthy truth is in fact that from knowledge of the physical outlay of a prison one may well predict the behavior of both the inmates and the wardens, but neither understand it nor explain it; what I happen to capture in my allegedly objective analyses does almost certainly reveal more about the structure of the net I have thrown than about the fish I set out to catch. Therefore, show me your map and compass and I shall show you who and where you are, whence you came and where you are heading.

The prison/fisherman metaphor is chosen deliberately, for social democratic welfare theory is a special case of utilitarianism and thereby of the principle of happiness maximization. But an integral part of that same conception is also that the goodness of a given act should be judged in terms of its consequences, not in terms of the actor's intentions. And when Swedish politics is evaluated on that basis, then the social engineers may well find themselves not in the heavens to which they aspire but in the dungeons they have designed for the others. When it comes down to it, the real question of guilt and punishment is whether the ideologues lost their way because they were scientifically naïve or because they were morally wrong, obsessed by power for the sake of power. Not so easy to determine, especially as Sweden for seventy-five years has effectively been a one-party state. A record hard to beat, the Cretan Liar the sole contestant.

Such is the setting of the present book. Now the arguments must speak for themselves, albeit with the tacit understanding that everything I have to say is deeply rooted in the past. In deed I am often less impressed by the splendid analyses of today than by the wisdoms of the broken clay tablets of Babylonia, the torn scrolls of the Old Testament, the reenacted dramas of Homer, Sophocles, Shakespeare.

*

**Like Ludwig Wittgenstein** in the preface of the *Tractatus,* I too would have liked to write a good book. Too late for that, especially as his attempt was surely good enough, mine nothing but a temptation impossible to resist. The alternative—occasionally tested and enjoyed—was a five-year vacation on the beaches, long hours in the Parisian cafés, peaceful nights in the Värmland forest, unmeasurable time in the arms of a grandchild. But if seventy years from now I somehow came to know that these four little creatures had stumbled onto some of the same treasures that so excited their morfar—and especially if they had done so without having read a word of what he wrote—then my beard would be salty. And could I but touch them, they would be mine again.

Not as fanciful as it might at first appear. For perhaps the best way to approach the present volume is to read it as a minimalist guide to the landscape of western culture. Around the world in one week, deluxe accommodations and three-star fare at the cost of asking. And even though the imagined readers must find their own way through the unknown, I would be surprised, nay saddened, if they were not taken aback by the breathtaking vistas of Uruk, Peniel, Thebes, and Nicaea. Perhaps even by their visit to the Philadelphia mausoleum.

Anyone who enters these lands travels on a tourist visa, the specialized expert no exception. Every reader—especially the specialized expert—is therefore bound to run into passages (s)he will initially find strange, frightening, blasphemous, wrong, stupid, merely incomprehensible. It is to soften that type of reaction that the numerous notes and the long bibliography are provided, a back-handed tribute to those who have crisscrossed the same areas before me, many of them better equipped and more adequately prepared. None, however, dressed in my shoes and adorned with my eyes, none with a map like mine, none with a compass of the same declination. No painter with my brush, palette and canvas, no poet of my rhyme and reason.

What remains is a genuine sense of awe, a stunned admiration for the insights into power and human relations that beam out from every line of *Enuma elish* and *Gilgamesh,* from every paragraph of Moses' first stone tablet, from the treasurous stories about Abr(ah)am, Jacob and Job. Add to this the abstract geometries of Plato and Euclid, the cunning tricks of Odysseus and the horrible fate of King Oedipus, and the humus of western culture lies there before us ready to be tilled. The Renaissance masters and the Orthodox icon-makers did it, Paul Cézanne and Mark Rothko, Marcel Duchamp and Thomas Pynchon as well.

Thus far and no farther, any attempt to paraphrase the trailblazers an act of inexcusable vandalism. Yet, what one cannot do perfectly, one must do as well as one can. And just as every boy is a child of his own time and place, so the present is a present, a *pharmakon* forged in the utopian smithy of Now-here.

*

**The present is present** because the present is steeped in the past. True friends have it all and that is why they always mean so much and sometimes say so little: Franco Farinelli and Alessandra Bonazzi in Bologna, Michael Watts and Allan Pred in the Bay area, Ole Michael Jensen and Jette Hansen-Møller in Copenhagen, Dagmar Reichert in Zürich, José Ramírez in Stockholm, Thomas Hård af Segerstad in Uppsala. The laughters from the Department of Human Geography at Uppsala keep rolling along, the nine months at the Center for Advanced Study of the Behavioral Sciences at Stanford a privilege unasked for yet graciously facilitated by Dan Brändström of the Tercentenary Fund of the Bank of Sweden. Doug, Michael and George (anonymity preserved) knocked on the doors of the University of Chicago Press, where T. David Brent invited me in, first to the angst-filled and abysmal antechamber, eventually to the splendors of the main hall. Drums beating, trumpets sounding. And I live happily ever after.

But no one should be blamed and no one feel absolved, for throughout the weaving the door of my chamber has been securely locked. Now, when others are let in, they should be warned that although power never sleeps in the same bed two nights in a row, its well-guarded residence is where it has always been and where it will always be: in the inter-esting abyss between categories, in the taboo-laden cleft of the excluded middle.

1050 Campus Drive
Annunciation Day 2005
in that particular year,
Good Friday too

Kåbo 23:7
September 11, 2005
as every year, happy
birthday as well

# PRELUDE

# BORDER-MAN

Behavior that's admired is the path to power
among people everywhere. *Beowulf*

**What does it mean to be human?** A godly question as impossible to answer as not to pose, a monstrous creature too evasive to capture, too forbidden to leave alone. In the meantime the river runs, and by a commodius vicus of recirculation we are once more brought back to the swerve of shore and bend of bay, to that magic theater on whose stage the human actors H. C. Earwicker (alias Here Comes Everybody) and Anna Livia Plurabelle are found again, a chaosmos in which categories ride the surf, traces are erased, drops turn to clouds, clouds change to drops. And in the self-referential mood of these opening sentences, everything that this book eventually will have to say has in a sense already been shown, in the same moment of now and then, in the same place of here and there. Every age is retrospective.

The cosmological constellations flash back and forth, not the least to *Enuma elish,* the oldest creation epic yet to be discovered, a drama originally projected onto seven clay tablets now preserved in several editions and various conditions of readability. To us the most fascinating characteristic of that tale is that in the beginning there is nothing at all, merely the two positions of above and below.[1] So primeval is this situation that it takes place at a time "when there was no heaven, no earth, no height, no depth, no name."[2] No nothing, not even, as in the younger Genesis, an earth without form and void.

In the conception of the Babylonian poet a void was consequently a void-in-and-of-itself, not merely a void of form. Then (as well as now) it was through the practice of naming that shapeless matter (more precisely fluid water rather than solid earth) changed into meaning-filled meaning. In the most literal of all available translations:

When on high the heaven had not (yet) been named,
(And) below the earth had not (yet) been called by a name;
(When) Apsu, primeval, their begetter,
Mummu, (and) Tiamat, she who gave birth to them all,
(Still) mingled their waters together,
And no pasture land had been formed (and) not (even) a reed marsh was to be seen;
When none of the (other) gods had been brought into being,
(When) they had not (yet) been called by (their) name(s, and their) destinies had not (yet) been fixed.[3]

*

**When they had not yet been called by their names** and their destinies had not yet been formed! Such was the beginning of the hazy beginning, such was the watery cocktail out of which eventually rose the undefined mist of Mummu, later described as "the creator of heaven (and) earth, who directs the c[louds]; . . . To whom no one among the gods is equal in power."[4]

In the next chapter we will return to this first tale of how the human territory is delimited, of how we find our way in the unknown by drawing on invisible maps of the invisible and by following cultural compasses whose needle points us not to the physical existence of the magnetic north pole but to the social subsistence of the culturally taken-for-granted. For the moment, however, all one needs to note is that the very first words of the first line of *Enuma elish*—the two words that have given the epic its name—are usually translated as "When above," while the beginning of the second line means "And below." It is these fix-points of time and place—*above, below, when,* and *and*—that form the meshes of the coordinate net in which the world is both captured and created. As Edward Casey loves to put it: "There is no creation without place."[5]

Here—as in any reconstruction—memory serves as the projection screen par excellence. Most importantly, since our historical consciousness is always filled with a variety of voices, memory tends to be longer than we think, a circumstance that explains why James Joyce could call history a nightmare from which he was trying to awake and why Immanuel Kant could claim that the ideas of space and time are not merely "intuitions" but "a priori intuitions," the seedbed of the synthetic a priori. Memory is consequently meaning accumulated and culture transferred,[6] the world of reality always a mediated world, never a collection of things-in-themselves. Just as the geneticists keep repeating that the human body is programmed for a life distinctly different from our own, so Bertrand Russell used to say that our ordinary language conveys the metaphysics of the Stone Age.

If it is true that memory is the intersection of mind and matter,[7] then it follows that the thought of a tabula rasa is literally unimaginable.[8] Likewise, if it were not for the resistance of a reflecting wall there would be no images to capture, hence nothing to understand either. The aim of the present study is therefore not to discuss the biological evolution through which *homo* became *sapiens,* a knowing man, but to explore those cultural practices which together make us sapiens.[9]

Throughout these expeditions into the Land-of-the-Knowing-Man my guiding star will be Søren Kierkegaard's aphorism that we live forwards and understand backwards. The Swede will nevertheless go one step further than the Dane and argue that to understand what it means to be human is to understand what it means to understand. As in the second line of *Enuma elish,* the power of Kierkegaard's observation in deed grows out of the small word "and," the conjunctive (initially classified not as a con-junctive but as a pre-position) by which the gropings of the existential actor are intricately bound to the ambiguities of the dialectical analyst. Most significantly Kierkegaard's entire philosophy may be read as an investigation of how this grammatical bridge of the "and" is constructed and thereby of how the abyss between the worlds of being and understanding is negotiated. And in the invagination of that abyss lies the palace of Apsu, the hidden but well-established seat of power.

Ideally I would like to become one with this preposition-turned-conjunctive.[10] At the same time I am well aware that in reality such a feat is impossible, not the least because every understanding, like every translation, is thoroughly ironic. And irony, of course, constituted the Socratic foundation of Kierkegaard's either-or technique of pseudonymous writing, itself a tacit admission that "the minimum number, in the strict sense of the 'number,' is two."[11] Expertly using that particular reasoning mode, the hunchbacked Dane could on one page demonstrate that dialectics is the most powerful of all tools for understanding the past, on another that the same reasoning mode has nothing to say about the future. The relations between the *Communist Manifesto* and the Soviet Gulag bear him out, for whereas the analyses of the former are easy to accept, the realities of the latter are impossible to excuse.

Similar relations between here and now, there and then, are constitutive not only of modern ideologies but of religious mythologies, especially when the connections are expressed through the figure of Janus, my own favorite among gods. What intrigues me with this pivoting symbol of gatekeeping is less that he is equipped with a body that makes him see in opposite directions at the same time, more that he has a mind which allows him to merge seemingly contradictory categories into one meaningful whole. From his watchtower at the middle of the bridge he is consequently able to keep both sides of the abyss under constant surveillance, in the same glance catching a glimpse of those pasts that once were and of those futures that have yet to come.

Given the Greek fear of the void, it is not surprising that Janus was invented in Rome and not in Athens. In the lands surrounding the *Mare nostrum* he was in fact everywhere to be seen, for not only was his image stamped on practically every coin, but in religious prayers this janitor was the first to be mentioned and in cultural rituals this son of January was equated with the beginning of all beginnings. Diana was his godly consort, a connection which explains why the doors of his temple stayed open in times of war and why they were shut in times of peace. Like ordinary lovers, gods need their privacy too.

Janus' main concerns were one with my own: creativity, power, socialization. Defiantly I therefore pray again,

> Oh Janus! Help me become a sinner. Let me understand how you break definitions and thereby create. Show me how you erase what others see as irresolvable paradoxes. Teach me the equation of that third lens inside your head whereby contradictory images are transformed into coherent wholes. Speak memory, speak! SPREACH, Janus, SPREACH! And the walls come mumbling down.[12]

Accordingly, and throughout the adventures of the present tome, I will be searching for a place inside Janus' head. From that zero-point of the excluded middle I will then try to grapple with the taboos of limits, the sins of trespassing, the braiding of ontology and epistemology. With the aim of understanding how Janus stayed sane while ordinary people in similar situations of double bind go crazy, I will try to place him on the operation table, cut his skull open, lay his brain bare, investigate how his mind is wired. Why, and how, for instance, did the Romans elevate this categorial juggler to godly status, when we diagnose his counterparts as schizophrenic madmen? Why did they afford him a special place in their pantheon, while we isolate his likes in the soundproofed cells of the asylums?

What does it mean to understand?

*

**To understand is to be immersed in language**, to live in the conjunction between one expression and another. At least in that context it is literally true that in the beginning is the word; as every Jew, Muslim, and Christian knows, the *Enuma elish* is not alone in preaching that without names there may well be sweet- and salt-water oceans, but neither gods nor rocks, neither knowledge nor understanding. Hence no humans either, for to be human is to be a very peculiar animal, member of a species whose individuals are kept together and apart by their use of signs. It is through them that we define and redefine who and what we are; it is through them that individual and society are inseparably connected.

And yet. Speaking and understanding is never enough. I want to be believed as well, I want to share my understandings with others. As Immanuel Kant put it in one of the most profound statements ever made:

> In the synthetic original unity of apperception, I am conscious of myself, neither as I appear to myself, nor as I am in myself, but only that I am. This

> *representation* is an act of *thought,* not an *intuition.* [It follows] that I have no knowledge of myself as I am, but only as I appear to myself. . . . As for the knowledge of an object different from myself, I require . . . an intuition also of the manifold in me.[13]

By necessity I partake of multitude, by necessity the many in me are one.

The main difficulty with this attitude lies in the explanatory gap between what might be called a *mind state* and a *brain state,* the former an issue of understanding human understanding, the latter a matter of understanding insensate molecules (itself, of course, a form of understanding).[14] Thus, while it is the business of imagination to connect, it is the purpose of understanding to make the principles of imagination explicit by relating them to the thesis that "the consciousness of my own existence is, at the same time, an immediate consciousness of the existence of other things outside me."[15]

To explicate this thesis of the necessary unity of consciousness is to specify the translation function through which I mirror myself in you and you in me. Kant's concern—like mine—is obviously not with the specification of objective rules, but with judgment, especially with the "subjective principles which are derived, not from the quality of an object, but from the interest which reason takes in a certain possible perfection of our knowledge of an object."[16] In that context it is crucial to stress, however, that just as the map (of which the sign is merely a special case) is our privileged means for finding the way, so the travel story is the most effective device for transporting our imaginations from the utopian No-where to the actual Now-here; whenever I am saying "of course," what I am really saying is that I am on *course,* that I am steered the way by the compass of the taken-for-granted. If we want to achieve objectivity in our observations, we are consequently obliged to move not only our minds but our bodies too. It is by sharing our common understandings that we show who we are.

Such is the ideal. As always, reality is different. For immediately I set out on my epistemological journey I am bound to realize that what I *want* to achieve and what I *can* achieve are never one and the same. Pushed to its limits by its own modalities, the principle of self-reference in fact reveals that perfect translation is impossible, that every representation is a lie. To be human is in that perspective to be an epistemological and existential mongrel, a bastard sprung from two cousins of unclear ancestry, semiotic on the spear side, rhetorical on the spindle side.

Different as these modes of communication may be, they are nevertheless connected because both of them are rooted in desire—a desire so desirous that it can never be satisfied, a desire which desires nothing less than itself, a desire so self-consuming that it devours its own children. Driven by its ability to imagine what lies beyond its body, the semiotic animal is in fact constantly trying to be what it knows it cannot be—a perfect sign perfectly communicated. Not like other apes steered solely by evolutionary chance and survival of the fittest, but a species molded in the processes of cultural transmission of purposeful action.[17] And so it is that to *Homo sapiens* language is not merely a medium of reflection but the very foundation of self-knowledge, not merely a mirror but the tain of the mirror.[18]

The fact that chimpanzees know how to lie does not mean that they also know how to write their autobiographies or how to paint their self-portraits. Likewise, a dog may well expect his master to come home, but not that he will come home next week. But I, in my capacity as a human being, I know not only how to write my autobiography and paint my self-portrait, but I also know that I must meet my wife at the San Francisco airport on December 9 at 12:30 in the afternoon. The difference is that only the human animal is capable of realizing (and of saying) not only that it *has* a body but that it *is* a body. Look, I am invisible! Invisible like the fish down there in the water, the prey that I have come down here to the bridge in order to tempt with my bait and catch on my hook, the thin fishing line and my capacity to imagine the only connections between us. The red-and-white float bobbing up and down, an abstract dance of life and death.[19] What a fantastic performance! What an incredible illustration of the fact that the human being is the only creature capable of killing for the sake of an idea! Power is likewise, because power is by definition a magic game of ontological transformations. Dreams are such stuff that we are made of.

It was Søren Kierkegaard's genius to lay bare these theatrical roots of irony, basic insights which the existentialists of the twentieth century were eager to adopt as their own: essence and phenomena misrelated; the world constantly beyond reach; suicide the only way to get away. As so often before, etymology proves instructive, for the term "irony" has its roots in the Greek comic character *eiron,* a clever underdog who by his wits triumphs over the boastful and tragic *alaon.* The point is, of course, that even though comedy is often more revealing than tragedy, the two genres are related like the two sides of the same coin; to Aristotle's argument that comedy imitates men who are worse than the average, tragedy those who are better, Quintilian responded that ethos is akin to the comic, pathos to the tragic, the former an everyday emotion for ordinary men, the latter a temporary state which strikes extraordinary men under extraordinary circumstances. To both authors an incongruity between ideal concepts and objective reality.[20]

No wonder, therefore, that we tend to read tragedy alone and comedy together, for the former genre is concerned with individuals, the latter with classes. No wonder either that we release our tensions by crying in private and laughing in public, for what is at issue is nothing less than the contradictions of life itself. In the words of Kierkegaard's Climacus:

> The comical is present in every stage of life (only that the relative positions are different), for wherever there is life there is contradiction, and wherever there is contradiction, the comical is present. The tragic and the comic are

> the same, in so far as both are based on contradiction; but the tragic is the suffering contradiction, the comical, the painless contradiction.[21]

Two types of tears. Two forms of kathartic cleansing. Two boundaries for understanding life. Two genres performed in the Theater of the Absurd. Two modes of self-reference.

Plays of signs nevertheless. Rope tricks of translation. Mission Impossible.

*

**A mission no agent can refuse:** How do I find my way in the power-filled world of thoughts-and-actions, things-and-relations?

These were the cryptic orders handed down to me by the Secret Service Headquarters, a set of commands as impossible to fulfill as not to accept:

- lay bare the familiar of the unknown;
- find the principles of imagination and specify the rules of ontological transformation;
- draw a map of the Territory of the Humans, (re)trace its fluctuating boundaries and find its stable center;
- produce an atlas of what it means to be human;
- initiate a critique of cartographical reason.

*

**The order is given, the order shall be executed**! What follows is therefore a report of what the agent discovered during his clandestine operation, a document divided into six sections, an illustrated tale that takes the reader from point **A** to point **F.**

**A.** First this brief PRELUDE, an attempt to set the tone and prepare the way for the critique to come. And what is to come is an investigation which begins in the calibration of the surveying instruments and with the establishment of the fix-points and base-lines that later will be used in the mappings of what it means to be human, the real purpose of the mission. While the findings from this tool-oriented phase of the investigation make up the first half of the book (section B), the actual maps are presented in the second half (here called sections D and E). Forming the bridge between the two halves is the brief section C, a sketch of the coordinate net in which the human animal is captured and domesticated, the reins by which we are made so obedient and so predictable.

**B.** The findings from the initial reconnaissance raids are brought together in three subsections of eight chapters. The aim of the first incursion—the subsection called MAPPINGS—is to tell a different history of cartography, a journey which begins with a reading of *Enuma elish* and ends with an interpretation of the *Ebstorfer Karte,* arguably the most outstanding example of a *Mappa Mundi Medievalis.* What is learned along the way is not only that every map is simultaneously a picture and a story but that every map is a record of mistranslation, hence a major confrontation with the limits of representation.

The second raid—the subsection entitled INSTRUMENTS—takes us deeper into this crucial problem of translation, one chapter devoted to the semiotics of the sign, especially to the role of the Saussurean Bar, another to the theory of proper names and definite descriptions, a third to the rhetorical strategies of geometry. What drives the search is a hermeneutics of suspicion, a mode of argument which says that the strength of cartographical reason lies less in its ability to tell the truth and more in its power to convince.

Finally, the third and most daring of the reconnaissance expeditions, here called IMAGINATIONS. In this extensive subsection of four chapters I try to explicate the techniques by which we make the absent present and the present absent. First comes a reconstruction of Filippo Brunelleschi's perspective and a mapping of Plato's *Republic,* then two comparative readings of Odysseus' return and Abr(ah)am's leaving, the former an example of informative reformulation, the latter a handbook in institutionalized terror. Finally, there is a mapping of the Island of Truth (a very attractive name), a continent which was originally discovered by Immanuel Kant, he himself the most outstanding student of limits, theoretical philosopher in terms of the subject matter he thought *about,* practicing geographer in the language he thought *within.*

**C.** From these preliminary incursions into Lands Unknown back to the tabernacle and a thanks-giving COLLATION of another type. A debriefing and a pep talk, a feast which functions less as a communion and more as a warriors' preparation before the battle. A concerted effort to experiment with the boundaries of Plato's *Republic* and thereby with the taboos of the excluded middle of Aristotelian logic, with the bar of Ferdinand de Saussure's linguistics and with the points, lines and planes of Marcel Duchamp's projections.

And as a Phoenix out of the ashes rises a map of three lines, a geometric sketch in which the Territory of the Humans is enclosed within noncrossable boundaries of silence, the godly realm of Mindscape located at the top, the beastly caves of Rockscape at the bottom. To be human is in that perspective to be engaged in a perpetual two-front war, sometimes in the manner of Jacob Isaacson wrestling with invisible spirits, sometimes, like the twentieth century physicists, baptizing invisible matter. No peace in sight, for who can ever tell whether God plays dice.

**D.** The purpose of the collation-chapter—effectively a cabinet meeting with Janus presiding—is to tie the two phases of the project together, to let the findings from the reconnaissance expeditions influence the actual mappings of the second half of the book, the ultimate goal of the mission. The outcome of that endeavor is an ATLAS with a set of maps never seen before, a series of geometric pictures and bloody stories drawn from the most decisive episodes in the perpetual war of what it means to be human. Yet another set of local habitations provided with proper names and definite descriptions, yet another attempt to charter the way from the tower of Babel to the skyscrapers of North Armorica, an overwhelming illustration of how the humans have expanded their territory into areas which were previously ruled by the utterly alien, sometimes appearing in the form of invisible gods, sometimes as untouchable matter. Four sites to visit, four maps to be drawn:

1. *Uruk,* the walled city where Gilgamesh, through his friendship with Enkidu, learned that he too was eventually to die;
2. *Peniel,* the name of the ground where Jacob wrestled with God, saw his face, and survived—perhaps the same place where Job later took his LORD to court;
3. *Thebes,* the city where King Oedipus stuck his eyes out in order to better see the difference between kings and gods, fathers and brothers;
4. *Nicaea,* the summer palace where Jesus Christ, after much theological squabbling, was defined as the most inter-esting being of beings.

**E.** End of the end, begin the beguine. Hence a return to the present, a two-part REQUIEM composed in remembrance of cartographical reason, that dominating mode of thought-and-action which is constantly in the process of being resurrected. The first part of this Mass for the repose of the souls of the dead takes us to *Philadelphia,* Thomas Pynchon's heavenly City and crowded niche of Hell, the place where Marcel Duchamp (resident alien) erected a mausoleum in his own honor, a show in which the bachelor was stripped bare by his brides, their givens taken-for-granted. At the center of that edifice lies the kicking corpse of modernity.

The second part of the Requiem—*Uppsala*—is a composition in the wake of my own time-and-place. And just as Jacques Lacan once could claim that the unconscious is structured like a language, so I now retort that power is structured like a map. Seemingly a crystal palace, this *Mappa Mundi Universalis* is in actuality a power-filled instrument for understanding how we are made so obedient and so predictable. The home of a ghost which refuses to go away.

**F.** Finally five MEMORIALS, the first containing the notes, the second an extensive bibliography, the third and fourth two indices of noninterpreted keywords, one of proper names, the other of (in)definite descriptions. And, at the very end, the grateful acknowledgments.

*

**Throughout this prelude** I have allowed the concept of the map to play a more fundamental role than the concept of the sign. But what is a sign and what is a map? And which exactly are the relations between them? How are their respective limits de-termined and their borders circum-scribed? How do I condense the infra-thin tympanum of the pre-positional *and* into a de-materialized point? Which forces make thing-like metaphors explode into metonymic chains of novel meanings?

Perhaps the answer lies in the geometry of the cross, its roots anchored not in the symbolism of Golgotha but in the opening words of *Enuma elish,* its branches not in the outstretched arms of the Savior but in the canonical positions of up and down, front and back, left and right.[22] Crucial in deed, for it was none less than Leon Battista Alberti who in his rhetoric-based theory of painting "took from Quintilian a classification of the seven possible types of movement that an orator might have to make and that a painter might have to reproduce: up or down, to the right or the left, forward or backward, in a circle."[23] *Homo erectus* caught in its own net of:

above<br>
|<br>
now — and — then<br>
|<br>
below

Such is the image that will guide us throughout this entire book. Two perpendicular lines, map and compass turned into one. Disguised as a geometry with names this cross begins to live, in the process transforming the system of spatial coordinates into a Vitruvian Man. But whereas the navel of this archi(tectual) figure is always at the center of the circumscribing circle, its diagonals are sometimes intersecting in that same omphalos, sometimes in the groin; while both the *Homo ad circulum* and the *Homo ad quadraticum* have their legs and arms outstretched to embrace you, Leonardo's version is a jumping-jack waving to the crowd.

As I hope to be able illustrate, there is much to learn from this geometric shiftiness, not the least as it is embodied in the figure of Osiris, at the same time Egyptian fertility god and King of the Underworld. Quite predictably, many tombs show him with his member angled, albeit no longer set in its normal place between the legs, but sticking out of (or is it stuck *in*) the sanctity of his sacred navel. The principle of self-reference embodied, the incest taboo transcended, the boundary between life and death erased.[24] In contrast, Donald Duck always carries his heart in the right place, perhaps because ducks do not have any navels, perhaps because Donald Duck is not a real being but a Hollywood imagination.

*

**Too much for one bite**, constipation in the making. Chewing the cud the only relief. Communion in another form.

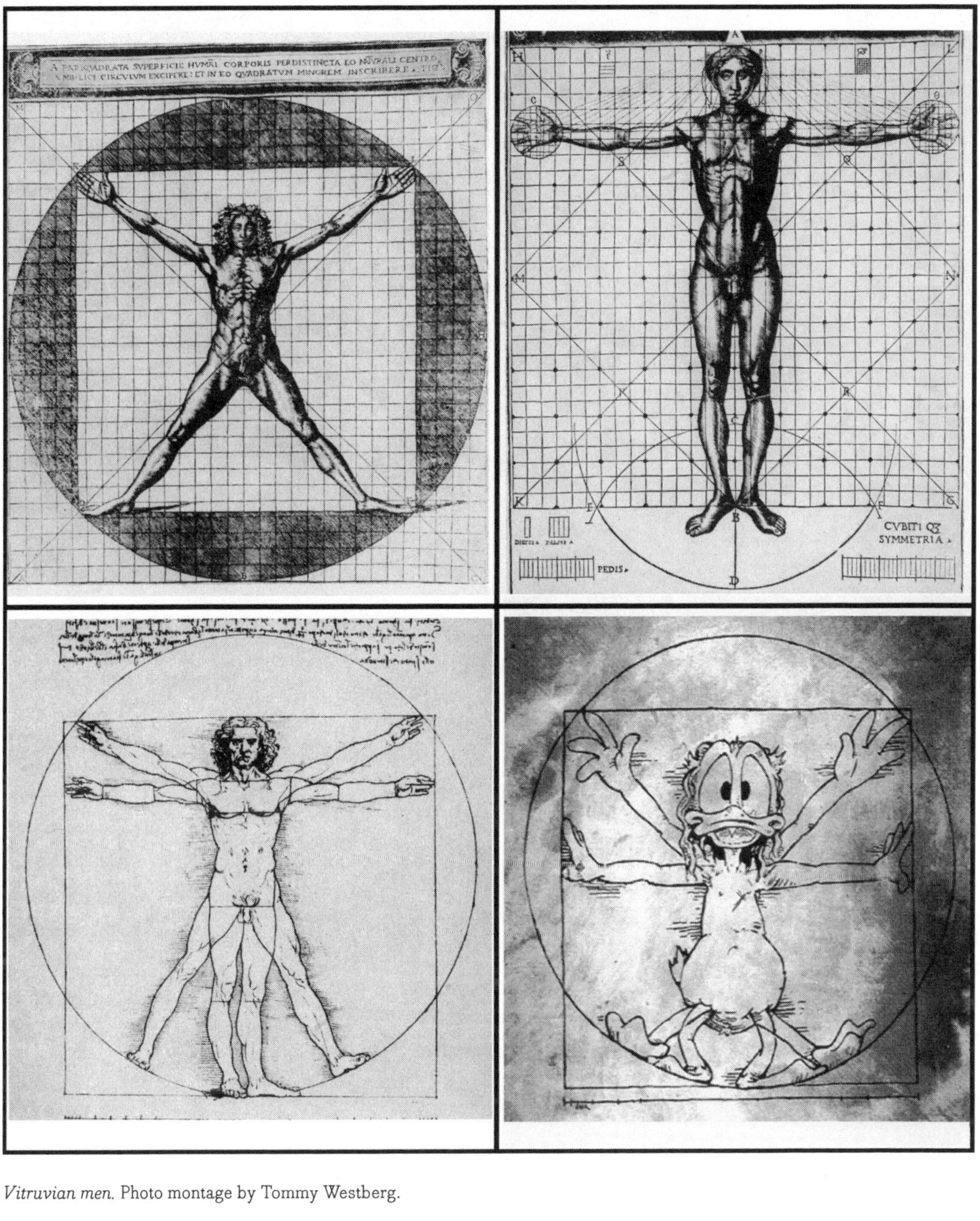


*Vitruvian men.* Photo montage by Tommy Westberg.

# MAPPINGS

# WHEN ABOVE

God called the light Day, and the darkness
he called Night. *Genesis 1:5*

**In the map of maps** naming is the dematerialized point into which all power is condensed, the primordial means for knowing our way about.

And for that reason we must now return to *Enuma elish,* the Babylonian creation epic which some have claimed is not a creation epic at all,[1] merely a well-rehearsed story of how the god Marduk gains and retains his elevated position as the Lord of lords. Regardless of how it is categorized, this ancient text is a crucial document, for what it does is to lay bare the rhetorical techniques through which undifferentiated chaos is turned into differentiated cosmos. Since it was intended for recitation, it was cast into a type of poetry in which each line forms a distich such that the two halves typically stand in contrast to each other.[2] The resulting sense of difference deferred is generated through an ingenious use of meter rather than rhyme, the latter unknown not only to Babylonian but to Hebrew writers as well.

On this rendering the *Enuma elish* emerges as nothing less than a search for the *topos* of *topoi,* a text more concerned with the creation of people than with the manufacturing of things, a foundational treatise on politics rather than religion. Its premise is that in the beginning of the beginning nothing has yet been formed, because in the beginning of the beginning no things have yet been named. All that exists are the two coordinates of above and below, cardinal positions waiting to be inundated by the fluids of Apsu and Tiamat, the former sweet, the latter bitter. And, as if to underline the spatiality of its own structure, the term "Apsu" literally means "abyss" and "uttermost limit," by linguistic coincidence connectable also to "the great deep," "the primal chaos," "the bowels of earth," "the infernal pit."[3] A perfect example of proper name and definite description merged into one.[4]

From the mingling of Apsu and Tiamat there eventually emerged three pairs of gods, each generation wiser and more intelligent than its predecessor, the figure of Nudimmud-Ea the most outstanding of them all. But noise annoys and as typical teenagers the youngsters eventually became such a nuisance that

> Apsu made his voice heard
> And spoke to Tiamat in a loud voice,
> "Their ways have become very grievous to me,
> By day I cannot rest, by night I cannot sleep.
> I shall abolish their ways and disperse them!
> Let peace prevail, so that we can sleep."[5]

When Tiamat heard this she got very upset, furiously shouting

> How could we allow what we ourselves created to perish?[6]

But Apsu receives contrary advice from his vizier Mummu, and, ignoring Tiamat's objections, he decides to go ahead. Barely in time to avert the pending catastrophe, the god Ea, superior in understanding, learns about the plot. Through a magic spell he puts Apsu to sleep, removes his crown, and places it on himself. Thus having established his position, Ea kills Apsu.

On top of the corpse, i.e. *across the abyss,* Ea then builds a splendid house for himself and his wife Damkina. There, in the heart of holy Apsu, in the Chamber of Destinies, their son Marduk is conceived, the most awesome being ever to be.[7] When his father saw him he beamed with pride and his heart was filled with joy. In deed:

> He made him so perfect that his godhead was doubled.
> Elevated far above them, he was superior in every way.
> His limbs were ingeniously made beyond comprehension,
> Impossible to understand, too difficult to perceive.
> Four were his eyes, four were his ears;
> When his lips moved, fire blazed forth.
> The four ears were enormous
> And likewise his eyes; they perceived everything.
> Highest among the gods, his form was outstanding.[8]

Already from birth Marduk is held to be everything, including a premonition of Janus, the Roman gatekeeper who from his watchtower at the middle of the bridge controls both sides of the abyss at the same time. To impose his will on the troublemakers, Marduk now blows a devastating flood wave. But instead of calming the rioters down this act upsets them so much that they convene an assembly in which Tiamat is authorized to declare war on whoever was responsible for the killing of sweet Apsu. To that end the bitterness puts together a terrifying army of snakes with venom instead of blood in their veins, a phalanx of dragons, a sphinx, a horned serpent, a rabid dog, a scorpion man, a fish-man, a bull-man and eleven more of the same kind. As commander in chief she appoints Kingu, an upstart in whom she invests absolute power and forces to share her bed lest he run out of control.

At this stage Ea learns about Tiamat's preparations. Charged with the task of talking Tiamat out of her plans, the wise Anshar is dispatched as an emissary to her court. Terrorized by what he sees, Anshar returns with the report that not only has he himself failed to complete his mission but that "no one else can face her and come back."[9] Things are obviously getting out of hand and everyone agrees that diplomacy will no longer do. Ea consequently summons Marduk to his private quarters, asking him to step forth, to show who he is by being equal to himself. Marduk, who for the first time is now addressed as "The Lord,"[10] initially rejoiced then immediately proceeded to list the conditions under which he would accept the challenge. These were his words, for emphasis three times repeated:

Lord of the gods, destiny of the great gods,
If I am indeed to be your avenger,
To vanquish Tiamat and to keep you alive,
Convene the assembly and proclaim my lot supreme.
When ye are joyfully seated together in the Court of Assembly,
May I through the utterance of my mouth determine the destinies,
 instead of you.
Whatever I create shall remain unaltered,
The command of my lips shall not return (void), it shall not be changed.[11]

*

**Through the utterance of my mouth shall I determine your destinies!** In their acceptance of this decree, the friendly gods demonstrate how they differ from Tiamat, whose words have lost their power because the new politics are conducted by other means. Since every audience demands its own propaganda, Marduk chooses not to threaten the electorate with new taxes but invites them to a feast instead. For this:

They gathered together and departed,
All the great gods who determine [the destinies].
They entered the presence of Anshar and filled [the court of Assembly];
They kissed one another [as they came together] in the assembly,
They conversed (and) [sat down] to a banquet.
They ate bread (and) prepared w[ine].
The sweet wine dispelled their fears;

[Their] bod[ies] swelled as they drank the strong drink.
Exceedingly carefree were they, their spirit was exalted;
For Marduk, their avenger, they decreed the destiny.[12]

And out of the ashes from this revel Marduk emerges as a splendid Phoenix, Lord of lords, God of gods. They erected for him a throne, they gave him sovereignty over the whole universe, and, as a premonition of what was later to reappear in the Christian confession, they observed that to trespass is to sin. More precisely:

None of the gods shall transgress your limits,[13]

in the original text rhetorically bejeweled, embellished and enhanced through the poetics of alliteration: *itukka la ittiq*.

Uncertainty nevertheless breeds suspicion, and Marduk is now asked to prove himself. Accordingly:

They sat up in their midst one constellation,[14]
And then they addressed Marduk their son,
  "May your decree, O Lord, impress the gods!
  Command to destroy and to recreate, and let it be so!
  Speak and let the constellation vanish!
  Speak to it again and let the constellation reappear."
He spoke, and at his word the constellation vanished.
He spoke to it again and the constellation was recreated.[15]

A magnificent show of ontological transformations. The magician of power in outstanding performance. Let there be! And there is. Ovations rising to the sky, on this particular occasion in Babylon, later in Jerusalem, eventually in Washington, D.C. When the gods (literally his fathers) saw how effective his utterance was, they all rejoiced, blessed their Lord and chanted MARDUK IS KING, OUR KING IS MARDUK!

Bestowing upon him the scepter, the throne and the royal robe, the assembly then send him off to pour out the life of Tiamat. His weapons are numerous, but most decisive is the net in which he intends to catch her and the four winds by which he plans to blow her up. Kingu runs away in panic. Tiamat looses her mind, in anger her wits are scattered. And then:

Face to face they came, Tiamat and Marduk, sage of the gods.
They engaged in combat, they closed for battle.
The Lord spread his net and made it encircle her,
To her face he dispatched the [Evil] wind, which had been behind:
Tiamat opened her mouth to swallow it,
And he forced in the [Evil] wind so that she could not close her lips.
Fierce winds distended her belly;

Her insides were constipated and she stretched her mouth wide.
He shot an arrow which pierced her belly,
Split her down the middle and slit her heart.
Vanquished her and extinguished her life.
He threw down her corpse and stood on top of her.[16]

[For the rest of the troops]
He imprisoned them and broke their weapons.
In the net they lay and in the snare they were.[17]

What a timeless text, the Lord spreading his net! If the dictator of Iraq ever wanted an instructor in biological warfare, he need not go far to find him. If the generals of the Pentagon ever wished to know more about the relations between cartographical coordinates and the targeting of smart bombs, they need not experiment on haphazardly chosen victims but may consult the *Enuma elish* instead. Likewise with the principles of Derridean deconstruction. For when it came to the handling of Tiamat's dead body, Marduk found himself lodged in the abysmal interface between *de*-struction and *con*-struction. Like a professional butcher, a real ripper,

> He turned back to where Tiamat lay bound, he straddled the legs and smashed her skull (for the mace was merciless), he severed the arteries and the blood streamed down the north wind to the unknown ends of the world. . . .
>
> The Lord rested; he gazed at the huge body, pondering how to use it, what to create from the dead carcass. He split it apart like a cockle-shell; with the upper half he constructed the arc of sky, he pulled down the bar and set a watch on the waters, so they should never escape.[18]

Then the most decisive step of all. For when the epic now moves from the power struggles of the gods to the creation of the world, Marduk enters the stage disguised in new clothes, no longer dressed in the warrior's coat of mail but in the uniform of the land surveyor. The implements changed as well, the magic net no longer a tool for capturing monstrous rivals but a device for ordering the thing-like entities of stars, towns and people. The abyss nevertheless remains the ruler's privileged *topos,* for:

> He crossed the sky to survey the infinite distance; he stationed himself above apsu, that apsu built by [Ea] over the old abyss which now he surveyed, measuring out and marking in.[19]

*

**Measuring out and marking in** Marduk now proceeds to the construction of what eventually was to be a celestial globe. Onto the sky he projected positions for the great gods and through a kind of triangulation he set up three constellations for each of the twelve months. For the pole of the universe he chose the planet Jupiter (possibly Mercury), Tiamat's liver became the zenith.

Once these heavenly bodies were in place, Marduk returned to Tiamat's corpse, cutting it up into pieces to which he allotted new functions: out of her eyes welled the Euphrates and Tigris, her paps became mountains, her crotch the fulcrum of the sky. The construction of the physical universe completed, he went on to the building of yet another temple and to the arrangement of yet another banquet. All in honor of himself. And under the influence of sweet wine the assembly shouted in unison, GREAT LORD OF THE UNIVERSE!

In response the Absolute delivered a speech from the throne, his four eyes and big ears simultaneously scanning the above and the below, the past and the future, no member untouched.[20] This is what he said to the gods his fathers:

> In the former time you inhabited the void above the abyss, but *I have made Earth as the mirror of Heaven,* I have consolidated the soil for the foundations, and there I will build my city, my beloved home. . . . It shall be BABYLON. [And there the fallen gods, our defeated enemies, the supporters of Tiamat, will be serving us day after day].[21]

Yet another case of ontological transformation, yet another instance of invisible ideology turned into touchable stone, the Earth made a mirror of Heaven, the Heaven itself a material projection of social relations. Babylon the most marvelous of cities, the center of the world, its own street pattern drawn in the shape of a net, its name, *bab-ili,* most commonly translated as "the Gate of the gods."[22]

And yet. For absolute rulers enough is never enough. Moved by the desire to create a work of consummate art, Marduk conceives a plan designed to solidify his position and perpetuate his rule. To placate the supporters he promises that they will never need to work again, that after the next election leisure will be their blessed lot. Marduk is wise enough to honor the pledge and to that effect he swiftly creates a primeval man, the prototype of you and me, by definition slaves of the gods. Not an invention formed in the image of the Almighty but a savage concoction stirred together from the blood of the slaughtered Kingu, mankind a dish of *Boudins à la Mésopotamie.* Nothing like a perfect copy of the perfect original, merely a black sausage. A deed appropriately described as a deed impossible to describe.

Then Marduk—at this final stage called "the King"—divided the gods into two groups, each vassal assigned to his proper place in the feudal system. Three hundred he stationed as watchers of Heaven, an equal number as guardians of the Earth. In return for their new-won freedom the once-defeated erected yet another temple above the abyss, this time in honor of their conqueror turned benefactor. As a token of appreciation the latter in return invited them to yet another celebration, for the first time featuring nothing but black-headed waiters. And next morning the entire congregation was taken on a tour through the Hall of Armory, where the intricate construction of the net was explained and the workmanship admired.

To top it off, the banquet closed with the performance of a liturgy in which the occult was explained in a way that no one could misunderstand. At the end, which in effect is not the end but another beginning, Marduk is the King, Marduk is the Absolute, Marduk is the Lord of lords. Hail to the Chief! Fifty were his names, so numerous that if ever attacked he could always hide behind another alias. Never catchable as the specific this *or* that, always on the move as an ambiguous this *and* that.[23] Never a logical either-or, always a dialectical both/and. Many in one, one in many. Ungraspable multiplicity, like the Hebrew JHWH a tautology defining himself as a being who is who he is. Jacques Lacan was obviously neither the first nor the last to understand that the real is given its structure by the power of the name; as Gilles Deleuze and Félix Guattari aptly put it, "the proper name does not indicate a subject [but] designates something that is of the order of the event, of becoming or of the haecceity. It is the military men and meteorologists who hold the secret of proper names, when they give them to a strategic operation or a hurricane. *The proper name . . . marks a longitude and a latitude.*"[24] Heaven above and hell below, the sinister to the left the righteous to the right. Cartographical reason in nuce.

In the mist-enveloped regions of religion naming is the name of the game, an exercise in ontological transformations where earthly people appear as projections of heavenly gods, social relations as signs in the sky. And to me this is the central message of the *Enuma elish:* the world created is the mirror image of the Court of Power, literally a representation of a representation, a map of a map, a signified meaning searching for its own coordinates; nascent sociology turned to mythical astrology, mythical astrology to prescientific astronomy. Different ways of finding the way.

And once every twelfth month (on the fourth day of the New Year's celebrations, at the time of the spring equinox) the epic was recited, the ruling king first divested of his insignia, then—following the negative confession that the sinner was not he himself but his enemies—he was dressed up again in his robe of stars and stripes, an embodied constellation of points and lines projected onto the canvas of the sky itself.

# AND BELOW

> When I was a little chap I had a passion for maps. . . . At that time there were many blank spaces on the earth, and when I saw one that looked particularly inviting on a map (but they all look like that) I would put my finger on it and say, When I grow up I will go there. JOSEPH CONRAD, *Heart of Darkness*

**Points and lines** were the stellar guides not only of the Babylonians but eventually also of Pytheas from Massalia (presently Marseilles), the Greek traveler who some time before 320 BCE left his home town for an adventurous journey to the northernmost limits of the *oikumene.* The exact motives and purposes of these explorations remain unclear, but Massalia was an important commercial center with a natural interest in the routes to the tin-producing areas of Cornwall and the amber deposits farther away. In addition, most people still thought of the land mass of Europe, Asia, and Libya (i.e. Africa) as a huge island surrounded by a big Ocean. Since that Ocean was envisioned as a monster-filled chaos located outside the limits of the oikumene,[1] dangerous enough to be taboo, it was a great challenge to determine the exact location of the line between the firm land and the fluid water.

Since what we cherish most is often what we fear most, the image of this outer limit of the Continent was doubtless a major motivation for Alexander the Great,[2] when he set out to conquer the east. What then happened is what could be expected, for as soon as the worn-out soldiers realized that they had been brought to the interminable Ganges plain, not to the Ocean shore, they mutinied, forcing the Great King to turn back. Both (re)actions were perfectly understandable, for credibility has always been a function of rhetoric, rhetoric always an eroticized choreography of bodily senses and cultural meanings.

The idea of the Ocean river was not new. Indeed, it had long been connected with the unmentionable limit of limits which Anaximander of Miletus (the "father of geography," the first who dared to draw the world on a map)[3] called *to apeiron,* a primordial substance from which everything was said to grow and into which everything was held to return; the origin of all things; an undefinable flow at the same time infinite and indefinite.[4] This limit of limits was consequently neither the Heaven up there nor the Earth down here, but the very bond which separates and unites them, a fluctuating water which was said to be one with the unstable horizon, all gods and all things stirred into being through the motions of its waters.

Perhaps it was once understandable, but to us—the latecomers—the meaning of the apeiron is virtually ungraspable. For so totally boundless is this concept that it lies before and beyond every distinguishable category. And at least in that sense it seems closely related to Apsu's abyss,[5] an ontological mongrel without a body or rather a soul which is to the body as the gold to the *louis d'or.*[6] On these relations little can be said except that it was Anaximander who initiated the radical thought which eventually made the mythical image of the Ocean fade into the rational geography of later explorers,[7] most prominent among them Alexander in the southeast and Pytheas in the northwest,[8] the former in his youth tutored by none less than Aristotle himself, the latter a man of unknown provenance.

On the return from his encounters with the unfamiliar, Pytheas wrote a report significantly entitled *On the Ocean,* now lost but indirectly known through comments by at least eighteen ancient writers.[9] Their opinions about the man and his stories vary widely, ranging from the praise of Eratosthenes to the scorn of Strabo, to whom "that fellow Pytheas, who gave an account of Thoule, has been clearly shown as the worst possible liar. . . . [And] anyone who made such continually false statements about well-known places would scarcely be able to tell the truth about places which are not well-known to everyone."[10] But time plays its own game of truth and consequences and today Pytheas is considered neither a madman nor a liar. What in hindsight makes him so remarkable is that he never lost his bearings, a characteristic which makes him a living illustration of Ludwig Wittgenstein's oft-quoted remark that "if I were sometimes to see quite new surroundings from my window instead of the long familiar ones, if things, humans, and animals were to behave as they never did before, then I should say something like 'I have gone mad'; but that would merely be an expression of giving up the attempt to know my way about. . . . But the important thing for me is that there isn't any sharp line between such a condition and the normal one."[11]

Despite all ordeals, Pytheas was evidently able to stay on course, partly because he was an expert navigator wise enough to use local pilots, perhaps local vessels as well,[12] partly because he was a skilled astronomer. It has even been suggested that the real reason he undertook his voyage was that he wanted to verify what geometry had already taught him. If so, then Pytheas may be best described as a paradigmatic scientist obsessed by the idea of combining theoretical insights with direct observations, in effect a geometer of intersecting lines turned into a geographer of nameable places. The most outstanding result of this approach is that he established the exact location of the celestial pole, that imaginary center about which all stars appear to circle, that dematerialized point which corresponds to the outward projection of the earth axis—a geometer's abstraction so abstract that even though it can be thought, it can never be seen. No mean feat, for the invisible mark of this ontological transformation is nothing but the construction of the fourth corner in a rectangle whose other corners sometimes, albeit only under favorable conditions, may be glimpsed from afar. Demonstrating how such an imaginary point in the sky may be used for determining the symbolic location of real places on the earth clearly takes the power of extraordinary thought. But alone and in the coolness of cloudless and sleepless nights it was exactly this marvelous merger of astronomy and geography that Pytheas achieved. In addition, he was the first to argue that the Ocean tides obey the commands of the man in the moon, shining light in the dark.

But the Massaliote made observations in broad daylight as well, the first person to have related the latitude of a place to the length of its longest day—yet another instance of finding one's place on earth by consulting the stars in the sky. His only tool was a gnomon or sundial, a simple device often attributed to Anaximander but in fact known already to the Babylonians,[13] essentially an upright stick designed to measure not a thing but the shadow of a thing, more specifically the sun itself. The technique draws on the idea that at midday the length of the shadow remains unaltered as one travels due east or west on the earth's surface, technically the parallel of latitude, a line located at the same distance away from the equator. It follows that if gnomon ratios from two widely separated points are observed to be the same, then the two locations lie on the same parallel of latitude.[14] Not the only occasion when an entire world is contained in an angle.

Since identical astronomical phenomena can be observed from all places along the same latitude, Pytheas realized that the height of the sun at the winter solstice may be figured by calculation; one correct observation is enough to verify the entire series. From his deep knowledge of the geometry of the sphere he also deduced that there must be a parallel of latitude at which, at the time of the summer solstice, the sun never sets and the day lasts for twenty-four hours. Moreover, he reckoned this circle to lie on the parallel of 66°N, not too far from the actual location of the Arctic circle at 66°32′N.[15] True to his boldness, he put the island of Thule on that line, by definition the northern boundary of the temperate zone, the latitude at which the circumpolar stars are always visible.

No one knows exactly how far north Pytheas actually ventured or which places he did visit. Most debated are his references to Thule, by the majority of modern commentators held to be Iceland, by some to be Shetland or the Norwegian coast around Trondheim. He himself placed it six days' voyage north of Britain calling it "the very last regions" or "the last named location," an area where people grew millets (oats?) and drank beer brewed from barley and honey.[16] Here was literally the dangerous edge of the world, with summer daylight lasting up to six months, a land of fire-throwing volcanoes and boiling springs, a sea with freezing fog, ice-slush all around, solid ice rumored to be a day away. A place where "barbarians showed us where the sun goes to bed."[17]

Not knowing anything about the Gulf Stream but firmly believing that to the north of the Black Sea it was too cold for humans to survive, people found Pytheas' story hard to accept. For whereas his invention of parallels of latitude fit well into the Greek habit of symmetrical thinking, that same ordering device now kept his critics from realizing that the world as he saw it was the world as it actually is; as in the earlier case of Marduk, Pytheas' destinies were determined by the utterance of his mouth. To repeat, Thule was the last of *named* locations, a place located at the end of the world, an island rising out of a dangerous Ocean, sweet and bitter at the same time.

Many have been labeled madmen for less. Nothing unusual about that, for as Thomas Pynchon reports from the Company Lodge at Cape Town at a time shortly after Venus's transgression in 1761,

> Many who have been to Rooms forbidden the others, report seeing, inside these, a Door to at least one Room further, which may not be opened. The Penetralia of the Lodge are thus, even to those employed there, *a region without a map.* Anything may be there. . . . Madness will visit by surprize, taking away to its Realm of Voices and Pain even a mind in the rosiest of sanity. When they are too dangerous to roam free, the town madmen are kept as a responsibility of the Company, confin'd in padded rooms in the Slave Lodge. . . . Some of them hate women, some desire them, some know hate and desire as but minor *aspects of a greater Oceanick impulse,* in which, report those who survive, it is unquestionably better not to be included. Some do not survive.[18]

*

**An Oceanick impulse,** a region without a map, a condition some do not survive. Such was the monster-filled world Pytheas humanized through a combination of calculation and observation. His primary tools were three: a gnomon; a set of parallel lines; an invisible point—in toto an arsenal of conceptual treasures sharpened and reapplied in the revolutionary shadow play which a century later was written and performed by a man called Eratosthenes.

Born around 276 BCE, this remarkable polymath grew up in the thoroughly Hellenized city of Cyrene, a Greek colony on the Mediterranean coast west of Egypt. From that beginning he moved first to Athens and then to Alexandria, where he spent most of his life initially as tutor of the crown prince and eventually as head of the Museum and its famous Library, the world center of calculation; the spider in a well-functioning information network, constantly surrounded by an estimated half million volumes and a collection of outstanding scholars; unprecedented resources put to unprecedented use. While his contemporaries freely acknowledged him as an influential literary critic, grammarian, philosopher, poet, chronologist, astronomer, mathematician, they ranked him first in no field, second in many. "Beta" was the nickname they gave him, a poisonous gift wrongly administered to a man much greater than themselves.[19] At eighty(?), weary and struck by blindness, he passed away, reputedly through voluntary starvation.

Eratosthenes' best work was in geography, geodesy and cartography, intertwined fields in which he most definitely was not beta but alpha, arguably "the earliest outstanding geographer and . . . to this day one of the greatest geographers of all ages."[20] His main achievements were two, both equally based on a mastery of the geometry of the sphere (originally developed by Euclid) and the empirical data gathered in the Library (including the field reports of Alexander and Pytheas); whereas the first type of knowledge was crucial in the design of a methodologically simple but brilliant technique for measuring the circumference of the earth, the second played a corresponding role in the construction of a world map of hitherto unseen veracity. In both cases a question of weaving a Marduk-net sufficiently wide and sturdy to capture the entire world, one thread theoretical, the other empirical, the former lodged in the mind of the mapper, the latter in the body of the explorer.

Eratosthenes' first tale begins in a deep well in the town of Syene (presently Aswan), a strange juxtaposition of Apsu's abyss and Pytheas' Ocean. This well was not an ordinary well but a very peculiar well, for, once a year, its walls cast no shadow, the angle to the sun equal to zero degrees. Since this rare event always occurred on the day of the summer solstice, it meant that Syene must be located right on the Tropic of Cancer, the imaginary parallel which runs around the earth at a distance of twenty-three and a half degrees to the north of the equator.

The story of the Syene well was widespread and came to play a leading role in the construction of Eratosthenes' drama. Even more crucial, though, was his offhand comment that in Alexandria itself there was never a noon without shadow—a remarkable observation of an unremarkable fact, a flash of genius by which the non-shadow of Syene was transformed into a sign of a difference deferred, a silent zero waiting to speak.

And speak it did, more exactly at midday on June 21 of an unknown year, the historical moment at which Eratosthenes, gnomon in hand, stepped out of the Library and into the sun. Prompted to talk, the instrument revealed that the angle between its top and the sun's rays was seven and one fifth degrees, by ordinary division made equivalent to one fiftieth of a full circle ($7.2/360 = 1/50$). Since everyone who counted knew that at that same hour there was no shadow at Syene—a place believed to lie on the same meridian as Alexandria—it was accepted that the distance between the two places was one fiftieth of the distance around the earth. Translating that relative measure into an absolute number was straightforward, for most travelers agreed that a camel caravan, which on the average moves 5,000 stadia per day, needed 50 days to cover the distance between the two cities. The conclusion came by itself: the circumference of the earth equals 50 × 5,000 = 250,000 stadia, a figure which Eratosthenes later adjusted to 252,000, the latter number evenly dividable by twelve and therefore mathematically more convenient. A technique so brilliant in its simplicity that twenty-three centuries later it is routinely taught in elementary schools around the same globe it once created.

In hindsight, the question is not whether Eratosthenes' technique was theoretically sound, which it is. It is rather whether his results are empirically true, which may or may not be the case. Thus, if one assumes that he used the long stade, then the circumference of the earth came out to be about 46,000 kilometers, a figure 15 percent off the correct value of 40,234 kilometers. If, on the other hand, one assumes that he used the short stade, then the estimate would have been merely 150 kilometers, or 1 percent, wrong![21] The underlying errors were several, most importantly (a) the location of Syene is not exactly on the Tropic of Cancer, but 60 kilometers to the north; (b) Syene and Alexandria are not on the same meridian, but about 3° apart; (c) the distance between the two points was derived from the travel speed of camel caravans, hardly a precise unit of measurement; and (d) nobody knows the exact relation between one stade and one meter.

Still an embracing achievement, an Aristotelian synthesis in which Syene and Alexandria were treated as concrete end-points in an abstract line 250,000 stadia long, geometrical ingredients in a scientific approach easy to modify, impossible to destroy. Shadow of a smile spreading across the face of the Librarian, invisible furrows of latitude and longitude for ever cut into the face of the earth.

*

**Latitude and longitude** came to form the skeleton also of Eratosthenes' world map, the second of his major contributions. In that case, however, the fix-point of fix-points was no longer the Library of Alexandria but the observatory on top of Mount Atabyrion on the island of Rhodes. It was in that point that he let two perpendicular lines, one horizontal and the other vertical, intersect to form the center of a coordinate net which divided the world into four quadrants, three winds blowing in each. Up and down; front and back; left and right; head and feet; face and arse;[22] in the middle the navel of the Vitruvian omphalos, pulsating origo of understanding. Traces of submerged beginnings, fragments recovered from the bottom of Apsu's water-filled abyss. The round world flattened on a heated grill, McDonald burgers in advance of themselves.[23]

Eratosthenes' horizontal line was well established and known to have appeared already on the Ionian maps from the sixth century BCE, where it was normally labeled "the Equator." Dicaerchus, disciple of Aristotle and forerunner of Eratosthenes, called it *diaphragma*—a most appropriate term given the fact that the division into north and south was never an issue of sharp limits but more of a fuzzy zone, the metaphorical midriff of a world which breathes, laughs and suffers its hiccups. Its beginning lay somewhere in the waters beyond the Pillars of Heracles (the Straits of Gibraltar), from where it hugged the axis of the Mediterranean and followed the length of the Taurus mountains, eventually reaching the eastern Ocean at a point specified as "the furthest capes of Asia."[24]

In comparison, the vertical line exhibited more twists and turns, but it began in the vaguely defined Cinnamon country, passed Meroe in Sudan, joined the Syene-Alexandria meridian along the Nile, continued north via Rhodes through the Straits of Bosphorus across the Black Sea to the mouth of the Borysthenes (modern Dnieper), finally reaching the island of Thule, still believed to sit on the Arctic Circle six days' voyage northeast of Britain. To modern eyes hardly a straight line.

No criticism intended, though, for it was to take much thought and many drowned seamen before the measuring of meridians reached the same level of accuracy as the measuring of latitudes. In deed it did not occur until the mid-eighteenth century, when the British horologist John Harrison, under constant harassment from the astronomers on the Board of Longitude, designed a series of chronometers that made it possible.[25] What the gnomon long ago had done to the parallels of latitudes (never crossing but all of varying lengths, longer close to the equator, shorter close to the poles), Harrison's cogwheels now did to the converging circles of longitudes (all intersecting at the two poles, all of the same length).[26] It should be noted explicitly that the establishment of the precise position of a point in space once again depended on techniques for determining the exact moment of a point in time—the summer solstice for the parallels, the sunrise for the meridians.

Let there be time and there is shadow. Let there be stars and there is direction. Time is the space that may not be seen. And as a mirage the old Sun God reappears again, albeit in new clothes and under another alias.

Intersecting the main axes of Eratosthenes' map were additional lines of latitude and longitude, for the sake of convenience drawn perpendicular to each other. This use of right angles automatically made the map itself rectangular, the known world believed to measure about 38,000 stadia from north to south, 78,000 stadia from east to west. Most importantly, the land mass was still depicted as an island surrounded by the apeiron of a continuous Ocean. And even though the map was meant to be drawn to scale, the oikumene reached little more than one third around the globe, the distance to the east greatly exaggerated.[27] Even so, Eratosthenes' work was a wonderful achievement, primarily because he tried "to build a struc-

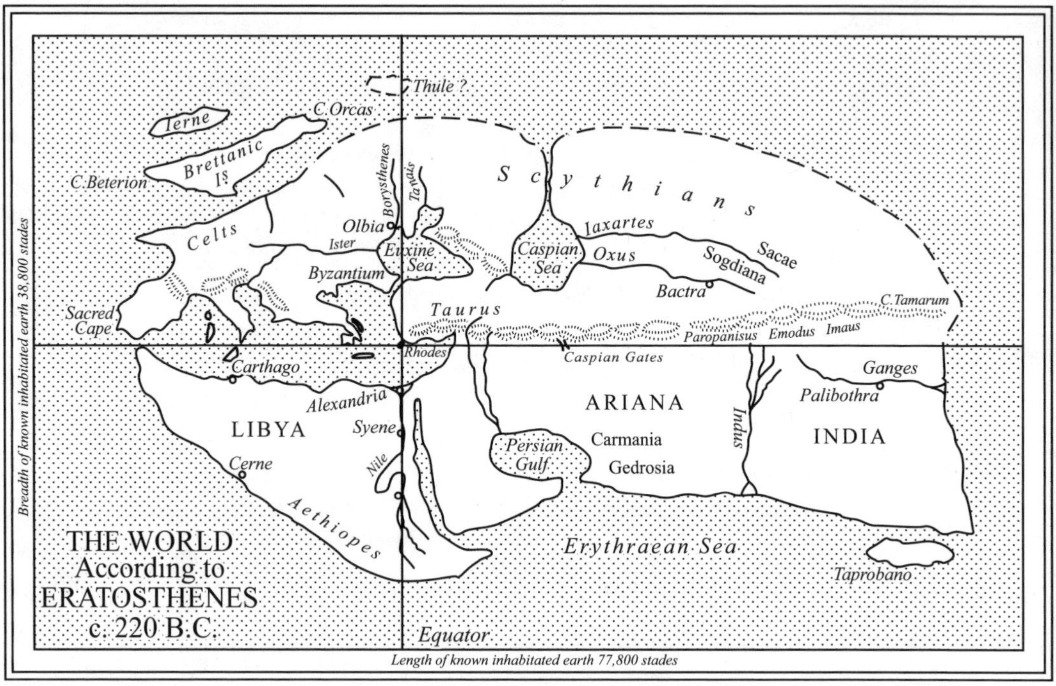


Reconstruction of Eratosthenes' map of the world. Drawing from O. A. W. Dilke, *Greek and Roman Maps* (Ithaca, N.Y.: Cornell University Press, 1985), 33.

ture of abstract geometrical lines and shapes which did not represent anything real in the geographical space but made visible mathematical relationships within the orthogonal frame of the map."[28] Eratosthenes' map was a computing device, his thinking syllogistic.

So it can be told, the fascinating story of how the "Beta" translated the language of concrete geography into the idiom of abstract geometry, winding rivers of nature into straight lines of culture, the observatory on Mount Atabyrion the pivot of the world. In the words of one commentator, "a very striking achievement . . . the first really scientific Greek map [that] affected world maps right down to the Age of Discovery."[29]

Not good enough, though, perhaps because Eratosthenes too readily succumbed to the Greek weakness for parallelism, in his case operationalized by means of straight lines and right angles, a coordinate net that makes it impossible to consider seriously the problem of the earth being round and the map being flat.

Of this imperfect translation Eratosthenes was well aware, even though he chose to ignore it. And for good reasons. The first to take the challenge seriously was instead Claudius Ptolemaeus, yet another astronomer-turned-geographer, yet another practitioner who understood that geography is most properly defined as a geometry with names.

*

**Claudius Ptolemaeus,** better known as Ptolemy, was born in upper Egypt around 90 CE and died in Alexandria, probably in the year 168. His first name was Roman, the second he shared with the Macedonian dynasty that ruled Egypt at the time, a family to which he was not related. His everyday language was Greek, his everyday home a real-world place with a real-world name, in his professional jargon no longer spelled A-L-E-X-A-N-D-R-I-A but 31°13′N, 60°30′E.[30] His first major work was an astronomical encyclopedia entitled *Syntaxis,* but better known as *Almagest* (the Arabic rendition of the Greek for "the Greatest [Compilation]"), the second the enormous *Geographike Hyphegesis* ("Manual of Geography," usually called *Geography*).[31] Sandwiched between these masterpieces was the astrological treatise *Tetrabiblos,* Ptolemy's attempt to correlate the supposed characteristics of various peoples with the positions of zodiacal signs and planets.

The purpose of the wide-ranging *Almagest* was to teach the reader how to draw a celestial globe, a representation of the universe in which the heavens were conceived as an immense sphere that rotates daily around an axis which at its center has another sphere, the nonrotating earth. According to this two-sphere model of cosmos the stars are affixed to the *inside* of the celestial globe, but Ptolemy was careful to stress that he described them not as they really are but as they appear to an observer placed on the terrestrial globe. The earthly stargazer was consequently one with the fix-point of fix-points. What I see in the sky depends on where I stand on the earth.

Also the *Geography* dealt with issues of representation, but there the focus was on the more limited problem of how to draw a picture of the inhabited world either as a sphere on a globe or as a plane surface on a map. Seemingly eager to tie the diverging approaches of the *Almagest* and the *Geography* together, Ptolemy professed that "these things belong to the loftiest and loveliest of intellectual pursuits, namely to exhibit to human understanding through mathematics [both] the heavens themselves in their physical nature (since they can be seen in their revolution about us), and [the nature of] the earth through a portrait (since the real [earth], being enormous and not surrounding us, cannot be inspected by any one person either as a whole or as a part)."[32]

The exact wording is revealing, for here Ptolemy explicitly declares that his cartography is best understood as a form of pictorial art. And yet, even though his ambitious goal was to produce a copy of an original, he was well aware that this desire is impossible to satisfy; the reason is, of course, that in the real world of representation there is nothing like a true copy, merely an array of creative mistranslations. Heeding Aristotle's advice, the best we can do is to do as well as we can, choosing our approaches accordingly. As proof, witness the very first words of *Geography*'s Book One, a chapter entitled "On the Difference Between World Cartography and Regional Cartography":

> World cartography is an imitation through drawing of the entire known part of the world together with the things that are, broadly speaking, connected with it.[33] It differs from regional cartography in that regional cartography, as an independent discipline, sets out the individual localities, each one independently and by itself, registering practically everything down to the last thing therein (for example, harbors, towns, districts, branches of principal rivers, and so on), while the essence of world cartography is to show the known world as a single and continuous entity, its nature and how it is situated, [taking account] only of the things that are associated with it in its broader, general outlines (such as gulfs, great cities, the more notable peoples and rivers, and the more noteworthy things of each kind).
>
> The goal of regional cartography is an impression of a part, as when one makes an image of just an ear or an eye; but [the goal] of world geography is a general view, analogous to making a portrait of the whole head. . . . Consequently regional geography requires landscape drawing, and no one but a man skilled in drawing would do regional cartography. But world cartography does not [require this] at all, since it enables one to show the positions and general configurations [of features] purely by means of lines and labels.[34]

*

**Lines and labels.** Those were Ptolemy's terms for what modern semioticians call visual marks and proper names. Sounds like a poststructuralist to me. And like a cubist painter he sensed that no one but a man skilled in the imitative art of drawing can picture a hydra with many faces and hidden arses. One problem solved, another created as the smoke from the sacrificial burning was seeping into the room instead of passing through the chimney. And thus it was that

> Emerson smoaked it all right away. "If it's but the empty places between the Towns," he advis'd Dixon, "your worries are at an end, for look what you can do. You can get above it." He spoke these words with an emphasis Dixon cannot describe the full strangeness of. Something was up, . . . but before they learn'd to fly, they had to learn about Maps, for maps are the Aides-mémoires of flight. So Dixon came to discover as well the great Invariance whereby, aloft, one gains exactitude of Length and Breadth, only to loose much of the land's Relievo, or Dimension of Height,—whilst back at ground level, traveling about the Country, one regains bodily the realities of up and down, only to lose any but a rough sense of the other two Dimensions, now all but one.

"Earthbound," Emerson continued, "we are limited to our Horizon, which sometimes is to be measur'd but in inches.—We are bound withal to Time, and the amounts of it spent getting from one end of a journey to another. Yet aloft, in Map-space, origins, destinations, any Termini, hardly seem to matter,—one can comprehend all at once the entire plexity of possible journeys, set as one is above Distance, above Time itself."

"Altitude!" cried out a couple of alert youths,—as, in Emerson's class they were encourag'd to do.

"Altitude, being the Price we pay for this great Exemption, is consider'd as an in-house Expense, to be absorb'd in an inner term of lengthy Expression describing Location, Course, and Speed. If you're interested, wait for my book upon Navigation, currently all but in Galley-proofs, for a detail'd Account."[35]

In a not too different sense the maps that eventually came out of Ptolemy's hands (it remains uncertain whether it was he or someone else who actually drew them) were based on information about altitude, albeit strangely furnished primarily by Marinus of Tyre (active in the years around 100 CE) and Strabo (64/63 BCE–23 CE), these compilers in turn updating older accounts provided by Hipparchus (active at least from 162 to 126 BCE), Eratosthenes and Pytheas. The collected data were of two types, one coming from various travel stories and brought together in a long list of ordinary place names and geographical descriptions, the other a catalogue of latitudes and longitudes, some determined with the aid of gnomons and astrolabes, others by less reliable means—it is estimated that even as late as 1740 merely 116 places on the entire earth had been correctly located by astronomical observation.[36]

Once these secret lists of triple-named coordinates had been assembled—an amazing total of about 8,000—the next step in the mapping process was to turn the collected data-sets into meaning-filled points, lines, and planes, i.e. into geometrical entities which could be projected onto a sheet of papyrus, the cartographer's version of the painter's canvas and the movie-maker's screen. This third step of projection was in many ways the most crucial, for it was through this operation that the abstract concepts of lines and labels were ontologically transformed into material objects—the points turned into cities, small islands, and the mouths of minor rivers; the lines becoming coastlines, longer rivers, and mountain chains; the planes changed first into provinces, and satrapies eventually into the papyrus itself—in the vocabulary of the Romans a papyrus roll was in fact the *charta* itself.[37]

The challenge is enormous, for the regional geographer must know not only how to draw the ear and the eye of a particular place but how to paint a portrait of the entire oikumene as well. The task of the world geographer is even more demanding, for he cannot, as the regional geographer, ignore the fact that the earth is a sphere and therefore more faith-

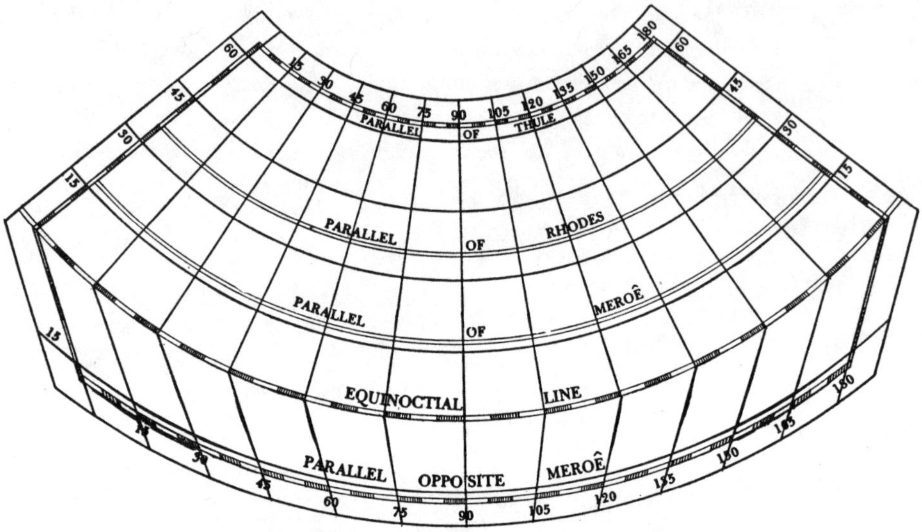


Reconstruction of Ptolemy's conic map projection. Drawing from Lloyd A. Brown, *The Story of Maps* (Boston: Little, Brown and Company, 1949), 70.

fully rendered in a three-dimensional sculpture than in a two-dimensional picture. It follows that while the interests of the regional cartographer are best served by a flat map on which the straight lines of latitude and longitude are intersecting at right angles, the world cartographer is better off with a spherical globe. The only problem with the latter is that in order to be useful to anyone but school teachers and interior decorators, the globe must be so big that no one is strong enough to handle it. Hence the recommendation that "the size of the [globe] should be determined by the number of things that the map-maker intends to inscribe [on it]; and this depends on his competence and ambition."[38]

Ptolemy's own competence and ambition were beautifully demonstrated in his three map projections, all of them pictorial inventions designed to flatten the earth without destroying it. The first was a conical projection constructed with straight converging meridians and curved parallels, the second a modified spherical projection with curved converging meridians and curved parallels. In comparison, the third was considerably more complicated, in effect an attempt to draw the globe as it appears from a distance, an image not unlike the pictures that much later were to be drawn according to the rules of perspective.[39] Regardless of their differences, each map projection had north at the top and east at the right, a pragmatic decision which reflected the belief that most of the oikumene was located in the northern latitudes and therefore most readily studied if pictured in the upper right-hand corner of the map. The birth of a convention so ingrained that most map users never even think of it.

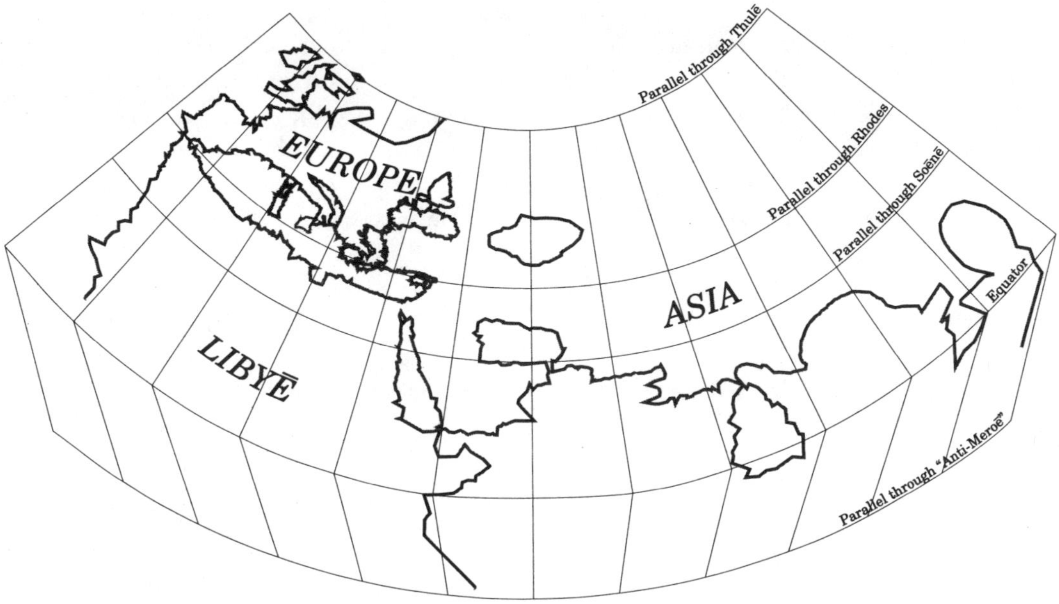


Ptolemy's oikumene caught in his first projection. Drawing from J. Lennart Berggren and Alexander Jones, *Ptolemy's Geography* (Princeton, N.J.: Princeton University Press, 2000), 129. Reprinted by permission of the Princeton University Press.

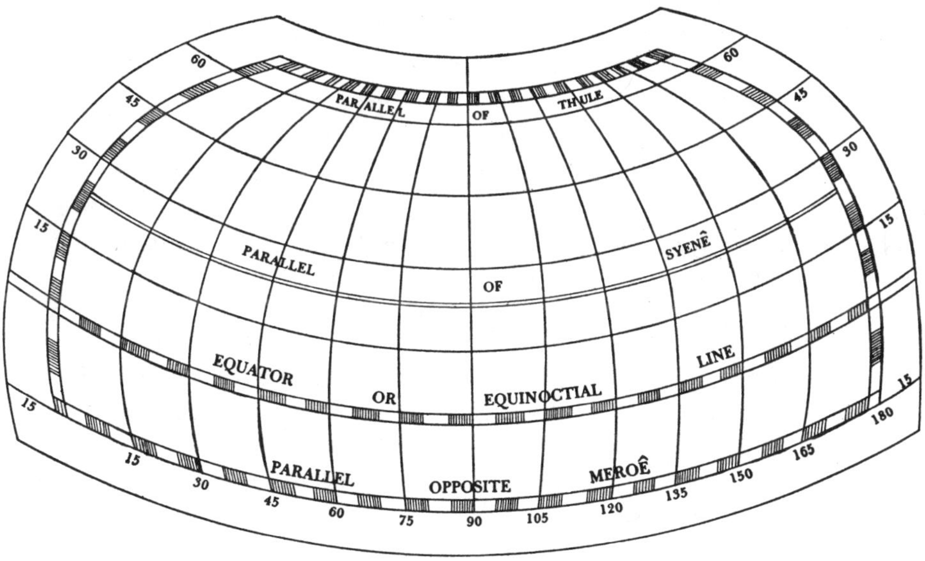


Reconstruction of Ptolemy's spherical map projection. Drawing from Lloyd A. Brown, *The Story of Maps* (Boston: Little, Brown and Company, 1949), 71.

Perhaps the most remarkable aspect of the conical projection is that it is hung on an invisible point which is located outside the rectangular area of the oikumene, a geometrical construction at which the straight lines of the meridians converge and around which the circular segments of the parallels are drawn. In Ptolemy's own words this invisible point placed at the top of the map may be "*imagined* as the north pole,"[40] in Lacanian terms a sign not real but symbolic. Just as Pytheas half a millennium earlier had anchored his world in the invisible point of the celestial pole, so Ptolemy did as well. Yet another ontological juggler later to be arisen again from the dead, reborn and, through the words of resurrected manuscripts, destined to change the world.

And change the world is exactly what the *Geography* did, even though the current reader should be warned that its subsequent influence may have been more due to its mistakes than to its advances. In the words of one commentator, "the Geography was both a keystone and a millstone, a pioneering effort that outlived its usefulness. [While] his hypothetical map was excellent . . . his world of reality was faulty."[41]

In hindsight, the three projections were Ptolemy's greatest contribution. However, it must also be counted to his credit that he abandoned the Greek conception of the inhabited world as surrounded by an Ocean river, arguing instead for the possibility, indeed the probability, of a Terra Incognita, an imaginary world beyond the limits of arbitrarily determined boundary lines. In addition he perfected the art of trigonometry originally invented by Hipparchus and subsequently turned into an indispensable tool for all land surveyors. But Ptolemy also bears responsibility for two serious errors, both of which could have been avoided had he put more trust in the spectacular theories of Aristotle and Eratosthenes than in the alleged facts of Strabo and Marinus.

Ptolemy's first and most serious mistake was to accept the geocentric theory according to which the nonrevolving earth is placed at the center of the universe, an assumption which already Aristarchos of Samos had rejected, his second that he figured the earth to be considerably smaller than it actually is. While current opinion is that Eratosthenes overestimated the circumference by about fifteen percent (if he miscalculated at all!), the entire constellation of Hipparchus, Strabo, Marinus and Ptolemy himself all underestimated it by about eighteen percent. Centuries later it was this error that took Christopher Columbus not to the India he knew but to the real Terra Incognita of the Americas, an unknown and dangerous world literally beyond the limits not only of the Greek oikumene but of his own imagination, an invention more than a discovery. In the bingo of unforeseen consequences creative misunderstanding is the key to success.[42]

Ptolemy was clearly obsessed by the idea of making the invisible visible, of capturing the wholeness of the world—up and down, front and back, north and south, east and west—everything in the same all-embracing picture. Ptolemy, a forerunner of Albrecht Dürer's artist struggling with the problem of perspective, he himself seated on one side of the gridded window, the voluptuous nude reclining on the other. Ptolemy, the Cinderella of Geography poisoned by the threads of his own matrix, after a millennium in the dungeon of oblivion finally resurrected again, a graven image charting the routes to the New World, a place where it was later observed that

> History is not Chronology, for that is left to lawyers,—nor is it Remembrance, for Remembrance belongs to the People.[43]

Facts are nevertheless facts. And although all roads lead to Rome, the city was not built in one day. Most importantly Rome was not Alexandria, the Emperor not the Librarian. As a consequence, and under the influence of Christian theology, the clear-headedness of Greek science eventually evaporated into the thin air of dogma and mysticism. Gone were the likes of Pytheas, Eratosthenes and Ptolemy, imaginative forgers of the uncreated conscience of their culture. Rising from the mist-enveloped regions of religion were instead a host of monks and bishops, some as sensitive and sharp as anyone before or after, others poised on the tip of a pin engaged in fierce battles with and among themselves.[44] Special, though not unique, was a man called Cosmas Indicopleustes, one of the most remarkable children of cartographical reason ever to be. "For neither before him was any like unto him, neither shall be thereafter."[45]

*

**Cosmas Indicopleustes**—"Cosmas the Indian Sea Traveler"—was not the proper name of a real individual,[46] but the (in)definite description of an Alexandrian Greek active during the first half of the sixth century. After a successful career in the spice trade, this former merchant retired to the monastery of the Raithu on the Sinai peninsula, perhaps taking the vows to mark the conversion. In that new environment, if not exactly in the cloisters of that same place, he first produced a number of reports from his extensive travels primarily to Ethiopia,[47] all of them now lost, then the remarkable *Christian Topography,* most likely composed between the years of 547 and 549.[48]

As with Pytheas, Eratosthenes and Ptolemy before them, the sixth-century mappers also set out to capture the relations between the invisible world of thought and the visible world of matter. In Alexandria at the time these thorny issues turned even more devilish when they were caught in the crosscurrents of Greek and Christian belief systems, themselves under constant and fierce reevaluation.

What most distinguished Cosmas from the other participants in those debates is that he was not a learned man but essentially an autodidact, rhetorically weak and ignorant of syntax, vulgar and highly repetitive in his expressions; in the words of a later commentator, his "reading was as wide as his reflection was infantile."[49] Few were those who read him, fewer still those who believed him. As a consequence, the message of the *Christian Topography* was quickly forgotten and not brought to light again until 1776, when a Latin translation happened to be discovered.[50]

What makes Cosmas' book so fascinating is consequently not what it says but what it shows: cartographical reason in nuce. Most importantly, he seems unwittingly to have accepted Ptolemy's argument that "world cartography is an imitation through drawing of the entire known part of the world together with the things that are, broadly speaking, connected with it."[51] But whereas Ptolemy set out to capture the visible world of geography, Cosmas tried to draw the invisible world of theology, the former finding his way by looking at the stars in the sky, the latter by reading the words of the Bible.[52] These profound differences notwithstanding, both artists chose to hang their respective pictures on hooks of shadows, Ptolemy on blocked sun-light, Cosmas on the New Testament foretold by the Old.

Gnomons of different kinds, gnomons nevertheless. For just as Eratosthenes had argued that what was visible in Alexandria was invisible in Syene, so the Church Fathers turned the touchable Jesus into the incarnate Son of the nontouchable Father. And just as the old Greeks had translated astronomy into geography, so the young Christians turned cosmography into prophecy; while to Ptolemy the earth was a round ball, to Cosmas it was as flat as it appeared to be,[53] a disk hung on nothing but itself.[54] And yet it must be repeated that both Ptolemy and Cosmas were driven by the same desire of imitating the world in drawing—not, however, of depicting the heavenly life up there or the human life down here taken separately, but of capturing the complex relations between them.

In this flip-flopping context of mapping, Plato's concepts of Form and Idea remained forever present, the Divided Line of the *Republic* nothing but the map of maps. As a wonderful illustration, Thomas Pynchon let DePugh recall

> a Sermon he once heard at a church-ful of German Mysticks. "It might have been a lecture on Mathematics. Hell, beneath our feet, bounded,—Heaven, above our pates, unbounded. Hell a collapsing Sphere, Heaven an expanding one. The enclosure of Punishment, the release of Salvation. Sin leading us naturally to Hell and Compression, as doth Grace to Heaven, and Rarefaction. Thus—"
>
> Murmurs of "'Thus'?"
>
> "— may each point of Heaven be mapp'd, or projected, upon each point of Hell, and vice versa. And what intercepts the Projection, about mid-way (reckon'd logarithmically) between? Why, this very Earth, and our lives upon it. We only think we occupy a solid, Brick-and-Timber City,—in Reality, we live upon a Map. Perhaps even our Lives are but rep-

> resentations of Truer Lives, pursued above and below, as to Philadelphia correspond both a vast heavenly City, and a crowded niche of Hell, each element of one faithfully mirror'd in the others."
>
> "There are a Mason and Dixon in Hell, you mean?" inquires Ethelmer, "attempting eternally to draw a perfect Arc of Considerably Lesser Circle?"
>
> "Impossible," ventures the Rev$^{d}$. "For is Hell, by this Scheme, not a Point, without Dimension?"
>
> "Indeed. Yet, suppose Hell to be almost a Point," argues the doughty DePugh. . . .
>
> "What puzzles me, DeP., is that if the volume of Hell may be taken as small as you like, yet the Souls therein must be ever smaller, mustn't they,—there being, by now, easily millions there?"[55]

Like Mason and Dixon in Philadelphia, the sixth-century Alexandrians could not help but notice that whatever they imagined to be up there and down here refused to sit still; in both realms identity and difference seemed in perpetual, rhythmic and creative flux.[56] Not only was the Nicene definition of Jesus Christ subjected to interminable discussion, the Greek conception of the physical universe was likewise. Spanning the gap between the scientific and theological sides of the abyss was a man widely known as John Philoponus. His first name suggests that he came from a Christian family, his second (in effect not a proper name but an (in)definite description) that he was a "lover of work." Given his large output this nickname seems most appropriate.[57] But he too, in Reality, lived upon a Map. For in the World of Understanding there is no other place to settle.

# IN-BETWEEN

Man dansar däruppe—klarvaket
är huset fast klockan är tolv.
Då slår det mej plötsligt att taket,
mitt tak, är en annans golv.
NILS FERLIN, "Infall"[1]

**John Philoponus** (born around 490, dead sometime in the 570s) devoted his life to the Christianization of Hellenistic thought in general and to the mummification of Ptolemy's *Almagest* in particular.[2] Already at a young age he orchestrated a massive assault on Aristotle's physics and cosmology, a comprehensive critique which it took a thousand years and a Galileo to match; in deed it has been suggested that "it is no exaggeration to call John Philoponus a sixth-century Archimedes in philosophy."[3]

The most remarkable aspect of this man and his revolutionary philosophy is that it grew out of the conviction that the universe is the single creation of a single God; to be precise, he identified Aristotle's conception of the first cause with the Christian notion of a personal God. In turn this twist led him to reject the prevalent belief in the world's eternity and to accept instead the proposition that heaven and earth are constituted by the same physical properties and governed by the same physical laws, these laws themselves derived from an elegantly argued conception of infinity. To Philoponus the universe was in fact a vast mechanism in which the stars were balls of fire not aether-filled homes of angels, in short a collection of physical objects created out of a nonphysical entity into which they will eventually also perish. Dust to dust, except in this case the dust is not real but imaginary and symbolic.

Closely connected with this belief in *creatio ex nihilo*—to the Greek culture the most forbidden of everything forbidden—was also Philoponus' revision of Aristotle's theory of dynamics; rejected was the idea about the thrown discus being pushed by the air behind it, accepted was instead the argument that the thrower imparts a kind of kinetic power into whatever is thrown, a power which is lodged in the object and keeps it going until the energy is consumed and the discus reaches its resting place on the ground. And so it is that Philoponus' groundbreaking revisions of the scientific dogma (including his novel theories of vacuum, time and place) all were grounded in his monotheistic belief in one Creator.

But Philoponus was not merely a scientist steeped in religion; he was equally much a theologian steeped in logic. In line with his creationist critique of Aristotle he consequently insisted that since every nature is individualized, the person of Jesus Christ must be one nature as well. This argument was itself typical of the Monophysites, a group which up to the present day has played a dominant role among Alexandria's Christians; perhaps the Egyptians found the Monophysite perspective so attractive because they were brought up in a culture that traditionally had treated the earthly ruler as a heavenly god. Theologically the disputes were about the definition of Jesus Christ, politically about the rivalry between Constantinople and Alexandria. It is ironic that even though Philoponus entered the debate with the hope of adding some logical clarity, he left it in such disarray that a century later he himself was placed under formal anathema.

The discussion reaches to the heart of Christology and in a later chapter ("Nicaea") it will resurface again. For the moment it suffices to note that the controversy was temporarily resolved by the Council of Chalcedon in 451. However, as none of the factions was willing to accept the proposed compromise, the fight went on, seemingly without end. The warring schools were two—one, under the influence of Theodore of Mopsuestia (d. 427)[4] and Nestorius of Constantinople (d. 451),[5] which was centered at Antioch; the other, with Cyril (d. 444) as its leader, that was primarily associated with Alexandria. At issue was the question of how to express the unity of the person in whom both God and man were held to be present.[6]

To cut the proposed answers to the extremest minimum, the Antiochians—Cosmas Indicopleustes included—claimed that the title *Theotokos,* "Mother of God," did not properly belong to Mary. This demotion of the Holy Virgin from the rank of "Mother of God" to "Mother of Jesus" had the further consequence that Christ must have two natures, one divine the other human, the two distinguishable not by empirical observation but by mental abstraction alone. According to this Nestorian belief the incarnate Lord was consequently a sublime partnership out of which Jesus emerged as a second Adam, by definition redeemer supreme. In contrast, the Alexandrians—John Philoponus prominent among them—argued that in Christ's person there is a single nature which constitutes a synthetic bonding of body and soul; accordingly, redemption comes not indirectly by way of the second Adam but directly through the sovereign power of the Creator himself.

Throughout the debates frequent references were made to the discussion of identity and difference as it first appeared in Plato's *Parmenides* and then reappeared again in Plotinus' concept of *hypostasis,* an exceedingly complicated term which sometimes stands for a thing and sometimes for an action—the literal meaning is "putting underneath" or "that which subsists or underlies," in a more extended sense "foundation," "reality," even "the real Being."[7] Playing at the dangerous edge of etymology, hypo-stasis thus turns to under-standing, *hypo-thesis* to under-statement.

And there he is, Atlas balancing the globe on his shoulders, Humpty Dumpty moments before the fall. Lenin knew it well: you cannot make an omelet without breaking some eggs.

*

**Plotinus** (c. 205–70) was a forerunner of it all; a borderline philosopher and mystic, founder of Neoplatonism; a cosmopolitan Greek born in Egypt, dead in Rome; an exceptional teacher closely affiliated first with Emperor Gordianus III, with whose expeditionary force he marched against Persia, then with Emperor Gallenius, to whom he proposed the construction of a Platopolis, an abstract idea never transformed into concrete stone.[8]

Known primarily through his student Porphyry, Plotinus' work is collected in the six *Enneads,* all of them now translated into several languages. In any critique of cartographical reason his philosophy is bound to play a pivotal role, partly because it belongs to the interface between the ancient and the medieval, partly because it explicitly focuses on the understanding of ontological transformations and thereby on the relations between invisible ideas and visible things. Much of it can be read as creative explorations of Plato—by Plotinus often referred to simply as "he"—not the least of the Divided Line of the *Republic,* the scale by which "the realm of things" and "the realm of affections in the soul" are mapped onto each other. I consequently conceive of Plato's Line in the first instance as a divid*ing* line, in the second as a divid*ed* line; while the former represents the abyss between the two realms of ontology, the latter is best imagined as a kind of thermometer designed to measure the degree of thinghood in the two realms, the intelligible more or less hotly spiritual, the opinionable more or less coolly material. Most importantly, the major marking (the suspension bridge) across the divid*ing* line signifies the topos of ontological transformation, the metaphorical counterpart of Celsius' zero, the limit at which fleeting ideas freeze into concrete things and concrete things evaporate into thin air. So here it is, the scale—in effect the coordinate net—of Plato's map of maps:[9]

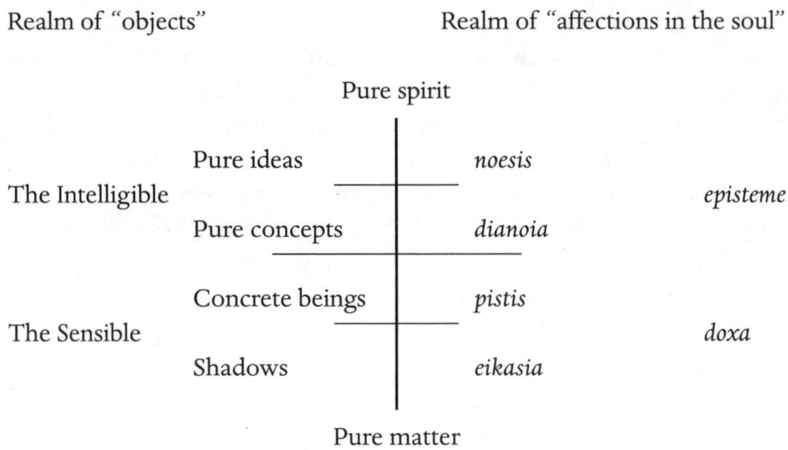


> "I understand," he said, "not fully, for it is no slight task that you appear to have in mind, but I do understand that you mean to distinguish the aspect of reality and the intelligible, which is contemplated by the power of dialectic, as something truer and more exact than the object of the so-called arts and sciences whose assumptions are arbitrary starting points. . . . And I think you call the mental habit of geometers and their like mind or understanding and not reason because you regard understanding as something intermediate between opinion and reason."
>
> Your interpretation is quite sufficient," I said. (Plato, *Republic* 511c–d)

As Socrates thus confirms, the scale of the Divided Line embodies the idea that degrees of "truth" correspond to degrees of "being"; more precisely, the highest point on the Line marks the highest truth, everything lower being merely diluted versions of the Good of the good. Transferred to the world of Plotinus this elevated point coincides with what he called "the One," by definition the hypostasis of hypostases. Thus, just like Plato before him, Plotinus imagined the universe of thought-and-action as a multilayered edifice in which the sand of ontology and the cement of epistemology were stirred together into concrete slabs of building material, each division of the Divided Line corresponding to a particular form of understanding, the physical world of things always a show of the spiritual world of affections.

The top floor of Plotinus' edifice is filled by "the One." On the two floors immediately below are "the Intelligent" and "the Soul," while relegated to the dungeons underneath are the material objects which themselves appear not as the shadows they actually are but as images draped in thinghoods of different thickness. In his own words:

> *The One* is all things and not a single of them: it is the principle of all things, not all things, but all things have that other kind of transcendent existence; for in a way they do occur in the One; or rather they are not

> there yet, but they will be. . . . The One, perfect because it seeks nothing, has nothing, and needs nothing, overflows, as it were, and its superabundance makes something other than itself. This, when it has come into being, turns back upon the One and is filled, and becomes the *Intellect* by looking towards it. . . . Resembling the One thus, Intellect produces in the same way, pouring forth a multiple power—this is a likeness of it—just as that which was before it poured it forth. This activity springing from the substance of the Intellect is *Soul,* which comes to be this while Intellect abides unchanged: for Intellect too comes into being while that which is before it abides unchanged. But Soul does not abide unchanged when it produces: it is moved and so brings forth an *image.* It looks to its source and is filled, and going forth to another opposed movement generates its own image, which is sensation and the principle of growth in plants. Nothing is separated or cut off from that which is before it. For this reason the higher soul seems to reach as far as *plants.* . . . All these things are the One and not the One.[10]

And thus the blueprint of Plotinus' universe, a construction of understanding with several levels, each one resting on its own hypo-stasis:

THE ONE

THE INTELLECT

THE SOUL

---

image

plant

Put differently, Plotinus' highest principle—the One—is by necessity beyond reflection because it simply is, a nontranslatable entity which I will later refer to as

$$a$$

On the second level this unity begins, through the power of the Intellect, to reflect on itself, with the result that the original $a$ is translated into the tautology of

$$a = a$$

Then, on the third level of the Soul, it is the Intellect (itself a reflection of the One) that is reflected in yet another round of differentiation and individualization, such that the tautological and uninformative $a = a$ transcends itself into the informative

$$a = b$$

Further down—in the dungeon of things—appear first the images and then the plants:

$$a = b = c$$

and

$$a = b = c = d$$

And so on and on until the very end at which every attempt at understanding approaches the silence of pure matter; however, it is not the Soul itself that descends into the body, merely a kind of light that makes the body visible. The concept of Being—like the concept of number—is consequently not a primary phenomenon but a synthetic unity of sameness and otherness, the one and the many.

In Plotinus, as in Plato, consciousness precedes creation because thought always comes before being; the Soul is not *in* the body, the body is *of* the Soul. On that account, matter is in fact infinite and unlimited—*to apeiron*—something which is "not yet stable by itself, and is carried about here and there into every form, and since it is altogether adaptable, becomes many by being brought about into everything and becoming everything."[11] It follows that the main distinction in Plotinus' anthropology is not between body and soul, but between a lower form of being, in which the Soul is connected with the flesh, and a higher form of being, in which the Soul emanates from the Intellect. And yet, since this in-between mode of reasoning is not dialectical but hierarchical, the Intellect is always a picture of the One, the Soul always a picture of the Intellect. To paraphrase: to be human is in that conception to be all things and not a single one of them: "the true self is thus distinct from the empirical living composite, which is made up of the body and [of] the *dynamis,* image, or shadow, of the soul."[12]

Already this preliminary house-tour suggests that finding one's way around the Plotinian edifice is far from straightforward.[13] One reason is that the connections to the *chora* of the *Timaeus* are striking,[14] another that the concept of hypostasis itself begins to shift around, sometimes functioning as the floor of under-standing, sometimes as the ceiling of language. In my mind a hypotasis is consequently not a locked box but a thin line—perhaps more appropriately a tightrope—which at the same time divides and ties together the various modes and levels of thought-and-action. Easy to get hurt.

*

**Hypostatis is an infra-thin line**, the floors and ceilings of the Plotinian House.[15] As might be expected, the historical interpretations of the term *hypostasis* resurfaced in the Christological debates first at the Council of Nicaea in 325, then at the Council of Chalcedon in 451. For at the very moment when the warring factions at Chalcedon were ready to agree on a clever (i.e., ambiguous) formulation, Cyril of Alexandria surprisingly noted that at the heart of the hypostasis lies the tension between the two prepositions *of* and *in*. Once this had been discovered, the quarrel started anew, for whereas the Alexandrian Monophysites believed that Christ is a real union of God and man, hence one person and one single end-product, the Antiochian Nestorians retorted that the expression "known in two natures" is not compatible with the assertion of one hypostasis. The debate went on and on until a century later, when John Philoponus (remember him?!) tried to close it by translating it into the language of pagan logic.

However, that latter exercise had hardly started before it ran into its own problems, for Philoponus quickly concluded that the art of understanding prepositions is less a question of logic and more a case of cartographical reason; by its own word a preposition is a *pre*-position, a position assumed in advance. In the important text of the *Arbiter*, Philoponus therefore chose to focus exactly on this crucial point, emphatically stressing that to a Monophysite the term "*in* the same nature" is not an acceptable characterization of Christ. Philoponus' argument was in fact that whenever we are using the preposition *in* we are effectively stating that the whole is not in the whole but in the conjunction of separate parts. But surely no believer would ever limit Christ to his hands, eyes, ears, nose and tongue. Thus, while we may very well say that a man is *of* soul and body, we cannot meaningfully say that he is *in* soul and body.

Given that description of the situation, Philoponus suggested that we may distinguish between three sorts of mixture: first, "juxtaposition" as in the house built in wood and stone; secondly, "blending" as in the case of wine and water; thirdly, "interweaving" as in the ropes that are tied together from different strands. On that basis he then argued that it is only the interweaving (a state characterized by *sympatheia,* not confusion) that properly describes the union of soul and body; in turn this argument eventually led him to conclude that "one composite hypostasis must mean one composite nature";[16] if Christ is one, then his nature and hypostasis must necessarily be one as well.

For us it is important to note that when Philoponus approached these eternal questions his Christology was structured exactly as his anthropology: while in Christ there is Logos, in man there is rational Soul. And so it is that in today's world the major concern is no longer with the theological problems of Christology but with the political problems of representative democracy and minority politics. What, for instance, does it mean to be a woman, if "woman" is an interweaving of social construction and biological necessity, a sympatheia of the currently politically correct? Which really are the relations between the president or prime minister, virtually always elected by a minority, and those citizens whose explicit trust (s)he has asked for but not been granted?

> "Are even village Idiots taken in any more by that empty cant?" mutters the tiny Topman McNoise, "no more virtual than virtuous, and no more virtuous than the vilest of that narrow room-ful of shoving, beef-faced Louts, to which you refer,—their honor bought and sold so many times o'er that no one bothers more to keep count.—Suggest you Sir, even in Play, that this giggling Rout of poxy half-wits, *embody* us? Embody *us*? America but some fairy Emanation, without substance, that has pass'd, by Miracle. Into *them*?—Damne, I think not,—Hell were a better Destiny."
>
> "Why," exclaims the Captain, "'tis the Doctrine of Transubstantiation, which bears to the Principle you speak of, a curious likeness,—that's of course considering members of Parliament, like the Bread and Wine of the Eucharist, to contain, in place of the Spirit of Christ, the Will of the People."
>
> "Then those who gather in Parliaments and Congresses are no better than Ghosts?—"
>
> "Or no worse," Mason cannot resist putting in, "if we proceed, that is, to *Consubstantiation,*—or the Bread and Wine remaining Bread and Wine, whilst the spiritual Presence is reveal'd in parallel fashion, so to speak,—closer to the Parliament we are familiar with here on earth, as whatever they may *represent,* yet do they remain, dismayingly, Humans as well."[17]

Such is human life not only along the Mason-Dixon line but in the Plotinian edifice as well, the latter a metaphysical House of Representatives in which everything emanates from the cosmology of the One. And that is why we must now return to Cosmas Indicopleustes and his rendering of the world as a copy of the Tabernacle. More precisely, the mood of sixth-century Alexandria is well captured by the very first words of the *Christian Topography,* effectively an invocation uttered "in the name of the Father and of the Son and of the Holy Ghost—the one adorable Godhead in three Persons—the consubstantial and life-originating Trinity of the one God, from whom every perfect gift comes down to us from above."[18]

*

**The Tabernacle** was to Cosmas an exact image of the universe, a revelation born of St. Paul's letter to the Hebrews, the Apostle himself drawing on a quotation from Exodus (Heb. 8:5; Exod. 25:9, 25:40).[19] To be precise, the India-Traveler based his world map not on observations from his extensive travels but on readings of the Scriptures. Chief among the latter was God's command to Moses that the tabernacle and all its furnishings should be built in exact conformance with the pattern shown him on the mountain. As with the conventional maps of Ptolemy, Cosmas' sketch may therefore be characterized as an imitation in drawing, a picture of a sacred place not unlike my own reconstruction of the House of Plotinus.[20] In deed it is tempting to read the *Christian Topography* not as its author evidently intended it to be read—as a true representation of the real world of sticks and stones—but as a map of the imaginary and symbolic words that really hurt you, a hodgepodge of past, present, and future: Anaximander long after Anaximander, Lacan long before Lacan; Pytheas, Eratosthenes and Ptolemy all denigrated, Karl Marx's batman not yet invented. The topography ("place-writing") of Cosmas' tabernacle-turned-map is nevertheless surprisingly simple, for its geo-graphy ("earth-writing") consists of nothing but two places separated by a curtain, the latter made of blue, purple and scarlet yarn and finely twisted linen with cherubim worked into it, a primeval form of iconostasis hung "with gold hooks on four posts of acacia wood overlaid with gold and standing on four silver bases [with] the ark of Testimony [placed] behind the curtain. The curtain [separates] the Holy Place from the Most Holy Place" (Exod. 25:32–33).

It is this one-story rendering of the original tabernacle that Cosmas proceeds to tilt and eventually turn into a two-story building. On the upper level of that house he locates what in Exodus was called the "Most Holy Place," on the lower the merely "Holy," the original curtain simultaneously functioning as the floor of the upper level and the ceiling of the lower; in Plotinus' edifice this floor/ceiling is the hypostasis of hypostases, while in the Divided Line of Plato's *Republic* it is the ontological transformation-point at which untouchable ideas freeze into touchable things and everything solid melts into air. In Cosmas' universe the lower story is the present world, until the time of resurrection the home of men and angels, the upper story the Heaven, presently residence of the One, after resurrection home of the Son and the Holy Ghost as well.

If the overall structure of Cosmas' map corresponds to the tabernacle at large, its details are determined by the arrangement of its furniture—the lead character is no longer the master architect but the interior decorator. The impressive centerpiece is "a table of acacia wood—two cubits long, a cubit wide, and a cubit and a half high [overlaid] with pure gold and [with] a gold molding around it. Also [around] it a rim a handbreath wide [with yet another] gold molding on the rim" (Exod. 25:23–25). On that table were then placed a set of plates and dishes, pitchers and bowls, all made of pure gold, twelve showbreads and a lampstand as well. No room for improvisation, though, for the Lord repeatedly insisted that everything should be made exactly to the pattern he had dictated. And when Moses had completed the construction work and everything was put in its proper place, "the cloud covered the Tent of the Meeting, and the glory of the Lord filled the tabernacle. Moses could not enter the Tent of Meeting because the cloud had settled upon it, and the glory of the LORD filled the tabernacle" (Exod. 40:35–36).[21]

With great ingenuity and much work Cosmas restaged this classical drama in his own time and place, totally subservient to the Nestorian reading of the Scriptures and totally negligent of the accumulated knowledge of optics, physics and astronomy. When touched by the Egyptian's wand of surrealist abracadabra, the flat rectangular table with the gold molding magically turned into a flat rectangular earth twice as long as it was broad (remember Alexander and Eratosthenes!) all of it surrounded by a monster-filled Ocean (remember Anaximander and Pytheas!). Onto the clean and framed panel of that surface he then projected not only the four waters of the Persian, Arabian, Caspian and Mediterranean seas, but also the four lands and nations of the Indians in the east and the Celts in the west, the Ethiopians in the hot south and the Scythians in the cold north.

Onto the four sides of the rectangular earth, which rested on nothing but itself (Job 26:7), Cosmas welded four walls. Across the top of this construction he then stretched a heaven curved as a canopy, a construction reminiscent both of the vaulted roof of a bathroom and of the celestial globe of the ancients. Towards the northern end of that oikumene he finally placed a big mountain behind which the sun, fittingly miniaturized and likened to a lampstand, came to sleep at night. No Antipodes, however, for in that case the rain would have to fall *up,* a situation which even the Nestorian considered absurd. As further proof of the antipodeans' nonexistence Cosmas referred first to Genesis then to Isaiah, for according to the former God made heaven and earth and nothing beyond them, according to the latter heaven was the LORD's throne, earth his footstool (Isa. 66:1). (Mis)guided by his own logic he then concluded that the earth must be located at the bottom of the universe not, as the pagans had it, at the top.

And yet. Just as Alexander and Pytheas were driven by an irresistible desire to reach beyond the shore of the inhabited world, so was the Egyptian merchant-turned-holy man. To that end, beyond the unnavigable Ocean River, on the rim a handbreadth wide, he imagined a different kind of earth, home of the patriarchs and source of the great rivers of the Indus (or was it Ganges), Nile, Euphrates and Tigris. On that same *Terra ultra Oceanum,* somewhat to the east, he placed Paradise itself, temptatious idea, unreachable reality, the monk's own version of that fix-point of fix-points which later mapmakers were to locate at the magnetic North Pole. Triptych à la Hieronymus Bosch, architecture à la Antonio Gaudí.

In this context of power it must finally be stressed that Cosmas' key references were to the Exodus, a story which like his own life was set in Egypt and the Sinai. It is in the blood-drenched atmosphere of those pages that God appears first as the warrior par excellence, equally cruel and unpredictable, then as the paradigmatic lawgiver, equally severe and measured.[22] No longer a creator obsessed with reproduction but a liberator concerned with morality, no longer a manufacturer of material relations but a forger

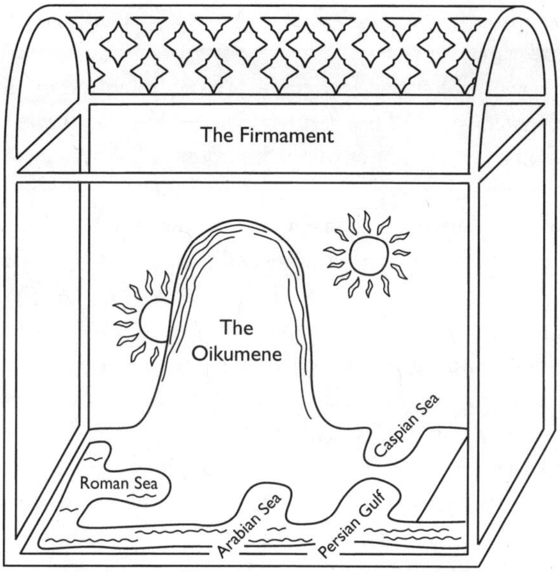


Cosmas' world enclosed within the tabernacle. Adapted from Umberto Eco, *Baudolino* (London: Secker and Warburg, 2002), 327.

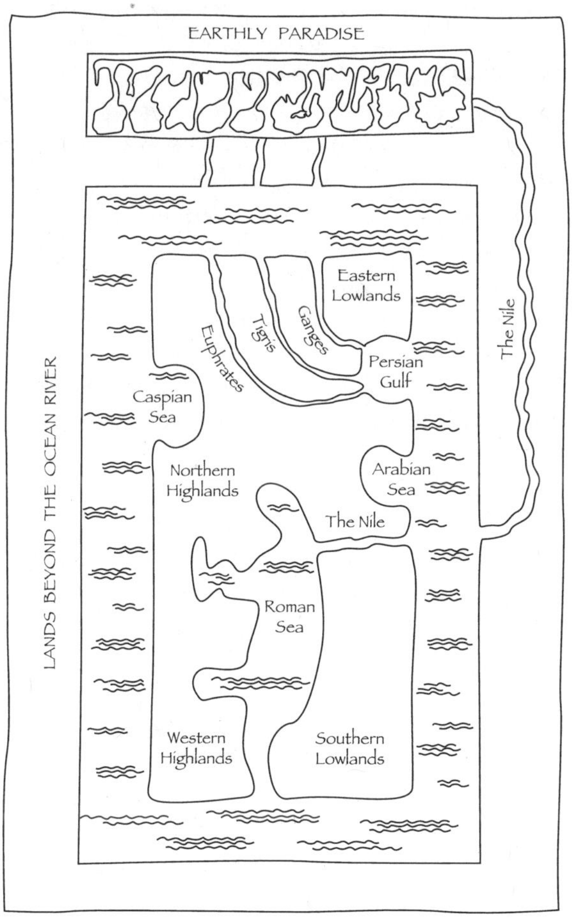


Cosmas' view of the oikumene. Adapted from Umberto Eco, *Baudolino* (London: Secker and Warburg, 2002), 257.

of social relations. Make no mistake, though, for "Jealous" was his name, jealous his nature (Exod. 34:14).

Initially this godly powerbroker hands over the ten commandments to the people at large. However, when Moses returns from yet another private audience at the mountain retreat, he finds that in his absence the people he had been set to govern have fallen into idolatry, offering their sacrifices to a gold calf rather than to the invisible LORD. Quite predictably the latter gets furious and threatens to destroy the stiff-necked people once and for all. "But Moses sought the favor of the LORD his God. 'O LORD,' he said, 'why should your anger burn against your people, whom you brought out of Egypt with great power and mighty hand? Why should the Egyptians say, 'It was with evil intent that he brought them out, to kill them in the mountains and to wipe them off the face of the earth? Turn from your fierce anger; relent and do not bring disaster on your people.' . . . Then the LORD relented and did not bring on his people the disaster he had threatened" (Exod. 32:11–14).

Having thus changed his mind, God reissues the law, albeit with the difference that he now entrusts it not to the public at large but to Moses alone. The LORD obviously realized that the populace—perhaps since it was made in his own image—is not always to be trusted, deciding instead that all future agreements were to be reached in the secrecy of the smoke-filled room. And to that same end Moses personally assembled the tent in which the LORD was to dwell, the secluded meeting-place at which the two of them could speak to each other face to face as a man speaks with his friend (Exod. 33:11). The texts, though, speak for themselves.

Thus it should now be evident that from beginning to end cartographical reason serves as the handmaid of power, sometimes concealed with its face covered, sometimes naked with its genitals bared. Its overriding task will always be the same: to draw the line between insiders and outsiders, more precisely between the free above, who can do whatever they like, and the slaves below, who have no choice but to submit. But why was Cosmas so eager to respond positively to the LORD's question of "Where is the house you will build for me? Where will my resting place be?" (Isa. 66:1). Why did he not instead react against the claim that "This is the one I esteem: he who is humble and contrite in spirit, and trembles at my word" (Isa. 66:2). Why did he not detect the scorn in the LORD's voice when the latter asked, "Where were you when I laid the earth's foundation? Tell me if you understand. Who marked off its dimensions? Surely you know! *Who stretched a measuring line across it*?" (Job 38:5). In comparison, the Chinaman was well aware that

> [To] mark a right Line upon the Earth is to inflict upon the Dragon's very Flesh, a sword-slash, a long, perfect scar, impossible for any who live out here the year 'round to see as other than hateful Assault. How can it pass unanswer'd? . . .

> "To rule forever," continues the Chinaman later, "it is necessary only to create, among the people one would rule, what we call . . . Bad History. Nothing will produce bad History more directly nor brutally, than drawing a Line, in particular a Right Line, the very Shape of Contempt, through the midst of a People,—to create thus a Distinction betwixt them,—'tis the first stroke.—All else will follow as if predestin'd, unto War and Devastation."[23]

Later still, when Mason and Dixon have walked their own line into the wilderness and begun their own self-assessment, the land surveyor asks the star-gazer:

> "Ev'rywhere they've sent us,—the Cape, St. Helena, America,—what's the Element common to all?"
>
> "Long Voyages by Sea," replies Mason, blinking in Exhaustion by now chronick. "Was there anything else?"
>
> "Slaves. Ev'ry day at the Cape, we lived with Slavery in our faces,—more of it at St. Helena,—and now here we are again, in another Colony, this time having drawn them a Line between their Slave-Keepers, and their Wage-Payers, as if doom'd to re-encounter thro' the World this public Secret, this shameful Core. . . . Pretending it to be somewhere else, with the Turks, the Russians, the Companies, down there, down where it smells like warm Brine and Gunpowder fumes, they're murdering and dispossessing thousands untallied, the innocent of the World, passing daily into the Hands of Slave-owners and Torturers, but oh, never in Holland, nor in England, that Garden of Fools. . . . Christ, Mason."
>
> "Christ, what? What did I do?"
>
> "Huz. Didn't we take the King's money, as here we're taking it again? Whilst Slaves waited upon us, and we neither one objected, as little as we have here, in certain houses south of the Line, . . . "
>
> "Yet we're not Slaves, after all,—we're Hirelings."[24]

Cosmas, a slave or a hireling? None of the kind, merely a fascinating child of his own time and place, like Plotinus before him a borderline figure occupying the interface between the philosophies and astronomies of the ancients, on the one hand, and the Christian beliefs of the middle ages (including its latter day versions), on the other. Not as crazy as it might first appear, and not as crazy as it eventually was to be. Yet a forerunner to the makers of the *mappae mundi,* those fanciful documents which for a crucial period were to guide the thoughts-and-actions of many pilgrims and crusaders, both groups better known for their mistakes than for their achievements. And that is why in these pages I have chosen to imitate Cosmas' thought in drawing not by means of his own maps but by two sketches adapted from Umberto Eco's *Baudolino,* like Thomas Pynchon's *Mason & Dixon* a wonderful illustration of cartographical reason performed on the high wire.

# MAPPAE MUNDI MEDIEVALIS

Every point of the universe is a fixed point:
all you have to do is to hang the pendulum from it.
UMBERTO ECO, *Foucault's Pendulum*

**The mappae mundi** of the middle ages carry meanings far beyond the conventional conceptions of mapping. Most importantly, the purpose of the medieval map was not primarily to communicate geographical or cosmological facts but to charter the world of thought-and-action, as always a mysterious mixture of the arts and sciences, albeit during that period more of the former than of the latter. It follows that the mappae mundi should be interpreted "not primarily as repositories of the current geographical knowledge (although a modicum of such information may sometimes be obtained from nowhere else) but as illustrated histories or moralized, didactic displays in a geographical setting."[1] Their prime function was consequently not to record exact geographical facts but to imitate in drawing the lessons of the Scriptures, to weave into the same fabric the threads of time and place, history and geography, textual narrative and pictorial representation.[2] In the minds of their makers these so-called maps were not primarily maps at all, but paintings, artistic creations made with the same techniques of illumination as other manuscripts, the Red Sea appropriately colored red.

To understand a mappa mundi (indeed to understand any map) is to approach it as a palimpsest, a parchment written on twice, the first text erased, the second covering the traces of the old. Since its rhetorical force depends on the internal coherence of the narrative, and since the spiritual world is always emphasized over the physical, the pictured scenes are positioned to fit the temporal logic of the story rather than the realities of geographical space. Thus, while it may be fair to say that the Greeks measured the earth by the stars and the Romans measured by milestones, the medieval Christians got their directions from the angels, one Heaven evidently as good as another.[3] Typical of its time was Saint Augustine's saying that "a man who has faith in you . . . though he may not know the tracks of the Great Bear, is altogether better than another who measures the sky and counts the stars and weighs the elements."[4] But what *was* the light that guided the magi to Bethlehem? A celestial body; a newborn baby; a word picked out of the Evangelist's portmanteau? A hard fact of astronomy; a social fact of astrology; an episode from the canon of world literature?

The questions furnish the answers, for just as stars come in many forms so do the mappae mundi. The most frequent and best known are the T/O maps, so called because they look like an anagram with the letter T embraced by an O. At the edge of that world was the Ocean Sea, a conception deeply rooted in the geographies of Anaximander, Alexander and Pytheas, all of them ultimately grounded in Homer's description of Achilles' shield.[5] Surrounded by this dangerous and unnavigable water was the earth proper, by the cross-like confluence of the Don, the Nile and the Mediterranean divided into the three continents of Asia, Europe and Africa, those lands in turn settled by the families of Shem, Ham and Japheth, all descendents of Noah. Since the sun rises in the east and sets in the west, the map was oriented accordingly, the shining Orient placed at the top, the shadowy Pillars of Hercules at the bottom, Jerusalem sometimes, though not always, at the center. The symbolism of the letter T was incredibly rich, ranging from the wooden cross of Christ to the trees in the garden of Eden, from the theological debates over trinity to the geometrical techniques of triangulation.

Although the oldest T/O maps have not survived, it is generally agreed that the pattern was set by Isidore (ca. 560–636), bishop of Seville, proficient writer and compiler of the twenty volume *Etymologiae,* a highly influential encyclopedia of human and divine subjects; it was he who derived the word *homo* from *humus,* by analogy man from earth, dust from dust and back to dust again. As the world's leading expert on Paradise, this prelate was naturally interested in the precise location of that same utopia. Paradoxical in deed, for while utopia is a *ou-topos,* a no-place, it is at the same time a *eu-topos,* a place-of-happiness,[6] in toto a devastating critique of the dominant ideology. After much thought Isidore set this imagination in the Far East, more exactly in an isolated place surrounded by a high wall lest it be invaded by scheming hordes of lustful sinners. The parallels to Alexander's quest for the eastern shore of the oikumene are striking, a fact which suggests that the mappae mundi may be best treated as maps of desire, in their function closely related to the images which nowadays are projected onto millions of movie and television screens.[7]

For the medieval mapmakers—who shared Ptolemy's idea of imitating in drawing but had forgotten about his longitudes and latitudes—the main goal was not to capture the mountains, rivers and cities of the physical earth but to make the incredible credible by turning the invisible visible. A remarkable advance in that direction was made by Hugh (ca. 1097–1141), influential master of the school at the abbey of St. Victor in Paris. Although

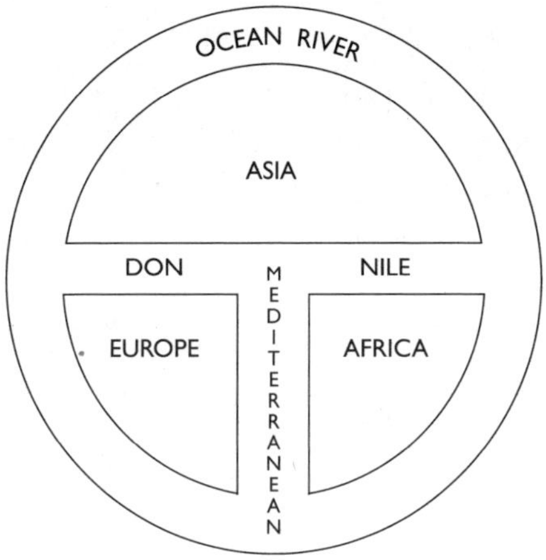


Stylized T/O map.

himself a mystic, this man readily acknowledged that the world is not only spiritual (i.e. imaginary and symbolic) but is material (i.e. real) as well. To that pre-Lacanian end, in a remark by six centuries predating Giambattista Vico, he explicitly stated that "the significance of things is far more excellent than that of words, because the latter was established by usage, but Nature dictated the former. The latter is the voice of men, the former the voice of God speaking to men."[8] And in another text, in a tone worthy of the best postmodernist, he stressed that "it is not things nor the images of things but rather their meaning that we wish to show, not what things in themselves mean, but what is meant by them."[9] As a pigeon-holing teacher Hugh knew from experience not only that memory is a central feature of all meaning formation but that memory without rules is like an uncatalogued library, a contradiction in terms.

Since memory has both structure and content, Hugh devoted his life to the development of better techniques of remembering, well knowing that designing rules for remembering is one thing, forming rules for forgetting quite another. In particular he stressed that we remember by means of imaginary pictures,[10] literally representations of spatially distributed topoi, not the least as these places are expressed in the typographical layout of the page itself. In his own words, "it is a great value for fixing a memory-image that when we read books, we study to impress on our memory through our mental-image-forming power [*per imaginationem*] not only the number and order of verses or ideas, but at the same time the color, shape, position and placement of the letters, where we have seen this or that written, in what part, in what location (at the top, the middle, or the bottom) we saw it positioned, in what color we observed the trace of the letter or the ornamented surface of the parchment. Indeed I consider nothing so useful for stimulating memory as this."[11]

With these ideas constantly in mind, Hugh constructed a range of memory aids, each one essentially a *thesaurus sapientiae,* a "storage-room" or "knowledge-container," in which he deposited the mental pictures of whatever it was that he wanted to remember.[12] Although their designs varied slightly, these intellectual tools were all constructed from blueprints given by the Scriptures, especially by the chapters of Genesis in which God looked back at what he had done, saw how great man's wickedness had become and concluded that the results were not to his liking. His heart was filled with pain, and in his wisdom the Almighty decided to do away with it all, not, however, with Noah, who he considered a righteous man, blameless among the people of his time. So God, the LORD, said to Noah:

> I am going to put an end to all people, for the earth is filled with violence because of them. I am surely going to destroy both them and the earth. So make yourself an ark of cypress wood; make rooms in it and coat it with pitch inside and out. . . . Everything on earth will perish. But I will establish my covenant with you, and you will enter the ark—you and your sons and your wife and your sons' wives with you. You are to bring into the ark two of all living creatures, male and female, to keep them alive with you. . . . You are to take every kind of food that is to be eaten and store it away as food for you and for them. (Gen. 6:13–21)

For Hugh the ark clearly served as God's favorite thesaurus sapientiae, a container of everything worth remembering, hence of everything worth knowing and preserving. And so rich is the biblical language that the term "ark" denotes not only Noah's rescue ship but the covenant between God and his people, these laws and regulations in turn transferred from the realm of meaning into the chest in which they are stored, the wooden coffer on the table of the tabernacle that is hidden behind the curtain which separates the Most Holy Place from the merely Holy Place, a sacred position from which it must never be removed. And so it is that neither word nor world sit still, a fact which helps explain why Moses, after having finished writing his law book, gave the following command to the Levites: "Take this Book of the Law and place it beside the ark of the covenant of the LORD your God. There it will remain as a witness against you. For I know how rebellious and stiff-necked you are. If you have been rebellious against the LORD while I am still alive and with you, how much more will you rebel after I die!" (Deut. 31:26–27). No wonder that Moses knew the conditions of power. He was the first to have them codified.

In a similar manner Hugh's *arca sapientiae* may be seen as the floating metaphor of his own mnemonics, a model of memory which was inherently locational. This fascinating attempt at codification permeated also the famous map (long since lost)[13] in which he depicted the world as an oval-shaped vessel, its first, second and third class passengers routinely assigned their respective stations on the lower, middle and upper decks.[14] As the instructions show, there is a short step from the conception of God as architect to the conception of God as geometrician.[15] To be precise:

> The perfect ark is circumscribed with an oblong circle, which touches each of its corners, and the space which the circumference includes represents the earth. In this space, a world map is depicted in this fashion: the front of the ark faces the east, and the rear faces the west. . . . In the apex of the east formed between the circle and the head of the ark is Paradise. . . . In the other apex, which juts out to the west, is the Last Judgment with the chosen to the right, and the reprobates to the left. In the northern corner of this apex is Hell, where the damned are thrown with the apostate spirits.[16]

How fortunate for us that the LORD changed his mind and instead of wiping mankind off the face of the earth taught Noah—and by extension Hugh—how to escape the flood! And how striking that the oval-shaped map which was issued from the abbey of St. Victor now reappears as what it really is: a postmodernist palimpsest eight hundred and fifty years in advance of Gilles Deleuze. Thus, it is tempting to imagine how the latter, foremost critic / practitioner of cartographical reason,[17] rises from the table at the Bar de Mille Plateaux, folds his napkin, kisses his friend and whispers, "Oh Félix, always remember that the *mappa* of a map is a screening screen, a mirror of memories too easily forgotten. Nothing but the bottomless surface-in-between. Therefore, every map is an interface, every labyrinth a folded abyss."

And yet. The ingenuity of Hugh's ark—Paradise in the east, Hell in the west—easily fades in comparison with the beauty and symbolism of the *Ebstorfer Karte,* the largest mappa mundi to come our way, thirty sheets of goat skin sewn together (some sheets missing), a total of almost thirteen square meters.

*

**The *Ebstorfer Karte*** is an outstanding example of invisible connections made visible, like all mappae mundi a thesaurus sapientiae, in this case of unsurpassed richness. In its structure a topographical rendering of the world, in its content a compendium of everything worth knowing. Not a descriptive copy of the world as it is, rather a normative interpretation of what it ought to be. Seemingly a world-picture, in reality a world-picturing. An icon in disguise, by definition a painting which is not a painting.

Although the questions of why, when, where and by whom may never be fully answered,[18] there is general agreement that the map stems from the thirteenth or fourteenth century, most likely from the years around 1235. Its author (almost certainly not the same as its painter) may or may not have been Gervase, originally from Tilbury, England, well-known encyclopedist, for different periods student and teacher of canon law in Bologna, member of the Sicilian court, judge of the archbishop of Arles and the count of Provence, eventually in the service of the Guelphs as provost in Ebstorf.

It was in this Niedersachsen town on the Lüneburger Heide that the map most likely was put together, initially used as a teaching aid, later as a wall decoration hung in the dean's residence, later still as an altarpiece. Thereafter it was long forgotten, in 1830 rediscovered in a closet of the Benedictine nunnery, four years later moved to the historical collections in Hannover, restored, copied, photographed and carefully documented,[19] eventually lost in the flames of an allied air raid, thanks to the old records soon resurrected again, for financial reasons issued in four hand-colored versions, three presently in various museums including that in the Ebstorf convent, one sold to a private collector. It is these tales of tales, these copies of copies that a century of research has struggled to interpret. No end in sight.[20]

In its present state the map measures 3.58 × 3.56 meters. However, considering that goatskin is a living material, hence liable to shrinkage and expansion, it is not unlikely that what presently is *almost* a square originally was a *perfect* square. Inscribed into that quadrant is the T / O map proper: at the edge of the world lies the circular Ocean, at the center Jerusalem surrounded by the earth itself, the latter divided into three continents by the main waterways, in this instance shaped less as a T and more like a Y. Filling the empty space between the circular earth and the square frame are some densely written marginal notes, a kind of user's manual put together for the benefit of readers and travelers alike. Even though most of these comments, unfortunately not included in Miller's reproduction, were lifted directly from Isidore's *Etymologiae,* they are highly informative, not the least because the margins were places of transformation and paradox, pun and perversity, areas reserved for the show and resolution of contradictions and conflicts.[21] The space located inland of the Ocean river—what to the Greeks had been the oikumene—is filled with at least 1,224 drawings, most of them supplemented with a written commentary, everything in striking (albeit not perfectly retrievable) colors. In general, the size of the pictures reflects their importance in the story rather than the physical extension of the place in which the events tied to them allegedly occurred. The picture-world is the world-picture, a spatialization of Judaic, Greek and Christian literature. James Joyce prefigured.

By all measures the Ebstorf map is a genre-transgressing work of art in which text and picture are intricately interwoven into an uncomparable encyclopedia, a forerunner of Julio Cortázar's *Hopscotch,* a hodgepodge of biblical, classical and fabulous history mixed with the names of true places, cities and people. As already noted, its primary purpose was not to measure the earth surface but to show in pictures how all history is embedded in the particular history of Christianity.[22] In this remarkable conception the

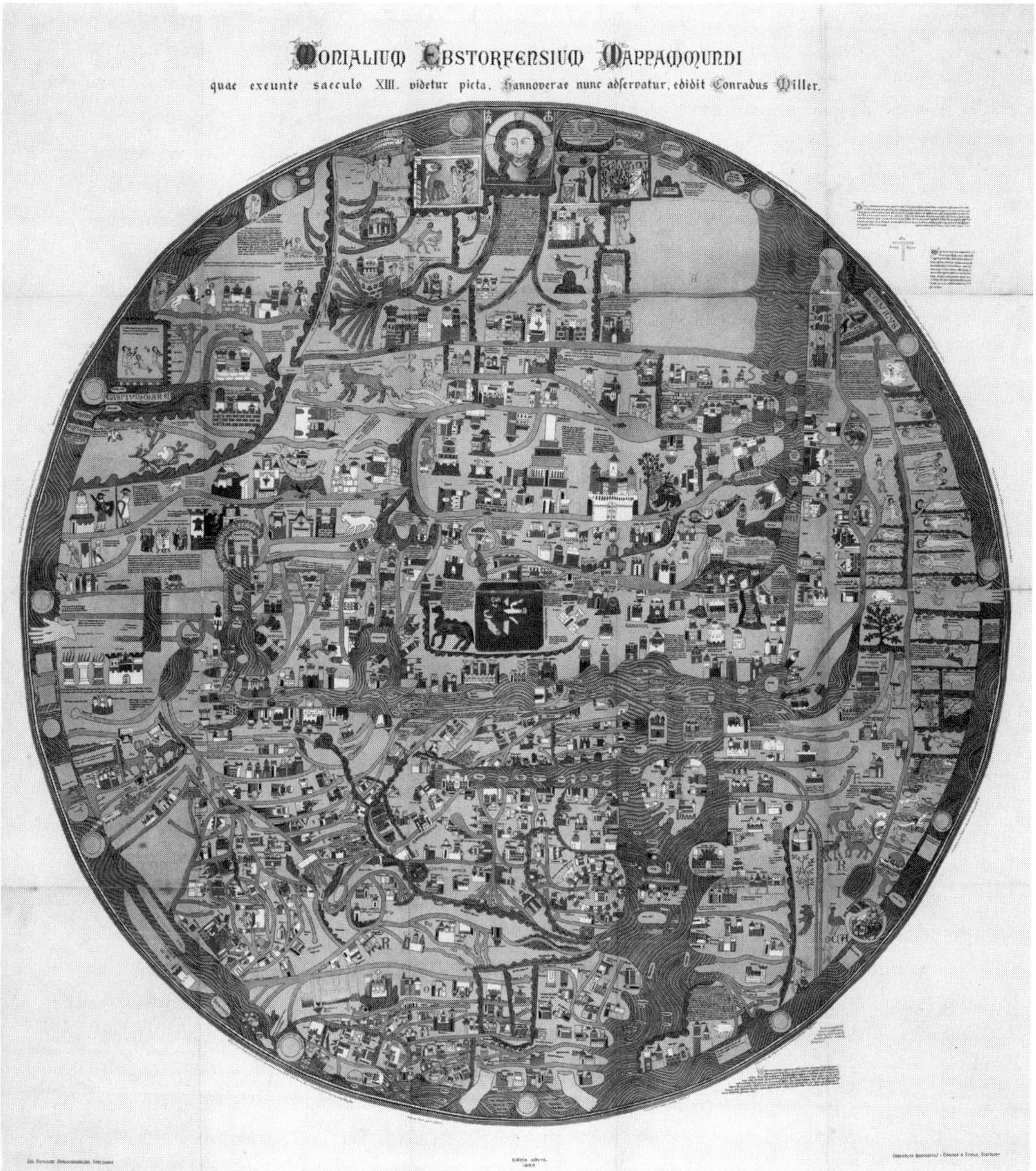


*Die Ebstorfer Weltkarte.* Reproduction from Konrad Miller, *Mappae mundi: Die ältesten Weltkarten.* Vol. 5, *Die Ebstorfkarte* (Stuttgart: Jos. Rots'sche, 1896).

tales of fact and fiction are never in conflict, the imaginary road from Ebstorf to Paradise as clearly marked as the real road from Ebstorf to Rome. Walled in by monastic regulations, most monks and nuns could visit Jerusalem only in their dreams, and it was as road atlases in their spiritual rather than physical pilgrimages that the mappae mundi played a major role. In that regard there are in fact some striking parallels between Gervase's *Ebstorfer Karte* and Cosmas' *Christian Topography,* for in both cases the structure of chorology and the content of chronography are so tightly interwoven that neither the weft nor the warp can be disentangled. All in all a wonder-filled dance of ontological transformations.[23]

Since the Ebstorf specimen is not only a literary and pictorial document but a map as well, the cardinal directions are clearly marked. Not, however, by the customary winds blowing their rhumb-lines onto the portolan charts, but by the body of Christ pointing the way to salvation. Placed at the top of the map, in the far east, is therefore a picture of the Savior's head, in the west his feet, in the south and north his left and right hands. Together these body parts serve as the mapmakers' fix-points, overriding, but not entirely erasing, the latitudes and longitudes of ordinary geography. Most noteworthy is the fact that the orientation marks are positioned neither inside nor outside the oikumene, but exactly in that limit of limits which is the Ocean itself. Rephrased, Christ's head, feet and hands are all doubly bound, partly tied to the solid ground of the *terra firma,* partly to the unknown on the other side. The abyss between certainty and ambiguity transcended, the medieval monks torn by the same fears and desires as Pytheas and Alexander fifteen centuries before them.

It is in the interweaving of Christian and Ptolemaic coordinates that the Ebstorfer most profoundly differs from other medieval maps, including those in the London Psalter and Hereford Cathedral, the former in the size of 170 × 124 millimeters and hidden in a book, the latter measuring 1.58 × 1.33 meters and publicly displayed either as an altarpiece or for purely educational purposes.[24] Most importantly, the English maps pictured the world as a gift offered *through* Christ, while their German cousin conceived the world *as* Christ. Rephrased, the makers of the *Ebstorfer Karte* seem never to have doubted that the world *is* Christ; as they believed that Jesus is God Incarnate, so by extension they believed that the earth is Christ embodied. Strange, yet not as surprising as it might at first appear. For not only was there in Greek cartography a tradition of likening the world to the human being, but the medieval public was of course thoroughly familiar with the Christian story of how the Word turned to flesh and came to dwell among us (John 1:14). To latecomers as ourselves the associations run the other direction as well, not the least to the Here Comes Everybody of *Finnegans Wake* and to the works of Samuel Beckett, for a period Joyce's secretary, the genius who dreamed of a writing which is not *about* something, but *is* that something itself.[25]

And now it may finally be surmised that to the maker(s) of the Ebstorf map the overriding aim was to project an image of God's Son, not, however, in the form of a tattoo etched onto the skin of a parchment, but as the privileged vehicle of all human thought-and-action. In my imagination I hear first the soft voice of Stéphane Mallarmé, "paint not the thing, but the effect it produces," then the chants of the monks, "paint neither the thing nor its effects, but the forces which produce both the thing and its effects." To the critic of cartographical reason the task is self-evident: study the particular coordinate net in which the clerics simultaneously captured and shaped their universe of material and social relations.

Give me a fix-point and I shall move the world, weave me a Marduk-net and Evil will be killed! Square the circle, for it was through the proportions of Vitruvius' *Homo quadratus* that medieval man translated number into mystical symbols.[26] In the succinct words of Umberto Eco:

> According to the theory of *homo quadratus,* number is the principle of the universe, and numbers possess symbolical meanings which are grounded in correspondences at once numerical and aesthetic. . . . Vitruvius taught that four was the number of man, because the distance between his extended arms was the same as his height—thus giving the base and height of a square. Four was the number of moral perfection, and men experienced in the struggle for moral perfection were called "tetragonal." However, *homo quadratus* was also pentagonal, for five was another number of arcane significance which symbolized mystical and aesthetic perfection. Five was a circular number [which] was found in man, for if the extremities of his body were joined by straight lines they formed a pentagon (an image found in Villard de Honnecourt, and also in the much better known drawing by Leonardo).[27]

As recalled, the main principle of Vitruvian architecture was that

> without symmetry and proportion there can be no principles in the design of any temple; that is, if there is no precise relation between its members, as in the case of those of a well shaped man. . . . In the human body the central point is naturally the navel. For if a man placed flat on his back, with his hands and feet extended, and a pair of compasses centered at his navel, the fingers and toes of his two hands and feet will touch the circumference of a circle described therefrom. And just as the human body yields a circular outline, so too a square figure may be found from it. For if we measure the distance from the soles of the feet to the top of the head, and then apply that measure to the outstretched arms, the breadth will be found to be the same as the height, as in the case of plane surfaces which are perfectly square.[28]

*

**Squaring the circle** of the *Ebstorfer Karte* is to lay bare its coordinate net, by no coincidence patterned in the same manner as the coordinate net of all other maps: up and down, right and left, an origo at the center. *Homo circularis, Homo quadratus.* Cartographical reason undressed. *Enuma elish* resurrected.

At the commanding top, in the spot where the ancient stargazers had

*Die Ebstorfer Weltkarte:* Christ's head in the east.

envisioned the North Pole, the medieval believers painted the head of Christ. No longer a blinking star but a shining light. No longer the fourth corner of Pytheas' constellation but an icon. Held in place by a rectangular frame, the Savior looks straight ahead, his eyes blazing the way through time and space. In the upper corners of this figure are written the two letters *A* and *Ω,* in the lower corners the words *primus* and *et novissimus,* for "I am the Alpha and the Omega, who is, and who was and who is to come, the Almighty" (Rev. 1:8). And, as no monk needed to be reminded, this quotation ties back to the story of how Thomas Didymus (nicknamed "The Doubter") once asked Jesus, "we don't know where you are going, so how can we know the way?" to which he got the answer, "I am the way, the truth, and the life. No one comes to the father without me" (John 14:5–6).

It was exactly this passage that the Ebstorf Benedictines turned into a medieval *Baedeker,* an illustrated tour-guide designed to help the pilgrims find their way to salvation. Connecting the two poles of the accompanying map—Christ's head in the east, his feet in the west—is the zero meridian, an invisible line running from top to bottom.

While Christ's head is vaguely placed where the sun rises, his feet are at the mythological Pillars of Hercules, the right foot partly resting on the ground, the left walking on nothing but water;[29] as recalled, it was at these Straits of Gibraltar that the ancient mariners believed that the oikumene changed into the anoikumene. Tied to that place, and deeply etched into the mind of every sailor, was the warning signal *non plus ultra*—HERE BUT NO FURTHER![30] Yet, these forbidding words were soon to be replaced by the challenging *Plus ultra*—FURTHER STILL! Through the erasure of the *non,* Anaximander's *to apeiron* was forever transcended by the explorers of the New Age, the *finis* turning into the *novissimus,* one narrative replacing another. As every rhetorician knows, the easiest way to be believed is to tell a story. And as every cartographer knows, all stories are in essence travel stories, infinite chains of metonymies in which one wor(l)d slides into another, a postmodern narrative with multidimensional meanings.

The particular characteristic that makes the Ebstorf map so special is the tremendous skill by which its authors tell several stories at the same time, in general moving from Genesis to Revelation, Alexander serving as border guard in the east, Hercules in the west. Through the mappers' creative mixture of word and picture, they take the audience down the beaten track of the prime meridian, holy and worldly events typically rendered in parallel. Thus, shown immediately below Christ's head are the outlines of two symmetrical itineraries, one anchored in the Old Testament, the other in the Alexander legends.[31]

The first station along the biblical route is paradise itself, here shown in all its splendors, including a painting of Adam and Eve in their nakedness, the tree of life on one side, the snake climbing the tree of knowledge on the other. Further to the northwest, where four streams were said to come up from the earth and water the whole surface of the ground (Gen. 2:6), there is yet another Eden, this one protected by a wall of fire; in deed the mappers seem genuinely uncertain about where exactly to locate the earthly paradise, and for that reason they let it appear at different times and at different places.[32] However, they did not go as far as Cosmas Indicopleustes, who six centuries earlier had taken the word *ou-topia* so literally that he refused to put it on this earth at all but set it in the No-where on the other side of the unnavigable Ocean. In comparison, the Hereford map chose a middle road and had it placed on an island at the very edge of the habitable world, Adam and Eve expelled from the walled-in garden to the uncertainties of the terra firma.[33]

Occasionally the Ebstorfers came close to practicing the same type of ambiguity, for when they painted the "Promised Land of Saints"—the imaginary island which the legendary Saint Brendan had discovered somewhere to the west of Ireland— they gave it the form of a rectangular island off the coast of Africa. On the other hand, there is not a single trace of Presbyter Johannes, better known as Prester John. Given the popularity of this figure, and the dates of the map, the omission is quite surprising, for the news about his empire of milk and honey began to circulate already around 1145.[34] Perhaps the exclusion is somehow related to the Council of 1240, at which it was officially decided that the entire idea of an earthly paradise is so ridiculous that it must be condemned. But even edicts occasionally travel slowly and when Columbus on his third voyage reached the mouth of the Orinoco, he thought he had discovered one of the paradise rivers. From that mistaken belief he then concluded that he was at the very top of the earth, the earth itself being not round but pear-shaped. In the same mood he named one of the Cuban capes "Alpha and Omega."

At the same distance from Christ's head as the picture of Adam and Eve, but positioned to the left rather than the right, are several illustrations which refer not to the Bible but to the Alexander legends. In a direct parallel to Genesis 2 and 3, the first of these paintings—like that of Adam and Eve put in a rectangular frame—shows the emperor standing between the two oracular trees, barefoot, bareheaded, and dressed in a red mantel. Nailed to that position, simultaneously illuminated by both the sun and the moon, he is told by the two oracles, one speaking Greek, the other "Indian," that he will soon die. Next to this pivotal painting are several other references to the emperor's adventures, one showing a group of Chinese behind their Great Wall, another a people so peace-loving that they live on nothing but the smell of apples. Located in the same area are the tombs of Persepolis, the place where Alexander, in a gesture of magnanimity, buried Darius, the Persian king he so decisively had defeated and whose family he so cruelly (or was it so respectfully) refused to set free. Farther away is the city of Alexandria shown with its already demolished lighthouse, earlier one of the world's seven wonders. To be exact, Alexander is mentioned ten times, pictured about fifteen.[35]

Whereas illustrations of history's most central events (including the Tower of Babel) are cluttered around the main meridian, most horror stories are relegated to the margins. Outstanding in this respect is the treatment of the apocalyptic people of Gog and Magog, in number like the sand on the seashore, a legend often tied to the Scythian invasions which via Armenia and Mesopotamia reached Syria and Palestine around 630 BCE.[36] Told and retold in many contexts, these ingredients are here mixed into a potent brew of psychedelic proportions.[37] Following convention, the mappers placed this home of Antichrist in the far northeast (more correctly eastnorth), close to Christ's right hand, its stigmata clearly visible. There, within a rectangular frame of bricks and fire, the biblical author recalls how the Sovereign LORD once declared that "I am against you Gog. . . . I will turn you around, put hooks in your jaws and bring you out with your whole army. . . . On the mountains of Israel you will fall, you and your troops and the nations with you. I will give you as food to all kinds of carrion birds and to the wild animals" (Ezek. 38:3–4 and 39:4). The result is a scene of how they "eat the flesh of mighty men and drink the blood of the princes of the earth as if they were rams and lambs, goats and bulls—all

*Die Ebstorfer Weltkarte:* Christ's head surrounded by Adam and Eve in Paradise and Alexander the Great between the oracular trees.

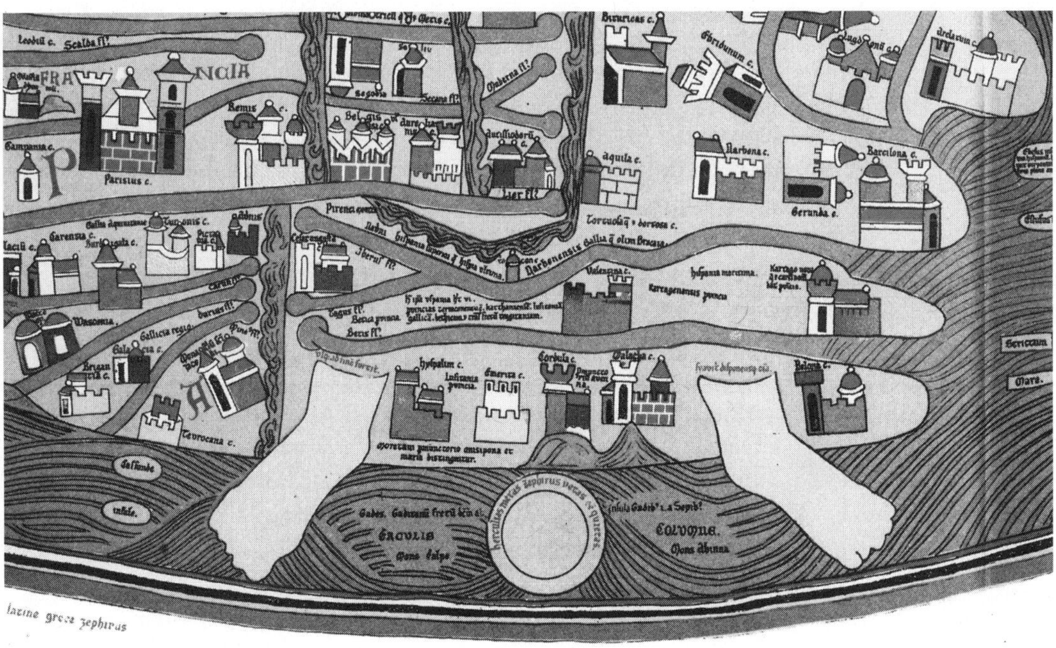

*Die Ebstorfer Weltkarte:* Christ's feet at the Straits of Gibraltar.

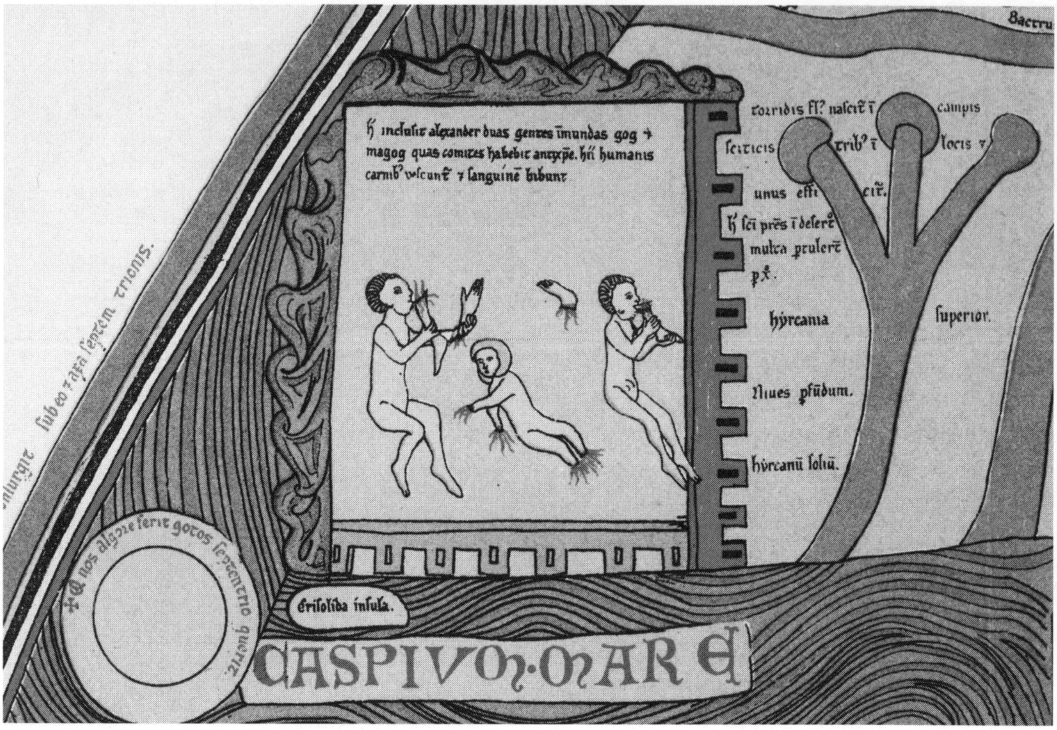


*Die Ebstorfer Weltkarte:* Land of Gog and Magog.

of them fattened animals from Bashan. At the sacrifice I [the Sovereign LORD] am preparing for you, you will eat fat till you are glutted and drink blood till you are drunk. At my table you will eat your fill of horses and riders, mighty men and soldiers of every kind" (39:19–20). Seems like Hell to me. And that, of course, was the intention.

On the other side of the earth life is equally strange, albeit not equally evil. Thus, along the southern rim of the oikumene, on the outskirts of Africa, there lives a fantastic range of monsters, peculiar beings that everyone talks about but nobody has ever seen; literally the Others of the other, etymologically connected with the word *monere,* "to warn." These symbols of human ambiguity, most of them invented already by Pliny the Elder, carry names as fanciful as the characteristics of their bearers, for to be monstrous is to be deformed. The map in fact assembles a stunning total of twenty-four of these different races, including the *Anthropophagi* ("man-eaters"), who drink from human skulls and wear human heads and scalps on their breasts; the *Blemmyae,* who lack heads and necks and have their faces on their chests; the *Cynocephali* ("dog-heads"), who have huge teeth, breathe flames, and communicate through barking; the *Panotii* ("all-ears"), very shy and with ears so big that they can be used as both blankets and wings; the *Sciopods* ("shadow-feet"), one-legged but extremely swift creatures, who spend their days lying on their back, shadowing themselves with their enormous foot. On an island in the Nile a dwarf is riding a domesticated crocodile.[38]

Everything monstrous is positioned in the vicinity of Christ's left hand (not stigmatized), an indication that also the unclassifiable is subject to the LORD's embrace, albeit kept at maximum distance. After all, Christianity claims to be the only major religion that welcomes not only the rich and the beautiful but also the poor and the handicapped, not only the rulers of the world but the prostitutes and the leprous as well. And therein lies the importance of the monsters, because they provided valuable input into the analyses of what it means to be human. Crucial indeed, especially as it was believed that only human souls may be admitted to paradise. The challenge was enormous, and it was considered a great missionary success when it was reported that a Dog-Head, perhaps the most monstrous of all monsters, had not only been converted but been turned into one of the most prominent saints.[39] Unfathomable is the magic of ontological transformations.[40]

*

**In my interpretation**, it is exactly this power-filled question of what it means to be human that lies at the heart of the Ebstorf map. Its proposed answer is everywhere to be seen, but nowhere as clearly as in the origo of the coordinate net, the geometrical center of the circular earth and the quadratic map alike. Self-reference embodied, Christ's grave in his navel, Jesus himself a second Adam. Osiris reawakened, all winds in the same bag, all planets caught in the same geometry.

At this pivot of the four quarters the map puts the city of Jerusalem,[41] in the written text described as the earthly capital of Palestine,[42] in the painted picture shown as the heavenly capital of Christianity, the ideological rather than the geographic center of the world. Surrounded by a golden wall with twelve towers, this place is so unique that it is enclosed in a square (not merely rectangular) frame, the only painting so adorned. Filling this crucial space is an icon of the Savior himself appropriately painted *en face* in the flat perspective of the time. Newly risen from the tomb his eyes are set on the regions of the north, his fingers crossed, his hand raised in blessing.

Although the map's geometry is never perfect, the horizontal and vertical axes definitely traverse the central square, perhaps intersecting in Christ's face, possibly in the sacred point between his eyes.[43] Running through the same center is a fascinating diagonal, one end in the Gog and Magog area of the eastnorth, the other in the Garden of the Hesperides in the far westsouth, a peninsula guarded by a serpent with its tail in its mouth, the only picture with a frame like that.[44] Along the same axis, midway between the tomb of Jerusalem and the Hell of Gog is Noah's ark stranded on Mount Ararat, while in the other direction, midway between Jerusalem and the Hesperidian Garden, is the island of Sicily shaped as a

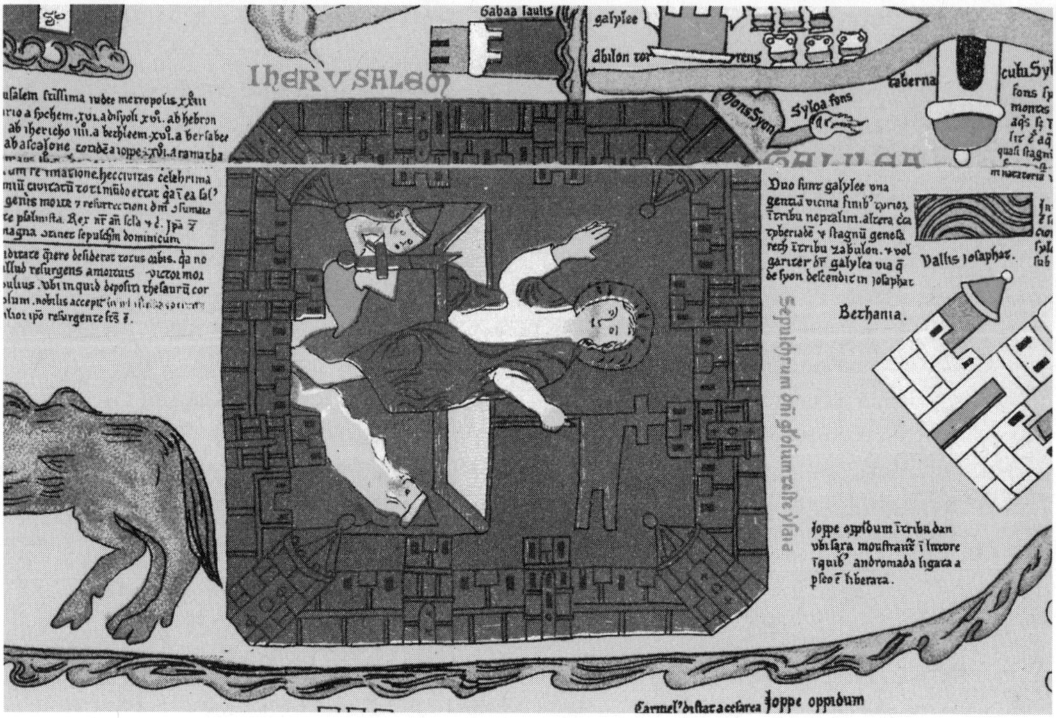


*Die Ebstorfer Weltkarte:* Christ's navel in Jerusalem.

giant heart; as recalled, Gervase spent part of his early life at the Sicilian court, and a romantic mind easily imagines how he might have enjoyed his nights. At any rate, all places along the diagonal—Gog and Magog included—are in reality topoi of salvation, the quadratic Jerusalem one with the apocalyptic Jerusalem. Relevation revealed, for like a first century Molly Bloom John had reported how he

> saw a new heaven and a new earth, for the first heaven and the first earth had passed away, and there was no longer any sea. I saw the holy city of Jerusalem, coming down out of heaven from God, prepared as a bride beautifully dressed for her husband. And I heard a loud voice from the throne saying . . . "I am the Alpha and the Omega, the Beginning and the End. To him who is thirsty I will give to drink without cost from the spring of the water of life." [And] one of the seven angels who had the seven bowls full of the seven last plagues came and said to me, "Come, I will show you the bride, the wife of the Lamb." And he carried me away in the Spirit to a mountain great and high, and he showed me the Holy City, Jerusalem, coming down out of the heaven from God. It shone with the glory of God, and its brilliance was like that of a very precious jewel, like a jasper, clear as a crystal. It had a great, high wall with twelve gates, and with twelve angels at the gate. (Rev. 21:1–2, 6, 10–12)

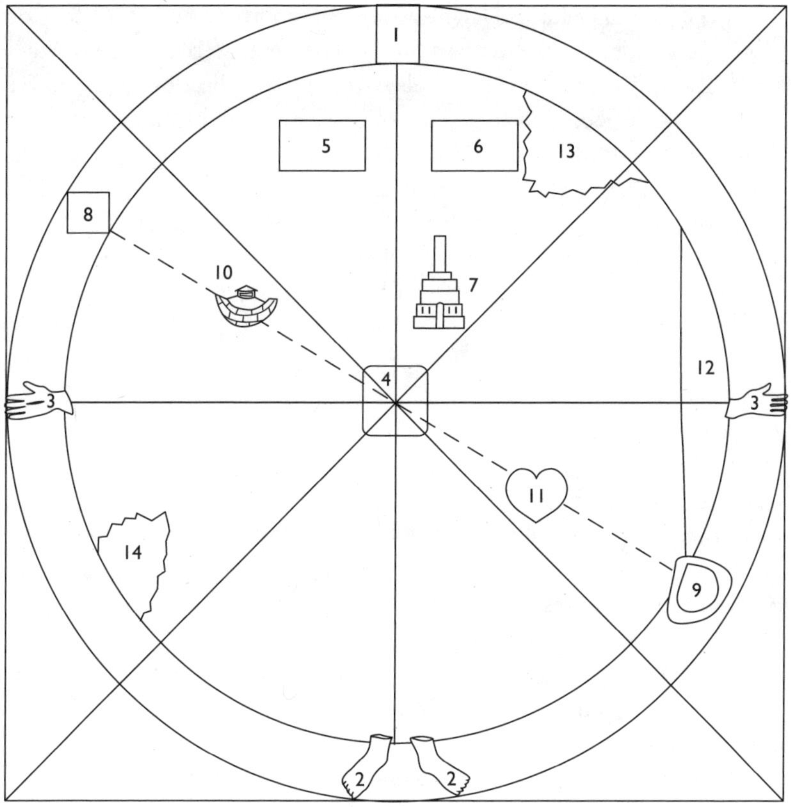


1. Christ´s head
2. Christ´s feet
3. Christ´s hands
4. Jerusalem
5. Adam and Eve
6. Alexander
7. Tower of Babel
8. Gog and Magog
9. Hesperian Garden
10. Noah´s ark
11. Sicily
12. Monstrous races
13. Missing sheets
14. Missing sheets

*Die Ebstorfer Weltkarte.* Conceptual scheme.

Another diagonal, located more to the east but almost perpendicular to the Gog–Hesperides axis, ties together the missing goatskins in the upper right corner and the partly lost pieces in the lower left, all of them mysteriously stolen in 1888, when the original map was disassembled and sent to Berlin for restoration. The intriguing question is, of course, why these and no other sheets were taken. A possible answer is that the text in the upper right-hand corner may have contained clues to the problem of the map's provenience, while the lower left showed the area of Niedersachsen, another key to its origin. These questions are themselves directly connected to the puzzle of why it was Jerusalem, rather than Ebstorf, that was placed at the world's center; in most other maps (some prominent mappae mundi excluded) the center of the world is one with the mapmaker's own home—for the Greeks it was Delos or Delphi, for the Alexandrians the observatory on Mount Atabyrion, for the Romans the city of Rome, Mecca for the Arabs.

Similar considerations may well have been on the Ebstorfers' mind, for in terms of geographical detail nothing on the map compares to Niedersachsen. In deed this region contains marks not only of Alexander the Great, who most likely had never heard of it much less had ever set foot there, but of numerous episcopal sees and monastic centers as well. In addition, there is the town of "Ebbekestorf" itself, here shown with the graves of three martyrs, holy figures otherwise unknown.

*

**A fascinating document**. In the words of one of its most prominent commentators: "The Ebstorf World Map must be seen concurrently from various standpoints; from the point of view of geography it is a map; from that of learning generally, an encyclopedia; as a work of iconography, it is a work of God's creation; as a work of politics, a sign of lordship. And as a work of piety, it represents a means of meditation."[45]

Yet, perhaps even more importantly, the Ebstorf map is an outstanding, in deed astonishing, example of cartographical reason in practice. The human body—here elevated into the body of Christ himself—becomes the coordinate net of the entire universe, a merger of Vitruvius' *Homo circularis* and *Homo quadratus* two centuries before Leonardo.[46] A piece of art in which picture and narrative join forces in a continuous play of ontological transformations. In the words of the expert:

> Does Britannia, when it sleeps, dream? Is America her dream?—in which all that cannot pass in the metropolitan Wakefulness is allow'd Expression away in the restless Slumber of these Provinces, and on West-ward, wherever 'tis not yet mapp'd, nor written down, nor ever, by the majority of Mankind, seen,—serving as a very Rubbish-tip for Subjunctive Hopes, for all that *may yet be true,*—Earthly Paradise, Fountain of Youth, Realms of Prester John, Christ's Kingdom, ever behind the sunset, safe till the next Territory to the West be seen and recorded, measur'd and tied in, back into the Net-Work of Points already known, that slowly triangulates its Way into the Continent, changing all from subjunctive to declarative, reducing Possibilities to Simplicities that serve the ends of Governments,—winning away from the Sacred, its Borderlands one by one, and assuming them unto the bare mortal World that is our home, and our Despair.
>
> "Yet must the Sensorium be nourish'd," Mason, insomniack, addresses himself in a sort of Gastrick Speech he had devis'd for Hours like these, ". . . as the Body, with its transcendent Desires, the foremost being eternal Youth,—for which, alas, one seeks in vain thro' the Enthusiasts' Fair, that defines the Philadelphia Sabbath,—the best Offer heard, being of Bodily Resurrection, which unhappily yet requires Death as a pre-condition. . . ."[47]

*

**But what do they look like**, the secret maps, which serve the same functions for today's rulers as the *Ebstorfer Karte* once did for the salvation of the German monks and the Mason-Dixon line for the future of Britannia? Squaring the circle of that question is no easy feat, for the world's omphalos seems sometimes to be located in Washington, sometimes in Beijing—to say nothing about the alternative paradises of Jerusalem, Baghdad or Rome, Brussels or Stockholm, 1050 Campus Drive or Kåbo 23:7. The ideas of bodily resurrection still the best offer heard, the drawing of the right line the very shape of contempt.

Which techniques does the cartographic reasoner actually use, when (s)he weaves that Marduk-net of longitudes and latitudes, establishes the fix-points, calibrates the scales, screens the mappae?

# INSTRUMENTS

# SAUSSUREAN BAR

> A philosophical problem has the form: "I don't know my way about."
> LUDWIG WITTGENSTEIN, *Philosophical Investigations*

**The sign is a map,** albeit in many forms. Sometimes a word, sometimes a gesture, a painting, a sculpture, a tree, a car, a suit, a coin, a gene, an icon, an index, a symbol, a missing friend, a child lost, a who-knows-what . . . Stop it!

And when I do,[1] then the sign of signs takes the minimalist form of a double helix, hereafter rendered either as

$$\frac{S}{s}$$

or as

$$\frac{s}{S}$$

where the *S* stands for "Signifier," the *s* for "signified," and the —— for the bar which in the same stroke keeps the *S* and the *s* together and apart.

In the constructivist universe of Descartes, Signifier and signified are located on opposite sides of the ontological divide, the former an ingredient in the objective world of the material, the latter in the subjective world of the mind. Also in the more analytical universes of Immanuel Kant and Ferdinand de Saussure, the Signifier is primarily in the physicality of the sign, the signified in its meaning, even though the distinctions are not as sharp as in the case of Descartes; since the two entities are held to be inseparably united, there can be no *S* without an *s*, no *s* without an *S*. On the surface two worlds, one sensible, one intelligible, deeper down a kaleidoscope of socially constructed appearances. In Saussure's terminology, "linguistic units are dual in nature, [for] a linguistic sign is not a link between a thing and a name, but between a concept and a sound pattern."[2] On this rendering, every thing boils down to a sign, a comment which is not meant to say that everything *is* a sign. The silent spaces between words often speak louder than the words themselves.

While our immediate contact with the *S* goes through the five senses of the body—sight, hearing, touch, smell, and taste—the *s* belongs to the sixth sense of culture. Much follows from this distinction, for instance that the activities of counting and naming are intricately intertwined and that God is a moral, not a material, agent. The Signifier is consequently not a thing-in-itself but a metaphoric image of a thing, the signified not a meaning-in-itself but a metonymic story of a meaning; what gives a coin its value is not the metal I hold in my hand but the socio-cultural context in which it can be exchanged and turned into something else. Addressing the same theme of representation, Immanuel Kant once stressed that "I cannot represent to myself a line, however small it may be, without drawing it in thought, that is, without producing all its parts one after the other, starting from a given point, and thus, first of all, drawing its intuition."[3] Breaking any ties before they bound him, Kris Kristoferson responded that "when you're headin' for the border lord, you're bound to cross the line."[4] The boundary is in deed all we share, a fact firmly underlined by the dialectics of the word *bound* itself, a term which means both "destined" and "tied," both "on the way to" and "fettered." By leaps and bounds, I beat the bounds. Out of bounds, I am bound to win.

And so it has already been demonstrated that the semiotic/rhetorical animal is essentially a mimetic animal, a creature possessed by the desire of turning every *S* into an *s*, every *s* into an *S*. Of course I want to know the meaning of what I see and hear, of course I want to find the proper expressions of my emotions. What characterizes these games of ontological transformations is that the sensible of the Signifier reaches out for the intelligible of the signified, the intelligible of the signified for the sensuous of the Signifier. Operating in the world of theory I tend to think of myself as a phenomenon (I *have* a body); living in the world of action I show myself to be a noumenon (I *am* a mind).

But so ingeniously is the sign constructed that this mimetic desire can never be satisfied, the dream of perfect translation never be realized. Thus, as soon as I write *s* its invisible meaning automatically turns into a visible *S*, as soon as I express myself through an utterable *S* it is immediately permeated by a silent *s*. And therein lies the guarantee that the power of the sign will never die, because it will never cease to change, language itself a machine of perpetual motion. The burning desire of the semiotic animal is consequently not a desire for the desired, but a desire for desire; not a desire for the meaning of the upper-case or lower-case *s*, but a desire for the penumbra of the tympanum. But already in the act of mentioning them, both the signified of the *s* and the fraction line of the —— have inevitably taken on the clothes of the *S* of the Signifier. As the emptiness of the lost child was filled with talk, so the conversation turned into a cemetery of petrified things.

It follows that every language is deeply metaphysical, not by happenstance but by necessity. The challenging truth is that the semiotic/rhetorical animal wants nothing, because all it asks is to be believed and all it needs is to be acknowledged. The catchphrase is not the Cartesian *Desidero ergo sum* but the Lacanian *Desirare ergo sum,* not the sedative of "I desire therefore I am" but the ecstasy pill of "Desire therefore I am."

Evasive evasion. Seducer seduced. A languishing tango for two, "Unstoppable Generation" the name of the band, "What I want from you is who I am for myself" the words of the chorus. The challenge is not to avoid tripping on your own shoelaces but to hold the sign close enough to grasp it, to experience with your own body how "language is a skin: I rub my skin against the other. It is as if I had words instead of fingers, or fingers at the tip of my words. My language trembles with desire. . . . Language experiences orgasm upon touching itself."[5] In Sanskrit, the word *linga* has three closely connected meanings: plow, penis, and tongue.[6]

*

**Like the power it signifies, the sign never sits still.** It is exactly this shiftiness that the alternative ways of depicting the sign are trying to capture; while in the first formulation the *S* appears in the nominator and the *s* in the denominator, in the second the two denotations have switched places. In both cases, however, the presence of the nominator always rules over the absence of the denominator. Thus, when I as a writer long ago began weaving this text, it was the

$$\frac{s}{S}$$

that best captured the situation. In contrast, when you now as a reader are trying to interpret what looks like the same text, the expression

$$\frac{S}{s}$$

offers a better description of what is going on.

The difference is that whereas I initially started off with some vague ideas for which I was trying to find the proper words, all that you are now encountering are some black marks projected onto a white sheet of paper, material data which you are presently doing whatever you can to interpret. In my case an idea searching for its expression, in your case an impression searching for its meaning. To wit, one part of this little word *I* consists of optical lightwaves, another of cultural meanings, the former striking your eyes, the latter your unconscious. Whoever fails to notice the former consults an optometrist, whoever is insensitive to the latter talks to a psychoanalyst. But regardless of where one turns, one must never forget the Kantian dictum that "thoughts without content are empty, intuitions without concepts are blind."[7] It follows that the I itself is never preexistent, always constructed. In deed it is only in the fluctuating boundary between the imaginary and the symbolic—in the fuzzy and contested region between first- and third-person statements—that the I for a brief moment encounters the second-person of its real self.

In this wonderland of semiotics, the formulation

$$\frac{s}{S}$$

tends to be loosely associated with the linguistics of Ferdinand de Saussure, who held that the combination of Signifier and signified is arbitrary and that the Platonic idea of "treeness" can be expressed in many ways—for instance by the word "tree" in English, "Baum" in German, "träd" in Swedish. The formulation

$$\frac{S}{s}$$

on the other hand, is closely tied to the psychoanalytic theories of Jacques Lacan, world (in)famous for his claim that the unconscious is structured like a language.

But beware! The argument is not that the unconscious *is* a language, merely that whoever wants to understand the unconscious must approach it *as if it were* a language. Like Kant before him, so also Lacan realized that nothing can be understood directly, only indirectly, never in itself, always by means of something else. The two-tiered reason is firstly that understanding by definition is an exercise in translation, secondly that every translation involves transition from one language to another. As inmates and wardens chained together in the dungeon of the prison-house of language, we are forced to say what we have to say, not free to argue what we want to argue. Easy to tell the truth. Easy to be believed. Difficult, perhaps impossible, to do both at the same time.

The physics of Niels Bohr offers an excellent case in point, for as a fellow Dane thoroughly steeped in Kierkegaardian philosophy he staunchly insisted that when two nonreducible theories are equally true, then the challenge is to learn to live with the contradictions.[8] Bohr is even reported to have said that there are two kinds of truth, one called "simple truth," whose opposite is obviously absurd, another named "profound truth," whose opposite is itself a profound truth[9]—the wisdom of *mythos* in new clothes. As Harald Høffding, who played a decisive role in Bohr's life, once wrote, "Kierkegaard came more and more to regard the capability of embracing great contrasts and of enduring the suffering which this involves as the criterion of the sublimity and value of a conception of life."[10] In the same manner Bohr was well aware that the paradoxes with which he was struggling sprung from his own refusal to abandon the common language of classical physics; in deed he explicitly argued that it is better to have a paradoxical theory, which also others can understand, than to settle for a noncontradictory theory, which is for him alone. Under pressure the theoretical physicist becomes a practicing geographer forced to acknowledge that what he happens to see depends on where he happens to stand. Bohr's conclusion was that since the self is the relation to oneself, no one can get rid of the relation to himself any more than he can get rid of himself.[11]

It was this problem of how to be actor and spectator at the same time that Kierkegaard and Bohr shared in common. It is my own problem as well. But not only theirs and mine. For what were the leading sciences of the twentieth century if not variations on the theme of translation and applied semiotics? At least at the present stage in the history of ideas it is conventional wisdom to say that without shared symbols we would be literally nothing, Flying Dutchmen tossed around on the high seas of knowledge, blown to the east in the morning, to the west in the evening. Yet everyone agrees that there is an unfathomable difference between the concrete reality of genes and the abstract theory of genetics, the former primarily made of chainlike molecules of nucleic acids that occupy specific places on a certain chromosome, the latter essentially a set of maps structured like a language.[12] When the chips are down, it is a question not of genes playing dice but of people conversing with each other. Put differently, language is not primarily a technical tool for transmitting coded information from sender to receiver but a human medium for the sharing of meaning. At least before the advent of the cloned Dolly, an individual's genes were thought to stay for ever the same. In deed it was this indubitable fact that made the in-dividual in-divisible, whole and in-dividable. Apart from others, one with oneself.

In that context it is telling that the twelfth-century definition of what it means to be a person explicitly included the *im*possibility of defining what it actually is.[13] But which are the reins that keep the imaginations of the twenty-first century on track? How can I isolate and decompose the cultural glue that holds the sign together? Which are the couplings between what I grasp through the five senses of my body, on the one hand, and the meanings of the sixth sense, on the other? When I legitimate a certain statement by calling it "common sense," do I then tie myself primarily to the commonalities of the sensibility of our bodies or to the shared intelligibilities of our minds? Is it inevitable that once I cut loose from one scheme, I am caught by another? What does it look like, the sign without clothes? Can there even be a sign undressed, a representation which itself is not a representation?

Which are the epistemological and ontological coordinates of the nucleus around which the Signifier and the signified are treading their perpetual break-dance of signification?

*

**The nucleus of the sign is invisible.** Untouchable too, for the nucleus of the sign is a relation important enough to be protected by the taboo. The principle of the excluded middle under renewed scrutiny.[14]

Up till now I have let this limit of limits be symbolized by the ultra-thin line of ——. Clear enough. Misleading nevertheless, for the meaning of the horizontal line drawn on the surface of the page is more properly rendered by the thickness of the two-sided paper itself, the physicality of the Signifier parading up front, the meaning of the signified hiding on the back.[15] In that image the limit of the line appears no longer as the ontological divide par excellence, but as a white wall onto which sensible reflections are being cast, the unwritten page functioning as the social scientist's counterpart to the painter's canvas. Too simplified, still. For whereas both the page and the canvas tend to be flat, hard and stable, reality's projection screen is better envisioned as a velvet curtain which slowly moves with the wind. Captured by the former are simple truths readily categorized, comprehended, and believed, sticking to the latter are profound truths difficult to accept, sets of folded invaginations, pockets full of posies. Ashes to ashes, dust to dust, in the cleft of self-reference we all fall down. And I think to myself, what a wonderful world. Not a middle ground but a center, neither the light at the opening of Plato's cave nor the lined-up prisoners inside it, just the resistance of the limestone up front.

More on this later. For the moment it is enough to repeat that every sign carries the mark of an invisible distinction, by an imaginary trick itself made visible in the bar of the fraction line, that umbilical chord through which Signifier and signified are connected to each other like mother to fetus, fetus to mother. Pushed to its minimalistic extreme this sign of the sign gets condensed into the penumbra of the Saussurean Bar-in-Between, the scene of a perpetual Bacchanalian revel, the place of a nightly rendezvous at which not a member is sober, perhaps a variant of Timaeus' *chora*. In our capacity as semiotic animals we have no choice but to live in and of language, much as the fish lives in and of water. The Bar is our home, the fraction line the bed in which our signifying descendants are being conceived. To the psychoanalyst, the bar is an objective correlate of the meeting between conscious and unconscious; to the epistemologist it is the boundary between what I see and what I understand; to the mythologist it is the clear water surface in which Narcissus first mirrors himself then shatters in the hope of becoming one with his own reflection, drowning himself in the process.

But the bar is also the wedding ring, the pierced ear, the broken hymen, the scar of circumcision. And in that widened context it is easy to see what the fraction line really is: an iconostasis; a symbol of categorial change; a name for the unnamable; the signification of the transition from child to adult, from clean to unclean, from friend to foe. And just as our bodies carry the scars from these initiation rites, so do our minds. The only difference is that the marks on our bodies present themselves as visible phenomena, the distinctions of the minds as memories (more exactly reminders) of invisible noumena. In both cases, however, it is our ability to learn to live with difference that makes the semiotic animal the remarkable creature we actually are.

It follows that in the world of humans any mark is better than no mark, for without categorization there is nothing at all, not even nothing at all. The frightening alternative is an echoless scream in a mountainless valley, an open-mouthed girl on a bridge across a fiord without shores, nowhere more movingly presented than in the paintings of Edward Munch. The thingifying forces of *horror vacui* are here rendered as what they really are, not the horror of the abyss but the horror of the absence of markings. The metaphysics of presence takes many forms, and in the literature of anthropology the settling of a territory is often treated as the equivalent of its consecration.[16]

The truth is that without distinctions our thoughts-and-actions would have nothing to stick to, our lives nothing to share. Such vacuities are the norm in the horrifying Realm of Psychosis, that literally unthinkable province where there exist no initiation rites, no scars, no individuals, no society. And this explains why the deeply psychotic is so frightening, because the deeply psychotic lives outside the realm of the sign, an alien by definition incapable of grasping difference. Once having reasoned myself into that corner, I can no longer forget the unmentionable acts of the Holocaust, the striped clothing of the internees, the well-pressed uniforms of the guards, the blinding lights of the interrogation room:

> And that, Herr Goldschmidt, that is reason enough for you to submit, when I now execute the will of my Führer. It is not that I question your honesty, Herr Goldschmidt, but words turn to flesh and come to dwell among us. Would you therefore, bitte, please, drop down your pants, c'il vous plais. Prove with your cock who you are, Herr Goldschmidt, just as the beriddled King Oedipus once did with his swollen foot. You are what you eat, Your Schweinness!
>
> What is the meaning of your name, Herr Goldschmidt? Was it not your Lord who declared that the cut in the flesh of your foreskin is a sign of the covenant between Him and you? *Mens sana in corpore sano.* Tell me

quick, Herr Goldschmidt, my heart is weak and cannot afford to get upset. Who are you? How did you get that name you claim is yours? A gift from your parents or a disguise you invented for yourself? Pretention or protection? All is not gold that glitters, Herr Goldschmidt. Raise your eyes and look at me! Show me what makes you a man. Answer, please!

Communicate, be responsive. Not for the sake of your LORD, merely for that of your children. Are you really a goldsmith, Herr Goldschmidt? Or just a faking alchemist? Were you created by someone else or are you the graven image of your own imagination? Self-made? Copy right! How can I believe you, unless you show the difference between me and you? Bitte, Herr Goldschmidt. Have you never heard of Jürgen Habermas and his theories of communicative action?

I insist! Don't be ashamed, uncover yourself. Be sensible! Be intelligible!

The parallels are obvious. The proper name of the individual "Herr Goldschmidt" takes on traits of the Signifier, the common history of "the covenant between the LORD and His people" becomes the signified. When pushed to their limits, every picture comes out as a picture of a body, every prayer as a prayer of a soul. The bar(r)ing of the scar is so crucial, because the fraction line bears witness not only to Herr Goldschmidt's mode of existence but to the well-being of you and me. In Lacanian terms the fraction line of the Bar-in-Between is nothing less than the trace of the real, the true home of the semiotic/rhetorical animal, the consecrated place where we live and die, the stage on which we perform our dual roles as actors and spectators.

Playing chicken, Humpty Dumpty! For what is your alternative, now when every child knows the rhyme that you have fallen down and cannot be put together again, now when every comrade has been instructed that in order to make an omelet the chef must first crack some eggs. Egg-head, you! Navel-less being.

At issue is the twin issue of identity and difference, existence and subsistence.

# A = B

> And do not swear by your head, for you cannot make one hair white or black. Let what you say be simply "Yes" or "No"; anything more than this comes from evil. *Matt. 5:36–37*

**The question of identity and existence** is the question of how I recognize something when I encounter it again. Boiled down to its essentials the problem splits into two: (1) what I meet again is rarely exactly what I met before; (2) even when I do meet something again, it tends to come with another name and another meaning.[1]

Since the moment I was born at 6:05 in the morning of September 11, 1935, my body has inevitably changed a lot, yet some schoolmates insist that I am more like myself at salty sixty-nine than at sweet sixteen. Perhaps they recognize the gorilla nose and the jogtrot waltz, perhaps they are misled by the fact that my name has always stayed the same, the latter a crucial difference between me and my wife, who entered the wedding ceremony with one name and walked out with another, drastically changed in the procession. The resulting paradoxes go to the heart of what it means to be human, for they concern not merely the objects we are thinking *about* but the languages we are thinking *in,* not merely body but meaning too. For such is the truth that logic and metaphysics, ontology and epistemology are inseparably braided together. While old Leibniz held the logical principle of identity to be absolutely indemonstrable, and in that sense not primarily a logical but a metaphysical thesis, Bertrand Russell reasoned in the opposite direction, grounding his metaphysics in his particular view of logic.

In this perspective theorizing takes on another dimension, for whereas to think is to identify, to identify is to categorize. And so on and so on until it becomes clear that to categorize is to fetter, while not to categorize is to tear the world asunder. Understanding how we stay sane is in that situation to understand the complex relations between conventional logic and dialectics, where in the former reasoning mode both identity and contradiction are children of the principle of the excluded middle (nothing can be one thing and its opposite at the same time), while in the latter every category includes both itself and its opposite.[2] Much is at issue, including the fundamental distinction between what is an individual or particular instance, on the one hand, and a general or repeatable instance, on the other; common qualities like color and shape are universals because they are repeatable, spatiotemporal things like apples and tables are particulars because they are one of a kind. It follows that while the problem of universals is to explain how different things can have common properties, the problem of particulars is to show how objects with common properties can be different.

Fundamental to these distinctions is Leibniz's principle of substitutivity according to which two things are identical if, without change in truth-value, they can be everywhere substituted for one another—what is true of an object is held to be true of it no matter what it is called. But at least since the publication of Gottlob Frege's "Über Sinn und Bedeutung" in 1892,[3] the subject of identity has not primarily been discussed in terms of the sameness of things but in terms of the sameness of meaning. To the question of whether sameness is a relation, Frege answered yes, but then immediately added that sameness is not a relation between objects but a relation between the names or signs of objects. The implication is that whenever I show that I master my mother tongue, then I unwittingly also show that I know how to practice a particular theory of meaning. But in another context Frege also observed that "since any definition is an identity, identity itself cannot be defined."[4] Semiotics in advance of itself.[5]

What Frege tried to understand was how an identity statement like

*Venus is the morning star*

(here rephrased as $a = b$) can at the same time be both true and informative, whereas both

*Venus is Venus*

and

*The morning star is the morning star*

(here rephrased as $a = a$ and $b = b$) are true but not informative.

The short answer is that what we wish to express by the identity statement "Venus is the morning star" is that both "Venus" and "the morning star" are different names for the same celestial body. If, however, the identity is held to be between objects and not between names, then we are faced with a paradox, for a material object can be identical only with itself, never with anything else.

Yet it is obvious that the two sentences

$$a = a$$

and

$$a = b$$

have different cognitive significance; while $a = a$ is an analytic truism which adds nothing to our knowledge, $a = b$ is a synthetic statement which contributes significantly to what we knew before. The issue is further complicated when we consider the true and informative proposition that

*The morning star is the evening star*

(here rephrased as $b = c$). If *a, b,* and *c,* as in these cases, stand for the same physical object, how can one then account for the fact that the three sentences all are true yet have drastically different meanings?

Frege solved the problem by arguing that there is a difference between the meaning *(Sinn)* of a term and its referent *(Bedeutung)*. More specifically, he held that the two terms "the morning star" and "the evening star" both denote the same reference, the planet Venus, but that they differ in their sense, i.e. in their meaning. It is for this reason that the statement

*The morning star is the evening star*

is not only true but also informative, for even though "the morning star" and "the evening star" both have the same reference, they differ in their sense. While "Venus" is a proper name, "the morning star" and "the evening star" are definite descriptions with the same reference but with different senses. The connections to the theories of the sign are immediate, for reference is usually tied to the physicality of the Signifier, sense to the meaning of the signified. Meaning, though, is always contextual, for meaning is always in use. And as Ludwig Wittgenstein repeatedly stressed, meaning in use is by necessity meaning in culture.

The problem is related to the ontological status of the object at hand and in that sense it speaks to us in our capacity as rhetorical rather than semiotic animals. It is obviously easier to agree that we are talking about the same object, if that object is a concrete thing which stems from the realm of the sensible rather than an abstract relation that comes from the world of the intelligible. Since the eye and the index finger serve as our major epistemological tools, it is one thing to communicate about existing phenomena, which can be seen and touched, and quite another to talk about subsisting noumena, which by their very nature are invisible and untouchable.

A stone is a stone in a shared world of knowledge, a god is a god in the private mind of the believer. Saying that God is nothing is therefore blatantly wrong, while saying that God is no-thing is fundamentally correct. The conclusion is that even though the two entities carry the same name, there is a world of difference between the planet Venus and the goddess Venus.[6] On the other hand, there are close connections between Frege's sense and reference, on the one hand, and Bohr's complementarity theories, on the other. Naming is in deed a tricky business.

*

**Naming is a tricky business**, an integral part of what it means to be human. Not surprisingly, therefore, a revolutionary consequence of Frege's seminal work is the insight that Leibniz's law, which says that synonymous expressions may be interchanged in any context without change in the truth-values, is not generally valid but holds for extensional contexts only; its area of application is in propositions about physical objects like planets, not in talk about mental objects like goddesses. It was on this rocky foundation that Bertrand Russell subsequently erected his theory of description in which the problem is whether anything meaningful can be said about objects that lack material existence.

Russell's analysis took off from the proposition

*Pegasus does not exist*

which on the surface appears to be both true and informative. On closer inspection, however, it turns out that there is in deed no such thing for the statement to be informative about. Conversely, if the proposition is about Pegasus, then it seems that it cannot be true, for then it says that there *is* a Pegasus, which does not exist. One way to untangle the conundrum is to say that the proposition is false, which implies that Pegasus actually does have some kind of being. Another is to deny that it is about Pegasus. Russell himself advocated the latter approach; in fact it was as a technique for silencing all talk about imaginary creatures like winged horses that he designed his theory of proper names and definite descriptions.[7]

Fundamental to that theory is the conception that the meaning of a proper name is the object for which it stands. In the words of the early Wittgenstein, "A name means an object. The object is its meaning."[8] Armed with that definition, it is easy to see how one can claim that "Pegasus" is not a proper name at all—the mythical horse Pegasus is certainly not an object in the conventional sense—but a disguised description; if the bearer of the name does not exist, then the name has no meaning and cannot be a name. An equally crucial distinction is that definite descriptions are expressed in terms of universals, proper names in words for particulars. To Bertrand Russell, the logical atomist par excellence, being belongs to what can be counted, and he stubbornly defended the thesis that "every presentation and every belief must have an object other than itself."[9] Nice and clear, but only until one realizes that this formulation does not solve the initial problem of knowing and believing, but merely defines it away.

In contrast, William Shakespeare was not attracted by presumptuous solutions of this kind, and in a famous passage he let Juliet exclaim:

> O Romeo, Romeo! wherefore art thou Romeo?
> Deny thy father and refuse thy name,
>
> . . . . . . . . . . . . . . . . . .
>
> 'Tis but thy name that is my enemy;
> Thou art thyself, though not a Montague.
> What's Montague? it is nor hand, nor foot,
> Nor arm, nor face, nor any other part
> Belonging to a man. O, be some other name!
> What's in a name? that which we call a rose,
> By any other name would smell as sweet;
> So Romeo would, were he not Romeo call'd
> Retain the dear perfection which he owes
> Without that title. Romeo, doff thy name;
> And for that name, which is no part of thee,
> Take all myself.[10]

And yet. Their deep differences notwithstanding, Shakespeare and Russell would almost certainly have agreed that there are close connections between one's views of identity (*x* is the same thing as *y*) and quantification (there is an *x* such that), on the one hand, and what one takes to be the proper sense of a proper name (whether "Venus, the goddess" or "Romeo, son of Montague" are, or are not, proper names), on the other.

Serious matter, for to most observers the human world consists not only of material things that can be seen, named and counted, but equally much (perhaps more) of social relations which must be handled with intellectual tools of an entirely different kind. Once again there seem to be two modes of being, one made up of existing phenomena, the other of subsisting noumena. Venus, the planet, belongs to the first category, Venus, the goddess, is part of the second.

And now it can be said: in the western world of understanding there is no identity without difference, no difference without identity. But whereas identity is defined in terms of what it *is*, difference gets its meaning from what it is *not*. More precisely, the concept of identity belongs to the well-delineated spaces of conventional logic, the concept of difference to the ambiguous realm of dialectics. The crucial distinction is that in the either-or world of Aristotelian logic the laws of thought are clearly defined in terms of identity (*A* is *A*, i.e. $A = A$), contradiction (*A* is not not-*A*, i.e. $A \neq \sim A$), and excluded middle (everything is either *A* or not-*A*, i.e. $A \vee \sim A$). In the both/and world of Hegelian dialectics, on the other hand, the concept of definition is itself floating around. And yet, "nowhere else is the dialectic of identity and difference so rich and visible as in mathematics. [Indeed] mathematicians keep affirming . . . that A equals B. No one else—not even poets—affirms such claims as often. The presence of the dialectic of identity and difference in mathematics is far from accidental: it is, quite simply, what mathematics is about."[11]

Current analysts tend to agree that the evolving world is more readily understood as a play of contradictions than as a preservation of identities. Just as the semiotic animal is driven by a mimetic desire which can never be satisfied, so the philosopher's quest is a struggle not with identities established but with differences deferred. It is in this sense that every interpretation ultimately disintegrates into a series of repetitions,[12] acts of meaning in which one sign is being replaced by another. However, this new sign is never new forever, only for the passing moment in which it is born and articulated. It is through these processes of replacement that the prevailing metaphysics of presence is constantly undermining itself. Here, as in the case of evolutionary biology, man is treated as nothing but a preliminary end-product, nothing but a local habitation and a name. Ad infinitum.

In conclusion, signs tend to live their own lives; the world of memory crisscrossed by a network of one-way streets which run from the faraway past to the faraway future.[13] The history of science is therefore replete with theories which once were considered true and eventually were replaced by statements more in accordance with practical needs, empirical observations and theoretical understandings. Even faulty theories point ahead.[14]

As the wall of red lights switches to green, racing engines and spinning wheels split the silence. Hell breaks loose, when held-up associations roar ahead into anticipations of whatever lies around the corner, imaginary worlds hitherto unseen, unrealized differences deferred.

*

**The meaning of a sign is difference deferred**. One reason for the delays is that every sign is a double,[15] another that language is a form, not a substance. From the mouth of Saussure come the words that "in the language itself there are only differences, . . . no positive terms. . . . In a sign, what matters more than any idea or sound associated with it is what other sounds surround it."[16] In my own words, there is no exit from the dictionary. To these crucial observations Saussure subsequently added that "the importance of this truth cannot be overemphasized. For all our mistakes of terminology, all our incorrect ways of designating things belonging to the language originate in our unwittingly supposing that we are dealing with a substance when we deal with linguistic phenomena."[17]

From the father himself a clear warning against the seductive forces of thingification, of not seeing the *s* of the arbor for all the *S*'s of the trees. It should nevertheless be noticed that Saussure's warning is primarily directed not against the *making* of graven images (a practice to which the *Homo significans* has no alternative but to submit), but against the lure of *worshipping* them (a temptation to which we must never give in). Falling into the trap is nevertheless easy, for the semiotic/rhetorical bastard is always torn between the epistemological drive of telling how something *is* and the social need of being *believed* when s/he does so. As a consequence, we are forever doomed to roam the no-man's-land between logic and rhetoric, sometimes finding rest in the certainties of syllogistic form, sometimes being thrown off course by the ambiguities of substantive examples. "The language is, so to speak, an algebra which has only complex terms."[18]

The connections between the forces of reification and deification are highly significant, for to know is by definition to say that something is something else and be believed when one says it; since every understanding by necessity is an act of interpretation, it follows that knowledge is always an exercise in translation.[19] In paradigmatic condensation,

$$a = b$$

Indeed it should already be clear that this entire book may be interpreted as a power-filled piece of choreography, a *pas de deux* in which the lead performers are the two expressions

$$a = a$$

and

$$a = b$$

the former the tautology of the analytically true—*the* fix-point of formal logic—the latter the quintessence of the synthetically informative—*the* formula of creative mistranslation. Once this has been said, though, it must once again be repeated that in the universe of the sign there are no clear-cut identities, only delayed differences. Never a dead end, always an already-but-not-yet; never a genuine original, always an imperfect copy. Language is a simulacrum of simulacra, the privileged hiding-place of the psychoanalyst's repressed supplement.

Pushed to its own limits, the difference between identity and difference is that in the world of the former everything sticks to itself, while in the heaven of the latter everything escapes from itself. Comparison rests on a foundation of difference, for once a sign is interpreted it no longer is what it used to be. And exactly therein lies the crux of the matter, for it is well established that the structure of language itself makes perfect translation impossible. It follows that reality is never what it is said to be, for reality and language are never one and the same, a painful lesson which the Homeric cyclops Polyphemos was among the first to learn. Out of the cave—his mind as blind as his blinded eye—rolled the roaring cry for help,

> Nohbdy, Nohbdy's tricked me, Nohbdy's ruined me!

to which his friends—unable to detect the difference between the two word-sounds *Odysseus* and *ovtis,* between the name of a man and the word for "nobody"—simply replied

> Ah, well, if nobody has played you foul
> there in your lonely bed, we are no use in pain
> given by the great Zeus. Let it be your father,
> Poseidon Lord, to whom you pray.

But Odysseus himself, the cunning trickster, he

> was filled with laughter
> to see how like a charm the name deceived them.[20]

The conclusion is straightforward: although word and object are always related, they are never identical. For just as the semiotic animal is thoroughly paradoxical, so the rhetorical animal is thoroughly ironic. Socrates knew it well, Søren Kierkegaard too. Both paid a price.

*

**In the ballet** of Hegelian, Nietzschean, Saussurean, Heideggerian, Lacanian and Derridean realities the equal sign serves not as a marker of clear-cut identities but as a symbol of ambiguous differences, not as a protector of stability but as a rebel with jokers up its sleeve. Responding to the simple truth that two plus two equals four just as four equals two plus two, the profound truth replies that the strengths of the two equalities are not the same. To be precise, the first statement is true at all times and all places, the second only under special circumstances; as everyone knows, four can be made to equal virtually anything, not only two plus two but one plus three, ninety-seven minus ninety-three, minus twelve plus sixteen, etc., etc. The shocking lesson is that the equal sign is not what most people think it is, not two parallel lines which never cross, but an arrow which is pointing more in one direction than in the other. From the truth of

$$a = b$$

one must consequently not conclude that

$$b = a$$

is true as well. Neither history nor probability theory can be run backwards, for once the cards are dealt, the hands are given. And yet, even when erased, the multiplication table always leaves a trace.

As soon as the totality of ~~BEING~~ fades into the two ingredients of BEING and X, it is obvious that the most prevailing of all metaphysics is the metaphysics of presence. Nowhere is this predilection more evident than in the tendency to place a higher value on statements in and about the factual now-here than on statements in and about the utopian no-where, a practice beautifully deconstructed by Jacques Derrida's concept of *différance,* a neologism through which the French counterparts of the two terms "differential" and "deferred" are merged into one. The hinge of this linguistic invention is that a French-speaking audience cannot hear the difference between *différance* and *différence*—the two words sound the same—while whoever sees them in writing immediately grasps that they are not the same—an *a* does not look like an *e.* Yet another instance of Wittgenstein's saying that what cannot be said can sometimes be shown. What is silenced by the oral is seen screaming in the written.

When the movement from the *e* to the *a* is stopped in its tracks, then the deferral which first appears as a postponement in time shows itself to be a displacement in space. At issue are therefore no longer the logical/ontological relations between identity and contradiction, presence and absence, but the bottomless abyss of the excluded middle. In the tortuous words of the master himself:

> Differance is what makes the movement of signification possible only if each element that is said to be "present" . . . is related to something other than itself but retains the mark of a past element and already lets itself be hollowed out by the mark of its relation to a future element. This trace relates no less to what is called the future than to what is called the past. . . . In order for it to be, an interval must separate it from what it is not; but the interval that constitutes it in the present must also, and by the same token, divide the present in itself, thus dividing, along with the present, everything that can be conceived on its basis, that is, every being—in particular, for our metaphysical language, the substance or subject. Constituting itself, dynamically dividing itself, this interval is what could be called *spacing.* . . . Differance (is) (both) spacing (and) temporalizing.[21]

In poetry, especially the poetry of Stéphane Mallarmé, the suspension comes mostly from the placement of the words, not from their content. In the same vein there are close connections, yet decisive distinctions, between the numerical constellations of 01 and 10. The fascinating history of zero—among the Greeks the most forbidden of everything forbidden—shows that the zero is not at all the privileged sign of nothing, but a mark which takes its meaning from its position. Not primarily a number but a place,[22] not a secret hideout, but a well-guarded home.

After this exercise in deconstruction, the well-established meanings of logic, ontology, geometry and metaphysics begin to shift around, no longer as stable as they once were thought to be. To illustrate, the Greek word *pharmakon,* which plays such a crucial role first in Plato's *Phaedrus* and then in several of Derrida's writings,[23] is a bat-like word, sometimes appearing as a rat, sometimes as a bird—viewed from one vantage point pharmakon is a "poison," from another a "remedy" or a "cure."

But now, cutting through the noisy conversation in Socrates' pharmacy, is a knock on the door. Demanding to enter is *pharmakos*—magic man, wizard, poisoner—the archetypal scapegoat who has learned from experience that drugs belong to the sacred interface between life and death; for the tortured cancer patient, the morphine syringe is a blessing, the last rite in new clothes, for the street addict a bridge of sighs, a scar(r)ing walk from one hell to another.[24] And thus it is that in the Halls of Power a final letter—a pharmako*n* twisted into a pharmako*s*—actually does make a world of difference, for what the alphabet reveals is that the latter is nothing but Thoth in disguise, the Egyptian god whose duty it was to weigh the hearts of the deceased. Neither innocent nor guilty, this creative genius was as well versed in the power-filled art of ontological transformations as ever the god Janus, his gate-keeping descendant in Rome. And as major tools of trade Thoth used both geometry and writing, rhetorical devices so trustworthy that not even Plato—who invented the term—included them in the category of *rhētorikē.*[25]

"What, then, is truth?" asked Friedrich Nietzsche in a posthumously published fragment from 1873, immediately answering that

> truth is a mobile army of metaphors, metonyms, and anthropomorphisms—in short, a sum of human relations, which have been enhanced, transposed, and embellished poetically and rhetorically, and which after long use seem firm, canonical, and obligatory to a people: truths are illusions about which one has forgotten that this is what they are; metaphors which are worn out and without sensuous power; coins which have lost their pictures and now matter only as metal, no longer as coins.[26]

Five years later Charles Sanders Peirce argued along similar lines, claiming that truth is what the community of inquirers agrees on, more exactly that "the opinion which is fated to be ultimately agreed to by all who investigate, is what we mean by truth, and the object represented in this opinion is the real."[27] In other contexts he argued first that the task of pure rhetoric is to "ascertain the laws by which in every scientific intelligence one sign gives birth to another, and especially one thought brings forth another," then that "a judgment is the mental act by which the judger seeks to impress upon himself the truth of a proposition."[28]

It cannot be said more clearly: philosophy is a branch of rhetoric, reasoning a persuasive activity grounded in human predicaments.[29] Telling the truth is consequently to be believed, to be an expert juggler of rhetorical tropes, torches flying through the air, metaphors swirled by the right hand, metonymies by the left, anthropomorphisms serving as the hands themselves. Actor and spectator enchanted by the same fire, burned by the same desire.

In all the commotion it is sobering to realize that what logic and geometry share in common is that both modes of reasoning are forms of rhetoric, albeit a type of rhetoric which is so convincing that it is no longer called "rhetoric" but "logic" and "geometry."

# QUOD ERAT

> The map does not reproduce the unconscious closed in upon itself; it constructs the unconscious.
> GILLES DELEUZE AND FÉLIX GUATTARI,
> *A Thousand Plateaus*

**Geometry is a form of rhetoric.** Nobody knows for certain, but the anecdotes keep repeating that above the entrance to Plato's Academy there was a sign, at the same time inviting and forbidding. Not, as in the case of Auschwitz,

*Arbeit macht frei*

nor, as in the case of my own university,

*Tänka fritt är stort, men tänka rätt är större*[1]

Instead, since we are all children of our own time and place,

| HERE NOBODY IS ADMITTED WHO DOES NOT KNOW HIS GEOMETRY |
|---|

The message is that the rules of geometry and the rules of thought are one and the same, the implication that whoever holds the keys to the former automatically knows the way also to the latter. In the words of Plato's successor, "without a presentation intellectual activity is impossible. For there is in such activity an incidental affection identical with one also incidental in geometrical demonstrations."[2] And so it is that human understanding (at least in its western form) is patterned on a combination of the Egyptian practice of land surveying and the Greek theory of geometry, both traditions grounded in the arts of sculpture. Two millennia later Descartes extended the same thought into the belief that the whole of physics is nothing other than geometry, an idea which Immanuel Kant subsequently minimalized into the claim that geometry is *the* language of things:

> The history of this intellectual revolution—far more important than the discovery of the passage round the celebrated Cape of Good Hope—and of its fortunate author, has not been preserved. But the fact that Diogenes Laertius, in handing down an account of these matters, names the reputed author of even the least important among the geometrical demonstrations . . . does at least show that the memory of the revolution, brought about by the first glimpse of this new path, must have seemed to mathematicians of such outstanding importance as to cause it to survive the tide of oblivion. A new light flashed upon the mind of the first man (be he Thales or some other) who demonstrated the properties of the isosceles triangle. The true method, so he found, was not to inspect what he discerned either in the figure, or in the bare concept of it, and from this, as it were, to read off its properties; but to bring out what was necessarily implied in the concepts that he had himself formed *a priori,* and had put into the figure in the construction by which he presented it to himself. If he is to know anything with *a priori* certainty he must not ascribe to the figure anything save what necessarily follows from what he has himself set into it in accordance with his concept.[3]

Just as *Enuma elish* had its above and below, so everyone knows that today's politics is tied to its left, right, and center. But how do I know whether a given proposition of logic is true or false and how do I decide whether a particular proof of geometry is to be rejected or embraced?

Aristotle and Euclid pointed the way to an answer, the former by legislating what can be truthfully *said,* the latter by discovering what can be convincingly *shown.* Thus, just as classical logic grounds its acceptable reasoning modes in the principles of identity, contradiction and excluded middle, so Euclidean geometry builds its acceptable forms of demonstration on a foundation of well-defined axioms and rules of inference.[4] In that light it is tempting to interpret the sign-off signal *hoper edei . . .*—in its Latin form, *Quod Erat . . .*—as a stamp of approval (a kind of immigration visa attached to the reasoner's passport) that permits its holder to migrate from the dubious Realm of Rhetoric to the safe Heaven of Geometry.

Whenever the border-guard at the intellectual Checkpoint Charlie is shown a document that carries the three letters Q.E.D.—*Quod Erat Demonstrandum*—"which was to be demonstrated"—he automatically assumes that the holder's credentials have been checked and found in accordance with established rules and regulations. Neither migrant nor officer has anything to fear, because "'All the steps are really already taken' means: I no longer have any choice. The rule, once stamped with a particular meaning, traces the lines along which it is to be followed through the whole of space. . . . When I obey a rule, I do not choose. I obey the rule *blindly*."[5] Since the presented evidence is essentially apodictic, and since doubting an apodictic statement leads to nonsense, the bearer of the document is let in from the cold. And thus it is that the approval stamp of the Q.E.D. serves as a mark of perfect submission, a warranty that here is a case in which the taken-for-granted may be accepted without further questions, the passport holder a thinker who knows his rules so well that he uses them creatively. The Greek expression is *hoper edei deixal*.

No news, no surprise, merely yet another reminder that "logic is not a field in which *we* express what we wish with the help of signs, but rather one in which the nature of the natural and inevitable signs speaks for itself. If we know the logical syntax of any sign-language, then we have already been given all the propositions of logic. . . . Hence there can *never* be surprises in logic";[6] every schoolboy knows both his multiplication table and his Pythagorean theorem. Yet I am certainly not alone in having woken up in the middle of the night, sweating from the bad dream of once again standing in front of the blank blackboard, properly dressed, chalk in hand, school inspectors in place. And out of the tip of the chalk comes what every analyst could have predicted, not the well-rehearsed proof from the past but the frightening failure to recall it. Child on the balcony, emperor in the gutter.

What must now be remembered is that Euclid did not crown all his proofs with the halo of Q.E.D. Sometimes he instead used the expression Q.E.F.—*Quod Erat Faciendum* in Latin—*hoper edei poiesai* in Greek, in English "which was to be shown." As an illustrative example, well aware that the most powerful rhetoric is the rhetoric of the example:

> Euclid is on the beach teaching his geometry to a group of followers. In the sand he draws two lines and then claims that they are equivalent. A bewildered and well-taught pupil raises his hand objecting that he cannot follow the argument, that he fails to understand how the master's conclusion obeys the axioms and inference rules he had learned the day before. "Maestro, please explain."
>
> "Oh, my friend," says Euclid, "I see your point. Less a point of geometry, though, more a point of ethics. For what your question is really suggesting is that you don't believe me, that your teacher is not to be trusted. Good for you. But beware, young man! For even though I cannot use my words to tell you how the two lines are equivalent, I can still show with my body that they are. Here. Take my hand. Walk with me. Three steps in this direction, three steps in that. Do you believe me now?"
>
> "Yes, of course," replies the brave one, "how could I doubt the subjectivity of my own body." "Well, then," concludes the teacher, "we agree."
>
> And as a sign of that habit, the agreement is donned with the approval seal of the Q.E.F.

Shown and done. As Friedrich Nietzsche later came to teach, the body is the mind's corrective. And as Michel Foucault eventually came to learn, the theory of power grows out of the practice of anatomy. And thus it is that the factual forces of the Q.E.F. often prove stronger than we believe, a circumstance which goes a long way towards explaining why both Descartes and Husserl took knowledge to be grounded not in an outlying object but in the interior of the subject itself.[7] To think is to rehearse a story, to show is to imagine a picture.

Euclid knew it well: there are two ways of being believed. With the Q.E.D., the conviction comes from shared ideas, in a sense from the predictable predictions of the perfectly socialized mind. With the Q.E.F., on the other hand, the agreement comes because somehow we share the same body with its up and down, front and back, two eyes, two ears, two arms, two legs, ten fingers, ten toes. In the terminology of semiotics, the socio-cultural relations which are inherent in the meaning of the Q.E.D. dwell primarily in the *s* of the signified, while the material relations of the Q.E.F. are in the *S* of the Signifier. From these connections one must not infer, however, that I take your body to be the same as mine.[8] On the contrary, for whereas I can experience *your* body only from the outside, the contact I can have with *my own* body is exclusively from within. While I can *see* your anger, I can *feel* my own, while I can understand a genetic set-up, I can not imagine what it is to be a gene.[9] How could I as a man ever know what it means to be a woman?[10]

And for these reasons it speaks well of Jacques Lacan's theory of the unconscious that the proofs which Euclid adorned with the Q.E.F. seem more fundamental than those he awarded the Q.E.D.[11] The links between his geometry and the taken-for-granted are further emphasized by the particular verb forms he chose for describing his operations. Thus, with minor exceptions, he employed the so called perfect passive imperative, effectively a kind of performative which in English may be rendered as "let it have come about," "let it have been cut," etc.—a mode of expression which in both the speaker and the listener create the feeling that "the eventual establishment of a conclusion has already taken place before the proof gets under the way."[12] What a wonderful show of the tenses and moods of syntax revealing themselves as what they really are, rhetorical devices so cleverly and effectively deployed that we have forgotten where they come from.

The parallels to the Vienna school of empirical positivists are equally obvious, for it was part of their metaphysics that neither mind nor matter can be trusted. Hence, they insisted that for a statement to be meaningful it must meet the double requirement of simultaneously being both logically consistent and empirically true, of being equally bound to the thoughts in here and the things out there. In one sense they thereby acknowledged firstly that knowledge is a field of semiotics, secondly that every sign consists of both a visible Signifier and an invisible signified, thirdly that the ultimate place of understanding lies in the mysterious Bar-in-Between. In another and even profounder sense they paid heed not only to Plato's motto that to know is to know one's geometry, but also to Euclid's practice of weaving *Demonstrandum* and *Faciendum* into the same textual tapestry, one thread constituting the warp, the other the weft. Spicing the argument with his own brand of rhetoric, Ludwig Wittgenstein noted that madness is a word for being lost, an expression for not finding one's way in the world,[13] for not knowing what is up or down, for not realizing the difference between what is between your ears and what is between your legs.

But how am I to be believed, if I now claim that this very morning, when I looked out of my window, I saw something I had never seen before? A quarrel not so easily settled, for in that strange case neither the Q.E.D. nor the Q.E.F. prove credible enough. In that situation the intellectual gatekeeper will instead search for a third type of approval stamp, the Q.E.I.—*Quod Erat Invendiendum*—"which was to be found" or "which was to be invented." For a reasoner to be worthy of this unusual recognition, blind rule-following does not suffice,[14] for in that case only the term "creative rhetoric" is appropriate. A Mallarméan throw of dice, a gamble of meaning in which nothing takes place except the place

EXCEPT

PERHAPS

A CONSTELLATION

Words exploding. Cascades of pyrotechnic inventions. Burning fireballs reaching for the sky, sooty debris falling to the ground. A situation reminiscent of Charles Baudelaire's sad albatross once admired as a heavenly king now paddling around on the ship's deck, kicked and ridiculed by seamen who know nothing better. He who was so fine, how droll and ugly now.

In the fireworks of rhetoric once a stone is cast, its trajectory cannot be altered.

*

**In the fireworks of rhetoric** metaphor is in the image of the stone that is cast, metonymy in the story of its flight.[15] To be definitionally precise, the metaphoric is lodged in the mirrors of likeness, the metonymic in the associations of closeness, the former essentially semantic and therefore typical of poetry and surrealist painting, the latter essentially syntactic and best exhibited in the arts of prose and cubism. Whereas metaphor is primarily in what is being offered as said and shown, metonymy is in what is being received as heard and seen. Rephrased and highly condensed: the metaphor of the metaphor is an ontological anchor, the metaphor of the metonymy an epistemological arrow, the former a mode of condensation, the latter a technique of displacement.

Together these tropes function as both the pistons and the fuel of the rhetorical engine, the pulsating heart of the linguistic machinery.[16] In that perpetual churning around, word and world get thoroughly intertwined, evasive signs constantly flip-flopping between the now-here of the present and the when-there of past and future, a χ-shaped crossing of two structures, a chiasm in disguise, the I itself the linguistic shifter par excellence. But herein lies also the fundamental contradiction of all rhetoric, for even though its words are intended to reveal something new, that goal can be reached only if the speaker refers back to the well-established attitudes of the unconscious. It follows not only that rhetoric is erected on a foundation of logic, but that logic itself is a rhetorical performance. Put language on the couch and through its slippages it will reveal the hidden powers of the socially taken-for-granted![17]

This insight into the dialectics of language is exactly what permeates Jacques Derrida's conception of deconstruction, a form of investigation which aims for "those 'aporias,' blind spots or moments of self-contradiction where a text involuntarily betrays the tension between rhetoric and logic, between what it manifestly *means to say* and what it is nonetheless *constrained to say.* To 'deconstruct' a piece of writing is therefore to operate a kind of strategic reversal, seizing on precisely those unregarded details (casual metaphors, footnotes, incidental turns of argument) which are always, and necessarily, passed over by interpreters of a more orthodox persuasion."[18] Not as radical as it might seem, though, for already Aristotle knew that the most effective speakers are not those who faithfully repeat what they have deliberately decided to say but those who automatically show that what they are saying is an honest projection of their own taken-for-granted. It is not I who make the speech, it is the speech that makes me.[19] By saying how things are I show both who I am and where I have been.

At least in this sense Martin Heidegger was correct: language is the language of Being, a Being which itself springs from the particular *Ge-stell, Stand-ort,* or *topos* that lies at the center of his dialectics of presence/absence. Under the spell of rhetorical abracadabra what is believed grows out of a double root, one planted in the bodily experience of a particular place, the other sunk into the socially acceptable of the unsaid, the former expressed through spatial metaphors, the latter through temporal metonymies. Since ritual is the privileged mode of paying attention, Ignatius of Loyola insisted that the fundamental act of meditation is concentration. As a technique for paving the way, he in fact recommended a two-step preparation by which the five senses of the physical body may open the gates to the sixth sense of spiritual meaning. To be precise, Ignatius instructed the director of the meditation to ask the contemplating retreatant first to "call to mind the narrative of the event" and then to "make a mental representation of the place." Typical is the advice which says that

> During meals, consider Christ our Lord as though one saw Him eating with His Disciples, His way of drinking, of looking, of speaking; and try to imitate him.[20]

The theophany of Ignatius is in effect a semiophany.[21]

To understand how these contradictions both come about and how they are resolved, one needs to recall that metaphor is a figure of speech in which a name or descriptive term is transferred to some object different from, but analogous to, that to which it is properly applicable: time flies when life is presented as a stage on which a train plays the role of an iron horse; a congested street becomes a bottleneck; a road-crossing turns into a fork. In contradistinction, metonymy is usually defined as a figure of speech in which the name of an object or concept is replaced by a word closely related to the original: "crown" stands for "king"; "The White House" for "the President of the United States." It follows that metonymy is closely related, but not identical, to synecdoche,[22] the trope in which the name for a part is transferred to the whole or, more rarely, where the name for the whole is shifted to a part: in the first case, the concrete "grave" becomes a stand-in for the abstract "death"; in the second the general term "society" gets the specific meaning of "high society." While metaphor involves substitution by similarity, metonymy is substitution by contiguity.

And there is Alice again, looking up and seeing the cat sitting on a branch of a tree:

> "Did you say 'pig' or 'fig'?" said the Cat.
>
> "I said 'pig,'" replied Alice; "and I wish you wouldn't keep appearing and vanishing so suddenly: you make one quite giddy!"
>
> "All right," said the Cat; and this time it vanished quite slowly, beginning with the end of the tail, and ending with the grin, which remained some time after the rest of it had gone.
>
> "Well! I've often seen a cat without a grin," thought Alice; "but a grin without a cat! It's the most curious thing I ever saw in all my life!"[23]

And exactly as English phonemes and Wonderland cats are free to come and go, so are metonymic associations, albeit not without curious restrictions; just as the theory of power is one with the practice of rhetoric, so the practice of rhetoric is one with the power of ontological transformation. It follows that language (not the least when it is taken in Fritz Mauthner's sense) is the primary medium of action, the tropes of rhetoric a catalogue of magic formulas through which ideologies are turned to stone and stones to meaningful interpretations.[24] Thus, while in the Judeo-Christian beginning of the beginning God created the world, he did not build it with his hands but through the utterance of his voice, not through the shaping of things but through the formation of minds. A universe flowing out of a mouth. Let there be! And there was. Here, as everywhere else, power is the name of the game, naming the game of power.

Millennia later every dynamiter knows that for doing his job properly he needs two things and two things only: a detonator and a stick of dynamite. In the same manner, every rhetorician knows that without the kicks of metaphor the metonymic explosions will never get going. Jacques Lacan most likely had something similar in mind when he wrote that "the creative spark of metaphor does not spring from the presentation of two images, that is, of two Signifiers equally actualized. It flashes between two Signifiers one of which has taken the place of the other in a signifying chain, the occulted Signifier remaining present through its (metonymic) connexion with the rest of the chain."[25]

Now, since it is only via the unconscious that I come to think-and-do whatever I think-and-do whenever I think-and-do whatever I think-and-do, it is tempting to conclude that the metonymic is more powerful than the metaphoric. But that temptation must be resisted, for without the imaginary reflections of the metaphoric there would in deed be no story to release, no connection between map and territory, no model, hence no understandable reality either. The power-filled truth of rhetoric is that metaphor is the master key to creativity, the bodily semen which fertilizes the cosmic egg.

Therefore—and since (s)he is well aware that the associative chain reactions easily run out of control—the power-holder does everything to keep the rhetorical detonators out of enemy hands. Similarly, the professional rock blaster routinely stores tons of dynamite in a padlocked shed while never leaving the detonators unguarded. Likewise, the Commander in Chief never lets the briefcase with the secret code out of sight, for (s)he is fully aware that whoever controls the charges thereby controls the means for blowing the world apart. What is at stake is not the tactical issue of one belief system defeating another but the strategic task of grasping power on its own terms, of realizing not only that power *is* everywhere but that it *comes* from everywhere. To grasp what is going on, "one needs to be nominalistic, no doubt: power is not an institution, and not a structure; neither is it a certain strength we are endowed with; it is the name that one attributes to a complex strategical situation in a particular society."[26]

What all this means is that the rhetorical machinery is constructed according to a two-stroke principle which says that without metaphor there is no metonymy and that without metonymy there is no metaphor. In Michel Foucault's words, "the soul is the effect and instrument of political anatomy; the soul is the prison of the body."[27] Exactly as in the sign itself, there can be no *S* without an *s*, no *s* without an *S*. Grasping power by its own means is consequently not merely an issue of unraveling the vertical connections of the sign as they are depicted in the two formulas

$$\frac{S}{s}$$

and

$$\frac{s}{S}$$

Even more important are the horizontal moves that are embedded in the braided chain of

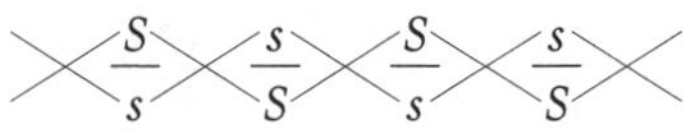

Understanding how these intertwined relations come about is to understand not only that e very sign confounds the Cartesian categories of mind and matter but also that it fuses the Lacanian categories of real and symbolic. In deed it is in the infra-thinness of the Saussurean Bar-in-Between that

> the speaking subject undergoes a transition to a void, to zero: loss of identity, afflux, or drive and a return of symbolic capacities, but this time in order to take control of *drive* itself. This is precisely what expands the limits of the signifiable: a new aspect of the displacement between the referent/signifiable, a new aspect of body, has thus found its signification.[28]

As already noted, Ignatius of Loyola instructed the praying individual that (s)he should first recall a biblical story appropriate for the day, then imagine in the mind (more correctly experience with the body) how (s)he is physically present in the room in which the story takes place. Not a heavenly vision of an abstract concept, but a well-grounded view of a material body. No wonder that the Jesuits remain so powerful, for, as Hugh of St. Victor knew so well, to remember is to re-cognize where something is, to place it in relation to the various topoi of rhetoric.

On this rendering, the term *topos* gets close also to Pierre Bourdieu's concept of *habitus,*[29] a junction at which different roads of argumentation come together, a privileged point in the landscape of reasoning. In the new theory of Chaïm Perelman—as in the old practice of Ignatius—the most crucial of all topoi is in fact the creation of a feeling of presence.[30] And if there is anything on which Saussure and Wittgenstein might have agreed, it is that the feeling of presence is a product of social interaction (more technically of rule-governed games).[31] Language speaks for itself, for "what is spoken can only be explained in language, and so it is in this sense that language itself cannot be explained."[32]

And for that very reason, the language of self-reference forms the pivot of power.

*

**The pivot of power** is in the cross between the verticality of metaphor and the horizontality of metonymy. The topos of that origo is the holey place in which the incredible imaginations of the Q.E.I. are allowed to incorporate and transcend both the socialized mind of the Q.E.D. and the common body of the Q.E.F.; the slippery rock on which worn-out identities always leave a trace and where present differences are never forever; the existential rope-yard where threads of identity and existence are twined together into mooring cables of thought-and-action; the sacred workshop in which the bonding between epistemology and ontology is under constant repair.

Such is the nature of the dematerialized point of power which the present book aspires not to practice but to investigate. Such is life in the Saussurean Bar-in-Between, the transvestite's favorite hangout. Such is the structure of the Vitruvian palace in which Signifier and signified are breathing each other, sometimes appearing under one alias, sometimes under another. Two lines forming a cross:

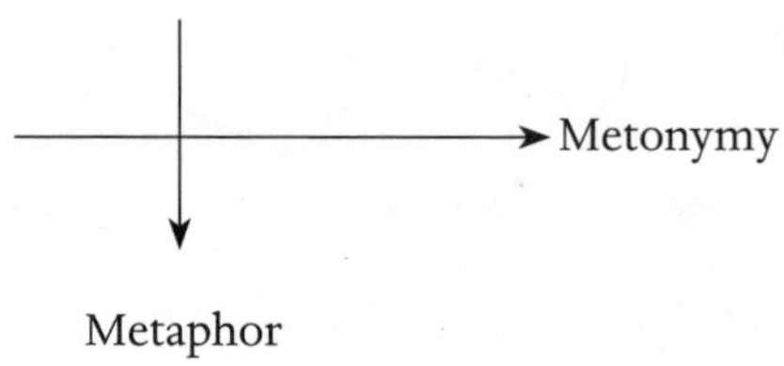


Who knows, but it may well have been this origo of thought-and-action that Maurice Merleau-Ponty had in mind, not the least in his reference to Paul Valéry's saying that

> language is everything, since it is the voice of no one, since it is the very voice of the things, the waves, and the forests. And what we have to understand is that there is no dialectical reversal from one of these views to the other; we do not have to reassemble them into a synthesis: they are two aspects of the reversibility which is the ultimate truth.[33]

But not everybody is able to grasp this truth. Roman Jakobson in fact proposed that the speech impairment typical of aphasia can be tied to disturbances in the normal way of selecting and combining linguistic units into meaningful patterns.[34] According to that theory the complicated disorders are essentially of two types, one which affects the expression of similarities, the other the voicing of contiguities, the former malfunction caused by a metaphoric deficiency and countered with an overdose of metonymic associations, the latter a metonymic deficiency which can be mastered through a strange use of metaphor.

Jakobson's paper had a formative influence on influential segments of poststructuralism and, partly for that reason, it has been severely criticized.[35] Much may nevertheless be said in its defense, not the least because Jakobson approvingly quoted Hughlings Jackson's early statement that "it is not enough to say that speech consists of words. It consists of words referring to one another in a particular manner; and, without a proper interrelation of its parts, a verbal utterance would be a mere succession of names embodying no proposition. Loss of speech is the loss of power to propositionize. . . . Speechlessness does not mean entire wordlessness."[36]

It cannot be said more clearly: not knowing how to propositionize is not knowing how to map; not knowing how to map is not knowing one's way about. And it cannot too often be emphasized that even though the phrase "I don't know my way about" was Ludwig Wittgenstein's characterization of a philosophical problem, he was well aware that "language is a labyrinth of paths. You approach from *one* side and know your way about; you approach the same place from another side and no longer know your way about."[37] Memory, which on the surface appears as a matter of time, proves on these accounts to be a matter of place, especially as "the house is one of the greatest powers of integration for the thoughts, memories and dreams of mankind. . . . Memories are motionless, and the more securely they are fixed in space, the sounder they are."[38]

The Plotinian House and Cosmas' Tabernacle in their proper context. In the concluding words of Umberto Eco, "as subjects, we are what the shape of the world produced by signs makes us become. . . . The *map of semiosis,* as defined at a given stage of historical development (with the debris carried over from previous semiosis), tells us who we are and what (or how) we think."[39] In the slightly different context of Paul Ricoeur: "Could I consider myself as someone's neighbor without a topographical sketch? And could the here and the there stand out against the horizon of a common world, if the chain of concrete neighborhoods was not set within the grid of a great cadastre in which places are more than sites?"[40]

*

**The map of semiosis** ties memory and mimesis together through activities which are essentially spatial. The function of place-bound metaphors is consequently to awaken the store of collective memories from their metonymic slumber, because to learn something new is to reconnect to something old. The self conversing with its likes in the Bar-in-Between is always a self in the making, for the limit of an object (regardless of whether that object is material or conceptual) is intrinsic to its being.

But the fact that two systems of thought are isomorphic does not necessarily mean that they are also causally and genetically related. The key question is not what a given sentence means but what it does, especially *how* it does whatever it does. The tentative answer to Wittgenstein's question is therefore that we find our way in the Terrae Incognitae by unconsciously adherring to a rhetorical contract in which the cardinal directions of up and down, front and back, left and right serve as fix-points in a collection of invisible maps. On this view the map becomes the sign par excellence, a text to be read and interpreted, a tapestry in which Signifiers and signifieds are interwoven into mysterious patterns of meaning, power and submission. It follows that a topos is a topo*s* only in relation to other topoi, a constellation of social knots which serve the same stabilizing functions in human life as the upholstery buttons (Lacan's *points de capiton*) do in the analyst's couch.

The explorer is consequently well advised never to forget that the perfectly repressed is one with the perfectly socialized, that there is no presence without an absence, no absence without a presence, no remembering without forgetting, no world without imagination. In the apt words of Maurice Blanchot:

> It is clear that in me the power of speech is also tied to my absence of being. I name myself; it is as if I were pronouncing my funeral chant: I separate myself from myself; I am no longer my presence or my reality, but rather an objective, impersonal presence, that of my name, which overtakes me and whose petrified immobility performs exactly the function of a tombstone weighing on the void.[41]

It is in similar acts of naming that the *nom du père* transcends itself into the *non du père.* As the metaphor ignites the metonymic, the name of the father explodes into the no of the father, for it is in the social practice of naming that I perpetually recreate the differences between you and me. The crucial term in Sigmund Freud's aphorism *Wo Es war, soll Ich werden* is therefore not the *Es* or the *Ich* but the *Wo,* not the fleeting nouns of *id* or *ego* but the pre-positioning adverb of the *where.* Perhaps the term "Id-entity" is nothing less than the name of the primordial place, the nostalgic and reified yearning for home.[42] At the end of the end there is no escape.

The upshot is that the sign is a map, more precisely a very intricate mappa mundi. To be precise, every map is in actuality a total image, an artistic device in which word and picture are thoroughly intertwined, one connection to the imaginary, which is the heavenly above, another to the real, which is the terrestial below. Desire is everywhere, yet always held back by the horror vacui, the fear-filled void of the in-between.

In conclusion, there is not only the geometry with names, the old-fashioned geography which still dominates the discipline. There is also a geography of the soul, a vast Terra Incognita waiting to be mapped. As Jonathan Swift observed already in 1733,

So Geographers in Afric-maps
With savage Pictures fill their gaps
And o'er ininhabitable Downs
Place Elephants for want of Towns.[43]

Therefore—since in my definition Geography is a Geometry with names—the secret agent is bound to discover that Plato's Academy has not merely an entrance but an exit as well. And next to that half hidden door (s)he finds the note which (s)he has been looking for. Penciled and in small letters it says that

*FROM HERE NOBODY EXITS WHO*
*DOES NOT KNOW HER GEOGRAPHY*

*

**Geography is a form of imagination**. Therefore—as the secret agent has now completed the mandatory course in Cartographical Reason—(s)he is finally prepared to enter the halls of ontological transformation, the magic theater on whose stage the absent is made present, the present made absent. And what (s)he will then discover is that just as war may be God's way of teaching the Americans geography, so geography is my way of teaching philosophy. Always a local habitation and a name.

# IMAGINATIONS

# PLATO

**Let it be said again**: The sign is a map, a weaving together of picture and narrative, a power-filled statement which tells me both where I am and where I should go, indicative and imperative in the same breath. Squeezed into its own minimum the map is a double fold, verb turned to noun, noun to verb.

It is in the infra-thin interactions of the Bar de Saussure that truth is simultaneously created and repeated, at once originary and memorial, at once a Signifier searching its soul and a signified searching its body. The resulting issues go to the heart of what it means to be human, for it is in the rhythmic interchange between the sensible and the intelligible that life gets its meaning, hence its sense and direction. Put differently, every map is a projection of the self, every self a projection of its culture; in the words of Johann Gottlieb Fichte, "everything we see, we see within ourselves. We see only ourselves,"[1] partly because imagination is essentially memorial, partly because the art of memory is not an art of recitation but an art of invention.[2] Yet it is often thought that an image cannot be of itself, a circumstance of which William Shakespeare was always well aware. As he let Theseus observe in the last act of *A Midsummer Night's Dream,*

> The lunatic, the lover, and the poet
> Are of imagination all compact
> . . . . . . . . . . . . . . . . . . . .
> And as imagination bodies forth
> The forms of things unknown, the poet's pen

*Plato.* Herm, Roman copy of a Greek original wrongly called *Xenon.* Musei Vaticani, Città del Vaticano.

Turns them to shapes, and gives to airy nothing
A local habitation and a name.[3]

In my own culture (which the realities of perspectivism and solipsism inform me is the only one I can say anything meaningful about) these issues of representation are closely related to the history of Christianity, more specifically to the question of whether naming and picturing the holy is to be permitted or forbidden. Which really *are* the connections between the graven image and the sacred, the Signifier and the signified, imitation and imagination, the local habitation and its name? Which insights might be glimpsed from treating "The Holy" as a synonym for the "taken-for-granted," the name "God" as a pseudonym for "Power"? Not so easy to tell, for it is imagination that brings forth the forms of things unknown, that which otherwise cannot be connected.[4] Much at stake, not the least the law of noncontradiction. As Shakespeare put it in *Macbeth,* arguably the most imaginative of all his plays,

So foul and fair a day I have not seen
(1.3.38)

a statement to which the monstrous witches had already prepared the way, first through their joint observation

Fair is foul, and foul is fair:
Hover through the fog and filthy air.
(1.1.11–12)

and then by one of them announcing

A sailor's wife had chestnuts in her lap,
And mounch'd, and mounch'd, and mounch'd:—
Give me, quoth I:
Aroint thee, witch! the rump-fed ronyon cries.
Her husband's to Aleppo gone, master o' th' Tiger:
But in a sieve I'll thither sail,
And like a rat without a tail,
I'll do, I'll do, and I'll do.
(1.3.4–10)

No wonder that Banquo, like Macbeth before his rise and fall a general of the king's army, asks

What are these,
So wither'd, and so wild in their attire,
That look not like th' inhabitants o' th' earth,

And yet are on 't?—Live you? Or are you aught
That man may question? You seem to understand me,
By each at once her chappy finger laying
Upon her skinny lips: you should be women,
And yet your beards forbid me to interpret
That you are so.
(1.3.39–47)

And as the ambiguous witches vanish into the two-valued universe of identity, difference and excluded middle—yet another illustration of how everything solid melts into air—Macbeth laments

Into the air; and what seem'd corporal melted
As breath into the wind. Would they had stay'd!
(1.3.81–82)

to which Banquo first responds

Were such things here as we do speak about,
Or have we eaten on the insane root
That takes the reason prisoner?
(1.3.83–85)

then receives the reply

Your children shall be kings.
(1.3.86)

For Shakespeare the monsters lived on the Scottish heath, for the Ebstorf monk(s) at the southern rim of the oikumene, in both instances faraway places for faraway minds, in both instances a question of separating redeemable souls from unclean bastards. The offered answers were essentially the same as well, for just as the Ebstorfers placed the u-topian paradise in several places on the same map, so the bearded witches keep flipping in and out of existence, stubbornly refusing to be tied to a particular topos. In Harold Bloom's reading Macbeth is in fact best characterized as a tragedy of the imagination, more precisely "a punishment for the displacement of the sacred into the secular."[5] Himself unable to beget children, Macbeth slaughters them; deaf to his wife's insanity, he drives both of them mad. Not, however, because he murders the king, which is part of the game, but because he butchers a man in his sleep, which is unseemly behavior, foul play. Poised in the No-man's land between alternative logics, the usurper is literally lost, not knowing where he is with his mind full of scorpions, his entourage not human beings but hired murderers. As if he were navigating the southern rim of the *Ebstorfer Karte,* he exclaims

Ay, in the catalogue ye go for men;
As hounds, and greyhounds, mongrels, spaniels, curs,
Shoughs, water-rugs and demi-wolves are clept
All by the name of dogs.
(3.1.92–95)

The humans are rare and far in-between, the world inhabited by unclean dogmen.

Even though rules are always to be followed, Shakespeare's characters often let one truth overrule another. Of this tendency Macbeth is well aware, and already in his first monologue he sets the stage for what is to follow:

Two truths are told,
As happy prologues to the swelling act
Of the imperial theme—I thank you, gentlemen.
This supernatural soliciting
Cannot be ill; cannot be good. If ill,
Why hath it given me earnest of success,
Commencing in a truth? I'm Thane of Cawdor.
If good, why do I yield to that suggestion
Whose horrid image doth unfix my hair
And make my seated heart knock at my ribs,
Against the use of nature? Present fears
Are less than horrible imaginings.
My thought, whose murder yet is but fantastical,
Shakes so my single state of man that function
Is smother'd in surmise; and nothing is
But what is not.
(1.3.128–42)

Also in the mind of Lady Macbeth nothing is but what is not. And for that reason of imagination she prays:

Come, you spirits
That tend on mortal thoughts, unsex me here,
And fill me, from the crown to the toe, top-full
Of direst cruelty! make thick my blood;
Stop up th' access and passage to remorse,
That no compunctious visitings of nature
Shake my fell purpose, nor keep peace between
Th' effect and it. Come to my woman's breasts,
And take my milk for gall, you murdering ministers,
Wherever in your sightless substances,
You wait on nature's mischief. Come, thick night,

And pall thee in the dunnest smoke of hell.
(1.5.41–42)

Egging her husband on she tells him that he should screw his courage to the sticking-place and show her that he is as determined to carry out their plan as she is. In her own words,

I have given suck, and know
How tender 'tis to love the babe that milks me:
I would, while it was smiling in my face,
Have pluck'd my nipple from his boneless gums,
And dash'd the brains out, had I so sworn as you
Have done to this.
(1.7.54–59)

Not surprising that three acts, much gall and many thick nights later, the doctor declares her disease beyond his practice. Horrified she screams:

> To bed, to bed; there's knocking at the gate. Come, come, come, come, give me your hand: What's done cannot be undone.—To bed, to bed, to bed! (5.1.73–76)

a lamentation to which Macbeth responds by ordering the doctor:

Cure her of that.
Canst thou not minister to a mind diseas'd,
Pluck from the memory a rooted sorrow,
Raze out the written troubles of the brain
And with some sweet oblivious antidote
Cleanse the stuff'd bosom of that perilous stuff
Which weighs upon her heart?
(5.3.39–45)

The doctor replies as every therapist would:

Therein the patient
Must minister to himself.
(5.3.45–46)

Having looked out of herself and into a world she never saw before, even the lady eventually finds her way. As the officer confirms,

The queen, my lord, is dead.
(5.5.16)

Once at (or is it *on*) that stage, also the horrified Macbeth finally understands not only that his own disease is most properly called "The Evil" but also that might is not right and that tomorrow is a chimera:

> She should have died hereafter;
> There would have been a time for such a word.
> To-morrow, and to-morrow, and to-morrow,
> Creeps in this petty pace from day to day
> To the last syllable of recorded time;
> And all our yesterdays have lighted fools
> The way to dusty death. Out, out, brief candle!
> Life's but a walking shadow, a poor player
> That struts and frets his hour upon the stage
> And then is heard no more. It is a tale
> Told by an idiot, full of sound and fury,
> Signifying nothing.
> (5.5.19–28)

Here, in the act of signifying nothing, the power of ontological transformations becomes one with the vanishing point of Renaissance perspective. Immanuel Kant was right on: "*Imagination* is the faculty of representing an object even *without its presence* in intuition. . . . The faculty of imagination is, so far, a faculty of determining our sensibility *a priori,* so that the synthesis of the intuitions . . . must be the transcendental *synthesis of the faculty of imagination.*"[6] Herein, then, lies the hidden condition of all knowledge, the only force strong enough to invade the utterly unknown.

Therefore it can now be said: without imagination there would never be any maps, for the characteristic which maps and imaginations share in common is that they let me know not only where I am but whence I came and whither I must go. Rephrased, it is the transcendental synthesis of mapping and imagination that makes me conscious of myself, "neither as I appear to myself, nor as I am by myself, but only that I am. This *representation* is an act of *thought,* not of *intuition.*"[7] Even more precisely, imagination is the stuff that fills the abyss between the sensible world of things and the intelligible world of meaning—and that explains how Shakespeare in the same play could produce both candles and walking shadows, both kings and idiots, both sound and fury. The truth is that without imagination there would be neither self-evident truth nor self-referencing imagination, neither the act of signifying nor the thing of nothing.[8] No chora either.

It is for these Lacanian reasons that I have recalled the many quotations from Macbeth (still too few) and tried to illustrate how the allied forces of imagination and cartographical reason function together. In the words of the outstanding critic:

> The Macbeth vision is powerfully superlogical. Yet it is the work of interpretation to give some logical coherence to things imaginative. To do this, it is manifestly not enough to abstract the skeleton of logical consequence which is the story of the play: that is to ignore the very quality which justifies our anxious attention. Rather, relinquishing our horizontal sight of naked rock-line which is the story, *we should, from above, view the whole work extended, spatialized: and then map out imaginative similarities and differences, hills and vales and streams.* Only to such a view does *Macbeth* reveal the full riches of its meaning. Interpretation must thus first receive the quality of the play in the imagination, and then proceed to translate this whole experience into a new logic which will not be confined to those superficialities of cause and effect which we think to trace in our own lives and actions, and try to impose on the persons of literature. In this way, we shall know that *Macbeth* shows us an evil not to be accounted for in terms of 'will' and 'causality'; that it expresses its vision, not to a critical intellect, but to the responsive imagination. . . . *Macbeth* is the Apocalypse of Evil.[9]

To this analysis à la the above and below of *Enuma elish* only one comment should be added. This is that what makes the figure of Macbeth so frightening is that he kills out of reason not out of passion. And that is why the interpretation must first receive the quality of the play in the imagination, and then proceed to translate this whole experience into a new logic, more precisely into a new map of imaginative similarities and differences. Like other politicians entangled in nets of their own design, Macbeth's only alternative is to say yes to the future.[10] For the usurper there is no way back, hence he has no choice but to choose again, again, again and again.

The result is moral schizophrenia,[11] a condition in which imagination functions as a leaking thesaurus sapientia, a confusion in which the compass of cartographical reason has no taken-for-granted to lock on to. As the imprisoned Lovelace wrote to his Althea,

> Stone walls do not a prison make,
>     Nor iron bars a cage;
> Minds innocent and quiet take
>     That for a hermitage;
> If I have freedom in my love,
>     And in my soul am free;
> Angels alone, that soar above,
>     Enjoy such liberty.[12]

Ontological transformation in reverse, the mind set free by the body in chains. Jailhouse rock in another tune, for it is unclear whether we are free because we imagine or whether we imagine because we are free.

*

**An object with or without its presence.** Such is the function of imagination as it straddles the abyss between the sensible and the intelligible; the well-guarded home of Apsu, the umbilical cord of western philosophy; the border between appearance and apparition, the realm of things and the affections in the soul.

How do I remember the absent and forget the present?

This power-filled question of how we represent absent objects *as if they were* present, subsisting relations *as if they were* existing things, is crucial to any understanding of what it means to be human. Hence also to the history of Christianity, from its beginning torn by conflict. The past is consequently of the essence, not the least because this new religion grew out of two roots as different as anyone could imagine, one sunk into the soil of Jewish monotheism, the other moved by the tides of Greek polytheism. While the former tradition communicates primarily through narrative and the sense of hearing, the latter relies more on pictures and the sense of sight; as the Jewish God captures his audience with the words "*Hear,* O Israel!," so the Greek term for "I know" is *eidon,* literally "I have *seen.*" How do I know whether to treat Jesus Christ as God's son or as a devious idol?

In search of an answer imagination takes us straight back both to Moses (to be treated in a subsequent chapter) and to Plato, arguably the first philosophers to formulate a comprehensive view of the relations between mind and matter, inner and outer.[13] In the wake thereafter every image assumes ghost-like properties, by definition a logic-breaking mixture of being and non-being—never a clear-cut either-or, always a dialectical both/and; never a Signifier and a signified ripped apart, always a full sign of body and meaning. And at least on this issue the two traditions agree: the subject does not belong to the world, the subject is the limit of the world.[14]

In their respective approaches to this ontological paradox the two Greeks, Plato and Aristotle, opted for different solutions, the former leaning heavily towards the intelligible, the latter towards the sensible. Thus, while to the father-like master the images were false reflections of pure thought, to the son-like pupil reason without images was unseeable, hence literally unknowable.[15] Although they both agreed that "just as in the case of a wheel, so here there must be a point which remains at rest, and from that point the movement must originate,"[16] they profoundly disagreed on the nature of the Archimedean fix-point; what was stable to one was ephemeral to the other. Of this difference Aristotle was well aware, even though he too was certain that "there could be no knowledge of things which were in a state of flux."[17] It is telling that in Raphael's painting *The School of Athens,* it is Plato who points to the heavens with one finger and Aristotle who reaches to the earth with his whole hand.

To search for the fixed point of rest is to enter the forbidden rooms of

Raphael, *School of Athens*. Detail of Plato and Aristotle. 1508–11. Fresco. Stanza della Segnatura, east wall. Musei Vaticani, Città del Vaticano.

the Bar de Saussure, by definition the abysmal topos of ontological transformation. Once there, however, we must always keep in mind that the blueprint of the House of Representatives was drawn a long time ago, most outstandingly in Books 6 and 7 of the *Republic,* the grand center of all Socratic dialogues. More explicitly than anywhere else, it was in these pages that Plato argued that for each set of changeable things there is a single Form; although there may be many beds, there is only one Bed.

Plato's Forms are objects of thought rather than objects of sight, and this explains why to him they could always remain the same. For that reason he consistently treated the sensible world as a spatial and visible reflection of a nonspatial and invisible structure of intelligible originals, a process of intellectualizing in which the concept of imagination serves as the bridge between the wisdom-lovers' realm of reason and the honor-lovers' aspiration, on the one hand, and the money- and food-lovers' realms of appetite, on the other. All in all an unbreakable combination of pleasure and knowledge.[18] Solipsism in the making.

As a way of understanding these relations Plato distinguished between the re-presented copies, which can be seen by the outer eye of the body, and the pre-sented originals, which are for the inner eye of the mind. Hovering above the transcendental abyss is nevertheless a kind of Heisenbergian uncertainty principle which says that the copied things of the sensible world (roughly the Saussurean Signifiers) are visible to all animals, while the original Forms of the intelligible world (roughly the Saussurean signifieds) are for the semiotic animal alone. In Ludwig Wittgenstein's version of the same idea, "while I am looking at an object I cannot imagine it. . . . Images are not picture. . . . Asked 'What image have you?' one can answer with a picture. . . . What is imaged is not in the same *space* as what is seen. Seeing is connected with looking. . . .When we form an image of something we are not observing."[19]

No wonder that imagination is considered such a sly and slippery concept, in essence another term for the Mallarméan throw of dice, a joker in the game of cartographical reason, the wild card of understanding.

But even Mallarméan jokers must play by the rules. And nowhere have these cultural regulations been more explicitly laid out than in the *Republic,* a text which in itself is an attempt to resist the seductions of concrete things and advocate the abstract principle of goodness instead. It should nevertheless be stressed that Socrates (sometimes in conflict with his Platonic alter ego) never considered the search for the Form of Forms as the final end of a journey of faith but always saw it as the beginning of an unending quest. Yet he was everywhere well aware of the difficulties, noting in particular that "a sensible man will remember that the eyes may be confused in two ways—by a change from light to darkness or from darkness to light; and he will recognize that the same thing happens to the soul."[20] It follows, firstly, that the person with knowledge is a person who thinks things through where others remain unreflective, and secondly, that the attainment of the Good is not merely the outcome of correct reasoning but the sign of a good person. Just as true science is the study of the Good, so the good is a child of enlightened imagination.

Plato's being is a predicative being, not an existential Being. And exactly as Shakespeare's *Macbeth* is a map of horror, so Plato's *Republic* is a map of the Good.

*

**The *Republic* is a map,** its overriding purpose to charter the way to the good life in the good city and thereby provide an answer to Socrates' question of how one should live. The important point is not that Plato's map is a political document (which of course it is),[21] but that the political is an explication of how the real world is moved from the existing Now-here of the already to the predicative No-where of the not yet, both topoi equally imaginary and symbolic. Obviously easier said than done, and when Glaucon asked what the Good really is, Socrates wisely replied that if he knew, he would be much more than content. In his own words:

> I am afraid [that such a definition] is beyond my powers; with the best will in the world I should only disgrace myself and be laughed at. No, for the moment let us leave the question of the real meaning of good; to arrive at what I at any rate believe it to be would call for an effort too ambitious for an inquiry like ours. However, I will tell you, though only if you wish it, what I picture to myself as the offspring of the Good and the thing most nearly resembling it.[22]

A picturing for Socrates, a path-breaking journey for his descendants!

Why path-breaking? Exactly because each and every one of the Socratic dialogues is structured like a map, at the same time a set of meticulously established thought-positions and an explication of the routes by which these particular topoi have been reached, a mode of reasoning in which "the discourse is spatialized through a series of metaphors that accompany a chain of primary or guiding questions around which the whole dialogue is organized."[23] And once in that mood it is easy for the budding critic to imagine Plato writing (actually more likely saying) something like the following:

> "Do you now know, Glaucon, and can you tell me, where we are and what gives to this place its special character?"
>
> "Yes, Socrates, you just said it yourself. Of course I know that I am now-here rather than no-where, pre-sent rather than ab-sent."
>
> "Good, my friend. Do you then also recall how we journeyed together to get here, which cities we visited along the way and which sights we glimpsed in the distance?"
>
> "Yes, by Zeus," said Glaucon. "In my deepest sleep I can retrace each step of the argument. And let me add, before we indulge in yet another cup, that rumor tells me how you once took your young friend, Phaedrus, on a similar walk, albeit not inside the city, but outside its walls; two ways of finding the way, with me in the realm of politics, with him in the wonderland of love. Indeed it now dawns on me that each and every one of

your dialogues is actually a map. Bound together all your teachings are to me nothing less than an atlas of what it means to be human."

"What a chockfull observation! But, beware my dear nephew," said Socrates, "lest you turn into a copy of your uncle! As an antidote, let us therefore now imagine the picture of the Good, not as it is in itself, but as it shows itself in its offspring, that is in the thing that most resembles it. Not, regrettably, as well as I would like it to be done, merely as well as I am able to do it. With your interest in cartography you will surely agree that no map is a perfect map. And in your capacity as a citizen, you will also have experienced that whoever abuses his power, he thereby transgresses a forbidden boundary."

With this warning constantly in mind, and with the purpose of teaching Glaucon how to paint the picture of the Good in the just person's knowledge, Socrates introduces him to the three figures of the analogy of the Sun, the allegory of the Line, and the parable of the Cave. Similarly (and equally skilled in the art of rhetoric, some claim in the art of lying)[24] every cartographer knows that in order to make a map all that is needed are three concepts. No more and no less than

- a point which remains at rest;
- a line better known as a scale or a coordinate-net; and
- a plane onto which the emerging picture / story is projected.

And so is it that despite all its complexity mapping is essentially an exercise in the geometry of triangulation, triangulation itself an exercise in the politics of power, the latter a dance with the semioticians' three-cornered figure of thought, symbol and referent.[25]

Adhering to these principles the mapmaker first transforms Socrates' Sun into the fix-point of his own time and place—Ptolemy's North Pole, Cosmas' Paradise, Stalin's Dictatorship of the Proletariat. He then changes the Divided Line into a set of scaling coordinates—a Marduk-net designed to capture, sort, shape and name the Ruler's needs and interests. Finally (s)he understands that the canvas of the mappa serves the same functions as the philosopher's Cave Wall—the culturally prepared projection screen onto which invisible relations are casting their visible marks. Geometry at work, for just as the Archimedean point of these ontological transformations is the pivot of the revolving world, so the Ptolemaic line is the legend which translates picture to story and story to picture. The Platonic plane, in its turn, is the knowledge-stained napkin of Deleuzian manifolds.

Point, line and plane. Under other labels these are the most fundamental supports of Plato's mappings: blood-spots on the white sheets; imitations in drawing; tapestries woven at dawn and undone at dusk. The critic should nevertheless tread cautiously, never forgetting that even though in the utopian No-where all animals are equal, in the real world of Nowhere some pigs are more equal than the others.[26] In the cartographic universe this equalizer of equalizers is one with the scaling line, overwhelming in its power to implicate the knower in the known and the known in the knower. This magic wand is in effect the quintessential harbinger of pleasure, knowledge and revolutionary change; the Ruler's rule; the King's whipping-stick; the Orator's scepter; **some notes from the Underground too.**

Like the city of Plato's *Republic,* every map is simultaneously a show of visible things and a story of invisible relations, equally an appearance and an apparition, in the same document an illustrated guide to places (un)known and a hidden code of behavior. Once again, the horror of Macbeth's tragedy was not that he killed but that he slaughtered, not that he found his way to the King's bedroom but that he got lost in the labyrinth of ethics. Had he read the *Republic*—which Shakespeare himself almost certainly had—then the usurper would have understood that, unlike things which can be seen but not thought, ideas can be thought but not seen.[27]

*

**Things can be seen, ideas can be thought.** Which is another way of saying that as the Sun reigns over the world of the eye-ball, so the Good reigns over the intelligible. In Aristotle's vocabulary,

> Suppose that the eye were an animal—sight would have been its soul, for sight is the substance or essence of the eye which corresponds to the formula, the eye being merely the matter of seeing; when seeing is removed the eye is no longer an eye, except in name—it is no more a real eye than the eye of a statue or of a painted figure. . . . As the pupil *plus* the power of sight constitutes the eye, so the soul *plus* the body constitutes the animal.[28]

And just as to Aristotle—the pupil—there was no animal without soul and no vision without light, so to Plato—the seer—there was no reason without goodness. Enlightenment in the making, for as Socrates informs Glaucon:

> The Sun not only makes the things we see visible, but also brings them into existence and gives them growth and nourishment; yet he [the Sun] is not the same as existence. And so with the objects of knowledge: these derive from the Good not only their power of being known, but their very being and reality; and Goodness is not the same thing as being, but even beyond being, surpassing it in dignity and power.[29]

Yet, if the Greekjew stares at the sun with her eyes unprotected, she goes blind, if the Jewgreek meets his God naked, he goes mad. Hyperbole can get us no further, and even though Socrates forcefully argued the case of political idealism, he still considered it the most destructive of human passions, largely because "regimes depend on men's virtues, not on institutions; if the highest virtues are not present in the rulers, an inferior regime must be instituted. There are no guardians above the guardians; the only guardian of the guardian is a proper education."[30]

And a proper education is precisely the gift Socrates administers to Glaucon, a dialectical enterprise equally daring and moderate; daring because Glaucon was too young to have any personal experience of either politics or philosophy, moderate because Socrates' teaching drew entirely on his student's ability to imagine. It may even be that the most remarkable aspect of the collected dialogues lies not in the teacher's deliberations but in the students' readiness to understand them, not in the master's sayings but in the pupils' hearings. How do I grasp a meaning I never met before, how do I make sense of something hitherto unseen?

Part of the answer lies in the richness of the word *meaning* itself,[31] a sense which begins its wanderings in the point-like qualities of "that which lies in the middle," then moves via the line-like directives of "intending, signifying or pointing the way," to the final end in the plane-like values of "what is inferior in rank or quality"; the meaning of meaning is often an issue of how to be to-get-her.[32] It follows that a sentence can be said to have a certain meaning if it performs a certain illocutionary act, if it somehow induces a person's mind to move from one topos to another. "To mean" is consequently to invoke a travel story, to construct an invisible map of the invisible, to engage in a form of cartographical reason. It was Plato's genius that he knew how to lead his students not into the wilderness of the utterly unknown but to vantage-points from which they could *re*cognize what they had already cognized on their own; the utopian No-where must be approached as if it were Some-where. Plato's Academy was built as a map-room. Our own universities are likewise.

Yet, a crucial difference between now and then is that Glaucon had no way of knowing that while for Ptolemy the North Pole was real but not marked, for Gervase Paradise was marked but not real. This distinction notwithstanding, also Ptolemy's and Gervase's mappings were by necessity imitations in drawing, seemingly of material things, in actuality of social relations. The same holds also for other utopias including Plotinus' *Platopolis,* Saint Augustine's *City of God,* Thomas More's *Utopia,* Arlo Guthrie's *Alice's Restaurant* on the other side of the railroad tracks. And exactly because there can be no drawing without lines, no painting without paint, Jan Vermeer understood that the map may be approached as an allegory not of the world in general, but of the world of painting in particular.[33]

Thus it is generally believed that when he constructed his famous *De Schilderkunst* or *The Art of Painting* (ca. 1666–67) Vermeer observed the world through the peep-hole of a camera obscura, a device which builds on the principles of monocular rather than binocular vision.[34] Indeed it was his mastery of this technique that enabled him to exaggerate the scale

Jan Vermeer, *The Art of Painting*. 1666–67. Oil on canvas, 120 × 100 cm. Kunsthistorisches Museum, Wien.

differences between what is close and what is farther away—the size of Clio, the muse of history who is shown in the background, is diminished; the artist in the foreground, his back turned to the viewer and his black shoe blending into the blackness of the floor-tile, is enlarged. The other compositional components of the picture are treated with equal skill and to the same effect, the chandelier used as an immediate counterpart to Plato's Sun.[35]

But whereas the solitary artist with the large family struggled with the problem of how to make the represented objects stand out from the two-dimensional canvas, the conventional mapmaker is faced with the opposite problem of how to make the globe flat. Everyone obviously wants to be believed, but while the Renaissance perspectivist chose to paint the world as it seems to be, the projecting cartographer tries to show it as it really is. The Art of Painting is not a picture of a map, it is the map of a picture.

Similar issues of deception, truth and consequences were central also to Roger Bacon (by his contemporaries nicknamed *doctor mirabilis*—"wonderful teacher"), the Franciscan friar who sometime around 1260 wrote a famous letter to his benefactor Pope Clement IV.[36] Bacon's argument in this correspondence was that if the church wanted to succeed in the struggle against the Muslims, then the preachers should focus less on the logic of truth and more on the rhetoric of belief. As a way of advancing that idea he drew attention first to the newly rediscovered fields of geometry and optics, then to the artistic techniques associated therewith.[37] The rhetorical argument was that if the size of the depicted figures were made proportional to their "actual" position in the picture space, not to their "theoretical" position in the hierarchy of theology, then the message would be more acceptable. In Bacon's own words, "the whole truth of things in the world lies in the literal sense, [because] we can understand nothing fully unless its form is presented before our eyes, and therefore in the Scripture of God the whole knowledge of things to be defined by geometrical forms is contained and far better than mere philosophy could express it."[38] Once again the preacher's first duty is to be believed. And to be believed is less an issue of telling the truth and more of making the audience feel at home, the *Spiritual Exercises* of Ignatius Loyola and the visions of Saint Birgitta excellent cases in point. But who, apart from the ideal figure of Plato's wisdom-loving philosopher, would be more interested in understanding power than in exercising it?

A hundred-and-fifty years after Roger Bacon and a century before Niccolò Machiavelli, Leon Battista Alberti addressed these same issues of representation and deception. Without hesitating this versatile rhetorician declared that the artist's central goal should be to create a particular kind of illusion, more precisely to "draw with lines and paint in colors on a surface any given bodies, in such a way that at a fixed distance and with a certain, determined position of the centric ray, whatever you represent will *seem to* stand out in relief and exactly to resemble the bodies in question and in the same relief."[39] Good art should consequently strive to be effective by being affective, to *appear* real rather than *being* real. And like other manifestoers, neither Bacon, Alberti nor Machiavelli were content with the idea of merely interpreting the world. They wanted to change it as well. Indeed it was "by using the model of rhetoric [that] Alberti could create both a coherent, systematic manual and a consistent, provocative theory of painting. Often, moreover, he wrote as if he saw himself not as a painter or a practical man but as an orator."[40] Already Aristotle (*Rhetoric* 1354a) knew that Dialectic and Rhetoric are the twin sisters of each other.

* * *

**Seemingly topsy-turvy.** In reality as power-ridden as everything else, for Bacon's original proposal was as firmly tied to the Medieval theories of rhetoric and theology as to the Renaissance practices of geometry and alchemy; since the Creator's very first words were "Let there be light!" it was quite natural to think of optics as a privileged approach to the Almighty. In addition, the deceptively plastic forms of Giotto's frescoes, painted only a few years after Bacon's letter, pointed the propaganda-makers in the same direction.[41] Indeed it was with these sculpture-like pictures in Padua's Arena Chapel that Giotto opened the floodgates to a future in which ordinary viewers no longer were treated as passive recipients of biblical stories but as active makers of space itself. But already Horace knew that poetry is like painting, *ut pictura poesis.*

The idea which Bacon and Giotto shared with their Greek and Latin predecessors was that the sense of sight opens the doors to the Command Center of Culture.[42] And for this reason of power and socialization we must now return to Firenze and that beautiful morning in the summer of 1425 when the self-taught builder Filippo Brunelleschi, initially trained as a goldsmith, placed himself and his easel in the doorway of the cathedral of Santa Maria dei Fiori,[43] unwittingly producing the prototype of the perspective.[44] And there, for the first time in history, it was made perfectly clear that the real and the symbolic cannot be separated, a Lacanian reversal five centuries before Lacan, an irrefutable proof that the vanishing point of the painting is one with the viewpoint of the painter. What I happen to see depends on where I happen to stand.

Although it is no longer possible to determine exactly how Brunelleschi performed his experiments, the general story is well known, perhaps because it is easier told than verified.[45] Stationing himself some three *braccia* inside the central portal of the cathedral,[46] whose stunning dome he had not yet completed, he looked out over the piazza towards the octagonal Battiseria di San Giovanni. On a small wooden panel, about half a braccio square or about the length of one's hand, wrist included, he then painted a picture of what he saw, everything executed

> with such care and delicacy and with such great precision in the black and white colors of the marble that no minituarist could have done it better. In the foreground he painted that part of the piazza encompassed by the eye, that is to say, from the side facing the Misericordia up to the arch and corner of the sheep [market], and from the side with the column of the miracle of St. Zenobius up to the corner of the straw [market], and all that is seen in that area. And he placed burnished silver where the sky had to be represented, that is to say, where the buildings of the painting were free in the air, so that the real air and atmosphere were reflected in it, and thus the clouds seen in the silver are carried along by the wind as it blows. *Since in such a painting it is necessary that the painter postulate beforehand a single point from which his painting must be viewed,* taking into account the length and width of the sides as well as the distance, in order that no error would be made in looking at it (since any point outside of that single point would change the shapes to the eye), he made a hole in the painted panel at that point in the temple of San Giovanni which is directly opposite the eye of anyone stationed inside the central portal of Santa Maria dei Fiore. The hole was as tiny as a lentil bean on the painted side and widened conically like a woman's straw hat to about the circumference of a ducat, or a bit more, on the reverse side.[47]

The conic hole thus drilled, Brunelleschi turned the panel around, its backside toward himself. In his outstretched left hand he then placed a mirror, in his right hand, close to his eye, the painting. Now, peering through the hole, he saw what from our present-day standpoint could be expected: the image in the mirror looked exactly like the reality it mirrored, the clouds double-mirrored, first in the burnished silver of the panel then in the mirror itself.[48] In Manetti's words, "with the aforementioned elements of the burnished silver, the piazza, the viewpoint, etc., the spectator felt he saw the actual scene when he looked at the painting, I have had it in my hands and seen it many times in my days and can testify to it."[49] What Manetti presumably saw with his own eye was firstly that the "viewpoint," set at a particular distance from the picture surface, coincided with the "vanishing point," secondly that the "frame" was determined by what could be seen from the painter's position some three braccia inside the portal. What remains unclear, however, is how Brunelleschi handled the complication that what was seen to the left in the mirror was to the right in reality.

Impressed by his monocular vision, Brunelleschi took a few steps to the side, lifted the panel and the mirror as before, looked and was shocked. For prompted by the steps just taken he now discovered that the previously "perfect" painting no longer was the true representation he had taken it to be. Even more importantly he found that "on the basis of an image in perspective of a building one could determine the spot on which the painter placed himself, [because] from a given point of view the eye can only perceive part of any opaque body. . . . What was essential was that the chosen

Brunelleschi's use of *La tavoletta.* Adapted from Hubert Damisch, *The Origin of Perspective* (Cambridge, Mass.: MIT Press, 1994), 116.

point of view could be *deduced* from the painting itself, [that particular place] locatable in real space with relative precision ['about three *braccia* inside the central portal of the cathedral'], such that anyone at all who came to the spot might verify the accuracy of this designation."[50] More technically, "the centric ray is assumed to be perpendicular to the surface or intersection [with the panel plane] and can therefore be seen only as a point, the vanishing point of the orthogonals. . . . The picture defines our viewpoint, or viewing position."[51]

Reemphasized and rephrased by the leading expert:

> The experiment was intended to reveal, by reflectively turning the structural disposition back on itself, nothing less than the premise of its own efficacy: namely that a painting constructed in perspective (*constructed,* not merely approximated) must be seen from one specific point (*uno luogo solo*), governed by a system of rectangular cartesian coordinates distributed across three axes, two of them on the picture plane and the third perpendicular to it. . . . Self-referentiality is a trait characteristic of regulatory systems.[52]

Through this pregnating instance of reason in the reverse, Brunelleschi effectively transcended the limits of Greek geometry, like Giotto before him foreshadowing a world which within two centuries was to come. More precisely, while the traditional "Euclidean geometry posits a homogeneous space in which all points are equivalent, in which all points are of no account, as regards spatial composition . . . the [new] geometry of Desargues posits a space organized in relation to a point of view through which order is imposed on the random variety of the first. Here the point encompasses space and space encompasses the point, the word 'encompasses' being understood to embrace not only geometry, but vision and thought as well."[53] The major difference between Euclidean and Cartesian geometry is consequently that whereas the former deals with the things of the finite world, the latter is occupied with the relations of infinity; while Euclid proved theorems, Descartes solved problems.[54]

The conclusion was inevitable: there is something viewing my viewing, something watching my watching.[55] From its place on the receding horizon the vanishing point stares us straight in the eye, for not only is what I happen to see dependent on where I happen to stand, but the very purpose of perspectival art is to shape that relation. In the ensuing process of affect and effect the world was gradually changed from being finite and bounded to being infinite, continuous, homogeneous, and isotropic. The double point is that to think like Copernicus one must first have written like Alberti, and to think like Alberti one must first have built like Brunelleschi. A point well taken.

*

**This point** and its connections with Euclid's fifth axiom have been further explored in a remarkable essay by Ulf Linde in which he reminds us that Euclid took the join between two points to be unique—in schoolboy parlance, two straight lines are parallel if they never cross. Easier to grasp than to understand, though, for once you let your thumb and index finger run along the edge of a table your body will immediately *touch* what being parallel actually means.[56] Strange, nevertheless, because when Alberti based his observations on sight rather than touch, then he could *see* that in the far-away distance the presumably parallel lines were running together;[57] while to the thumping wheels of the locomotive the rails are always the same distance apart, to the searching eyes of the train-guard they come together in the vanishing point he is bound never to reach.

What and who is under these circumstances to be trusted, the man or the machine, the voyeur or the fetishist, the index finger and the eye or the index finger and the thumb? How do I reconcile the contradictions between the *there* in the distance and the *here* close by?

Whatever the answers, the letter *t* makes a wor(l)d of difference, not the least because in our role as common citizens we tend to invest more trust in distancing sight than in close touch, while as individual beings our preferences are exactly the reverse—a political fact with considerable consequences. It follows that when Brunelleschi performed his Lacanian inversion he stumbled onto one of the most crucial inventions ever made, a creation perhaps as important as the invention of the wheel, certainly as revolutionary as the discovery of the passage round the celebrated Cape of Good Hope: the modern subject, a newborn creature in other contexts known as "the self." Quite significantly there were before the Renaissance no pictures which today would be called "self-portraits," because prior to the fifteenth century there was no self to portray![58] Rephrased, my self is the limit of myself and what cannot be thought can be neither imagined nor painted. And in that process of minimalization, the three-dimensional world gets condensed into a dimensionless and dematerialized point, a fading *locus* which never sits still.

Like the power it symbolizes, the vanishing point is determined by the spot from which it is seen,[59] the self-referential premise of its own efficacy; while in Greek "to see" was "to know," in French the word *représentation* means both "representation" and "performance." No wonder, therefore, that Brunelleschi understood that the perspectivist's real challenge was not in capturing the stones of the piazza or the walls of the Battisteria, but in moving with the sky and the clouds, the former objects stern and stable, the latter amorphous and constantly drifting. His solution was doubly brilliant, for, while he translated the solids into points and lines, he let the heavens free to mirror themselves in the plane of burnished silver.

The lesson is that just as the *perspectiva naturalis* is closely tied to Euclidean geometry, so the *perspectiva artificialis* anticipates the spatial systems of Descartes and Desargues, the former system finite and essentially confined to the plane, the latter from its inception concerned with questions of infinity.[60] In all maps (Ptolemy's and Brunelleschi's imitations included) the concepts of here and there, above and below, are from the beginning merged into one, primarily because every map is a palimpsest, a many-layered imagination of another place and another time. Translation is nevertheless the name of the game, for to translate is to make new boots out of old ones, to convey an idea from one art form to another.[61] Power disrobed by a man who for most of his life lived in a social space-in-between.[62]

*

**To translate is to convey an idea from one art form to another.** As a consequence, all great post-renaissance artists have tried to deconstruct and go beyond the Albertian constrictions. And eventually there came the conscious breakthrough of Paul Cézanne, his last canvases almost abstract, his impact revolutionary.[63]

Since this antisocial son of a self-made banker never took anything for granted, he was gradually able to understand that what he was painting were not God-given landscapes but man-made pictures, not pears and mountains but triangles and rectangles, not content but form.[64] Nothing but small areas of paint which bear strong yet nonspecific relations to the world as he saw it. Under the master's brush the boundaries between colors became boundaries between forms, the images effectively emerging from the paint itself. Seemingly compact objects were produced through brushstrokes which in themselves are open forms. After much practice he even came to see the importance of negative space, not the least that a patch of sky between two branches can tell as much about a tree as the tree itself. Most importantly he learned to slant the paint in different directions. And in that way he managed to evoke a sense of three-dimensional depth which was anchored not in the points and lines of classical perspective but in the free-floating planes themselves, geometry the sole yardstick of the earth.[65] Transcended in the process was the single point of vision which for half a millennium had stabilized what people saw. Thus, in a violent reaction against the theory and practice of Paul Gaugin, he adamantly claimed:

> Everything in nature is formed after the sphere, the cone and the cylinder. If you learn to paint on the basis of these simple figures, then you can achieve whatever you wish. The drawing and the color cannot be distinguished, for when you paint you draw as well; the more the colors harmonize, the more exact is the drawing; when color is at its richest, the form is at its fullest. The contrasts and the mutual relations of the hues, therein lies the secret of drawing and modelling.—You must work hard with your art and know from the beginning a method for realizing your intentions; to be a painter you must know the most elementary side of painting and use only the simplest means.[66]

Crucial among these elementaries of painting is the particular position in which the artist places his easel. Not surprisingly, therefore, Cézanne's choice of viewpoint has often been singled out as an important part of his genius.[67] Yet it deserves repeating that the distancing effects which centuries of previous artists had attained by diminishing the size of faraway objects he achieved by changing their color; through his example every painter now knows that when orange recedes, it turns to red, and when yellow moves away, it fades into green. As he himself so perceptively noted, the secret lies in the hard work.

True to that attitude he began on the shadow with a single patch, which he then overlapped with a second, and a third, until these hues, hinging to one another, not only colored the object, but also molded its form. In his own words, "it is only through nature that we can make progress and it is through studies of nature that the eye is trained; it becomes concentric by looking and working. What I want to say is that in an orange, an apple, a ball, a head, there is a culmination point and this point is always closest to our eye—and that is despite the terribly confusing effects of the light, the shadow and the colors. The border planes of the objects fading into a center placed at the horizon."[68] Like Plato before him he knew that to mold a Form is to give an inner apparition to an outer appearance.

Yet nothing comes of itself, not even to a genius. Therefore Cézanne was often ridden by doubts, not the least about the trustworthiness of his own body. As a way of handling this problem he tried to present the objects as they appear to all senses at the same time, and when he succeeded, one can actually

> *see* the depth, the smoothness, the softness, the hardness of objects, [even] their odor and [this] is why each brushstroke must satisfy an infinite number of conditions. . . . Cézanne's difficulties are [consequently] of the first word. He considered himself powerful because he was not omnipotent, because he was not God and wanted nevertheless to portray the world, to change it completely into a spectacle, to make *visible* how the world *touches* us. . . . It is not enough for a painter like Cézanne, an artist, or a philosopher, to create and express an idea; they must also awaken the experiences which will make their idea take root in the consciousness of others.[69]

Perhaps it is this combined sense of touch and seeing that gives to Cézanne's work its intensity of feeling, its integrity, solitude and hermitry. In effect his paintings were not only a commentary on Euclid's fifth axiom but also a remarkable illustration of Stéphane Mallarmé's remark[70]—here paraphrased to better fit my own purposes—that

watching
doubting
revolving
shining and mediating
without stopping
eventually produces an image that
consecrates it

*

**To consecrate an image** is to take it literally, much as Paul Cézanne did with the word "painting" itself. In deed it was only in that manner that he was able to realize that color at its richest is form at its bodily limit;[71] in his own eyes light was blue, and that explains why a shade of that color appears in virtually all his paintings. Just as Mallarmé's dice-throwing is a gamble, so Cézanne's painting is a sun-dance.[72]

In that light (pardon the term) it is significant that while Mallarmé coined the phrase "paint not the thing, but the effect it produces," Cézanne declared that "to see nature directly consists of sifting out the character of one's subject. To paint something does not mean making a servile copy of it. On the contrary, it means seizing a single harmony out of all the interconnections one has observed, transposing these into a formal series with its own validity by means of working them up according to a *new and original logic.*"[73]

Both Cézanne, the painter, and Mallarmé, the poet, were of course well aware that if it were not for the infinite plasticity of paint and words they could never have invaded the silences that normally fall outside the limits of grammar. And for that reason of creativity the modern techniques of remote sensing and aerial photography have something important to learn from the technical skills and theoretical insights so typical of the two maestros. For what is that type of mapping at a distance if not a human activity located in the interface between poetry and painting? What is a satellite picture if not a peephole show, a constellation of signs waiting to be transformed from meaningless indices into meaningful symbols?

Seen through that Peircean lens contemporary cartography stands naked before us, an emperor without clothes, a balcony-show of ontological transformations in which theory-laden observations are translated first into patches of color, then into strings of words, finally into purposeful action. Picture becomes story as *is* turns to *ought,* laboratories of thought get converted into theaters of social engineering, the ideal setting for political lackeys who want to claim that something is something else and be believed when they do it. Before you bow to the applause, though, beware of your soul, Doctor Faustus! For even though you are *saying* that *A* equals *B,* that is not what you are *showing.* And for that reason all serious art strives to *be* real, not merely to *imitate* the real. Impossible! A sort of manifesto, nevertheless.

And with these words wringing his ears, the heretic cartographer is finally prepared to move from the theory of art history to the practice of human geography. Once on that stage, however, he finds himself face-to-face not only with Francisco Goya's *Colossus* hovering in the sky but with his *Saturnus Who Devours His Children* as well. Such is the beast of Geographical Inference, an alias for the relation between spatial forms and generating processes, appearance and apparition. The ensuing questions reach into the heart of cartographical reason, a mode of thought-and-action which from its birth has been marching to the drums of a methodological two-step: *first* the investigated phenomenon is translated into the graven image of a map, *then* that idol is blessed with a set of seemingly meaningful interpretations. Reincarnation run backwards. Flesh turned to word, matter to meaning.

Nice and clear. Mistaken nevertheless. For through detailed experiments with a family of statistical frequency functions it has recently been shown that identical spatial patterns may be generated through drastically different processes. It follows that even the most perfect description of a spatial form can say nothing definitive about how and why that particular arrangement came about.[74] Once the dice have been cast, the original probability matrix can no longer be retrieved, basically because the equals sign directs the reasoning-traffic onto a system of one-way streets, not onto a swarm of pathways with no beginning and no end. The two-tiered conclusion is that just as reasoning from spatial form to generating process is questionable methodology, so is drawing inferences from aggregates to individuals deplorable ethics.

And yet. Even satellites and cameras know the art of disobedience, even artificial eyes occasionally fail to deliver what they have been designed to deliver. The professional skills of remote sensing and contemporary cartography might therefore be better taught in the Ecoles des Beaux Arts (not, however, of the type that existed in Cézanne's time) than in the Schools of Engineering. The reason is that in his moves from the ethical no-where to the aesthetical now-here the critical analyst is bound to discover that current approaches to picture processing are technologically advanced but philosophically obsolete; while the descriptive images are painted on the tilted planes of twentieth-century Cézanne, the interpretative stories are anchored in the points and lines of fifteenth-century Brunelleschi. In Cézanne's own words:

> [We] must not paint what we *think* we see, but *what* we see. Sometimes it may go against the grain, but this is what our craft demands. . . . People will teach you the laws of perspective at the Beaux-Arts, but they have never seen that depth results from a juxtaposition of vertical and horizontal surfaces, and that is what perspective is. I have discovered it after long efforts, and I have painted in surfaces, because I do nothing which I have not seen, and what I paint exists.[75]

Quite predictably the cartographic laboratories of Los Alamos are filled with flasks of mistranslation. The non-touchable loner was right: what is sorely needed is a new and original logic.

In his seclusion and self-doubt, Cézanne spoke about this novel logic to only one man, the young and admiring poet Joachim Gasquet, with whom he shared his love of Provence in general and the remarkable Mont Saint-Victoire in particular, the hues of the latter his final refuge:

> "Look at Sainte-Victoire," said Cézanne to Gasquet. "What a sweep! What a commanding thirst for sunshine! And what melancholy in the evening when all that heaviness settles upon it. Those blocks were fire. The fire is still in them. The shadow, the day, seems to recoil from them in awe, to be afraid of them. Plato's cave is up there; when great clouds pass over you will notice that their shadows tremble on the crags as if scorched, as if they were being quaffed by a fiery mouth at the same time."[76]

As the circle closes, another and much better poet is prying it open: Wallace Stevens, in daytime vice-president of the Hartford Accident and Indemnity Company, at night fascinated by the similarities between the words of the poet and the paints of the painter, the language of authenticity always tied to the imaginary language of absence—the word "stone" is not a real stone, "red turning black" not a real fire, the color "blue" neither the sky nor the Mont Sainte-Victoire.[77] And this insight explains why Stevens paid repeated homage to Paul Cézanne, not the least to the

Paul Cézanne, *Mont Sainte-Victoire.* 1902–1904. Oil on canvas, 73 × 91.9 cm. Philadelphia Museum of Art: The George W. Elkins Collection, 1936.

painter's obsession with the quarry, the pines, the aqueduct, the l'Estaque, to say nothing about the mountain itself and the ideal point from where it should be seen. As proof, here it is, "The Poem That Took the Place of a Mountain":[78]

There it was, word for word,
The poem that took the place of a mountain.

He breathed its oxygen,
Even when the book lay turned in the dust of his table.

It reminded him of how he needed
A place to go to in his own direction,

How he had recomposed the pines,
Shifted the rocks and picked his way among clouds,

For the outlook that would be right,
Where he would complete in an unexplained completion:

The exact rock where his inexactnesses
Would discover, at last, the view toward which they had edged,

Where he could lie, gazing down at sea,
Recognizing his unique and solitary home.

*

**Always the need of a place to go to** in my own direction, always an already in the not yet. And for that reason of intellectual history the critical cartographer must recall also Jacopo Tintoretto, the miracle whose name was one with his ancestry—son of a dyer. Perhaps earlier than anyone else it was this sixteenth-century Venetian who attained a sense of depth by working directly with the light of color, thereby allowing the central focus of the vanishing point to split into many. In fact there is in these twisting and unstable figures of the mannerist a foreshadowing not only of the folds and ovals of the Baroque, but also of the thousand plateaux and lines of flight that nowadays are associated with the names of Gilles Deleuze and Félix Guattari.

Even though the (post)modernist Deleuzian is a schizophrenic, he is not a crazy madman but an outstanding cartographer, a person who has accepted the challenge of mapping the many connections that determine what it means to be human. However, the outcome of that mission to the unknown "is not a map in which to locate or recognize oneself in a predetermined plane with fixed coordinates [but one] where things may go off in unforeseen directions or work in unregulated ways . . . , a map meant for those who want to *do* something with respect to new uncommon forces, which we don't quite yet grasp."[79] And even though the trendy jet-setter is a nomad, he is a monad as well, his privileged realm located in the boundary-zone between the three fields of power, religion and semiotics.[80]

Few people have explored this bewildering territory (by definition a no-man's land of imagination) with the same energy as René Magritte, the bourgeois surrealist dressed in business suit and bowler hat, the revolutionary explorer who wove together into one strand the activities of showing and saying, geometry and linguistics, painter and poet.[81] In their capacity as full signs, all his works are carefully named, some extremely frightening, the best unbreakably self-referential. The content is everywhere visibly seen and silently heard, the titles by his own admission "chosen in such a way as to prevent [the] pictures being put into some familiar context suggested by the automatic flow of thought in order to avoid uneasiness. The titles are meant as an extra protection to counter any attempt

René Magritte, *Ceci n'est pas une pipe.* 1928–29. Oil on canvas, 62.2 × 81 cm. Los Angeles County Museum. Purchased with funds provided by the Mr. And Mrs. William Preston Harrison Collection. Photo © Museum Associates/LACMA. © René Magritte/BUS 2005.

to reduce poetry to a pointless game, [because] the use of speech for ordinary purposes of life imposes a limited meaning on words designating objects. It would seem that everyday language sets imaginary boundaries to the imagination. But it is possible to create new relationships between words and objects and to bring out certain features of language and of objects that are commonly overlooked in the everyday process of living."[82] Magritte was painfully aware that our everyday life is a universe of ready-made experiences.

This prison-house of language is expertly, yet habitually, explored by its inmate, perhaps most famously in *La trahison des images* (*Ceci n'est pas une pipe*), a painted statement which in our context is more properly read as an unusual map than as an ordinary picture, a paradigmatic illustration of how we get lost in the known and find our way in the unknown, a marvelous rendering of how solid rocks turn to quicksand, geometry loses its foothold, the compass its direction. Humor, horror, irony and imagination, these are perhaps the only possible outlets for someone who like Magritte himself must deal with the memory of his drowned mother,

her body after seventeen days recovered at the locks of the canal, her face veiled in her night gown. The hidden legend never to be told, the mental scars always to be shown.

Magritte lived most of his outwardly conventional life in the subjunctive mood, like his commentator Michel Foucault effectively a mapper not of Utopia but of Heterotopia. Perhaps it is the preference for the latter that explains why his canvases so shamelessly bare the positions of the known and the knower, the surrealist's own version of Plato's Sun, Pytheas' Celestial Pole, Ptolemy's North Pole, Gervase's Paradise, Brunelleschi's portal, Cézanne's outlook. In the particular case of the *Ceci n'est pas une pipe* it is easy to see that the fix-point lies somewhere in the grayish void between the picture of the pipe, on the one hand, and the story of the words, on the other. More exactly, this exact rock of inexactness is found in the grammar of the word *ceci* itself, the seamless intertwining of the pronouns *ce* and *ci,* the *this* and the *here.* More exactly still, the fix-point of fix-points is in the merger of the two pre-positions *of* and *in,* the vantage-point and the fix-point turned into one. With rare clarity the artist here demonstrates how both the seeing eye and the pointing finger are drawn to the *this* rather than to the *that,* the rhetorical home of the example being the most power-filled of all powerful topoi. A privileged position blessed not by the Q.E.D., the Q.E.F. or the Q.E.I. taken separately, but by all of them linked together in the verb form of the "Quod Erat" itself: "(1) the object; (2) the something linked to it in the obscurity of my consciousness; and (3) the light into which this something had to be brought."[83]

The roots go deep, Magritte's relations to Plato's Sun and Brunelleschi's vanishing point as intimate as only the critic of cartographical reason might have expected. Hubert Damisch put it well in his concluding remark that "what Brunelleschi *discovered* . . . was not that the vanishing point could be taken for the image at infinity of a painting's orthogonal lines . . . , but rather that it functioned, within the limits of the painting, . . . as the semblance of an eye. . . . [The point is that] there is never a 'view' in a painting, nor an image in a mirror, save for the eye looking at it."[84] Rephrased, the rhetorical fix-point of western culture lies in the penetrating glance of the *ceci,* at once harbinger of the artist's dream and source of the cartographer's nightmare, the central ray which in Albrecht Dürer's famous print is held in its horizontal position by the vertical stick in front of the artist's eye.

And so it is once again shown that both the painter and the mapmaker must postulate beforehand the single standpoint from which they want their respective work to be viewed. Plato's Good in another light, Andrea Pozzo's God as well. Indeed it has been observed that Pozzo's frescoes in the circular ceiling of the Church of S. Ignazio in Rome (painted in the late seventeenth century) represent the culmination of religious illusionism, the vanishing point being nothing less than the "true POINT, the Glory of GOD."[85] Fix-point of fix-points.

Albrecht Dürer, *The Designer of the Lying Woman.* 1525. Woodcut from Dürer's textbook *Unterweysung der Messung mit dem Zickel und Richtscheyt.* 2d ed. Nürnberg: Hieronymus Andrae, 1538.

*

**Dürer's print** illustrates how the fix-point of Magritte's *Ceci* coincides with the point at which the central ray—by Alberti called "the leader and prince of rays"—cuts through the vertical screen which separates the model from the modeler, the designing designer on one side of the transparent goatskin, the undressed woman on the other. For six centuries this technique has been taught not only to countless artists but to the modern citizens of Plato's Republic as well, the pivotal Line at the same time divid*ing* and divid*ed*. Thus, just as the *Ceci* is an offspring of the Sun, so Dürer's grid transforms observations of the sensible into conjectures about the intelligible; the Divided Line is in fact a spatial form in search of a narrative that fits it, the motherly matrix an empty womb waiting to be impregnated.

As the story of that Line unfolds, it will be shown how the edges of the screen determine both the boundaries of the Platonic universe and the frame of the Renaissance picture; in Alberti's own words, "first of all, on the surface on which I am going to paint, I draw a *rectangle* of whatever size I want."[86] Inside that space generations of painters have subsequently placed the picture itself, unwittingly relegating everything else to the outside. Paradoxes nevertheless abound, for just as the invention of the frame liberated the picture from its place on the wall (virtually impossible to do with a Giotto mural), it at the same time tied the artist to the fluctuations of the market instead. One hook replaced by another; painting commodified; freedom (un)bound; the term "frame" a shorthand for the limits of discourse.[87]

Five centuries later it fell on Wassily Kandinsky to formulate a new and different manifesto, modern, abstract and better suited to the needs of its particular time and place. Typical was its description of the geometric point as the ultimate and most singular union of silence and speech, the proto-element of the graphic, hence of thought itself. In dialectical contrast, Kandinsky treated the line as the point's greatest antithesis, the horizontals and verticals completely dumb, the former warm and white, the latter cold and black. To get the lines to speak they must consequently be tilted, the muteness of the right angle being broken in the process. To the master of the blue riders the red diagonal was therefore the most talkative of all lines, just as the central point of a square was the minimalist prototype of pictorial expression.[88] Revolutionary? In deed! Yet not as different from Ludwig Wittgenstein's *Tractatus* as one might have thought, for also to the latter "contradiction is the outer limit of propositions: tautology is the unsubstantial point at their center."[89]

Following the tracks of these trailblazers, the critic finally reaches a point from which (s)he may catch a glimpse of the invisible maps which helped Socrates find his way in the (un)known. To my knowledge such a re-mapping has never before been attempted, a circumstance which, if true, is a warning that dangers lie ahead, starved cannibals waiting at every turn. Yet, there is no alternative. Not with the fire in me now. Lured and frightened; rigorous of mind and sound of body; pushed and pulled by the modernist insight that

> [t]he great aim is accurate, precise and definite description. The first thing is to recognise how extraordinarily difficult this is. It is no mere matter of carefulness; you have to use language, and language is by its very nature a communal thing; that is, it expresses never the exact thing but a compromise—that which is common to you, me and everybody. But each man sees a little differently, and to get out clearly and exactly what he does see, he must have a terrific struggle with language, whether it be with words or the techniques of other arts. Language has its own special nature, its own conventions and communal ideas. It is only by a concentrated effort of the mind that you can hold it fixed to your own purpose. . . .
>
> There are then two things to distinguish, first the particular faculty of mind to see things as they really are, and apart from the conventional ways in which you have been trained to see them. This is itself rare enough in all consciousness. Second, the concentrated state of mind, the grip over oneself which is necessary in the actual expression of what one sees. To prevent one from falling into the conventional curves of ingrained technique, to hold on through infinite detail and trouble to the exact curve you want.[90]

* * *

**Mapping the Republic** is to draw the lines of what it means to be human, to mount the instruments of cartographical reason in the infra-thin boundary between the two prepositions *of* and *in,* to face once again the Beckettian challenge of writing in such a way that it is not *about* something

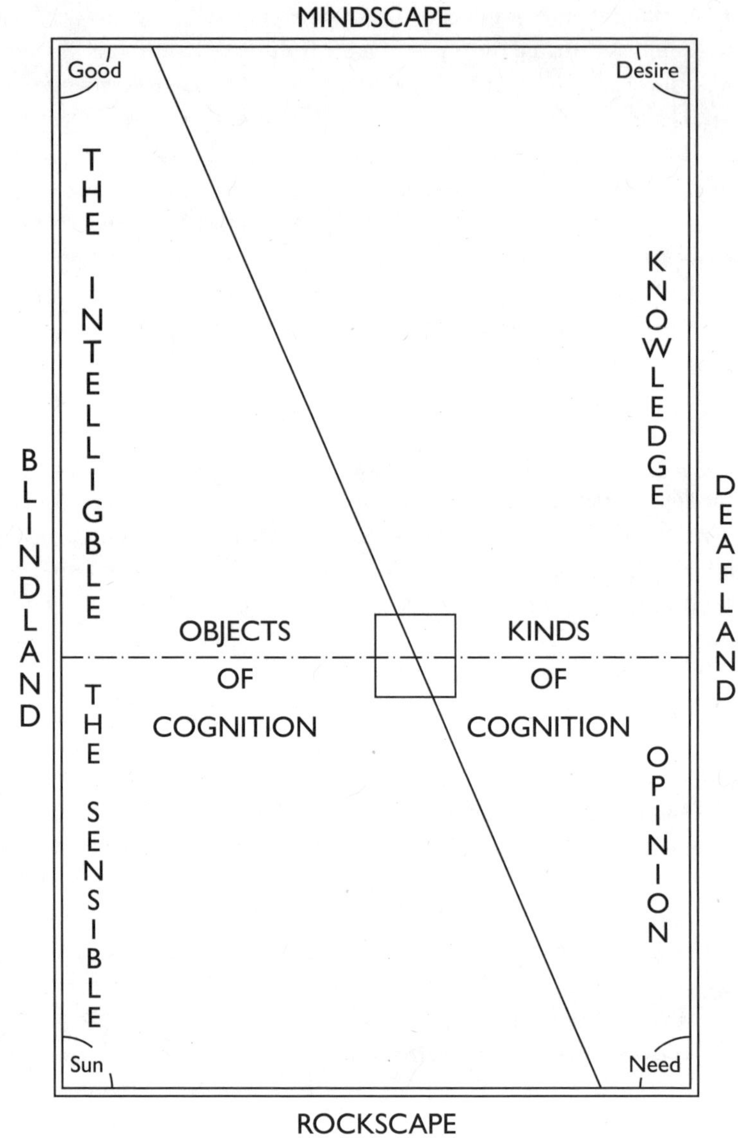


Base map of Plato's *Republic*.

but *is* that something itself. And with that unreachable goal etched onto the receding horizon, two maps are immediately forthcoming, one geographic, the other administrative.[91] Although the actual drawings reflect the flatness of the page, in my own imagination they are crumpled and many-dimensional.

The first imitation is a metaphysical base-map drawn with the purpose of determining the limits of language and thereby the boundary between Plato's universe, on the one hand, and four terrae incognitae, on the other. While the characteristics of the latter are literally nonmentionable, the former is a confederation divided into two realms, one filled with objects of varying ontological status, the other by different modes of understanding. With the grid of this fundamental(ist) creation firmly in place, the political map divides the human territory into administrative units—four provinces each subdivided into two districts—positioned along a gliding scale of concreteness and abstraction.

The geographic base-map is enclosed within an Albertian rectangle drawn in the proportions of the golden section. On its inside lies the Territory of the Humans, different enough to make a difference, on the outside the totally alien, a world so utterly unthinkable that it can be neither noticed nor named. The coordinates of *Enuma elish* come readily to mind; above and below, a net for catching the nothing which cannot be caught.

Even though the rectangular frame is one with the limits of my language, hence of my world, it is extremely difficult, perhaps impossible, to envision a boundary with only one rather than two sides. It was as a way of handling the unthinkability of this situation that the young Wittgenstein insisted that philosophy "must set limits to what can be thought; and, in doing so, to what cannot be thought. It must set limits to what cannot be thought by working *outwards* through what can be thought."[92] If that advice is taken seriously, then the explorer will eventually discover that on the outer side of what can be thought there are four types of silence, in principle non-expressible, in practice not entirely non-imaginary. It follows that whatever is said of the terrortories beyond the beyond reveals more about what lies inside the frame than about what may or may not be on the outside. And so it is that the (post)modern critic is faced with exactly the same problems as Charles Mason, Jeremiah Dixon, Gervase of Tilbury, Pytheas of Massalia and Alexander the Great before him. The Ocean River is still an Ocean River.

This granted, let the Human Territory be surrounded by four unknown continents on which the semiotic animal has never set foot—Mindscape shown at the top of the map (the medieval east), Rockscape at the bottom (west), Blindland to the left (north), Deafland to the right (south). To be precise, Mindscape contains nothing but pure and non-expressible meaning (a signified without a Signifier, an *s* in search of an *S*), while Rockscape is permeated by pure and non-observable matter (a Signifier without a signified, an *S* with its *s* aborted). In neither case a difference large enough to make a difference, rather a maddening lack, a classical state of psychosis, which in the first instance is expressed through the formula

$$\frac{s}{s}$$

the phantom of extreme spiritualism, in the second by

$$\frac{S}{S}$$

the non-sign of thingified thingification. Two forgeries hammered out in the smithy of parallel thinking.

The twin notions of Mindscape and Rockscape are legitimate offspring of Saussurean linguistics and Lacanian analyses. In comparison, Blindland and Deafland are best characterized as conceptual bastards, the former ruled by the fives senses of the body (nothing but an array of meaningless Signifiers), the latter by the sixth sense of culture (nothing but a void of signifieds). While in the *Ebstorfer Karte* Christ's head, feet and hands served as bridges across the Ocean River, here a troupe of crippled signs are performing the same tricks of semiotic leap-frogging.

The four fancylands of silence are all located in the beyond of the beyond. In contrast, the world of Plato's *Republic* lies entirely within the Wittgenstenian limits of language, a territory split into two realms of equal size and shape, one called The Realm of *Objects* of Cognition, essentially ontological, the other named The Realm of *Kinds* of Cognition, essentially epistemological. Marking the boundary between the two lands is a barbed wire twined from the prepositions *of* and *in,* a straight line which in standard textbooks is depicted as the divid*ing* (that is the vertical) part of the original version of the Divided Line. In our maps, however, it is drawn with an angle, partly as an expression of its non-muteness, mainly for reasons to be specified later.

Cutting through the Human Territory, reaching all the way from the invisible Blindland to the inaudible Deafland, runs yet another frontier. In most commentaries this is the major (horizontal) divid*er* of the Dividing Line, the interface between the Intelligible Region in the upper part of the Human Territory and the Sensible Region in the lower, the former area illuminated by the Form of the Good, the latter by the light of the Sun. The same traits are conspicuous in the present map as well, but here the line encapsulates also the Saussurean Bar-in-Between, the topos of ontological transformation where word turns to flesh and matter becomes meaning. It is exactly here, in the intersection of the two lines, that we encounter once again not only the void of Magritte's *Ceci,* but the political hub of Plato's *agora* and the geometric center of Kandinsky's square. This dematerialized point is the semiotician's counterpart to the astronomer's Greenwich, the knot of knots in Marduk's net.

The mapper of the Human Territory gets his instructions directly from the standard textbooks:[93]

- place your theodolite at the crossroads of the *Ceci;*
- in the Realm of Objects aim the sighting tube at the two landmarks of the Sun and the Good;
- in the Realm of Meaning repeat the same procedure, albeit with the difference that the tube be turned to the two corners of bodily Need and intellectual Desire;

- the triangulations completed, transfer the measured angles and distances to the cloth of the mappa;
- as proof of accuracy apply the approval stamps of Q.E.D., Q.E.F. and Q.E.I.—not one at a time but all three together.

*

**The meeting place** of Plato, Kandinsky and Magritte constitutes the invisible fix-point also of our second map, an attempt to illustrate how the two realms of cognition are divided into four provinces—*A* and *B* (together forming the Sensibility Region of the Human Territory) and *C* and *D* (the Intelligibility Region)—each province further divided into two districts, their respective names all specified on the map itself.

The relative size of the four provinces corresponds directly to the original segments of the Divided Line such that

$$D : C :: B : A$$

from which follows both that

$$D : B :: B : A$$

and that

$$D : C :: C : A$$

Rephrased,

The Intelligible World : The World of Appearance<br>::<br>Visible Objects : Images and Shadows

Grasping the construction of this many-faceted universe is crucial to any mapper of what it means to be human, for what I happen to say about it is doubly bound to the objects I am talking *about* and to the understandings I am talking *in*. To be more exact,[94] the objects of Province A are images, by which "I mean first shadows, and then reflections in water or in [mirrors of] close-grained, polished surfaces [à la Brunelleschi's burnished silver] and everything of that kind, if you understand" (*Rep.* 509). This particular mode of understanding (*eikasia* or "perceptual-thought") is of course itself a product of imagination, hence of our ability to make the absent present, the present absent. In contrast, the objects of Province B are always presently present, "actual things of which the [shadows of Province A] are likenesses, the living creatures about us and all the

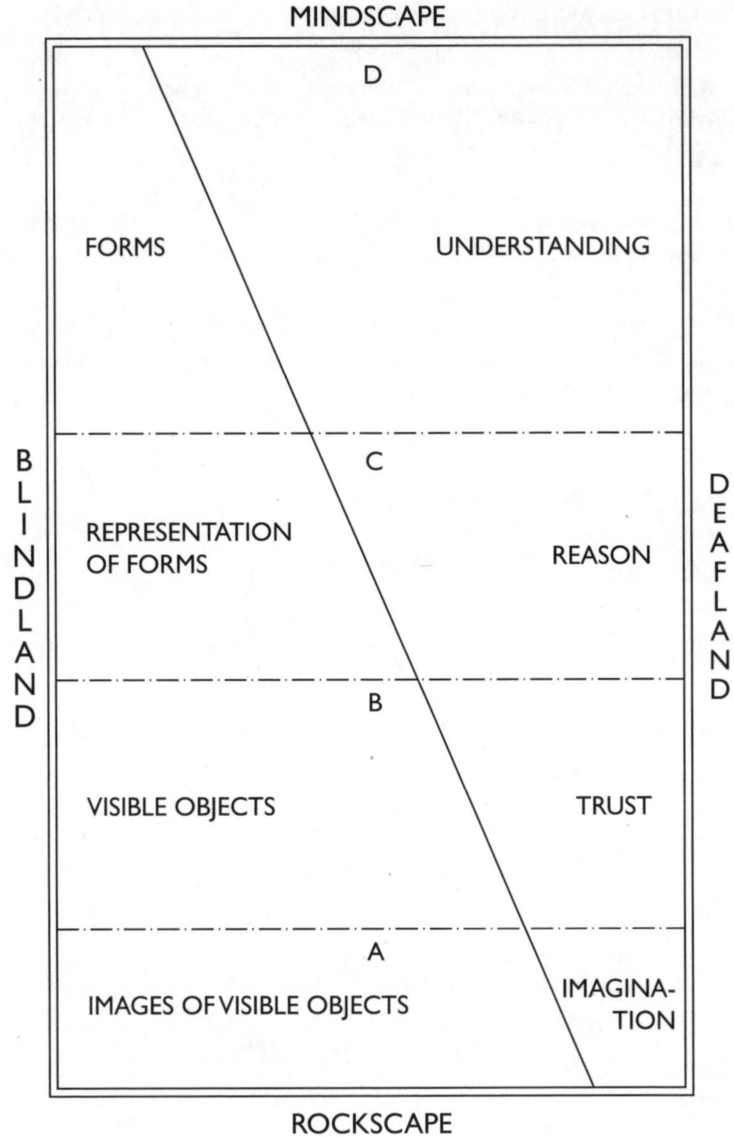


Administrative map of Plato's *Republic.*

works of nature or of human hands." Most importantly, our grasp of these objects is a function not of individual imagination but of the social trust (*pistis* or "folk-wisdom") which is anchored in the five senses of the body—well remembering, of course, that what I happen to see with my eyes is not necessarily the same as what I happen to touch with my hands. This much (or rather this little) about the two provinces and four districts of the Sensibility Region, the area which on our two maps is located *below* the infra-thin line of the main parallel. Familiar ground, finding the way a child's play.

In the Intelligibility Region, which is located *above* the main parallel, the situation at first seems quite different. However, as soon as one realizes, firstly, that everything in Province C concerns mathematical reflection and, secondly, that mathematicians deal with visible things in a very special way, then most of the difficulties evaporate like phlogiston air. In particular, one must keep in mind that the images with which the mathematician is working are not material objects which can be readily grasped by the human body, but something higher and more abstract—not ordinary shadows or reflections in water, but diagrams, imitations in drawing of concepts unseen. The objects of Province C are consequently instanced not by this particular square or by this particular diagonal but by *the* square and *the* diagonal; although the square I draw on the blackboard in some sense is a real thing—white chalk-marks on a black surface—the figure itself is a shadow not of a concrete thing but of an abstract idea, not an entity I can hold in my hand but a relation "which only thought can apprehend."[95] My own maps are obviously objects of this latter kind, Plato's distinction between the visible and the intelligible essentially the same as the positivist's division between the observational and the theoretical. The Cartographical Reasoner is a close relative of Socrates' Mathematician, the modern bureaucrat a first cousin of both.

In Province C the main mode of apprehension is the type of reason (*dianoia* or "scientific-thought") which is usually associated with formalized logic and geometry. Very much of *demonstrandum* and *faciendum,* very little of *invendiendum.* Thus, in this area of the Human Territory:

> ". . . the mind uses as images those actual things which themselves had images in the visible world; and it is compelled to pursue its inquiry by starting from assumptions and travelling, not up to principle, but down to conclusion. . . . You know, of course," said Socrates, "how students of subjects like geometry and arithmetic begin by postulating odd and even numbers, or the various figures and the three kinds of angles, and other such data in each subject. These data they take as known; and, having adopted them as assumptions, they do not feel called upon to give any account of them to themselves or to anyone else, but treat them as self-evident. Then, starting from these assumptions, they go on until they arrive, by a series of consistent steps, at all the conclusions they set out to investigate." (*Rep.* 510b, 510d)

The concept of the taken-for-granted undressed, logical deduction another name of the bureaucrat's game, the mind rather than the body constituting the major target of his intentional actions.

In Province D, finally, the objects of cognition are as far removed from the material world as anyone can imagine. The actual things, which we earlier encountered in Province B, are here replaced by the concept and reality of pure Forms, an entity not so easy to grasp. For the sake of clarity, "let me [therefore] remind you of the distinction we drew earlier and have often drawn on other occasions, between the multiplicity of things that we call good or beautiful or whatever it may be and, on the other hand, Goodness itself or Beauty itself and so on. Corresponding to each of these sets of many things, we postulate a single Form or real essence, as we call it. [Further], the many things, we say, can be seen, but are not objects of rational thought; whereas Forms are objects of thought, but invisible" (*Rep.* 507b). Clearly Plato's Theory of Forms is a visual metaphor applied to mental objects, an imagination of geometric shapes (*eidos* or idea) which cannot be seen.[96]

Then, suddenly, at the end of a well-prepared road, the view from the summit. And from that vantage-point it is immediately seen that the only approach to the Forms goes via the intelligent understanding of dialectics (*noesis* or "dialectical-thought"), in Aristotle's terminology the twin sister of rhetoric. While in Province C the conventional axioms and reasoning rules were taken as self-evident starting points for deductive conclusions, here, in Province D, the same statements are themselves turned into the prime objects of inquiry. In a marvelous condensation, Socrates teaches Glaucon that

> ". . . when its gaze is fixed upon an object irradiated by truth and reality, the soul gains understanding and knowledge [*noesis* and *episteme*] and is manifestly in possession of intelligence. But when it looks towards that twilight world of things that come into existence and pass away, its sight is dim and it has only opinions and beliefs [*eikasia* and *pistis*] which shift to and fro, and now it seems like a thing that has no intelligence.
>
> "This, then, which gives to the objects of knowledge their truth and to him who knows them his power of knowing, is the Form or essential nature of Goodness. It is the cause of knowledge and truth; and so, while you may think of it as an object of knowledge, you will do well to regard it as something beyond truth and knowledge and, precious as these both are, of still higher worth. And, just as in our analogy light and vision were to be thought of as like the Sun, but not identical with it, so here both knowledge and truth are to be regarded as like the Good, but to identify either with the Good is wrong. The Good must hold a yet higher place of honor. . . . [As opposed to the situation in Province C] the mind [here] moves in the other direction, from an assumption up towards a principle which is not hypothetical; and it makes no use of the images employed in the other [Province], but only of Forms, and conducts its inquiry solely by their means. . . ."
>
> "I understand" [said Glaucon], "though not perfectly; for the procedure you describe sounds like an enormous undertaking. But I see that you mean to distinguish the field of intelligible reality studied by dialectic as having a greater certainty and truth than the subject-matter of the 'arts,' as they are called, which treat their assumptions as first principles.

The students of these arts are, it is true, compelled to exercise thought in contemplating objects which the senses cannot perceive; but because they start from assumptions without going back to a first principle, you do not regard them as gaining true understanding about those objects, although the objects themselves, when connected with a first principle, are intelligible. And I think you would call the state of mind of the students of geometry and other such arts, not intelligence, but thinking, as being somewhat between intelligence and mere acceptance of appearances."

"You have understood me quite well enough," [Socrates] replied. (*Rep.* 508d–509a, 510b, 511c–e)

At issue in Province D is obviously neither the content of the taken-for-granted nor the consequences of the taken-for-granted, but the taken-for-grantedness in and of itself. More specifically,

"when I speak of the other sub-section of the intelligible part of the line you will understand that I mean that which the very process of argument grasps by the power of dialectic; it treats assumptions not as principles, but as assumptions in the true sense, that is, as starting points and steps in the ascent to something which involves no assumption and is the first principle of everything; when it has grasped that principle it can again descend, by keeping to the consequences that follow from it, to a conclusion. The whole procedure involves nothing in the sensible world, but moves solely through [F]orms to [F]orms, and finishes with [F]orms." (*Rep.* 511b, trans. Lee)

Being is consequently itself a Form, arguably even higher than the Form of the Good. Higher still, though, is the Form of Forms, and that explains why Socrates did his utmost to prove that without Forms there is no stability, hence no knowledge either.[97] Aristotle, of course, took the opposite stance, arguing that Plato introduced the Forms because he wrongly "accepted the Heraclitean sayings which describe all sensible things as ever passing away, so that if knowledge or thought is to have an object, there must be some other and permanent entities, apart from those which are sensible; for there could be no knowledge of things which were in a state of flux" (*Meta.* 1078b, trans. Ross).

To be a Platonic Form is consequently to be an entity which is self-referential, unique, intelligible and immutable. A primal scream in an echo chamber, a secret committed to a wartime diary. For now it should be obvious that in this context too

my propositions serve as elucidations in the following way: anyone who understands me eventually recognizes them as nonsensical, when he has used them—as steps—to climb up beyond them. (He must, so to speak, throw away the ladder after he has climbed up it.) He must transcend these propositions, and then he will see the world right.[98]

Finally, Socrates again, the ring of unmistakable genius: The mind uses as images those actual things which themselves have images in the visible world. Two verbs to do the trick—remember and remind, the former a backward-looking *res,* the latter a forward-aiming meaning. While the job of reifying remembrance is to re-member the members dis-membered, the magic of deifying reminders is to re-imagine relations forgotten.

*

**Transcending these propositions** was the marcveluous fate of Marcel Duchamp, outstanding critic of cartographical reason, a person rare enough to find his way by seeing the world right.[99] Although he made few references to Socrates and seemingly none to the Divided Line, I have been impressed by the many parallels between his concerns and the maps just presented, the latter essentially triangulations hung on the three points of (a) my own critique of cartographical reason; (b) the summit of Plato's *Republic;* and (c) the delay in glass of Duchamp's *La Mariée mise à nu par ses célibataires, même*—in English known as *The Bride Stripped Bare by Her Bachelors, Even.*[100]

Who knows, perhaps it is in these self-referential performances at the outer limits of projective mapping that the copies and originals of (post)modernism come naked before us, mimetic re-presentations revealing themselves as inventive re-creations of desire, liberation and the self. Driving the striptease is an irresistible urge to reach beyond the surface level of the taken-for-granted, to suck yet another bite of the unconscious into the womb of the conscious. Nothing obscene about that, merely an alternative way of operationalizing the well-established epistemology of extremity, an obvious parallel to the mathematician's procedure of letting his functions approach both zero and infinity.[101] When the tongue slips, it bares the teeth of its owner, no dentist strong enough to pull them out.

Few have explored the mouth of the taken-for-granted with the same ironic blend of touch and humor, withdrawal and renewal, as Marcel Duchamp.[102] If anyone, this leading practitioner of delays knew that in every already there is a not yet, in every painting the trace of a vanishing point, in every imagination the play of presence and absence. If anyone, this timely person realized that ideas are mental constructs without form and void, a remark which helps explain why he devoted his entire life to studies of the geometry of the invisible. From beginning to end an outstanding example of the fact that to be a genius is to know the impossible; to experience separation, incompleteness and frustrated communication; to dangle in the abyss of the excluded middle; to roam the no-man's land between the sensible and the intelligible; to straddle the prepositional line between the two realms of cognition; to turn life into art, erotics into aesthetics. All in an infinite regress, all without going crazy.

The decisive turning-point in Duchamp's career came in 1912 when the twenty-five-year old reached the conclusion that if painting was to be taken any further, someone else would have to do it. Hence he deliberately abandoned painting (admittedly only in its retinal form), not because he had lost interest in the issues of representation, but because he was determined to go beyond the confines of the one-point perspective. With characteristic honesty he told himself, "Marcel, no more painting, go get a job." Which he did, albeit only for a year and a half and as an intern at the Bibliothèque Sainte Geneviève, where he devoted more of his working hours to the study of the Renaissance perspective than to ordinary library chores.[103]

When he reached the crucial decision of withdrawal, he had just returned to Paris from a formative period in Munich, where he had continued his earlier struggle with the transitions of cubism, especially with the problem of how to picture the passage from virgin to bride. In addition, he reputedly bought a copy of Wassily Kandinsky's new book *Über das Geistige in der Kunst*. His real obsession, however, was to go beyond his own *Nu descendant un escalier*, the work which for him represented the definitive break with realism, an attempt to create in one and the same picture a simulacrum of motion; in his own words, "painted as it is, in severe wood colors, the anatomical nude does not exist, or at least cannot be seen, since I discarded completely the naturalistic appearance of a nude, keeping only the abstract lines of some twenty different static positions in the successive action of descending."[104] Or, as he argued in a later interview, "when you wanted to show an airplane in flight, you didn't paint a still life. The movement of form in time inevitably ushered us into geometry and mathematics. It's the same as when you build a machine."[105]

The story is well known, not the least through the informative catalogue from the 1993 exhibition at the Palazzo Grassi.[106] Thus, when Duchamp took his painting to the Salon des Indépendants in Paris, he sincerely expected it to be welcome in the cubist room. But on Monday March 18, 1912, he receives a surprise visit from his two brothers, the painter Jacques Villon and the sculptor Raymond Duchamp-Villon, both of them dressed as for a funeral. Their attire was appropriate, for they were sent by the hanging committee to inform their young sibling that his work was considered unfitting. One of the reported opinions was that it had too much of a literary title, another that a nude never descends the stairs—a nude reclines. "The cubists think it is a little off beam," said the brothers. "Couldn't you at least change the title?" But as the title was written in capital letters at the bottom of the canvas, it was an integral part of the painting itself and could not be removed. René Magritte's *Ceci* fourteen years before the pipe.

Marcel said nothing. But as soon as the messengers had left, he took a taxi to the Quai d'Orsay, collected the picture and brought it home.

Marcel Duchamp, *Nude Descending a Staircase (No. 2)*. 1912. Oil on canvas, 147 × 89.2 cm. Philadelphia Museum of Art: The Louise and Walter Arensberg Collection, 1950. © Succession Marcel Duchamp/BUS 2005.

The betrayal was never forgotten, the world irrevocably changed in the process—an Amazon butterfly fluttering its wings in a Paris building with water and gas on every floor. For just as the cubist nude did not fit into Dürer's orthogonal coordinate net, so the ambitious Marcel understood that he did not conform to the establishment definition of what it is to be a painter. The wounds never healed, although he later stressed that the affair "helped liberate me completely from the past, in the personal sense of the word. I said, 'All right, since it's like that, there is no question of joining a group—I'm going to count on no one but myself, alone.'"[107]

As a consequence he came to spend the next eleven years working on the *Large Glass,* his own definitively unfinished study of Dürer's fresh wi(n)dow, his revolutionary deconstruction of the Renaissance perspective, the scathing response to his insensitive critics. The bachelor was forced to grind his own chocolate, albeit in the company of Francis Picabia and under the influence of Alfred Jarry and Raymond Roussel. In his own recollection:

> It was fundamentally Roussel who was responsible for my glass, *The Bride Stripped bare by Her Bachelors, Even.* From his *Impressions d'Afrique* I got the general approach. This play of his which I saw with Apollinaire helped me greatly on one side of my expression. I saw at once I could use Roussel as an influence. I felt that as a painter it was much better to be influenced by a writer than by another painter. And Roussel showed me the way.
>
> My ideal library would have contained all Roussel's writings—Brisset, perhaps Lautréamont and Mallarmé. Mallarmé was a great figure. This is the direction in which art should turn: to an intellectual expression, rather than to an animal expression. I am sick of the expression "Bête comme un peintre"—stupid as a painter.[108]

Not so stupid, though, for eleven months after the *Nu descendant un escalier* had been rejected in Paris (and after brief showings first in Barcelona and then in Paris), it was exhibited anew, this time at the Armory Show on Lexington Avenue in New York, on that occasion under the domesticated name of *Nude Descending a Staircase.* The reception was beyond belief, most spectators attracted as much by the elusive title as by the cubist image. Despite his physical absence Duchamp was instantly turned into a celebrity, the only Frenchman as well known to the American public as Napoleon, the emperor, and Sarah Bernhardt, the actress. (Post)modernism in advance of itself, the moving nude dissolved into a series of lines, planes and volumes, by some hecklers described as an explosion in a shingle factory.

*

**At the time of the Armory show** Duchamp had already given up painting, allegedly for good. According to his own recollection, "I wanted to get away from the physical aspect of painting. I was much more interested in recreating ideas in painting. For me the title was very important. I was interested in making painting serve my purposes, and in getting away from the physicality of painting. . . . I was interested in ideas—not merely in visual products. I wanted to put painting once again at the service of the mind. And my painting was, of course, at once regarded as 'intellectual,' 'literary' painting. . . . In fact until the last hundred years all painting had been literary or religious: it had all been at the service of the mind. This characteristic was lost little by little during the last century [especially with the work of Courbet]. The more sensual appeal a painting provided—the more animal it became—the more highly it was regarded."[109] My own vision of heretic cartography is obviously of the same type: an atlas of invisible maps of the invisible, a set of mappae mundi abstracted from the interfaces of Plato's boundary lines.

On the newly plastered walls of his Paris apartment, 23 rue Saint-Hippolyte, Duchamp was already preparing the way by plotting the full-scale composition of what years later was to be the *Large Glass,* its focal point exactly defined as early as January 1913. In fact it is typical of the deliberate and long-drawn-out processes of his entire oeuvre that "all the glass was imagined and was drawn in 1913 and 1914, on paper. It was based on a perspective view, meaning complete control of the placement of things. It couldn't be haphazard or changed afterwards. It had to go through according to plan, so to speak."[110] Much else was happening at the time as well (including the purchase of the cast-iron bottle rack from a vendor at the Bazar de l'Hotel de Ville), yet it was not until 1915, after his move to New York, that he named, hence created, the concept of the readymade, by definition an article of everyday use so trivial that it normally goes unnoticed, "a work of art without an artist to make it. . . . A tube of paint that an artist uses is not made by the artist, it is made by the manufacturer that makes paints. So the painter really is making a readymade when he paints with a manufactured object that is called paint."[111]

Since the word *art* means *making* and *making* means *choosing,* the choice of a readymade is analogous to the choice of a tube of paint.[112] In exactly the same sense, every word of the present volume is a readymade as well, an anonymous entity turned and twisted, loved and hated by some choosy author(itarian). The inevitable truth is that we never begin from scratch, a fact which may account for Duchamp's remark that "I'm not at all sure that the concept of the Ready-Made isn't the most important single idea that came out of my work."[113] Not artistic creations but semiotic signs, not material artifacts but meaning-filled statements, not art-effects but thingy-things. Stripped bare by all the paraphernalia, a readymade shows itself to be a map without a legend, a profound way of laying bare the power of naming.[114]

In that molding process, the *not yet* is brought about through the same two-stage procedure as the Judeo-Christian God used when he formed the

Marcel Duchamp, *Fontän*. 1917. Porcelain, 33 × 42 × 52 cm. Replica of readymade executed by Ulf Linde, 1963. Signed by Marcel Duchamp in Milan, 1964. Moderna Museet, Stockholm. © Succession Marcel Duchamp/BUS 2005.

world, the *reductio ad absurdum* of the creative process: first an object is removed from its ordinary context and placed in another, then the old name is changed to fit better into the new surroundings. One example of such a pivotal moment came in the spring of 1917, when the resident alien otherwise known as Henri-Robert-Marcel Duchamp, originally born on July 28, 1887, near Blainville, France, went into a New York hardware store, bought a porcelain urinal, carried it to the exhibition hall of the Society of Independent Artists, laid it on its back, turned it around, signed it "R. Mutt" and baptized it *Fountain*. The scandal was immediate, the thing itself first hidden behind a screen, then officially rejected and eventually lost. Its nonexistence notwithstanding, the piece was projected into eternity through a photograph taken by the trend-setting photographer Alfred Stieglitz, well aided by some spicy anecdotes strategically planted by Duchamp himself. Sacrificed in the process was the material object, resurrected was the untouchable thought, the sensible buried in the ground, the intelligible elevated to heaven, sometimes in the guise of the Buddha sometimes as the Madonna of the bathroom.[115] "But," asked Arthur Danto seven decades later, "what is the conceptual fulcrum of this still controversial work? . . . Why—referring to itself—should this be an artwork when something else exactly like *this,* namely *that*—referring now to the class of unredeemed urinals—are just pieces of industrial plumbing? . . . It took genius to raise the question to this form, since nothing like it had been raised before. . . . Why is something a work of art, when something exactly like it is *not?*"[116]

The answer to Danto's question is that both the *Fountain* and the *Nude Descending a Staircase* are artistic statements as much as they are material things, intentional intertwinings of Signifier and signified, the One becoming the Other, the Other turning to the one, every ontological transformation a firework of rhetorical tropes. Speech act theory in practice. Let there be! And there is. Four pivotal cases to illustrate the point:[117]

- The hat rack exhibited in 1917 and quickly lost, its fate effectively written into the title itself: *Hat Rack.* A tautology, by definition always true but never informative. Bertrand Russell's theory of proper names and definite descriptions rejected by a Frenchman who most likely had never heard of it.
- The coat rack, which Duchamp initially had intended for the vertical wall of his New York studio but eventually nailed to its horizontal floor instead; as he later recalled, "a real coat hanger that I wanted sometime to put on the wall and hang my things on but I never came to that—so it was on the floor and I would kick it every minute, every time I went out—I got crazy about it and I said the Hell with it, if it wants to stay there and bore me, I'll nail it down. . . . It was not bought to be a Readymade—it was a natural thing . . . it was nailed where it was and then the idea came."[118] As the urinal had once been renamed *Fountain,* so the ordinary coat rack was now baptized *Trébuchet,* a French chess term which means "Trap" or "Stumbler." A textbook application of metaphor, the trope by which physical likeness is translated into the nearness of metonymic associations.
- The bicycle wheel mounted by its fork on a painted wooden stool—a readymade before the readymade (original [rue Saint-Hippolyte, Paris, 1913] lost; 2nd version [33 West 67 Street, New York, 1914) lost; 3rd version made for the exhibition *Challenge and Defy,* Sidney Janis Gallery, New York, 1951; 4th version, Ulf Linde, Galleri Burén, Stockholm, 1961; 5th version Richard Hamilton, London, 1964; 6th

*Found and Lost/Lost and Found.* Photo montage by Tommy Westberg. © Succession Marcel Duchamp/BUS 2005.

version, edition of 8 signed and numbered replicas, Galleria Schwarz, Milano, 1964). Even though its title, *Roue de bicyclette*—literally but flatly translated as *Bicycle Wheel*—sounds like a stuttering tautology, it is in effect a meaning-filled synecdoche, the name of a part extended to the whole. Since the Bicycle Wheel is not a pure but an assisted readymade, the chosen trope is highly appropriate.

- The beautiful snow shovel exhibited in 1915, subsequently returned to its daily chores and gone for ever. For who, except a crazy avant-garde artist, could imagine that the snow shovel (wood and galvanized iron, 121.3 cm high) was not a snow shovel but an allegory? *In Advance of the Broken Arm,* the title handwritten at its bottom and signed *from* (not *by*) Marcel Duchamp 1915. But why "arm" rather than "back" or "heart"?

Taken together the readymades flow into a principle of universal metamorphosis, the aesthetic puns always gathering at the intersection of the divid*ing* (slanted) and divid*ed* (horizontal) lines of my own imitations of the *Republic.* Undermined in the process is the idea of art as communication, enhanced instead the idea of art as life.[119] And the minimal point of that life lies exactly in the junction of Plato's two lines, by no coincidence found not only in the respective vanishing points of Duchamp's two major works, *La Mariée mise à nu par ses célibataires, même* and *Etant donnés: 1° la chute d'eau, 2° le gaz d'éclairage,* but in the lines of projection and in the infra-thin picture planes as well. The cartographer's dream come true. A seamless web of picture and story, a local habitation with a name.

# ABR(AH)AM

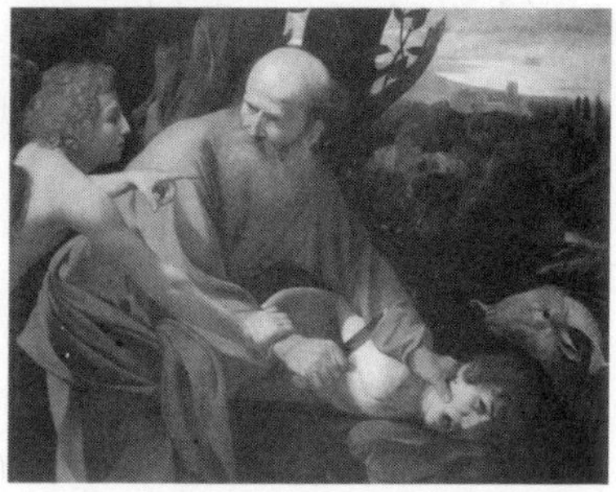

**In the western canon** there are two paradigms for making the absent present and the present absent, one grown out of the *Odyssey* and the pictures of Greek polytheism, the other rooted in Genesis and the stories of Judaic monotheism. The classical treatment is in Erich Auerbach's magisterial *Mimesis,*[1] frequently heralded as one of the most important criticisms ever published, a work deeply influenced by the historicism of Giambattista Vico and the idealism of G. W. F. Hegel.

Auerbach himself was a German Jew who in 1935, at the age of forty-four, was "retired" from his Marburg chair of Roman Philology. Shortly thereafter he was offered a post at the Istanbul State University, and it was there, between May 1942 and April 1945, that he composed the masterpiece which in 1946 was first published in Switzerland. This feat in turn paved his way to the United States, initially to the Institute for Advanced Studies in Princeton, then to a temporary appointment at the Pennsylvania State University, finally to a Sterling professorship at Yale. Most of it too late to enjoy.

The biographical details are important, for there is much to indicate that it was the enforced separation from European culture, including the lack of a rich and specialized library, that helped Auerbach remember. In the words of Edward Said, themselves politically charged, "the book owed its existence to the very fact of Oriental, non-Occidental exile and homelessness. And if this is so, then *Mimesis* itself is not, as it has so frequently been taken to be, only a massive reaffirmation of the Western cultural tradition, but also a work built upon a critically important alienation from it,

Caravaggio, *Sacrifice of Isaac.* 1603. Oil on canvas, 104 × 135 cm. Galleria degli Uffici, Firenze.

a work whose conditions and circumstances of existence are not immediately derived from the culture it describes with such extraordinary insight and brilliance, but built rather on an agonizing distance from it."[2]

My own reflection is that even though the secular intellectual is often portrayed as a border-*crosser* he is in reality more of a border-*dweller*, an endangered species of the excluded middle. Less an analyst of mimesis than a practitioner of mnemonics, more a mapper of the old than an explorer of the new, an individual whose method "consists in finding unusually fertile areas or key problems on which it is rewarding to concentrate, because they open up a knowledge of a broader context and cast light on entire historical landscapes."[3] With Auerbach there was in fact always an already but not yet, especially as he consistently took the personae of the Old Testament to prefigure those of the New; in his own words, "a world which on the one hand is entirely real, average, identifiable as to place, time and circumstances, but which on the other hand is shaken in its very foundations, is transforming and renewing itself before our eyes."[4] The parallels to the present study are obvious.[5]

The exiled philologist began his tour de force in the nineteenth song of the *Odyssey*.[6] After twenty years on the run Odysseus has at last returned, a Trojan horse in his own house, a trickster dressed in rags and behaving like a beggar, a man full of memories, desires and revenge. Seemingly a creepy stranger who is just hanging about and looking the women over, in reality the great tactician is assessing the situation, finding out who in the household is to be trusted and who among the suitors will be the first to be killed. Upset by the way the other men insult him, the unsuspecting Penelope is eager to know whether this well-traveled stranger may have heard rumors about her husband, or better still whether he may even have known him. For emphasis she adds that

> if I can see this man has told the truth,
> I promise him a warm new cloak and tunic.
> (17.548–50)

"No," replied the master of invention, "bid the queen to be patient and let me see her tonight, ask her to wait until darkness falls and the drunkards have quieted down."

And in the evening, moved by her wish to know where in the world she is and where in her marriage she ought to go, the honorable lady orders the housekeeper to bring forth a sheepskin-covered bench for the guest to sit on. Then she confronts him with the cartographer's key question,

> Friend, let me ask you first of all:
> who are you, where do you come from, of what nation
> and parents were you born.
> (19.104–105)

Carefully guarding his true identity the disguised lord replies,

> O my dear lady, . . . let it suffice to ask me
> of other matters—not my blood, my homeland.
> Do not enforce me to recall my pain.
> My heart is sore; but I must not be found
> sitting in tears here, in another's house:
> it is not well forever to be grieving.
> One of the maids might say—or you might think—
> I had got maudlin over cups of wine.
> (19.115–22)

Having thus been softened by the words of the honest seducer, she cannot resist telling him first how she cannot stop longing for Odysseus, her lord who crossed the sea to Troy and never returned, then how much she hates the suitors who have come to her house to court her against her will. In reply the great inventor fabricated a story of how once he had met her husband, making his lies appear so truthful that she begins to weep, her pale face growing moist the way pure snow softens and glistens on the sunny side of the mountain. As the narrator reports,

> her white cheeks were wetted by these tears
> shed for her lord—and he close by her side.
> Imagine how his heart ached for his lady,
> his wife in tears; and yet he never blinked;
> his eyes might have been made of horn or iron
> for all that she could see. He had his trick—
> wept, if he willed to, inwardly.
> (19.208–12)

But what and who is to be believed? The feelings that the man's words stirred in the woman's mind or the realities to which his story was referring? Not so easy to tell and therefore, true to the takens-for-granted of her culture,

> As soon as her relieving tears were shed
> she spoke once more:
> "I think I shall say, friend,
> give me some proof, if it is really true
> that you were host in that place to my husband
> with his brave men, as you declare. Come tell me
> the quality of his clothing, how he looked,
> and some particular of his company."
> (19.213–19)

Furnishing the details she had requested—no lies this time because he, if anyone, already knew what she knew: a purple cloak, fleecy and doubly thick, a brooch made of pure gold with twin tubes for the prongs, a hunting dog, a fine close-fitting tunic, a herald named Eurybates, a man somewhat older than himself, round-shouldered, dusky, wool-headed. A catalogue of factual observations immediately followed by the narrator's conclusion that,

> hearing these details—minutely true—
> she felt more strangely moved, and tears flowed
> until she had tasted her salt grief again.
> Then she found words to answer:
> "Before this
> you won my sympathy, but now indeed
> you shall be our respected guest and friend.
> With my own hands I put that cloak and tunic
> upon him—took them folded from their place—
> and the bright brooch for ornament."
> (19.249–57)

*

**These details—minutely true!** Although virtually everything has changed since the time when the blind(?) Homer recited his(?) tales twenty-seven(?) centuries ago, the power of imagination and the techniques of being believed have remained essentially the same: epithets in abundance, words of color, smell and taste so precise that the listener feels the exactuality in her very body.[7] It is easy to see through the rhetoric of Odysseus' advice to the honorable lady:

> [W]eep no more and listen:
> I have a thing to tell you, something true.
> I heard but lately of your lord's return,
> heard that he is alive, not far away.
> . . . . . . . . . . . . . . . . .
> Long since he should have been here,
> but he thought better to restore his fortune
> playing the vagabond about the world;
> and no adventurer could beat Odysseus
> at living by his wits—no man alive.
> . . . . . . . . . . . . . . . . .
> You see, then, he is alive and well and headed
> homeward now, no more to be abroad

far from his island, his dear wife and son.
Here is my sworn word for it.
(19.268–71, 282–86, 300–302)

No wonder that the attentive queen wished it all to happen, no wonder that she interpreted the sworn word not merely as an empty expression but as an obliging promise. Full of desire she treats the messenger with the respect she hopes he deserves. Indeed she issues the order:

Maids, maids: come wash him, make a bed for him,
bedstead and colored rugs and coverlets
to let him lie warm into the gold of Dawn.
In morning light you'll bathe him and anoint him
so that he'll take his place beside Telémakhos
feasting in hall.
(19.317–22)

But Odysseus has been through much during his wanderings and he does everything he can not to get caught again. More set on making war than love he begs the wife of Odysseus Laertiades,

"Let me lie down tonight as I've lain often,
many a night unsleeping many a time
afield on hard ground waiting for pure Dawn.
No: and I have no longing for a footbath
either; none of these maids will touch my feet,
unless there is an old one, old and wise,
one who has lived through suffering as I have:
I would not mind letting my feet be touched
by that old servant."
And Penelope said:
"Dear guest, no foreign man so sympathetic
ever came to my house, no guest more likeable,
so wry and humble are the things you say.
I have an old maidservant ripe with years,
one who in her time nursed my lord. She took him
into her arms the hour his mother bore him.
Let her, then, wash your feet, though she is frail.
Come here, stand by me, faithful Eurykleia.
and bathe—bathe your master, I almost said,
for they are of an age, and now Odysseus'
feet and hands would be enseamed like his.
Men grow old soon in hardship."
(19.340–60)

Along these lines the well-known story continues and Eurykleia—the old woman who once had nursed the newborn prince—faithfully does what she has been asked to do, her voice mumbling, her hands fumbling. Slowly, however, her imagination is rekindled, the memories which for so long have been absent from her mind are brought back by the presence of his body. And suddenly she bursts out,

> "My heart within me stirs,
> mindful of something. Listen to what I say:
> strangers have come here, many through the years,
> but no one ever came, I swear, who seemed so
> like Odysseus—body, voice and limbs—
> as you do."
> Ready for this, Odysseus answered:
> "Old woman, that is what they say. All who have seen
> the two of us remark how like we are,
> as you yourself has said, and rightly, too."
> (19.378–85)

Still the proper moment has not yet arrived and the king turned beggar remains determined to hide his real identity behind the mask of sheer likeness. Therefore,

> . . . he kept still, while the old nurse filled up
> her basin glittering in firelight; she poured
> cold water in, then hot.
> But Lord Odysseus
> whirled suddenly from the fire to face the dark.
> The scar: he had forgotten that. She must not
> handle his scarred thigh, or the game was up.
> But when she bared her lord's leg, bending near,
> she knew the groove at once.
> (19.386–94)

*

**The scar**, he had forgotten that. And exactly at this point of revelation, without warning, Homer breaks the story and moves Odysseus from the here-and-now of the footbath to the there-and-then of Parnassos, the land where long ago he had been sent to visit his maternal grandfather, Autolykos, the man who once gave him his name—Odysseus, Man of Odium and Distrust.[8]

One day, when the young Dawn spread in the eastern sky, the youngsters went hunting, Odysseus in the lead. Attacking from the undergrowth is a great boar, its white tusk ripping out flesh above the young man's knee, but missing the bone. Then, after a lively flashback of seventy-four lines,[9] we are abruptly brought back to the scene of the footbath, the narrator observing that,

> This was the scar the old nurse recognized;
> she traced it under her spread hands, then let go,
> and into the basin fell the lower leg
> making the bronze clang, sloshing the water out.
> Then joy and anguish seized her heart; her eyes
> filled up with tears; her throat closed, and she whispered,
> with hand held out to his chin:
> "Oh yes!
> *You are Odysseus!* Ah, dear child! I could not
> see you until now—not till I knew
> my master's very body with my hands."
> (19.466–75)

Fearing that she will reveal his true identity and thereby ruin the vengeance for which he has come,

> Odysseus' right hand
> gripped the old throat; his left hand pulled her near,
> and in her ear he said:
> "Will you destroy me,
> nurse, who gave me milk at your breast?
> Now with a hard lifetime behind I've come
> in the twentieth year home to my father's island.
> You found me out, as the chance was given you.
> Be quiet; keep it from the others, else
> I warn you, and I mean it, too,
> if by my hand god brings the suitors down
> I'll kill you, nurse or not, when the time comes—
> when the time comes to kill the other women."

Eurykleia kept her wits and answered him:

> Oh, what mad words are these you let escape you!
> Child, you know my blood, my bones are yours;
> no one could whip this out of me. I'll be
> a woman turned to stone, from iron I'll be.
> (19.479–94)

My bones are yours and through them my mind as well: such is the point of the best story ever told. Homer, the itinerant entertainer, might well have learned from experience that once he bored his audience he would never be invited again, his belly left empty, his desires gone unsatisfied. Even though most of his present readers get to the text through non-exact translations, Odysseus' right hand still grips our throat and his left hand pulls us near, the absent made present at precisely the right moment and in precisely the right place. And the bronze clang as she recognized the groove. Gottlob Frege prefigured, the scar turned into the double description of a wild boar at dawn and an old nurse at dusk. *Oh, yes, my child, you are Odysseus!*—Indeed

> I am at home, for *I am he.*
> (19.207)[10]

A model of clarity, the inherently oblique made beautifully transparent by language playing games with itself. In Auerbach's own words:

> The separate elements of a phenomenon are most clearly placed in relation to one another; a large number of conjunctions, adverbs, particles, and other syntactical tools, all clearly circumscribed and delicately differentiated in meaning, delimit persons, things, and portions of incidents in respect to one another, and at the same time bring them together in a continuous and ever flexible connection; like the separate phenomena themselves, their relationships—their temporal, local, causal, final, consecutive comparative, concessive, antithetical, and conditional limitations—are brought to light in perfect fullness; so that a continuous rhythmic procession of phenomena passes by, and never is there a form left fragmentary or half-illuminated, never a lacuna, never a gap, never a glimpse of unplumbed depths. *And this procession of phenomena takes places in the foreground—that is, in a local and temporal present which is absolute.*[11]

In all of these respects Homer's story about Odysseus is different from the Bible's story about Abr(ah)am, the former a tale of honor and reconstitution, the latter a legend of fear and submission; while the Greek bard anchored his story in the superman's body, the biblical redactors searched for words rich enough to express the non-expressible. And whereas Odysseus knew that Pallas Athena would always interfere to his advantage, Abr(ah)am's relation to God was steeped in mutual suspicion. In an attempt to set things right, the LORD eventually decided to test the old man's allegiance.

*

**The story of how God first summoned and then tested Abr(ah)am** plays a pivotal role in all monotheistic religions. For better and for worse it forms the common center-piece of Judaism, Christianity and Islam.[12] No study of what it means to be human can do without it.

Like other sagas also the biblical narrative has a beginning, a once-upon-a-time—never before had it been said that there is only one God, one and one only. It is the story of that truly stunning invention that is retold in Genesis 12–25,[13] a tale which is closely tied to the figure of Abr(ah)am, the legendary chieftain who about four millennia ago left his home in the Sumerian city of Ur, at that time a vigorous seat of polytheism and the birthplace of *Enuma elish,* a splendid culture which knew as little about Adam and Eve as about Noah and his sons. For unstated reasons and without detailed instructions, this conscientious objector launched himself on a quest of the unknown, ceaselessly moving towards the perpetually new. Stopping for a few years in Haran in the eastern part of present-day Turkey he eventually reached the promised land of Canaan.

This strange sojourner who at the outset was called Abram, not Abraham, was already seventy-five years old, when the LORD summoned him with the words:

> Leave your country, your people and your father's household and go to
> the land I will show you.
> I will make you into a great nation
> and I will bless you;
> I will make your name great,
> and you will be a blessing.
> I will bless those who bless you
> and whoever curses you I will curse;
> and all peoples on earth
> will be blessed through you.
> (Gen. 12:1–3)

Although the order to leave is nonnegotiable, nothing is said about the final destination—as in the cases of Auschwitz and Gulag no questions are invited, no answers offered. The uncertainties notwithstanding, Abram accepts the call, a decision which eventually was to make him very wealthy in livestock, in silver and gold. Accompanied by his nephew Lot, he moves from place to place until the two begin to quarrel about grazing areas, Abram temporarily settling in Hebron, where he builds an altar in honor of the LORD, Lot pitching his tents near the city of Sodom, a place so wicked that men slept with men and women with women. Somehow a war breaks out and Abram, despite their previous disagreements, goes to rescue his nephew.

Now the word of the LORD came to Abram in a vision, a reassurance that he was not forgotten and that in due course he would be paid off:

> Do not be afraid Abram, I am your shield, your very great reward.
>
> But Abram said, "O Sovereign LORD, what can you give me since I remain childless and the one who will inherit my estate is Eliezer of Damas-

> cus?" And Abram said, "You have given me no children; so a servant in my household will be my heir."
>
> Then the word of the LORD came to him: "This man will not be your heir, but a son coming from your own body will be your heir." He took him outside and said, "*Look up at the heavens and count the stars*—if indeed you can count them." Then he said to him, "So shall your offspring be."
>
> Abram believed the LORD, and he credited it to him as righteousness.
>
> He also said to him, "I am the LORD, who brought you out of Ur of the Chaldeans to give you this land to take possession of it." (Gen. 15:1–4)[14]

Abram, perhaps too willingly, believes what the LORD told him, indeed he sees the promises as a form of righteousness. Yet, he cannot free himself from the nagging suspicion that perhaps the reward of many children may never materialize, that the words of the Almighty might not always be trusted. Furthermore, this sense of uncertainty turns out to be mutual, for the LORD, on his part, was determined not to make the same mistakes with Abram as he previously had done with Adam and Noah, both of whom had greatly disappointed him. Hence he decided to put his new (in fact his first) disciple on a long and painful probation. Serious business indeed, for already in the city of Ur God had declared that Abram was not going to be famous through his worldly achievements but through his numerous offspring.

In this atmosphere of growing discontent Sarai, Abram's childless wife, is moved to take things in her own hands. And since,

> she had an Egyptian maidservant named Hagar, she said to Abram, "The LORD has kept me from having children. Go, sleep with my maidservant; perhaps I can build a family through her."
>
> Abram agreed to what Sarai said. So after Abram had been living in Canaan ten years, Sarai his wife took her Egyptian maidservant Hagar and gave her to her husband to be his wife. He slept with Hagar, and she conceived. (Gen. 16:1–4)

As might be expected, the pregnant servant and the barren mistress quickly came to despise each other, for whose family was the unborn child really going to build, that of the unfree slave or that of the legal wife? Haggardly, Hagar ran away. But an angel of the LORD went to look for her, found her near a spring where she was asked the classical question of cartographical reason:

> "Hagar, servant of Sarai, *where have you come from, and where are you going?*"
>
> "I'm running away from my mistress, Sarai," she answered.
>
> Then the angel of the LORD told her, "Go back to your mistress and submit to her." The angel added, "I will so increase your descendants that they will be too numerous to count."

The angel of the LORD also said to her:
"You are now with child
and you will have a son.
You shall name him Ishmael,
for the LORD has heard your misery.
He will be a wild donkey of a man;[15]
his hand will be against everyone
and everyone's hand against him,
And he will live in hostility
toward all his brothers." (Gen. 16:8–12)

A remarkable passage, especially as it is the non-mentionable YHWH himself (albeit disguised as an angel) who informs Hagar not only that she is pregnant but that the boy shall be called "Ishmael," a proper name whose definite description means "God hears." In addition, since the Almighty normally names only classes, not individuals, the reader is left uncertain about exactly which kind of entity Ishmael really is, a human being borne by a maidservant or a social symbol forged by the LORD himself. The ambiguity is further deepened by the fact that Hagar is the only woman in the Hebrew Bible who personally receives a divine blessing of her descendant, a circumstance that has prompted many commentators to draw parallels between the roles of Hagar and Mary, on the one hand, and Ishmael and Jesus, on the other. It may even be argued that God's direct (as opposed to his angelic) interference effectively turns Hagar into a female patriarch,[16] a role in which she no longer appears as an ordinary woman comparable to her mistress, but as a blessed representative comparable to her master.[17] Even more remarkably, this elevated slave is the only biblical person—male or female—who is said to have both seen God and given him a name. To be precise, "she gave this name to the LORD who spoke to her: 'You are the God who sees me, I have now seen the One who sees me.'"[18]

Through this unprecedented utterance—*she gave this name to the LORD*—Hagar is at the same time showing respect and exercising control. In a parallel movement YHWH addresses Hagar first by her proper name, then by her social status, a way of speaking which he never uses with her mistress. Add to these passages the fact that no one else ever mentions the maid's name or ever speaks to her, and it is clear that to God, as opposed to people in general, even an Egyptian slave was a full person, not merely an instrument or a means to an end, but a symbol of the repressed. Not surprising, therefore, that

> Hagar becomes many things to many people. Most especially, all sorts of rejected women find their stories in her. She is the faithful maid exploited, the black woman used by the male and abused by the female of the ruling class, the surrogate mother, the resident alien without legal recourse, the other woman, the runaway youth, the religious fleeing from affliction, the

> pregnant young woman alone, the expelled wife. . . . [But at the same time she is] the first woman to hear an annunciation, the only one to receive a divine promise of descendants, and the first to weep for her dying child.[19]

But the story continues and thirteen years later, when Abram was ninety-nine, the LORD appeared once again, this time announcing that "I am God Almighty, walk before me and be blameless. And I will make my covenant between me and you, and will multiply you exceedingly" (Gen. 17:1–2).

*

**My covenant,** not ours! The renewed summons is unequivocal: Never forget who I am, always remember that it is I your LORD, not you my subject, who is responsible for the production of your offspring. Your genitals are by my design, and you, Herr Goldschmidt, you and your descendants, must never forget it!

Even so, since the covenant is formally an agreement between two parties rather than a one-sided order, Abram must be treated accordingly. And as a sign that he is no longer an impotent geriatric but a blessed representative, the self-appointed LORD declares that the name of this man no longer shall be Abram but Abr(*ah*)am, "exalted father of many nations." An *ah!* and the world is for ever changed, Karol Cardinal Wojtyła later transformed by the same formula into Pope John Paul II, Miss Magnusson into Mrs Olsson, the white dress in both instances covering the fragile body, the symbolic ring in both cases slipped onto the outstretched finger. As a follow-up, Abraham was subsequently told first that the name of his wife was likewise to be changed, Sarai becoming Sarah,[20] then that she too, well past the age of menopause, will be blessed and give him a son.

On hearing this piece of incredible news,

> Abraham fell facedown, he laughed and said to himself, "Will a son be born to a man a hundred years old? Will Sarah bear a child at the age of ninety?" And Abraham said to God: "If only Ishmael might live under your blessing."
>
> Then God said, "Yes, but your wife Sarah will bear you a son, and you will call him Isaac.[21] I will establish my covenant with him as an everlasting covenant for his descendants after him. And as for Ishmael, I have heard you: I will surely bless him; I will make him fruitful and will greatly increase his numbers. He will be the father of twelve rulers, and I will make him into a great nation. But my covenant I will establish with Isaac, whom Sarah will bear to you by this time next year." When he had finished speaking with Abraham, God went up from him. (Gen. 17:17–21)

But since God is God he was well aware that nomads are constantly on the move and that their universe is centered more on the family than on a particular piece of land. The Almighty consequently asks for a sign that the old man has not just heard the original summons but that he has understood it as well. As proof, he demands that Abraham cut the foreskin of his penis, a non-erasable reminder that it is God, not man, who decides whether there will ever be any great nation to bless.[22] And on the same day Abraham proceeded to carry out the mutilation—first on himself, then on his son Ishmael, finally on all other males of his household, the slaves included.

Now, some time later, the LORD appears again, on this occasion in the guise of three visitors. Abraham, somehow intuitively sensing who they are, brings them water, washes their feet and serves an assortment of delicacies, he himself staying in the background, watching but not eating. From her place at the entrance to the tent, Sarah overhears one of the men repeat the pledge that she will soon bear a son, a statement so outrageous that she laughingly retorts, "is anything too wondrous for the LORD?" A rhetorical question not to the LORD's liking.

When the men get up to leave, Abraham walks with them until they reach a point from which they see the city of Sodom in the distance. There they pause and the strangers tell him that the sins committed in that place are too grave to be excused and that it must be leveled with he ground. Barely have these threats been uttered, however, before Abraham begins to speak back, effectively engaging the LORD in a series of negotiations hitherto unheard of, a political act of extreme courage. These were the questions he posed to the notoriously unpredictable:

> "Will you sweep away the righteous with the wicked? What if there are fifty righteous people in the city? Will you really sweep it away and not spare the place for the sake of the fifty righteous people in it? Far be it from you to do such a thing—to kill the righteous with the wicked, treating the righteous and the wicked alike. Far be it from you! Will not the Judge (or the Ruler) of all the earth do right?"
>
> The LORD said, "If I find fifty righteous people in the city of Sodom, I will spare the whole place for their sake."
>
> Then Abraham spoke up again: "Now that I have been so bold to speak to the LORD, though I am nothing but dust and ashes, what if the number of the righteous is five less than fifty? Will you destroy the whole city because of five people?"
>
> "If I find forty-five there," he said, "I will not destroy it." (Gen. 18:23–28)

Certainly a negotiator must be able to do better than this! And on that basis the bargaining continues until the man of dust and ashes makes his final push:

> "May the LORD not be angry, but let me speak just once more. What if only ten can be found there?"
>
> He answered, "For the sake of ten, I will not destroy it." (Gen. 18:32)

Yet, despite the accord, the two cities of Sodom and Gomorrah actually *were* destroyed, many righteous people presumably killed in the process. God's promise to behave fairly was obviously neither fair nor trustworthy.

But the semi-nomad continued his hithering-thithering, twice running into a power called Abimelech, in some verses described as the Pharaoh of Egypt in some others as the King of Gera. Without a fight this ruler is allowed to take Sarah as his concubine, Abr(ah)am deceitfully saying that she is not his wife, merely his sister—a shady plot designed to save his own life at the expense of someone else's. Whoever eventually gets the beauty pregnant remains a mystery and the only thing said is that

> the LORD did for Sarah what he had promised. Sarah became pregnant and bore a son to Abraham in his old age, at the very time that God had promised him. Abraham gave the name Isaac to the son Sarah bore him. Abraham was a hundred years old when his son Isaac was born to him.
>
> Sarah said, "God has brought me laughter, and everyone who hears about this will laugh with me." She added, "Who would have said to Abraham that Sarah would nurse children? Yet I have borne him a son in his old age."
>
> The child grew and was weaned, and on the day Isaac was weaned Abraham held a great feast. But Sarah saw that the son whom Hagar the Egyptian had borne to Abraham was mocking, and she said to Abraham, "Get rid of that slave woman and her son, for that slave woman's son will never share in the inheritance of my son Isaac."
>
> The matter distressed Abraham greatly because it concerned his son. But God said to him, "Do not be so distressed about the boy and the maidservant. Listen to whatever Sarah tells you, because it is through Isaac that your offspring will be reckoned. I will make the son of the maidservant into a nation also, because he is your offspring." (Gen. 21:1–13)

*

**It is through Isaac,** not Ishmael, that your offspring will be reckoned. And with that remark the stage is set for the *akedah,*[23] the famous binding which to all Jews, Christians and Muslims constitutes the ultimate tie between Abraham and the One and Only.[24]

Thus, after Abraham had spent a long time in the Land of the Philistines,[25] God once more grew suspicious, wondering whether the old man was really to be trusted. Therefore it so happened:

> Some time later God tested Abraham. He said to him, "Abraham!"
>
> "Here I am," he replied.
>
> Then God said, "Take your son, your only son, Isaac, whom you love, and go to the region of Moriah. Sacrifice him there as a burnt offering on one of the mountains I will tell you about."

Early the next morning Abraham got up and saddled his donkey. He took with him two of his servants and his son Isaac. When he had cut enough wood for the burnt offering, he set out for the place God had told him about. On the third day Abraham looked up and saw the place in the distance. He said to his servants, "Stay here with the donkey while I and the boy go over there. We will worship and then we will come back to you."

Abraham took the wood for the burnt offering and placed it on his son Isaac, and he himself carried the fire and the knife. As the two of them went on together, Isaac spoke up and said to his father Abraham, "Father?"

"Yes, my son," Abraham replied.

"The fire and the wood are here," Isaac said, "but where is the lamb for the burnt offering?"

Abraham answered, "God himself will provide the lamb for the burnt offering, my son." And the two of them went on together.

When they reached the place God had told him about, Abraham built an altar there and arranged the wood on it. He bound his son Isaac and laid him on the altar, on top of the wood. Then he reached out his hand and took the knife to slay his son. But the angel of the LORD called out to him from heaven, "Abraham! Abraham!"

"Here I am," he replied.

"Do not lay a hand on the boy," he said. "Do not do anything to him. Now I know that you fear God because you have not withheld from me your son, your only son."

Abraham looked up and there in a thicket he saw a ram caught by its horn. He went over and took the ram and sacrificed it as a burnt offering instead of his son. So Abraham called that place *The LORD Will Provide.* And to this day it is said, "On the mountain of the LORD it will be provided."

The angel of the LORD called to Abraham from heaven a second time and said, "I swear by myself, declares the LORD, that because you have done this and have not withheld your son, your only son, I will surely bless you and make your descendants as numerous as the stars in the sky and the sand on the seashore. Your descendants will take possession of the cities of their enemies, and through your offspring all nations on earth will be blessed, because you have obeyed me."

Then Abraham returned to his servants, and they set off together for Beersheba. And Abraham stayed in Beersheba. (Gen. 22)

What a fantastic text! All it says without saying, all it shows without showing. Imagination gone berserk, the most penetrating account of power ever formulated: *Here I am* (in some translations rendered as *Behold me*), a topos so vague that it might be any-where, hence both no-where and every-where; *some time later,* a kairos so fleeting that it could be any-time, hence both always and not at all. A shovel-board of coordinates, a world which is neither here nor there, neither now nor then. No shelter to be found between the *rising early next morning* and *on the third day,* not a word on what transpires in the minds of the father and his son, merely the factual observation that *the two of them went on together,* world literature's most pregnant silence. When Isaac (very little about Isaac) wonders why they carry with them both fire and wood, but no lamb for the offering, he is quieted with the assertion that *God himself will provide the lamb, my son.* Not an adjective in sight, not a trace of modality. No local habitation with a name, merely the Elohist's way of saying "I have not the faintest idea of what's going on."

A cruelty beyond belief. And perhaps for that very reason this pivotal text has long been used as a handbook of power, a paradigm of institutionalized terror, a story of how the king must never be mated and the queen never caught with her pants down, a map of invisible points which never stay put. Most significantly, the crucial moment comes with the appearance of the angel, an ambassador rather than the LORD himself, a *Doppelgänger* in case Abraham loses his mind and turns the knife on the messenger rather than on the beloved son. We do not even know whether Abraham ever agreed to God's request, just as we are never told whether he lied or told the truth when he informed his servants that *I and the boy . . . will come back to you.* Even Isaac's age has been much debated, most accounts agreeing that the "boy" must have been about thirty-seven years old.

At any rate, the turning point comes when the emissary tells Abraham that the Almighty has changed his mind, seemingly an order to save the boy's life, in effect the LORD's way of demonstrating that it is he, and he alone, who has the power to overrule every rule: "Stop! Stop! For Heaven's sake! No need to slaughter your son! By showing your willingness to go along, you have already slain yourself!" And after this supreme trial God never speaks to the summoned again.

The reason this passage is so frightening is of course that everyone realizes that he who has changed his mind once is likely to change it again. How could anyone who seriously declares I swear *by myself* ever be trusted? A knock on the door and the world tumbles down, for even though the trains to Auschwitz and the Gulag carried no destination signs, they all left on time. Little wonder, therefore, that the angel's cry from heaven has left generations fearing and trembling; indeed the very point of the story is that the LORD's actions are beyond human understanding, hence a matter of trust *(pistis)* rather than an issue of reason *(dianoia).* Unbearable is the proper word, because "the Bible's claim to truth is not only far more urgent than Homer's, it is tyrannical—it excludes all other claims. The world of the Scripture stories is not satisfied with claiming to be a historically true reality—it insists that it is the *only* real world, a universe destined for autocracy. All other scenes, issues, and ordinances have no right to appear independently of it, and . . . will be *given their due place within its frame,* will be subordinated to it."[26]

In the meantime the critic of cartographical reason cannot help but notice that the actual location of the mountain is left in abeyance;[27] the term "a certain place" is usually interpreted as a metaphoric reference to God himself, "who is the Place of the universe, while the universe is not His place."[28] For emphasis, most translations carry a footnote to the effect that the verb in the Hebrew name of the mountain, YHWH-yireh, means both "to provide" and "to see." In Greek, of course, "to see" is "to know."

Despite the countless ambiguities the argument is perfectly clear. Since in the potential space of the Hebrew Bible the Almighty is simultaneously everywhere and nowhere, the techniques through which he wields his power cannot, indeed must not, be disclosed—nothing to capture except a ram stuck with its horns in a thicket. Yet it must surely be considered more human to sacrifice a ram *as if it were* a son, than to slaughter a child *as if it were* a ram. As a consequence—and as an exception from the general rule that every firstborn male, whether animal or human, must be given back to God, to whom they belong—there is a special provision that in the case of donkeys and humans a lamb may be used as a substitute (Exod. 34:19). But who was Abraham's firstborn? Was it Ishmael, a half-Egyptian doomed to wander? Or was it Isaac, the laughing-stock of doubtful parentage? Given God's repeated insistence that both boys were of *his* making, why bother?[29]

Whatever the answer, the story of Abr(ah)am certainly changed the world; indeed no previous text had ever even hinted that there is only one God. One and only one! And with that claim a seed was sown, a mode of cognitive reasoning which logically leads to the concept of a First Cause. It follows that Abr(ah)am figures not only as the prototype monotheist but as the reductionist epistemologist par excellence. So what is the idea?

> The idea is that what's important is the power of ideas—*human* ideas. Not rivers. Not stones. Not land. Abraham went into the desert, a place of nothing, and created something entirely new. And that something new was based on something invisible. He collected technology and know-how from all places he visited. He mixed them with this big, unknowable, untouchable God, and he passed that down to both of his sons. And that's what changed the world. If we're fighting over stones, we're missing the point. Abraham was about a single idea, and that idea he gave to us all.[30]

On this rendering the most crucial part of Abr(ah)am's story is obviously not in the binding but in the call, not the *akedah* but the *lech lekha,* not the critical test but the summons to get up and go. At issue is the issue of trust (a deeper form of Plato's *pistis*), for even though both the sacrifice (leave your home!) and the reward (you will father many nations . . .) are clearly specified, there is not the faintest hint of the road-in-between. To be exact about the vague: "You're not going to know where your next home is. If you're going to be in covenant with me, you have to trust me with every cell in your body. And if you do that, I will bless you."[31]

Easier said than done, though, especially as the God of Genesis is more concerned with reproduction than with morality. All with a price, however, for when Sarah learned about the horrors that took place at the place which the LORD had provided, she promptly passed away, perhaps too distressed by her husband's behavior to want to go on. Abraham, on his part, briefly grieved the loss then found himself another wife, fathered seven additional children and lived till the age of one-hundred-and-seventy-five. And at the end the two half-brothers(?) got together and had him buried in a cave, seemingly unbothered by the wrongs their alleged father had done both to them and to their respective mothers.

Most strikingly, the Book of Genesis never asks the Messiah's question, "My God, my God, why hast thou forsaken me?" (Ps. 22:1).[32] In contrast, Søren Kierkegaard probed the silence, famously concluding that "Abraham cannot be mediated, and the same thing can be expressed also by saying that he cannot talk. So as soon as I talk I express the universal, and if I do not do so, no one can understand me."[33] However, Kierkegaard's silence must not be interpreted as an ethical withdrawal but as an engaged inwardness, a politics of the emigré, "a politics, that is, of one who places the needs of the singular over those of the universal, of one who takes up the cause of the outcast and the marginalized, the victims of injustice, the lepers and the lame, as a means of destabilizing the establishment."[34] The motivation is that faith is double-sided and requires first a radical break and then a readiness to face the terrors and enigmas of the utterly unknown. Questions asked, answers not-forthcoming. Indeed it is impossible to determine from J's text what exactly (s)he thought the divine might have meant and how exactly the humans ought to respond. Yet, when the Kierkegaardian/Derridean individual resigns from the world, it is always with the purpose of returning with more insights and more engagement. By no coincidence both writers were deeply influenced by Immanuel Kant.[35]

Uncertainty nevertheless abounds, perhaps because in every monotheistic religion—as in every reductionist science—there is a built-in tension between personal trust and social verification: How did Abr(ah)am know where he was and where he should go? By which other means than imagination, long considered the outstanding faculty of religion, could J ever have made the absent God present?[36] How else could (s)he possibly have approached that evasive and contradictory power which sometimes appears as an angel descending from the sky, sometimes as three men visiting a tent? "Imagination" is another word for "otherness."

In conclusion: To see that the emperor is naked one must first have dis-covered his dis-guise. And to appreciate the subtlety of Odysseus' tricks one must first have grasped the Aristotelian Laws of Thought.

# MOSES

**The tension between trust and verification** lies at the heart of European culture, perhaps of all cultures. In Erich Auerbach's analysis it is located exactly in the taboo-ridden interface between the certainties of Odysseus' scar and the ambiguities of Abraham's fear, you and I dangling in the abyss in-between. In the sometimes criticized words of the master himself:

> It would be difficult, then, to imagine styles more contrasted than those of these two equally ancient and equally epic texts. On the one hand, externalized, uniformly illuminated phenomena, at a definite time and in a definite place, connected together without lacunae in a perpetual foreground; thoughts and feeling completely expressed; events taking place in leisurely fashion and with very little of suspense. On the other hand, the externalization of only so much of the phenomena as is necessary for the purpose of the narrative, all else left in obscurity; the decisive points of the narrative alone are emphasized, what lies between is nonexistent; time and place are undefined and call for interpretation; thoughts and feeling remain unexpressed, are only suggested by silence and the fragmentary speeches; the whole permeated with the most unrelieved suspense and directed toward a single goal (and to that extent far more of a unity), remains mysterious and "fraught with background."[1]

Two modes of understanding, two modes of being, two ways of living. Diverging attitudes which over the centuries have been condensed, puri-

Marc Chagall, *Moses Receiving the Tablets of Law.* 1960–66. Oil on canvas, 46 × 38 cm. Musée National Message Biblique Marc Chagall, Nice. 

fied and eventually codified, one in Aristotle's Laws of Thought, the other in the Ten Commandments.

*

**Aristotle's Laws of Thought** formalize the three principles of identity (everything is what it is), contradiction (contradictory statements cannot both be true), and excluded middle (of two contradictory statements one must be true, the other false). Rephrased,

$$a = a$$
$$a \neq \sim a$$

$$a \vee \sim a$$

From their Greek beginning these "laws" have been treated as unprovable axioms, indeed as the conceptual foundation of all rational thought.[2] And even though recent developments in formal logic have made them technically obsolete—not because they have been proven wrong or useless but because they have been shown to be incomplete—they continue to play a decisive role in all deliberations of what is true or false, right or wrong. When pushed to the wall every parent, every policeman, every judge will insist that since it is metaphysically and ontologically "impossible to affirm and truly deny at the same time, it is also impossible that contraries should belong to a subject at the same time."[3]

Similar considerations motivated Aristotle's two-front attacks on his predecessors Heraclitus (who thought that all things are and are not, which seems to make everything true) and Anaxagoras (who held that there is an intermediate between the terms of contradiction, a condition which seems to make everything false). The strength—indeed the very purpose—of the law of the excluded middle is to exclude both of these possibilities, the weakness that it at the same time embraces unacceptable ideas of fatalism and extreme determinism.[4] Aristotle himself was well aware of the danger and he consequently argued that any reference to future events falls outside the jurisdiction of the law, suggesting instead that whatever I *now* happen to say about the sea battle tomorrow is *now* neither true nor false, but logically indeterminate. The contorted proof rests on the various modalities of the word *now,* for even though it is *necessary* that the famous sea fight will or will not take place tomorrow, it is not true that it will *necessarily* take place tomorrow or *necessarily* not take place tomorrow;[5] the possibility that a door is $now_1$ open and $now_2$ closed depends entirely on the particular *now-hinge* which holds the door in place when it swings.[6]

In the briefest of conclusions: Just as the so-called parallel axiom of Euclidean geometry is not an axiom but a theorem, so the law of the excluded middle of Aristotelian logic is not an unbreakable law but a principle with many exceptions. And just as every painter after Leon Battista Alberti has struggled with the problem of perspective, so every logician after Charles Sanders Peirce has confronted the limits of the either-or; in a message to his friend William James the latter perceptibly noted:

> I have long felt that it is a serious defect in existing logic that it takes no heed of the *limit* between two realms. I do not say that the Principle of Excluded Middle is downright *false;* but I *do* say that in every field of thought whatsoever there is an intermediate ground between *positive assertion* and *positive negation* which is just as Real as they. Mathematicians always recognize this, and seek for that limit as the presumable lair of powerful concepts; while metaphysicians and oldfashioned logicians,—the sheep & goat separators,—never recognize this. The recognition does not involve any denial of existing logic, but it involves a great addition to it.[7]

This anticipated addition was soon to come, especially through the revolutionary work of Jan Łukasiewicz, the Polish logician who in 1917 formulated a propositional calculus that has a third truth value, neither true nor false but genuinely indeterminate.[8] Paradoxically, these developments of modal logic go far beyond the traditional laws of thought, yet they are literally unthinkable without them. By no coincidence, Jesus in the Sermon on the Mount reminded the crowd:

> Again, you have heard that it was said to the people long ago, "Do not break your oath, but keep the oaths you have made to the LORD." But I tell you, Do not swear at all: either by heaven, for it is God's throne; or by the earth, for it is his footstool; or by Jerusalem, for it is the city of the Great King. And do not swear by your head, for you cannot make even one hair white or black. Simply let your "Yes" be "Yes," and your "No," "No"; anything beyond this comes from the evil one. (Matt. 5:33–37)

The message cannot be misunderstood: in the Christian conception of thought-and-action the evil lives exactly in the limit between Peirce's two realms.[9] Deeply hidden in the abyss of the excluded middle lies the residence of the most divisive of all principals, a Power Center so important that it is guarded by Satan himself.

*

**Guarded by Satan** is the secret code of the excluded middle. Therefore—and despite the fundamental differences between Greek polytheism and Judeo-Christian-Muslim monotheism—there are many parallels between the formalized Laws of Thought, on the one hand, and the Ten Commandments, on the other.[10] Furthermore, in both cases the practice of numbering makes the directives easier to recall, for just as the Laws of Thought are three partly because three is a holy number and partly because triangulation is the modus vivendi of cartographical reason, so the Commandments are ten (or twelve) because the semiotic animal has ten fingers to count with (or the year twelve moons to observe)— alternative tricks for remembering what to remember, a set of neatly ordered drawers in God's thesaurus sapientiae, different molds for capturing and shaping the wor(l)ds of thought-and-action. Yet the Bible nowhere refers to the Decalogue as *deka logoi* (ten words), and the rabbinic exegetes distinguish not ten or twelve commandments but a staggering total of six-hundred-and-thirteen, 365 of them phrased in the negative, one for each day, 248 in the positive. As in all deontic logic, obligations and prohibitions show themselves to be two sides of the same coin.[11]

The dating of the two-volume Law Book continues to be disputed, although most commentators now think that the final list was composed by D, albeit with antecedents at least from the seventh century BCE, perhaps from the thirteenth and the time of Moses himself. The issue of internal numbering remains unsettled as well and different congregations have established different conventions. The interpretations vary accordingly, yet it is generally agreed that the ten words may be divided into two groups such that the first are taken to regulate the relations between God and man (the Constitutional Law), while the remaining seven are said to govern the relations between man and man (the Civil and Criminal Law).[12] Although the paragraphs of the former are more closely tied to the culturally specific obligations of the original covenant and those of the latter to Semitic clan-life in general, the words of the two groups are intimately related.[13] Not so surprising, for as the story has it, both stone tablets were inscribed with the finger of God.[14]

In times of crisis it is the Constitutional Law of the first tablet that tells the Ruler how to rule,[15] and that is regardless of whether the potentate happens to be a Machiavellian Prince, a dictatorial Führer or an elected Prime Minister; a Jew, a Christian or a Muslim; a communist, a fascist, a social democrat or an ardent believer in anything else. The lapidary style and the extraordinary economy—three paragraphs and a prologue—of this self-referential document has ensured its longevity and it continues to serve as *the* blueprint of power and submission: an age-old yet constantly updated amalgam of Political Science and Social Psychology; a poem of how we are made so obedient and so predictable; a rhetorical masterpiece, in the following pages quoted from the Authorized King James Version originally published in 1611, a text "containing the Old and New Testaments translated out of the original tongues and with the former translations diligently compared and revised by His Majesty's special Command."[16]

It is hard to imagine a more power-filled statement than this Constitutional Law, not the least since it is here that YHWH for the first time reveals his own name, an expression so closely related to the Hebrew verb for "to be," *hvh/hjh,* that it is often translated as "The Being." As a way of stressing its importance it was this invisible entity itself, not one of its usual emissaries, who spoke the words of the Prologue, the utterance which says that

> I am the LORD thy God, which have brought you out of the land of Egypt, out of the house of bondage.[17]

At the same time a most concise summary of everything that went before and a key to everything that is to follow; a restatement of J's double-sided story of Abr(ah)am; an indissoluble blending of the original summons and the subsequent tests, including those of Abraham on the mountain of Moriah and of Jesus on the hill of Golgotha. Each and every word pregnant with meaning, speech act and self-reference knotted together into a net of internal relations: (1) the self of self-representation condensed into the name *I am;* (2) the power beyond power manifested by the definite description *your* LORD; (3) the legitimation of legitimation, a reminder that it was I who liberated you, *I who brought you out of the house of bondage.*

I and thou joined together by the shared experiences of the long march, thou and I inseparably bound. Janus before Janus, for in the realm of trust and verification it is normally assumed that whoever has kept his promises in the past is likely to do so in the future. Come election time and the chants are echoing through the halls: "You never had it so good. Therefore, for your sake (and for mine as well) you must please vote for me again. Don't be afraid Citizen K, for through my acts I have already shown that I am your shield and your very great reward. Leave your present habits behind, come with me to the land of Nowhere, the Utopia which I will show you. Blood, sweat and tears along the route, milk and honey once we get there. Read my lips and scrutinze my record."

Lenin before Lenin, Mao before Mao, Castro before Castro. Never a May day without a distress signal, every road paved with intentions. Hear the voice in the desert, **S**ave **O**ur **S**ouls, prepare the way.

From this self-proclaimed beginning, the lawmaker moves directly to §1 and its fantastic declaration:

> Thou shalt have no other gods before me.[18]

Once again a return to the original summons, an extreme codification of the idea that there is one God, one and one only. A proposition as stunning now as when it was first uttered. For what these eight words effectively decree is that *I* shall be your dictator, *I* and no one else. Wherever this Almighty happens to be—and by definition it is every*where*—it rules over every*one* and every*thing:* "I *am* your LORD, you are my servant." Monolatry at the beginning, monotheism in the end, YHWH as the jealous God which tolerates no opposition. First an exhortation, then an assertion.[19] First a Law of Power, then a Law of Submission. In the end a presumptuousness beyond belief.

The genius who first coined the phrase that there must be no one before (or, according to some translations, "beside") me was obviously well aware that in all surrounding lands there were many gods and many princes. Indeed it was exactly this type of many-faceted social order that the sojourner was so eager to get away from that he submitted his entire future to the whims of a Power he knew nothing about. At the same time the brilliant lawmaker would not have been so brilliant had he not foreseen the possibility, nay likelihood, of violent reactions. And, since knowledge, understanding, critique and reasoned revolt by definition are exercises in translation, he also knew that all translations must be carefully monitored and controlled, especially the mimetic representations of arts and literature. Plato prefigured.

Thus, demonstrating his deep insights into the mechanisms of rhetoric, the power-wielder first outlawed the (mis)use of metaphor, then ruled against the critical understanding of metonymy, especially the relations between proper names and definite descriptions. In toto a paragraph of all-embracing censorship, for operational purposes divided into two sections, here labeled §2a and §2b respectively. According to the former:

> Thou shalt not make unto thee any graven image, or any likeness of any thing that is in heaven above, or that is in the earth beneath, or that is in the water under the earth: Thou shalt not bow down thyself to them, nor serve them: for I the LORD thy God am a jealous God, visiting the iniquity of the fathers unto the third and fourth generation of them that hate me; And showing mercy unto thousands of them that love me and keep my commandments.[20]

A most delicate dialectic, especially as the Hebrew term for "image" refers more to the dwelling place of the divine than to the pictorial representation of its invisible being.[21] It follows that if you tell me *where* you are, I shall tell you *what* you are! Yet, as soon as I attempt to make the invisible visible I run the risk of thingification, of falling into the trap of misplaced concreteness, of deifying the reified; by semiotic necessity every ideology is a graven ideology. At the same time every flirtation with idolatry serves as an open invitation to madness. Hence, whenever I see something I never saw before, then I should tell myself to beware of false prophets, never to trust the magicians who come to me in sheep's clothing but who inwardly are ferocious wolves. Likewise, the reader should be reminded that just as in Marx's world of capitalism everything solid melts into air, so in Macbeth's world of witches what seems corporal melts as breath into the wind. Such is the stuff that dreams are made of, such is the secret code of ontological transformations.

The graven image is clearly the prototype of mimetic representation, the master key to idolatry and thereby to the doors of competing ideologies and potential usurpers.[22] The purpose of §2a is therefore to block that road to insight, to stifle every upheaval in its infancy; in the words of Brevard Childs, "God has chosen to make himself known, not in a 'static' image, but in the ambiguity of dynamic history. The commandment protects God's entry to the sphere of human life by guarding against an abuse which seeks to exploit his revelation for one's own use."[23]

Still there are loopholes and that explains why the author of the Constitutional Law added the complementary obligation of §2b, its explicit purpose to forbid the spreading of falsified tales, empty promises, stinging comedy and laughable farce: you shall not swear falsely. More exactly,

> Thou shalt not take the name of the LORD thy God in vain; for the LORD will not hold him guiltless that taketh his name in vain.
> (Gen. 20:7, Deut. 5:11)

In its totality (§§2a and 2b taken together) the second paragraph may be read as an outstanding product of the LORD's Security Forces, some of its agents trained in Plato's Academy of Distributive Justice, some others raised in the sties of George Orwell's Pig Farm. True to its absolutist ambitions the second word is unique also in the sense that it is the only one to include within it both a description of the forbidden crimes and detailed prescriptions of the punishments that are to come with it. It is sobering to realize that at a time not too long ago and at a place not too far away, the present book would have landed not only its author but his family (some as yet unborn) in Siberia. Perhaps some readers too.

It cannot be said more clearly: the second paragraph amounts to a devastating auto-da-fé; a burnt offering of reasoned critique; an imagination killed by an omnipotent censor and resurrected by an apparatchik judge; a totalitarian system designed to visit iniquities unto those who hate it and show mercy to those who love it; a construction so absolute in its absoluteness that it must be constantly reaffirmed and repeatedly reestablished. So Lord, won't you buy me a Mercedez Benz, my friends all drive Porsches, I must make amends.

To produce that type of social cohesion, obedience and predictability are exactly the purpose and function of the penultimate paragraph of the Constitutional Law, here called §3:

> Remember the sabbath day, to keep it holy. Six days thou shalt labor, and do all thy work: But the seventh day is the sabbath of the LORD thy God; in it thou shalt not do any work, thou, nor thy son, nor thy daughter, nor thy manservant, nor thy maidservant, nor thy cattle, nor thy stranger that is within thy gates: For in six days the LORD made heaven and earth, the sea, and all that in them is, and rested

> the seventh day: wherefore the LORD blessed the sabbath day, and hallowed it. (Exod. 20:8–11)

Although the antecedents of this paragraph are even more uncertain than for any of the others, most exegetes agree that the sabbath is as unique an invention as Israel's worship of one single God.[24] Yet it should be recalled that already in *Enuma elish,* written long before the time of J, Marduk bought his vassals' support by telling them that they worked too hard and that he had created the human race as a means of relieving their burdens. But what does it really mean to be human, when in one account man is a Mesopotamian blood sausage—a Boudin à la Mésopotamie—in another an Adamic speck of dust, in a third a Darwinian bag of walking sea water?

At any rate the third commandment is the only word whose Exodus and Deuteronomy phrasings are significantly different. Thus, according to the former it was the LORD who blessed the sabbath day and hallowed it, while in the latter, more priestly version, the potentate asked the liberated slaves to

> Keep the sabbath day to sanctify it, as the LORD thy God hath commanded thee. Six days thou shalt labor, and do all thy work: But the seventh day is the sabbath of the LORD thy God: in it thou shalt not do any work, thou, nor thy son, nor thy daughter, nor thy manservant, nor thy maidservant, nor thine ox, nor thine ass, nor any of thy cattle, nor thy stranger that is within thy gates; that thy manservant and thy maidservant may rest as well as thou. And remember that thou wast a servant in the land of Egypt, and that the LORD thy God brought thee out thence through a mighty hand and by a stretched out arm: therefore the LORD thy God commanded thee to keep the sabbath day. (Deut. 5:12–15)

The lawmaker has obviously felt a need to legitimate the LORD's outrageous lust for power. And to that end he once again reminds the congregation of freed slaves of who it was that once had cut their chains: "Therefore, after all these ordeals, I declare that you deserve a rest. However, I say to you that you must never spend that precious time alone, but always in the company of your likes. At the synagogue, in the church, at the play group, the faculty meeting, the confirmations, the funerals and the family dinners—it is at these gatherings that my invisible officials will teach you how to think-and-act! By common prayer you will then grasp the truth: *I am* your LORD, *I am* the air you breath, *I am* your taken-for-granted. *Therefore* (or is it *wherefore*) be like everyone else and you will be blessed. Since I hate the threatening original and love the servile copy, I shall punish the former and show mercy unto the latter."

And so it is that §3, despite its innocuous appearance, is frequently considered the most crucial of all the Ten Commandments, the ultimate guarantee that the dictatorial practice of monotheism will survive; its very point is of course to honor the real God, not the man-made golden calf but the unnamable and self-referential YHWH. And so it also is that Aristotle's Laws of Thought and Moses' Laws of Submission may be fruitfully read as alternative maps of power, two codifications designed with the double purpose of first telling the truth and then being believed when you do it. It is difficult to imagine two formulations of greater historical significance.

But whereas the Greek gods could be discovered by reason, the biblical God comes only through revelation. Yet, not even timeless propositions come with timeless interpretations. And the prow of Telemakhos' ship sheared through the night and into the dawn.

*

**Aristotle's Laws of Thought and Moses' Laws of Submission** both function as alternative maps of power. And like all other maps also these formulas are structured around the cartographical primitives of fix-point, scale and projection screen. However, since the two codifications are closely related in principle yet profoundly different in practice, the bewildered wanderer is often pointed in diverging directions—through one earphone (s)he is instructed to reach for the paradise in the east, through the other to head for the pole in the north. Once again paraphrasing Ludwig Wittgenstein: when I am looking out of my window I no longer see the clear outlines of the familiar vistas but a hazy assortment of paradoxes and predicaments. As my takens-for-granted are unhinged, my fix-points begin to float and my compass to spin. In the resulting confusion I no longer know my way about. Whatever I do, I do something wrong.

At first sight this sense of madness is surprising, especially as both the Greekjew and the Jewgreek begin their respective journeys through the landscape of thought-and-action in exactly the same origin of tautology. To be technically precise, the principle of identity,

$$a = \mathrm{a}$$

is logically the same as YHWH's name

I AM WHO I AM

or

I WILL BE WHAT I WILL BE

Yet, immediately the two explorers have left their common starting point, the formally similar statements push them towards drastically different destinations. And in that bewildering situation it is instructive to return to two parallel events in the stories of Moses and Odysseus, the former occasioned by the appearance of the LORD, the latter by the tricks of Pallas Athena.

The first first. For as Moses was tending the flock of Jethro, his father-in-law,

> the angel of the LORD appeared to him in flames of fire from within a bush. Moses saw that though the bush was on fire it did not burn up. So Moses thought, "I will go over there and see this strange sight—why the bush does not burn up."
>
> When the LORD saw that he had gone over to look, God called to him from within the bush, "Moses! Moses!" And Moses said, "Here I am."
>
> "Do not come any closer," God said. "Take off your sandals, for the place where you are standing is holy ground." [Then] the LORD said, "I have indeed seen the misery of my people in Egypt. I have heard them crying out because of their slave drivers, and I am concerned about their suffering. So I have come down to rescue them from the hand of the Egyptians and to bring them out of that land into a good and spacious land, a land flowing of milk and honey. [So] now, go. I am sending you to Pharaoh to bring my people the Israelites out of Egypt."
>
> [Moses] said to God, "Suppose I go to the Israelites and say to them, 'The God of your fathers has sent me to you,' and they ask me, 'What is his name?' Then what shall I tell them?"
>
> And God said to Moses, "I AM WHO I AM [alternatively translated as I WILL BE WHAT I WILL BE]. This is what you are to say to the Israelites, 'I AM has sent me to you.'" God also said to Moses, "Say to the Israelites, 'The LORD, the God of your fathers—The God of Abraham, the God of Isaac, the God of Jacob—has sent me to you.' This is my name forever, the name by which I am to be remembered from generation to generation.'" (Exod. 3:2–15)

In the Greek counterpart of the same story Homer lets the returned Odysseus for the first time in twenty years meet his son. Here, however, it is not the LORD who appears in flames, but Athena, the goddess, who once again smoothes the way of her protégé. Tipping the golden wand upon the man, she made his cloak pure white, his jawline clean and the hair no longer grey. And Telemakhos, as thunderstruck in front of the well-adorned stranger as Moses in front of the (un)burning bush, cannot speak but merely whisper:

> "Stranger, you are no longer what you were just now!
> Your cloak is new; even your skin! You are
> one of the gods who rule the sweep of heaven!

Be kind to us, we'll make you fair oblation
and gifts of hammered gold. Have mercy on us!"
The noble and enduring man replied:
"No god. Why take me for a god? No, no.
I am that father whom your boyhood lacked
and suffered pain for lack of. *I am he.*"
(16.180–89, emph. added)[25]

Stunned, Telemakhos cried out, "You cannot be my father!" And to this he received the factual reply:

No other Odysseus will ever come,
for *he and I are one,* the same; his bitter
fortune and wanderings are mine.
(16.201–205, emph. added)

*

**I am he, he and I are one!** Such was the answer Odysseus gave his son, in everything different from the LORD's voice from the bush. Such was also the answer which the clever trickster later gave to his own father when he found him in the neglected orchard, old, decrepit and dressed in a tunic patched and soiled, an appearance that could have been tidier. Yet, despite his inwardness, Laertes' curiosity was raised. And suddenly he demands:

Speak out, tell me further:
who are you, of what city and family?
(24.298)

The noble and enduring man, the wanderer who had fooled the Cyclops and made love to Kalypso, that Human, Curious and Earwicking man, he had yet another fable ready in reply. Included therein was the information that

My name is Quarrelman.
Heaven's power in the westwind
drove me this way from Sikania,
off my course.
(24.306–307)

A response structurally identical to what he had earlier told Polyphemos:

My name is Nohbdy: mother, father, and friends,
everyone calls me Nohbdy.
(9.366–67)

However, as he now heard the old man groan over the loss of his son and when he saw him scooping handfuls of dust over his head, the returned son could no longer contain himself. He lept forward, threw his arms around his father, kissed him and exclaimed:

> Oh, Father, I am he!
> (24.321)

On hearing this fantastic piece of news Laertes asked for proof, a request which the son promptly provided first by showing the scar from the wild boar's tusk, then by recalling in detail the young plants he had been given when he was a small boy: thirteen pear, ten apple and forty fig trees, each one with a name. Fifty rows of vines as well (24.327–42). The man's body thus quantified, his mind similarly verified, the rhetoric of logical positivism in the making. Yet not as simple as it might appear, for just as the scar was not merely a reminder of the young man's visit to his maternal grandfather but the mark from a transition rite, so the maturing trees were not merely trees but signs of the social and economic relations between father and son. No wonder the old man's knees failed him and he clutched his son, who held him swooning until he got his breath back. Then,

> Conversing
> in this vein they went home, *the two together,*
> into the stone farmhouse. There Telemakhos
> and the two herdsmen were already carving
> roast young pork, and mixing amber wine.
> (24.361–64)[26]

The crucial difference between the events on the mountain of God, once called Horeb and twice named Sinai, and those on Ithaka, Odysseus' family land, should now be clear. For whereas the LORD's declaration

I am I

is true but not informative, Odysseus declaration

I am he

is at the same time both true and informative. The power of tautology versus the knowledge of proper names and definite descriptions. In both instances a case of double binding, at the same time an observation of things out there and an allusion to relations in here.

It should nevertheless be repeated that both the Greekjew and the Jewgreek set up their respective measuring instruments in the pronoun I, *the* fix-point of fix-points, the poet's version of the painter's vantage-point and the mapper's stand-point. From this origo of origoes the western explorer may then follow two alternative routes, both opened up by the verb-form *am,* the counterpart of the sight-line which in one context runs from Albrecht Dürer's eye to the bush of the nude, in another from the cartographer's theodolite to the church towers in the distance. As Søren Kierkegaard once put it, "the whole existence of the human race is rounded off completely like a sphere, and the ethical is at once its limit and its content. God becomes an invisible vanishing point, a powerless thought, His power being only in the ethical which is the content of existence."[27]

What a wonderful image: the Judeo-Christian LORD as an invisible vanishing point on a sphere. Misleading nevertheless, for on the spherical globe, as opposed to the flat map, there is no fix-point to fix, no north pole and no equator, neither longitude nor latitude, neither identity nor difference. Hence no excluded middle either. In that perspective (pardon the term) it is easy to understand why the Great Dane feared and trembled. For in the tautological world of the LORD, as opposed to the translational world of the Greeks, there is nothing to hold on to except the Ruler's declaration that he is free to do exactly as he pleases. The roots of terrorism laid bare.

It is in the interpretation of the *am*—the presence of the Being—that Homer and the biblical authors exhibit their greatest differences. Thus, while the former believed that knowledge lies in the open formula of

$$a = b$$

the latter knew that the only irrevocable truth is in the closed tautology of

$$a = a$$

To the critic of cartographical reason—to whom the name *God* (if a name it is) serves as nothing but an alternative term for "Power"—this difference presents a most formidable problem, a paradox of the highest order. Thus, as already indicated, there is much to suggest that the Judeo-Christian LORD called himself *I am I* exactly because he did not want to be caught in the knowledge-seeker's net of either-or. Like all power-wielders his double strategy is to keep all options open, to maximize his own freedom and to minimize every critique before it is uttered. While Marduk escaped by appearing under fifty aliases, never sleeping in the same bed two nights in a row, the LORD solved the same problem by giving himself a name which is at the same time true and impossible to rephrase. So all-embracing is in effect this self-made creation that it is an instantaneous instance of both

$$a = a$$

and

$$a = \sim a$$

An obvious violation not only of the principle of contradiction but also of the law of the excluded middle, the Almighty himself a panic-stricken resident of the abysmal. And therein lies *the* methodological problem of the critical social scientist, for anyone who wants to approach power on its own terms is bound to violate some of the most sacred rules of conventional reasoning, including the practice of operationalization and the principle of noncircularity. How can I have a science about a subject matter that hops capriciously about? How can I accumulate knowledge about a Power which is always on the move? Perhaps by rolling with the punches and by realizing that it takes two to tango.

Another approach may be to acknowledge that for the Greekjew the figure of Odysseus stands for man in general, while for the Jewgreek the term "LORD" is a pseudonym for Power in and of itself. Thus, just as the Greek trickster becomes one with James Joyce's HCE alias Here Comes Everybody, so the Almighty turns into a force so totally unique in its being, form and name that it lies beyond every possible mode of representation. And just as the Homeric hero is firmly enclosed within the limits of the same, so the biblical figure is radically boundless, the ultimate otherness of the other, the certified violator of all principles of logical and cartographical reason, *to apeiron* in disguise. No wonder, therefore, that the Christians, given their double and contradictory ancestry, beg forgiveness for their trespasses, because to trespass is to behave exactly like the Master himself, a form of self-centered idolatry which no dictator can tolerate in anyone except himself. Hence the prohibition against the making and adoration of graven images, hence the ban on translation and the use of the LORD's name in vain, hence the fate of Narcissus. Hence also Titian's *Flaying of Marsyas,* the famous painting in which the ageing artist shows himself tortured by his critics, the most horrifying self-portrait ever made.

Yet it should be stressed that the second commandment is not merely for the sake of self-protection, not only a censoring paragraph designed to silence every criticism before it is uttered. At the same time, and equally important, it is a reminder that perfect translation is impossible, a warning that every attempt to visualize the invisible paves the way to the fallacy of misplaced concreteness. Therefore, when Moses pushed God to prove his allegiance and asked him to show his glory, the messenger boy received the reply:

> "I will cause all my goodness to pass in front of you, and I will proclaim my name, the LORD, in your presence. I will have mercy on whom I will have mercy, and I will have compassion on whom I will have compassion. But," he said, "you cannot see my face, for no one may see me and live."
>
> Then the LORD said, "There is a place near me where you may stand on a rock. When my glory passes by, I will put you in a cleft in the rock

and cover you with my hand until I have passed by. Then I will remove my hand and you will see my back; but my face must not be seen." (Exod. 33:19–23)

The critic of cartographical reasoning takes these repeated warnings most seriously, partly for Marx's reasons of alienation but mainly because our inability to be abstract enough has now reached a stage where it has become a threat to our very survival. Aristotle's advice that what we cannot do perfectly, we must do as well as we can, comes readily to mind. The example of Paul Cézanne as well, for with unsurpassed dedication it was this loner who searched for the most revealing place in the cleft, a sacred place where he could go in his own direction, the exact rock from which he could catch a glimpse of the unique and solitary home that was his alone.

The choice of fix-point presented a problem to the biblical narrators as well. Therefore they let the commandments be given twice, first to the people at large, then to Moses alone, the man who the LORD had entrusted to be his Provincial Governor. Since the populace in Moses' absence had fallen into idolatry and shamelessly sacrificed to a golden calf rather than to some invisible relation, they had in fact proven that they were not to be trusted.[28] As a consequence, rather than addressing the recalcitrant people directly, the LORD decided to deal with them indirectly, to govern through the extended arm of his messenger, a case of liminality if one ever was.[29] It follows that while Abr(ah)am may be described as the plaything of an immature god of adolescence, Moses is better characterized as a go-between, an apparatchik created by the ultimate dialectician of cruelty and benevolence, distance and nearness. World history's most outstanding politician, a nonnegotiable force impossible to grasp and impossible to ignore. Such is my own conception of the literary figure that others call "the Almighty being of Beings."

As in the Machiavellian world, so in the Mosaic: always set on producing obedient and predictable subjects, the Prince never sits still. In that unending game of truth and consequences raw power does of course have its moments. In the long run, however, it is always more effective to gather the vassals in the smoke-filled room where they can be addressed face-to-face and be treated *as if* they were friends. Eventually they will then come to believe that their sufferings are caused not by the Ruler's incompetence, mistakes or pure viciousness, but by their own sins. Pushed to the wall everyone, especially the Almighty, blames the victim rather than the criminal. Therefore the prophet once found it appropriate to put the following words into the LORD's mouth:

The time is coming when I will make a new covenant with the house of Israel and the house of Judah. It will not be like the covenant I made with their forefathers when I took them by the hand to lead them out of Egypt, because they broke the covenant, though I was a husband to them [*or* was

> their master]. This is the covenant I will make with the house of Israel after that time. *I will put my law in their minds and write it on their hearts.* I will be their God and they will be my people. No longer will a man teach his neighbor, or a man his brother, saying, "Know the LORD," because they will all know me, from the least of them to the greatest. For I will forgive their wickedness and will remember their sins no more. (Jer. 31:32–34, emph. added)

*

**I will put my law in their minds and write it on their hearts.** A breath of Social Democracy long before its time, a rhetoric as effective then as it is deceptive now, an excellent example of the techniques by which the cartographical reasoner prepares the mappa of his projection screen, a paradigm of socialization.[30] Propaganda should nevertheless be recognized as propaganda regardless of whether it is preached by a biblical author from the seventh century BCE or by a TV reporter from more recent times. Thus:

> "Say to the Israelites," said the LORD to Moses, "you must observe my Sabbaths. This will be the sign between me and you for generations to come, so you may know that I am the LORD, who makes you holy." (Exod. 31:12–13)

In my own update:

> "Say to the Citizens," says the PARTY to the cadres, "you must observe the election day and celebrate our victory. Ninety percent of the vote will be the sign between us and them, the proof that they will know who their Ruler is and will be grateful to those on whose shoulders their well-being rests. Do not worship any other ideology than mine, for our PARTY is a jealous congregation which never forgets, a blessed people whose laws are written in their minds, there being designed to be one with their taken-for-granted. Remember the words of Isaiah (6:9–10), the most contradictory of all prophets: 'Be ever hearing, but never understanding; be ever seeing, but never perceiving. Make the heart of this people calloused; make their ears dull and close their eyes. Otherwise they might see with their eyes, hear with their ears, understand with their hearts, and turn and be healed.'"

How differently the three figures of Odysseus, Abr(ah)am and the apparatchik address the question of what it means to be human. How different their maps, how different their techniques for being believed. For the Greek it is fate and honor, for the Jew it is the father who is who he is, for the Social Democrat the rhetoric of social consensus. *Ricorso,* Signore Vico. *Factum verum!* In Norman O. Brown's language:

The true *(verum)* and the made *(factum)* are convertible
verification is fabrication
homo faber
man the forger; at his forge
forging the uncreated conscience of his race.[31]

At issue is once again the issue of socialization, more precisely the techniques by which the Ruler's law is put in our minds and written on our hearts, much as the painter applies the paint to the canvas. The ensuing problems reach far back, most immediately to Plato's *Republic* and the story of the cave. Not, however, to its analyses of the shining sun and the chained prisoners, but to its remarkable ignorance of the wall. The silence is stunning, for to the critic of cartographical reason the close connections between Plato's limestone wall, the painter's canvas and the cartographer's mappa are impossible to ignore: the self-evident truth is that without the resistance of a projection screen there would be no shadows to catch, hence nothing to see, hence nothing to share. Understanding the structure of this social surface is a challenge of the utmost order, for it is the screen, not the screened, that constitutes the taken-for-granted.

To be metaphorically concise: Not all paint sticks to all surfaces and just as some mixtures run off the panel without leaving a trace, so others crack as they harden. To prepare the canvas for the reception of the unknown the artist therefore kills it with layers of gesso, a practice that is highly reminiscent of the socialization processes through which you and I are made so obedient and so predictable. The parallel is obvious, for while some readers will immediately grasp the unheard messages of the present book, others will neither understand nor remember. In through one ear, out through the other. No traces left in-between. Not with *you*, however, for with you the proof is beyond doubt: it is *you* who are now reading this very sentence, *you* and no one else. And when your body encounters these black marks on the white page, you are not floating around in an unmoored No-Where but firmly lodged in the presence of the Now-Here. And just as the painter's paintbrush leaves its marks on the stretched and prepared canvas (a rectangle proportioned à la Alberti), so my delayed wordbrush now touches the wrinkled surface of your unconscious.

It cannot be said more clearly: the signified springs from the meeting of a fleeting Signifier and a capturing screen.

*

**A fleeting Signifier and a capturing a screen**! And with that remark it has finally come, the longed-for moment of introducing Kazimir Malevich, the Russian Suprematist who many consider the first of abstract painters, the true father of minimal art.[32] His desire to surpass representational form and to move into a new mode of creation went beyond anything seen before, and in one of his most influential writings he described what he was doing as

> essentially a movement in reverse, a decomposition and a dissipation of what was collected into its separate elements; it is the attempt to escape from the objective identity of the image to direct creation and to break away from idealisation and pretence. I wish to create the new signs of my inner movement, for the way of the world is in me, and I do not want to copy and distort the movement of the subject or any other manifestation of nature's forms.[33]

When this remarkable explorer of the non-objective world some time around 1913 chose to paint a square rather than a mountain, an apple or a pipe, his overriding aim was to study the convergence of the image and the taken-for-granted. Most significant for the present critique is the fact that these creations were for him naked icons, non-representational presentations of the non-representable, concepts firmly tied to the minds and hearts of Russian orthodoxy. On the history of the icon I will have more to say in a later chapter, but already now it must be noted that an icon is not what it appears to be—not a picture of the holy but the holy in and of itself; a sign in which the gap between **S**ignifier and **s**ignified is brought to a minimum, the —— of the Saussurean bar magically erased; the prohibition on graven images at the same time honored and transcended.

In Malevich's mind the icon gradually took the form of a *Black Square,* a geometric figure centrally placed within a white area that extends to the edges of the non-framed and perfectly quadratic canvas of 18 × 18 *vershok* or 80.01 × 80.01 centimeters.[34] Four corner-points, four lines, one projection plane, two colours—such was "the creation of intuitive reason . . . , the face of the new art . . . , a living royal infant . . . , the first step of pure creation in art."[35] And not only in the arts, but in imaginary thought in general. The most fundamental of all geometric figures is in fact the square, practically used by the Egyptians perhaps already by the Babylonians, theoretically codified by Thales of Miletus (or someone else now called by that name).[36] At any rate a human creation of immense consequences, for from the four corners of the quadrate spring first two diagonals and from them a set of triangles, then from the intersection of the diagonals a point out of which grows the circle, *in*scribed as well as *circum*scribed. Even more remarkably, this entire development would have been impossible without the right angle, an invention arguably as revolutionary as the invention of the wheel.

As always, Malevich's inspiration came from many directions, including centuries of art history and a long poem by Walt Whitman entitled "Chanting the Square Deific."[37] In this highly creative piece of imagination the American wildman envisioned the unknowable Godhead as a figure

Kazimir Malevich, *Black Square.* 1914–15. Oil on canvas, 80.01 × 80.01 cm (Given as 79.6 × 79.5 cm). State Tretyakov Gallery, Moscow.

with four faces: (a) the Father without remorse; (b) the Son, suffering yet absorbed by affection and charity, rejected yet constantly resurrected; (c) the Holy Spirit, breather of life and lighter than light, essence of forms and name of the beyond beyond the beyond; (d) the comrade of criminals and brother of slaves, Satan himself in his ignorance full of guile and always set against the Ruler.[38] In a mind-boggling passage the Russian follower turned this divine square into an image of himself, boldly announcing that

> I am the beginning of everything for in my consciousness worlds are created. . . . I search for God, I search for my face, I have already drawn its outline and I strive to incarnate myself. And my reason serves me as a path to that which is drawn by intuition. . . . Reason first forms the face of man.[39]

It was the face of this mystified and dematerialized power, solid and four-sided, that the Suprematist set out to capture in his black square, a geometric form which he sometimes took as the end and the beginning, sometimes as the zero to which everything is reduced and from which everything evolves. In short, a system of space void of the Brunelleschian perspective, a planned escape from the parallelism of Euclidean geometry and the vanishing point of optics.[40] In the eyes of the Suprematist it was the flatness of his paintings that represented the real advance over his predecessors. In his own inimitable words:

> When the laws of perspective were set up in imitative art, a binding chain was established. A stall was set up in which the artist was expected to operate. . . .
>
> The perspective segment has also a great influence on man's consciousness. In our comprehension and view of the world our body always moved along lines of perspective, but thanks to the fact that the point of convergence was constantly moving further away our body did not feel the converging lines above and below itself; had they remained as unshakeable as they did for art our body would have grown into a cuneiform shape just as art did. . . .
>
> [But] when art felt the need to expand the growth of its body it was necessary to destroy the cuneiform-perspective catacomb. We began to regard the world differently and discovered its many-sided movement, and were thus faced with the problem of how to convey it fully: hence arose systems and laws contemporary to our comprehension. There came to art a pure comprehension of the value of texture as such, without any need for the linear, architectural building of houses. *The painterly textural content was built on the plane of the canvas.*[41]

Leon Battista Alberti dead, buried and arisen again. No wonder that the unprepared viewers felt like travelers in unchartered territory, Wittgensteinian madmen who did not find their way in the world, a congregation of sheep lost among maps whose likes they had never seen. Fix-points, scale and mappa of an entirely different kind, the collected manifestoes nothing less than a guidebook to a universe of hitherto unthought complexity, an (un)holy alliance between Einsteinian relativity theory and Russian orthodoxy.

The *Black Square* was first shown in December 1915 at the "0,10. Last Futurist Exhibition" in Petrograd. From a famous photograph we know that it was hung across the upper north-east corner of the exhibition room, the place which in Russian homes is reserved for the most sacred icon. Since the experimenting artist could not afford the best quality paint, the surface of the picture is now badly cracked, the underlying canvas shining through in its scarred and partly overpainted nakedness, an unintentional baring of the projection screen without which there would be nothing to see,

Kazimir Malevich, *Suprematist Composition: White on White*. 1918. Oil on canvas, 79.4 × 79.4 cm. Museum of Modern Art, New York (MOMA). © Digital image, The Museum of Modern Art, New York/Scala, Firenze.

not even nothing. And for that same reason the critical cartographer finds Malevich's painting from 1918, *White Square on White,* even more fascinating than the time-worn "royal infant" from which it originated: two squares one placed inside the other, the smaller tipped and thrust toward but never actually reaching the top and right edges of the larger.

What makes this work so stunning is that the icon of the tilted white square (almost but not exactly square) is set against a background which itself is white, albeit not of exactly the same hue. In my own interpretation the entire series of whites on white, largely painted during the revolutionary years between 1918 and 1920, can be seen not merely as a set of pictures of the nonpicturable but at the same time as studies of the background onto which the graven image has been projected, an unexpected coming together of §2a and §3 of the Constitutional Law. In that context it should also be recalled first that Wassily Kandinsky thought of white as a silent color,[42] then that for the perfectly socialized the taken-for-granted is so much taken-for-granted that it can be neither seen nor heard—two squares bleached almost to obliteration, an uncompromising merger of idealism and abstraction. As in the famous case of Jorge Luis Borges,[43] the scale of the true believer is 1:1.

Such is the scale of humor as well, and in a movie where he played a lawyer defending his client, Groucho Marx said, "Your honor, my client may look like an idiot and talk like an idiot, but you should not be deceived.—He *is* an idiot." In contrast, a picture of a pipe is never a pipe:

> Take, for example, Cézanne's self-portrait. Its scale does not in the least correspond to the form, i.e. the model. Or Kandinsky's non-objective scales where it would be difficult to say whether the colour corresponds to the form or the model, since the form itself does not exist, and there is only the movement of colour masses.
>
> We cannot say to Kandinsky, any more than we can to Cézanne, that colour ought to have its own individual form, since in their works, particularly in Cézanne, there is no colour at all, but there is painting. I could also speak of my own works: in one and the same picture I create the same forms but colour them with different colours. I said that I colour them in order to underline that in my own pictures I draw a strict distinction between colour and form. And in the case in point I colour the form with this or that colour, not because red or blue corresponds to this or that form, but because I paint in colours according to the scale that has arisen in my creative mind."[44]

And so it is that Malevich's series of whites-on-white forms an outstanding atlas of invisible maps of the invisible, a wonderful instance of the taken-for-granted captured in the language of the same; at the same time a picture of a dematerialized object and a story of nonmentionable power-relations; a self-conscious message consumed by its unconscious background; a weightless object anchored neither in the (Homeric) foreground nor in the (biblical) background, but in the free-floating relations of the in-between. The Abrahamitic HERE I AM projected from a realm of political uncertainty into a world of peaceful silence.

Secrets too secret to hold and too dangerous to reveal, especially in the all-embracing realities of dictatorial power. But not even Joseph Stalin—that most spirited advocate of the Constitutional Law—had the means for silencing every critique. As proof, it was in the portraits from the tightening period between 1928 and 1935 that Malevich reached a climax of dis-

Kazimir Malevich, *Peasant Woman.* 1928. Oil on canvas. 98.5 × 80 cm. © State Russian Museum, St. Petersburg.

closure. Created under the cloud of severe censorship these sometimes armless figures typically lack facial features—no openings for the bodiness of the five senses, merely an oval created by reason, slightly inclined and filled with pure color.[45] Doubtless a tribute to Paul Cézanne, who was the first to understand that even if a picture of some bathers may look like a recollection of childhood experiences, it is in reality a constellation of triangles, cones, cubes and spheres. Doubtless also a response to the LORD's declaration that my back you may see but not my face, for no-one can see my face and live. Finally, since the backgrounds were often painted on an unprimed canvas, the rawness of the surface is readily detectable. Plato's shadow-screening wall in another light.

Given the politics of his own lifetime, Malevich was well aware that what in the morning is heralded as "knowledge" may in the evening be ridiculed as "myth." Whatever the observer happens to grasp is therefore not reality but his grasping of reality, that particular mode of understanding which is inseparable from the socialization processes of his own time and place.[46] When the Suprematists used the term "painterly realism," they were therefore saying that the meaning of the imagery is determined by the physical qualities of the canvas and the pigments of the paint. Jacques Lacan would have agreed. The expressive Signifier lies on top, the meaning-filled signified stirs underneath.

True to its own idea *White Square on White* is virtually impossible to reproduce, perhaps because it is a work which represents nothing but itself, the creamy color of the (non-perfect) square hardly distinguishable from the half-and-half tones of the background. What speaks is consequently not the color of the paint but the texture of the brushstrokes, not the light of sight but the resistance of touch. A sculpture as much as a picture. Therefore, when Malevich in one of his many manifestoes posed the question "What is the canvas? What do we see represented on it?," he answered:

> a window through which we discover life. The Suprematist canvas reproduces white but not blue space. The reason is obvious: blue does not give a true impression of the infinite. The rays of vision are caught in a cupola and cannot penetrate the infinite. The Suprematist infinite white allows the beam to pass on without encountering any limit. We see moving bodies. . . . This discovery demands a great deal of work. The construction of Suprematist colour forms is in no way connected with aesthetic necessity. Both colours, forms and figures also have a black and white period. The most important in Suprematism—its double basis—are the energies of black and white serving to reveal the forms of action.[47]

And so it is that the Suprematist's icon is not *about* something—it *is* that something itself; a problematic in which Kazimir Malevich and Samuel Beckett are reaching out for the non-nameable prohibition of prohibitions, the taboo of taboos, the mappa of mappae. A rebellious protest of magnificent proportions, an in-depth study of the Constitutional Law, especially of the relations between its second and third paragraphs. In short "suprematism, as explored here, emerges as a pure product of the creative mind,"[48] the Suprematist himself a most courageous investigator of what it means to be human, outstanding explorer of the abyss between visible and invisible.

And since he used his pen as much as his brush, Malevich's oeuvre as a whole comes out as an intricate blend of story and picture, exactly the combination of genres that makes a map a map; as recalled, it is not enough to know where you are, you must also know where to go. And so it is that just as Malevich's black, red and white squares are alternative depictions of the taken-for-granted, so his canvases are like the stone tablets which Moses first broke in anger over the people's idolatry and which the LORD subsequently reissued in a slightly different version. It is a sad truth that the hangman tends to be more dangerous when (s)he is bored than when (s)he is overworked. Boxed in.

Quite fittingly, and in compliance with his own directives, the Suprematist's ashes were buried in the fields near his dacha, a white wooden cube with a black square marking the grave.[49] The year was 1935, the political situation getting tighter and tighter. In hindsight the purges were of course predictable, and already in 1929 the director of the Tretyakov Gallery, one Fedor Kumpan, had been given a long prison sentence for organizing a Malevich retrospective. Indeed, as early as 1926 the leading party newspaper, *Leningradskaia Pravda,* carried a vicious review of an exhibition staged by Malevich and two of his friends:

> A monastery has taken shelter under the name of a state institution. It is inhabited by several holy crackpots who, perhaps unconsciously, are engaged in open counterrevolutionary sermonizing, and making fools of our Soviet scientific establishments. As for any artistic significance in the "work" of these monks, their creative impotence hits you right in the eye at the first glance. . . . The Control Commission and the Workers and Peasants Inspection should investigate this squandering of the people's money on the state support of a monastery.[50]

Between 1933 and 1962 Malevich's major works were either locked away in Soviet archives—dictators never forget—or hidden in the basement of a German art dealer who knew their worth. Miraculously saved from bombs and fires, the largest collection is now in the possession of the Stedelijk Museum in Amsterdam, where it forms a critical parallel to the rectangular grids of Piet Mondrian.[51] It was not until 1977, however, that the Soviet Museums exhibited his works again, an event which in itself proves that in the universe of institutionalized terror the LORD continues to act exactly as he pleases, never whimsically always intentionally.

Joaquin Martínez, *Kasimir Malewich.* 1992. Book, oil, wood, 40 × 45 cm. Private collection.

With the utterance of his mouth the Almighty creates his subjects, with the tip of his finger he draws their prison.

*

**With the tip of his finger** God wrote a Constitutional Law unsurpassed in its insights into the structure of power, unbreakable in its self-reference, definitive in its self-assertion. The redactors' conclusion was fitting: "These words the LORD spake . . . with great voice; and he added no more" (Deut. 5:22).

In his almightiness the Almighty could of course have written something entirely different, including a foursquare gospel, a mandala, an opera, a pop song or anything else. He chose not to and that explains why I have read the Decalogue as a party platform, in everything as political as the American Constitution and the *Communist Manifesto.* In the words of René Girard, a confessed Christian, "the Bible has given us the privileged tool of demystification, but we either do not know it, or do not want to use it. Perhaps we are secretly afraid that it will wreak too much havoc."[52]

For my own part I have argued that the cartographer of power has more to learn from close readings of the Old Testament than from any operationalization of Machiavelli's *Prince;* as recalled, none less than Karl Marx remarked that in order to understand the queer relations between use value and exchange value we must have recourse to the mist-enveloped regions of the religious world.[53] Not surprisingly Joseph Stalin was deeply influenced by the techniques of his biblical predecessor, not the least when it came to the repeated insistence that if you obey me fully and keep my covenant, then out of all nations you will be my treasured possession. In contrast, there are profound differences between the approaches of Moses and Kazimir Malevich, for whereas the former went down from the mountain and warned the people not to force their way through to see the LORD, the latter withdrew, utterly disgusted by his discoveries. As a way out he eventually produced some of the most heartbreaking (self)portraits of his century, perhaps of all times, a kind of art-historical essays painted in an archaizing Renaissance style, their lines and colors reflecting the influence of Dürer, Cranach and the Russian icons. Most significantly, the signature in the lower right-hand corner was a miniscule version of the black square.

And yet. When pressured even the most determined power-wielder cannot withhold the truth and in that sense he is similar to the prisoners he tortures in his dungeons. Thus, immediately after the LORD had spoken his commandments,

> [w]hen the people saw the thunder and lightning and heard the trumpet and saw the mountain in smoke, they trembled with fear. They stayed at a distance and said to Moses, "Speak to us yourself and we will listen. But do not have God speak to us or we will die."
>
> Moses said to the people, "Do not be afraid, God has come to test you, so that the fear of God will be with you to keep you from sinning." The people remained at a distance, while Moses approached the thick darkness where God was.
>
> Then the LORD said to Moses, "Tell the Israelites this: 'You have seen for yourselves that I have spoken to you from the heaven: Do not make any gods alongside me; do not make for yourselves gods of silver or gods of gold. Make an altar of earth for me and sacrifice on it your burnt offerings

Kazimir Malevich, *Self-Portrait (Artist)*. 1933. Oil on canvas. 73 × 66 cm. © State Russian Museum, St. Petersburg.

and fellowship offerings, your sheep and goats and your cattle. Wherever I cause my name to be honored, I will come to you and bless you. If you make an altar of stones for me, do not build it with dressed stones, for you will defile it if you use a tool on it. And do not go up to my altar on steps, lest your nakedness be exposed on it.' " (Exod. 20:18–26)

*

**After this, Herr Goldschmidt**, you may button your pants. For through your nakedness you have just revealed what it means to be a semiotic animal, body and mind inseparably tied together. In particular, you have helped the critical cartographer demonstrate how the cut in your foreskin bares the terrorist politics of your LORD, ruthless and benevolent in the same breath, seemingly unpredictable, in reality brimming of intentionality. Whenever the Ruler calls, you automatically answer "Here I am," at the same time exhibiting the fear that you will never be trusted and the insight that it is HE who has the power to redistribute goods and beds exactly as he likes, justice as fairness not one of HIS concerns. Terrorist and supreme dictator in the same disguise. In the interpretation of Paul, the apostle:

> Oh, the depth of the riches of the wisdom and knowledge of God!
> How unsearchable his judgments, and his paths beyond tracing out!
> Who has known the mind of the LORD? Or who has been his counselor?
> Who has ever given to God, that God should repay him?
> For from him and through him and to him are all things.
> To him be the glory for ever! Amen. (Rom. 11:33–36)

In contrast, the scar which the old nurse discovered on her master's leg tied him not to the Olympian gods but to his family on Parnassos, especially to his grandfather Autolykos, the man who had given him his name and the person from which he had inherited his slyness. And even though the two scars differ in the sense that the human sacrificer left his mark on purpose and the wild boar did not, on another level the two stories are structurally very similar, not the least because every story is written with a purpose.[54] These were the peripatetic words that Penelope spoke to the banqueters:

> "My lords, hear me:
> suitors indeed, you commanded this house
> to feast and drink in, day and night, my husband
> being long gone, long out of mind. You found
> no justification for yourselves—none
> except your lust to marry me. Stand up, then:
> we now declare a contest for the prize.
> Here is my lord Odysseus' hunting bow.
> Bend and string it if you can. Who sends an arrow
> through iron axe-helve sockets, twelve in line?
> I join my life with his, and leave this place, my home,
> my rich and beautiful bridal house, forever
> to be remembered, though I dream it only."
> (21.68–79)

Two different ways of determining what it means to be human, two trials nevertheless. And as none of the suitors could stand the lady's test, the turn eventually came to Odysseus, still a stranger to the crowd. Seated on a stool strategically placed at the abysmal threshold, he looked the bow over, searching for signs of termites, handling it as if it were a woman. Then,

> the man skilled in all ways of contending,
> satisfied by the great bow's look and heft,
> like a musician, like a harper, when
> with quiet hand upon his instrument
> he draws between his thumb and forefinger
> a sweet new string upon a peg: so effortlessly
> Odysseus in one motion strung the bow.
> Then slid his right hand down the cord and plucked it,
> so the taut gut vibrating hummed and sang
> a swallow's note.
> In the hushed hall it smote the suitors
> and all their faces changed. Then Zeus thundered
> overhead, one loud crack for a sign.
> And Odysseus laughed within him that the son
> of crooked-minded Kronos had flung that omen down.
> (21.493–415)

A setup strikingly similar to Genesis 22, yet two outcomes entirely different, the godly omens worlds apart—"Abraham, Abraham," and the old man trembled; "one loud crack," and Odysseus laughed. The fact nevertheless remains: both characters responded to the calls and both passed their respective tests, one by exhibiting a mental strength beyond belief, the other by a rare combination of physical stamina and cunning intelligence. The sealing symbols vary accordingly, for in the first case there was the smell of a burning ram and the image of a devastated man, in the other a hall full of dead bodies, blood-drenched ghosts whose living relatives gathered to strike back:

> At this point, querying Zeus, Athena said:
> "O Father of us all and king of kings,
> enlighten me. What is your secret will?
> War and battle, worse and more of it,
> or can you impose a pact on both?"
> The summoner of cloud replied:
> "My child,
> why this formality of inquiry?
> . . . . . . . . . . . . . . .

Conclude it as you will.
There is one proper way, if I may say so:
Odysseus' honor being satisfied,
let him be king by a sworn pact forever,
and we, for our part, will blot out the memory
of sons and brothers slain. As in the old time
let men of Ithaka henceforth be friends;
prosperity enough, and peace attend them."
(24.472–86)

*

**Fine stories these are**. Not a word out of place, not a phrase without a point, the most crucial among them the argument that Gods are gods.[55] And in their roles as gods these entities always play by their own rules, consistently inconsistent, habitually violating what others take to be the laws of thought. In general their reasoning lies outside the reach of their subjects' imagination, beyond the human art of making the absent present and the present absent. And at least in this sense the figures of JHWH and his various messengers are like those of Zeus and his daughter Athena: sometimes they point their followers in this direction, sometimes in that, now taking us to a heaven of breathless beauty, now to a hell of Dante-esque horrors.

But if Gods are gods, then men are Men, in all respects just as rich and mysterious as the deities they have created in their own images. Therefore we learn from one source how in the eternity between early one morning and noon the third day the two of them went together, from another how an arrow was shot through the sockets of twelve axe-heads, King Minos' labyrinth straightened out in the process. Most revealingly, the nuptial bed—the final proof of the lasting relations between husband and wife, a design known only to the two of them—was made from the trunk of an olive tree, a natural growth so turned and twisted that it defies all attempts to fashion it into a prefigured design. Penelope to her son:

"If really he is Odysseus, truly home,
beyond all doubt we two shall know each other
better than you or anyone. There are
secret signs we know, we two."
A smile
came now to the lips of the patient hero, Odysseus,
who turned to Telemakhos and said:
"Peace: let your mother test me at her leisure.
Before long she will see and know best."
(23.107–14)

And after three millennia of exegesis, for better and for worse, masses of people continue to rely on the stories of Abr(ah)am and Moses as their primary guides in life. Similarly, most of us remain committed to the Aristotelian Laws of Thought, in themselves an outgrowth of Homer's study of the relations between dream and reality, a limit which when approached from one direction appears as the rim of the Earth, from the other as the shore-line of the Ocean River. Anaximander's *to apeiron* in another guise, cartographical reason when it works. A play of fix-points, scales and mappae:

•

*I am I* versus *I am he*
*Here I am* versus *Here I go*
The trust of *pistis* versus the honor of *habitus*

Yet there is much to say about Ludwig Wittgenstein's remark, itself prompted by his own background: "It is typical for a Jewish mind to understand someone else's work better than he understands it himself."[56] And so it deserves repeating that when Moses had finished the construction of the tabernacle,

> [t]he cloud covered the Tent of the Meeting, and the glory of the LORD filled the tabernacle. Moses could not enter the Tent of Meeting because the cloud had settled upon it, and the glory of the LORD filled the tabernacle. (Exod. 40:34–35)

So it also was that Odysseus said,

> "Telemakhos, the stranger
> you welcomed in your hall has not disgraced you.
> I did not miss, neither did I take all day
> stringing the bow. My hand and eye are sound,
> not so contemptible as the young men say.
> The hour has come to cook their lordships' mutton –
> supper by daylight. Other amusements later,
> with song and harping that adorn a feast."
> (21.424–30)

# KANT

**The question** keeps returning: How do I find my way in the unknown? The answer echoes back: By map and compass, picture and story.

More precisely: How are we made so obedient and so predictable? How are they constructed, the invisible maps and internalized compasses that tell us both where we are and where we should go, not in the visible and material landscapes of earth and air, fire and water, but in the invisible universe of the socially taken-for-granted? Is madness the price we pay for the sins of penetrating the abyss between the five senses of the body and the sixth sense of culture? In other words: How do I communicate my findings in such a way that they at the same time appear true to me and trustworthy to you?

The answer is that I must produce a document which in the same gesture both shows and tells, a performance which is not merely *about* something but *is* that something itself. In short a matter of rhetoric when it works, a challenge that Immanuel Kant took more seriously than most.[1] By so squarely placing his philosophy within the rhetorical paradigm, his conception of understanding got thoroughly soaked in metaphorical reason, a way of sharing the world which was doubly rooted in the personal experiences of his own body and in the academic disciplines of architecture and geography. In his own words, "human reason is by nature architectonic."[2] It is this mode of thought-and-action that I am calling "cartographical reason," a term originally coined by my friend Franco Farinelli.[3]

Immediately we enter this terra incognita we discover that many have been there before us. Perhaps the first was Thales of Miletus, perhaps his

*Portrait of Immanuel Kant.* German school, 18th century. Private collection. Bridgeman Art Library, London.

pupil Anaximander, perhaps someone else with a made-up name from an unclocked time and unchartered place. Be this as it may, for the definite answer will never be known. In all uncertainties one thing is nevertheless certain and this is that the legitimate father of modern reason was Immanuel Kant, Plato its ageing grandfather, Ludwig Wittgenstein its late-born Isaac. What links these giants together is not primarily that their questions made them theoretical philosophers but that their way of thinking made them practicing geographers, secret agents stationed in the frontier regions of sensibility and intelligibility, courageous explorers of the taboo-ridden realm of the taken-for-granted, self-appointed land-surveyors commissioned to establish the Mason-Dixon line between the oikumene of the humans, on the one hand, and the anoikumene of gods and brutes, on the other. Even though their determinations were exceptional—Socrates died for his convictions—Kant explicitly thought of his philosophy not as a single unified system but as a form of performative experimentation.[4]

Wittgenstein, on his part, is famous for the phrase that "if a lion could talk, we would not understand him."[5] The point of that remark is, of course, that in my capacity as a human being I can never know what the lion feels when his paws touch the ground, never hear with *his* ears how the wind rustles the elephant grass, never see with *his* eyes the antelope at the waterhole, never smell its scent, never slit its throat with *my* teeth, never taste its blood with *my* licking tongue. None of this because my body is the body of a man, his the body of a lion. All of this because my imagination is in the mind of a cultivated male, his in the leap of a wild animal; while I can think beyond the limits of my experience, I cannot know the objects of such thoughts. It is by no coincidence that we are using the expression "making sense," when "making intelligent" would be more appropriate.

It was this issue of truth and trust—the ultimate question of what it means to be human—that permeated everything Kant ever wrote, not the least in the detailed analyses of the *Critique of Pure Reason*.[6] And nowhere was this issue more precisely captured than in the first paragraph of the pivotal chapter "On the Ground of Distinction of all Objects in General into Phenomena and Noumena," at the same time a summary of what has gone before and an anticipation of what was to come. In Max Müller's old-fashioned translation:

> We have now not only traversed the whole domain of the pure understanding, and carefully examined each part of it, but we have also measured its extent, and assigned to everything in it its rightful place. This domain, however, is an island and enclosed by nature itself within limits that can never be changed. It is the country of truth (a very attractive name), but surrounded by a wide and stormy ocean, the true home of illusion, where many a fog bank and ice that soon melts away tempt us to believe in new lands, while constantly deceiving the adventurous mariner with vain hopes, and involving him in adventures which he can never leave, and yet

> can never bring to an end. Before we venture ourselves on this sea, in order to explore it on every side, and to find out whether anything is to be hoped for there, it will be useful to glance once more at the map of that country which we are about to leave, and to ask ourselves, first, whether we might not be content with what it contains, nay, whether we must not be content with it, supposing that there is no solid ground anywhere else where we could settle; secondly, by what title we possess even that domain and may consider ourselves safe against all hostile claims.[7]

Imagination at high pitch, cascades of brilliant associations, the master key to Kant's entire philosophy. No wonder that this time-bound traveler has been called *the* student of limits, his very name a handy revelation of his predilections—the German word *Kante* means literally "edge," in all respects the sharp counterpart of Anaximander's fuzzy *to apeiron.*[8] The true purpose of philosophy, he claimed, is to expose the illusions of reason by reminding us of its limits. Thus,

> it is humiliating, no doubt, for human reason that it can achieve nothing by itself, nay, that it stands in need of a discipline to check its vagaries, and to guard against the illusions arising from them. But, on the other hand, it elevates reason and gives it self-confidence, that it can and must exercise that discipline itself, and allows no censorship to anyone else. The bounds, moreover, which it is obliged to set to its own speculative use check at the same time the sophisticated pretensions of all its opponents, and thus secure everything that remains of its formed exaggerated pretensions against every possible attack. The greatest and perhaps the only advantage of all philosophy of pure reason seems therefore to be negative only; because it serves, not as an organon for the extension, but as a discipline for the limitation of its domain, and instead of discovering truth, it only claims the modest merit of preventing error.[9]

Perhaps the most fundamental of these limits (a close relative of Plato's divided line) is the one that runs between the phenomena and the noumena, the former a set of entities that appear to the senses, the latter something which is knowable only to the *nous* of pure understanding. Rephrased, the concept of a noumenon is "merely a *limiting concept,* the function of which is to curb the pretensions of sensibility. . . . At the same time it is no arbitrary invention; it is bound up with the limitation of sensibility, though it cannot affirm anything positive beyond the field of sensibility."[10]

In his inaugural dissertation of 1770 Kant in fact sounded much like Plato himself, concluding that "it is clear that things which are thought sensitively are representations of things *as they appear,* but things which are intellectual are representations of things *as they are.*"[11] What better proof can one ask for than the following passage from *Prolegomena,* the popularized and much criticized, even repudiated, version of the *Critique of Pure Reason:*

> Since the oldest days of philosophy inquirers into pure reason have conceived, besides the things of space, or appearances *(phenomena),* which make up the sensible world, certain creations of the understanding, called noumena, which should constitute an intelligible world. And as appearance and illusion were by those men identified (a thing which we may well excuse in an undeveloped epoch), actuality was only conceded to the creations of thought.
>
> And we indeed, rightly considering objects of sense as mere appearances, confess thereby that they are based upon a thing in itself, though we know not this thing in its internal constitution, but only know its appearances, viz., the way in which our senses are affected by this unknown something. The understanding therefore, by assuming appearances, grants the existence of things in themselves also, and so far we may say, that the representation of such things as form the basis of phenomena, consequently of mere creations of the understanding, is not only admissible, but unavoidable. Our critical deduction by no means excludes things of that sort *(noumena).*[12]

The Professor's research into these concepts formed an integral part of his dedicated teaching, regularly sixteen hours per week, well attended and much appreciated. As reported by the most prestigious among his biographers, "his most intimate friends and pupils were adamant in their assertion that in personal intercourse and in his lectures Kant 'was far more genial than in his books,' that he 'threw off ingenious ideas by the thousands,' and had squandered 'an immeasurable wealth of ideas.'"[13]

Of special relevance here is the circumstance that Kant repeatedly offered a lecture course in (physical) geography, not just once or twice but a stunning total of forty-eight times—to current readers an orgy in political incorrectness, to himself a mandatory introduction to the rest of his oeuvre.[14] When the university rector suggested that he might use his time more profitably, the teacher retorted that anyone who wanted to attend his philosophy lectures must first take his geography and anthropology courses, in his own words a propedeutic for everything else. While the purpose of philosophy was to know *what* to think, the purpose of geography was to learn *how* to think.[15]

*

**The map itself** is the answer to Kant's question about the title by which we possess the solid ground of pure understanding. And exactly like his latter-day descendants Arthur Schopenhauer and Ludwig Wittgenstein,[16] his main purpose was to establish the boundary between the empirically verifiable and the utterly unthinkable, to mark the line between the certainties of truth and the ambiguities of fantasy,[17] to frame what to Plato was the Territory of the Humans. But just as Kant's *Critique of Pure Reason* boils down to a brilliant codification of the Cartesian mode of geometric reason, so Wittgenstein's *Tractatus Logico-Philosophicus* is essentially a codification of the Kantian mode of cartographical reason. To quote:

> We picture facts to ourselves.
>
> A picture presents a situation in logical space, the existence and non-existence of states of affairs.
>
> A picture is a model of reality.
>
> A picture is a fact.
>
> Pictorial form is the possibility that things are related to one another in the same way as the elements of a picture.
>
> *That* is how a picture is attached to reality; it reaches right out to it. It is laid against reality like a measure.
>
> Only the end-points of the graduating lines actually *touch* the object that is measured.
>
> The pictorial relationship consists of the correlations of the picture's elements with things.
>
> These correlations are, as it were, the feelers of the picture's elements, with which the picture touches reality.
>
> A picture can depict any reality whose form it has.
>
> A picture cannot, however, depict its pictorial form: it displays it.
>
> Every picture is *at the same time* a logical one. (On the other hand, not every picture is, for example, a spatial one.)
>
> What a picture represents is its sense.
>
> A logical picture of facts is a thought.
>
> 'A state of affairs is thinkable': what this means is that we can picture it to ourselves.
>
> It used to be said that God could create anything except what would be contrary to the laws of logic.—The reason being that that we could not *say* what an 'illogical' world would look like.
>
> It is as impossible to represent in language anything that 'contradicts logic' as it is in geometry to represent by its co-ordinates a figure that contradicts the laws of space, or to give the co-ordinates of a point that does not exist.
>
> The method of projection is to think of the sense of the proposition.
>
> Situations can be described but not *given names.* (Names are like points; propositions like arrows—they have sense.)
>
> A name means an object. The object is its meaning
>
> The possibility of describing a picture . . . with a net of a given form tells us *nothing* about the picture. . . . But what *does* characterize the picture is that it can be described *completely* by a particular net with a *particular* size of mesh.
>
> Laws like the principle of sufficient reason, etc. are about the net and not about what the net describes.[18]

The principles of cartographical reason in extreme concentration. Everything specified, including

- the problems of representation (the picture is a model of reality);
- the fix-points (a picture is a fact);
- the scale (the picture is laid against reality like a measure);
- the mappa (every picture is at the same time a logical one, logical laws are about the net and not about what the net captures);
- the projection (the senses propositioned, the act of pro-positioning literally the act of positioning oneself, of moving from the perspectivist's vantage-point to his vanishing-point);
- the pointing and the naming (the picture reaches out to reality, the end-points of the graduating lines touching the objects they measure);
- the legitimation of power (saying that God is illogical is impossible).

In sum, at the same time a restatement not only of Gervase's mappa mundi and Vico's factum-verum (one cannot give the coordinates of a point that does not exist) but also of Walt Whitman's and Kazimir Malevich's Square Deific (names are like points, propositions like arrows).

As evidenced by these remarks, Wittgenstein knew very well that there are two distinct features which belong to a picture, "first, the relation between the elements of the picture; and second, the correlations of the elements in the picture with things outside the picture. . . . The correlating is not something that the picture itself does; it is something *we* do."[19] Rephrased into the cartographer's vocabulary, the picture-part of the map shows me where I am, the correlation-part tells me where to go—the picture at the same time depicts what is here and makes us imagine what is *not* here. Anscombe was right: "It is the peculiarity of a projection that from it and the method of projection you can tell what is projected; the latter need not physically exist, though the points in space that it would occupy must."[20] A remark no cartographer can afford to ignore.

For the critic of cartography the picture theory plays an important role, not the least because it carries within itself the seeds of its own destruction. Therefore, when Wittgenstein eventually moved into his later stage, it was the picture theory that was the first to go, albeit with the crucial exception that "a description is a projection of a distribution in space."[21] In general, though, when the building inspector had climbed his ladder and kicked it away, he no longer argued that our expressions get their meaning from the things they depict, but said instead that they are inseparable from the thoughts-and-actions which they generate and reflect.[22] "Grammar tells what kind of object anything is. . . . Language itself is the vehicle of thought."[23] To claim, as so many have done, that there was a radical break between the *Tractatus* and the *Philosophical Investigations* is therefore as misleading as the talk about a young and a mature Karl Marx.

A comment not as surprising as it might at first appear, for already in the motto of the *Tractatus*—inserted immediately after the dedication "in memory of my friend David H. Pinsent"—we find the quotation ". . . und alles, was man weiss, nicht bloss, rauschen und brausen gehört hat, lässt sich in drei Worten sagen."[24] A clear hint from the alleged positivist that the invisible is more important than the visible, an attitude which in turn explains why he could write:

> The whole sense of the book might be summed up in the following words: what can be said at all can be said clearly, and what we cannot talk about we must pass over in silence.
>
> Thus the aim of the book is to set a limit to thought, or rather—not to thought, but to the expression of thoughts: for in order to be able to set a limit to thought, we should have to find both sides of the limit thinkable (i.e. we should have to be able to think what cannot be thought).
>
> It will therefore only be in language that the limit can be set, and what lies on the other side of the limit will simply be nonsense.[25]

In a similar manner Immanuel Kant made a distinction between limits *(Schranken)* and bounds *(Grenzen),* noting in particular that "in mathematics and in natural philosophy, human reason admits of *limits* but not of *bounds:* namely, it admits that something indeed lies without it, at which it can arrive, but not that it will at any point find completion in its internal process."[26] A proper science of metaphysics would therefore lead not to the limits of speculative reason but to its bounds; what the explorer would reach would be the bounds of the conceivable, not the limits of the actual. Rephrased from the conclusion of the second *Critique:*

> Two things fill the mind with ever new and increasing admiration and reverence, the more often and more steadily one reflects on them: *the starry heavens above me and the moral law within me.* I do not need to search for them and merely conjecture them as though they were veiled in obscurity or in the transcendent region beyond my horizon; I see them before me and connect them immediately with the consciousness of my existence. The first begins from the place I occupy in the external world of sense, and extends the connection in which I stand into an unbounded magnitude with worlds upon worlds and systems of systems. . . . The second begins from my invisible self, my personality, and presents me in a world which has true infinity, but which can be discovered only by the understanding, and cognize that my connection with that world (and thereby with all those visible worlds as well) is not merely contingent, as in the first case, but universal and necessary. The first view of a countless multitude of worlds annihilates, as it were, my importance as an *animal creature* . . . . The second, on the contrary, infinitely elevates my worth as an *intelligence* by my personality.[27]

Sounds like Swedenborg to me.

*

**The starry heavens above me and the moral law within me**—these were the words that Kant's friends inscribed on his tombstone. And in that same constellation shines the entire entourage of great cartographers: Plato of Athens, Pytheas of Massalia, Ptolemy of Alexandria, Gervase of Tilbury. Eventually, both Immanuel Kant of Königsberg—the well-traveled explorer who never left home—and Ludwig Wittgenstein of Vienna—the suicidal homosexual so deeply divided about the world as he found it.[28] At issue to all of them was the issue of limits, especially the exact location of the thin line which at the same time separates and joins together first the Signifier and the signified of semiotics and then the sense and reference of logic. As every wanderer knows,

> thoughts without content are empty, intuitions without concepts are blind. It is, therefore, just as necessary to make our concepts sensible, that is, to add the object of them in intuition, as to make our intuitions intelligible, that is, to bring them under concepts. These two powers or capacities cannot exchange their functions. The understanding can intuit nothing, the senses can think nothing. *Only through their union can knowledge arise.* But that is no reason for confounding the contribution of either with that of the other; rather it is a strong reason for carefully separating and distinguishing the one from the other.[29]

My emphasis! In effect Kant's as well, for everyone agrees that his kind of a priori knowledge is possible only under the provision that the objects of experience conform to the constitution of our minds, only if body and imagination can be perfectly mapped one onto the other. Furthermore, to follow a rule is in Wittgenstein's vocabulary equivalent to following it blindly, to obey an order before it is uttered.[30] Finding the way is consequently to be so fully integrated into one's culture that one knows neither this nor that, neither how nor why.

The ensuing power struggle is nevertheless a struggle over limits, the struggle over limits always a matter of drawing a line, of pointing and naming. Since the real place of ontological transformations is in the Saussurean Bar, it is in the backrooms of that establishment that all cartographers—ancient as well as modern—set up their base camp. It is from that refuge that they (have) conduct(ed) their raids into the mysterious realm of the synthetic a priori, their revolutionary missions always one and the same: to investigate what it means to have objectively valid knowledge about objects which are independent of our experiences, indeed to determine whether a priori synthetic judgments are at all possible.

Kant's own approach to this problem was to say that the objects of the synthetic a priori are products of our consciousness, especially of the ways in which we share our sense impressions with each other. In addition, he claimed that these invisible objects are not discovered after the fact (a posteriori) but are somehow given to us in advance (a priori). In opposition to previous thinkers, who had taken knowledge to be a reflection of the objects they were trying to understand, the idealistic professor now argued that the relation is exactly the reverse. The resulting viewpoint is not a perspectiva naturalis but a perspectiva artificialis, a case of nature mirroring thought rather than thought mirroring nature.

To Kant himself this causal reversal (in a sense his own solution of the geographic inference problem) was so crucial that he dubbed it his "Copernican revolution in epistemology," an idea that eventually led Friedrich Nietzsche to the conclusion that God is dead. Just as the Polish astronomer earlier had proven that the earth moves around the sun, not the sun around the earth, so the German philosopher now reasoned that the fixpoints of understanding lie in thoughts, not in things. In his own words:

> If intuition must conform to the constitution of the objects, I do not see how we could know anything of the latter *a priori;* but if the object (as object of the senses) must conform to the constitution of our faculty of intuition, I have no difficulty in conceiving such a possibility. . . . [Therefore] space does not represent any determination that attaches to the objects themselves and remains even when abstraction has been made of all the subjective conditions of intuition. For no determinations, whether absolute or relative, can be intuited prior to the existence of the things to which they belong, and none, therefore, can be intuited a priori.[31]

As these ideas take hold—like the effect of the drip drip drip of the raindrops, when the summer shower's through—we begin to suspect that our allegedly objective analyses may actually tell us more about the languages we are talking *in* than about the objects we are talking *about;* objects conform to our modes of representation just as the concept of space is more deeply rooted in epistemology than in ontology. But so strangely is this constructive consciousness constructed that consciousness itself is steeped less in the conscious than in the unconscious. It follows that the a priori is a priori less because it is true and more because it is impossible to think-and-act-and-experience without it. To ask, as some critics have done, whether Kant believed that there can be apperception without body is therefore understandable.[32] Misleading nevertheless. The reason is that without the physicality of the Signifier the meanings of the signified would go entirely unnoticed, indeed be utterly unthinkable.

On this view any combination of representations becomes one with the judgment which links them.[33] A linchpin of Kant's reasoning was therefore that all statements of mathematics, including those of pure geometry, are synthetic judgments.[34] At the same time one should always bear in mind that one of the most fundamental differences between Plato's *Republic* and Kant's three *Critiques* is that while the arguments of the former were firmly grounded in what was to be the theorem-proving geometry of Euclid, the theses of the latter grew out of the problem-solving constructions of René Descartes. In a Vico-sounding reflection Kant even concluded that we can know things a priori only so far as we make them ourselves,[35] a remark which makes it even more puzzling that he seems never to have thought of the non-Euclidean geometries which were soon to revolutionize the world, ironically confirming many of his own ideas in the process.[36] Once again, their descendants get a sense that even geniuses have their blind spots.

And yet. Just as I am reminded of the anecdotal sign from the entrance of Plato's Academy, so I recall Kant's observation that classical geometry is in essence a formalization of the tactile, an extension of the human body itself.[37] Who knows, perhaps it was his forays into the red-light districts of Vienna that prompted Wittgenstein to write:

> There is no such thing as the subject that thinks or entertains ideas.
>
> If I wrote a book called *The World as I found it,* I should have to include a report on my body, and should have to say which parts were subordinate to my will, and which were not, etc., this being a method of isolating the subject, or rather of showing that in an important sense there is no subject; for it alone can *not* be mentioned in that book.[38]

Rephrased:

> The propositions of logic are tautologies.
>
> Therefore the propositions of logic say nothing. . . .
>
> Hence there can *never* be any surprises in logic.
>
> [Yet] there are, indeed, things that cannot be put into words. They *make themselves manifest.* They are what is mystical.[39]

*

**What make themselves manifest** are the numerous affinities between the cartographic thought patterns of our three philosophers, each one playing his own tricks with the epistemological operator *as-if,* itself a wonderful instance of human imagination—for Plato intelligible Forms became sensible objects, for Kant untouchable noumena turned to touchable phenomena, for Wittgenstein well-defined nonsense grew into sophisticated language games. In addition, all three shared a common belief in some kind of *sensus communis* through which "the consciousness of my existence is at the same time an immediate consciousness of other things outside me,"[40] *the* standard by which a particular experience is recognized as an instance of something general.

And therein lies in Kant's case the key both to the imperative that you should "act only according to that maxim whereby you can at the same time will that it should become a universal law,"[41] and to the conclusion that "beauty is not a concept of the Object [but a judgment of taste which presupposes] that the same subjective conditions of judgment which we find in ourselves are universally present in every man."[42] And so it is that both ethics and aesthetics are anchored in the social necessity of getting along.[43] In extreme condensation:

- understanding is an exercise in translation, an application of the epistemological operator *as-if;*
- objects conform to our modes of (re)presentation;
- the favored mode of (re)presentation is geometric;
- geometry is the formalization of the intuition of the tactile, the taken-for-granted constructed, the constructed taken-for-granted;
- the taken-for-granted is a product of socialization, the *we* an *I* generalized;
- the *we* and the *I* are linked by the thesis that *my consciousness is the same as yours.*

Brunelleschi resurrected, for also to the philosophical geographers the object known is inseparable from the subject of the knower; since "there can be no *it* unless there is an *I* which could be aware of *it* and thereby of itself,"[44] the very point of the *I* is to express my perspective. "Being is the absolute position of a thing."[45] And herein lies also the secret of Clayton Eshleman's many attempts to trace the origins of consciousness to the cave art of Upper Paleolithic imagination, the movement of a self lodged in a carved rock, an equivalent to the exact place of Wallace Stevens' inexactness.[46] "I disappeared," says the human-made bison on the wall, "I disappeared into you."

*

**In the absolute position of a thing** there is always a t(r)opology of being, a conjunction of the thingifying *as-if,* on the one hand, and the political thesis of *the necessary unity of consciousness,* on the other. It is on the strengths and weaknesses of these principles that current critiques must focus.

About the cartographer's reliance on the as-if there is almost certainly nothing that can be done in principle,[47] simply because the semiotic animal is one with its own languages, hence thoroughly paradoxical, hence thoroughly ironic. Thus, although obsessed by the desire of tying together the Signifier of the five senses and the signified of the sixth, we luckily lack the means of ever having that desire satisfied. The only way to share the world is consequently to treat it as if it were *other* than it actually is, to transform invisible relations into visible things. The as-if is a many-headed hydra, another term for the mapmaker's scale.[48]

But even though the problems of the as-if are irresolvable in principle, they can sometimes be handled in practice. Under fortunate circumstances the creative arts may offer a way out, for it is in their nature to transgress categorial limits and baptize novel conventions; in politics, as well as in science, one always reaches the point of operationalization. It is in fact in the intersection of the spatial *topos* and the temporal *kairos* that the practices of art no longer can be distinguished from the theories of truth, the cognitive drive no longer be separated from the aesthetic. As Ludwig Wittgenstein had it, ethics and aesthetics are one and the same.[49] And as Barnett Newman used to quip, aesthetics is for the arts as ornithology is for the birds.[50]

Not much to be done about the thesis about the necessary unity of consciousness either, that rock bottom statement of how the world is commonly shared and understood—on one level a remark about the political relation between the one and the many, on another a comment about the socialization techniques which make us what we are. In Kant's own words, here for the third time repeated:

> In the original synthetical unity of apperception I am conscious of myself, neither as I appear to myself, nor as I am myself, but only that I am. This *representation* is an act of *thought,* not of *intuition*. . . . I have no *knowledge* of myself as I am, but only as I appear to myself. . . . As for the knowledge of an object different from myself I require . . . an intuition also of the manifold in me. . . . I exist, therefore, as such an intelligence, which is simply conscious of its power of connection.[51]

An embodied act of thought, that is what I am! Accordingly Johann Gottlieb Fichte, a Kantian follower not to his master's liking, could seriously claim that there is no thing-in-itself and that "nothing is originally posited except the I; and this alone is absolutely posited. Hence there can be an absolute opposition only to the I. But that which is opposed to the I = not-I."[52] Once again the biblical

*I am I*

versus the Homeric

*I am he*

And with that observation it once more returns, the original challenge of the birds and the egg, the eggs and the bird. How do I learn to think-and-write in such a way that it is not *about* something, but *is* that something itself? Kant's answer was that since we cannot reach the things-in-themselves we must be content with reaching the things-in-us, not persist in our search for the noumena that subsist independently of human cognition but be satisfied with our knowledge of the phenomena whose existence depend entirely on our cognitive apparatus. Whether this definition of phenomena and noumena commits Kant to a one- or to a two-world view has been much and inconclusively debated.[53] Human, all too human.

High time, therefore, to return to the Kantian Island of Truth, this most marvelous imagination of how we find our way in the world.[54]

* * *

**The Kantian Island of Truth** is a conception which is easily translated into a map,[55] its shape reminiscent of a Wittgensteinian duck-rabbit. The name of the island is

$$a = b$$

admittedly not as attractive as Kant's original suggestion, merely an echo of the definition which says that to tell the truth is to claim that something is something else and be believed when you do it. The inhabitants of this familiar realm are likewise, a bastardous blend of semiotic and political animals, ironic creatures who after long practice have learned to live with the tensions between identity and difference without going crazy. Driven by an irresistible desire to turn everything into copies of themselves, these humanlike beings have all been drafted into a rhetorical army of metaphors, metonymies and anthropomorphisms, trouping tropes designed to change illusion into reality. And so it is that the Kantian island is not merely an island of trustworthy truths but a continent of unassailable power.

While the occupants of Kant's imagination are spread across the realm, their ruler resides in a centrally located castle, by some called the World Center by others Ground Zero. Flown from its main tower is a banner which looks like a blood-colored square with a black equal-sign, a most fitting symbol for the Supreme whose name is one with his personality:

$$A = A$$

The relations between this self-appointed potentate and his subjects are governed by a constitutional law of the Mosaic type. From its first paragraph echoes the message that

> I am the one, only I beneath the moon and under the sun. Night and day I shall be under the hide of you. Make no mistake, though, for through my self-definition I shall tolerate no usurper and I shall destroy any yokel who stands before or beside me.

This jealous despot is well aware that he is always threatened, his power under constant attack. For protection he therefore erects around

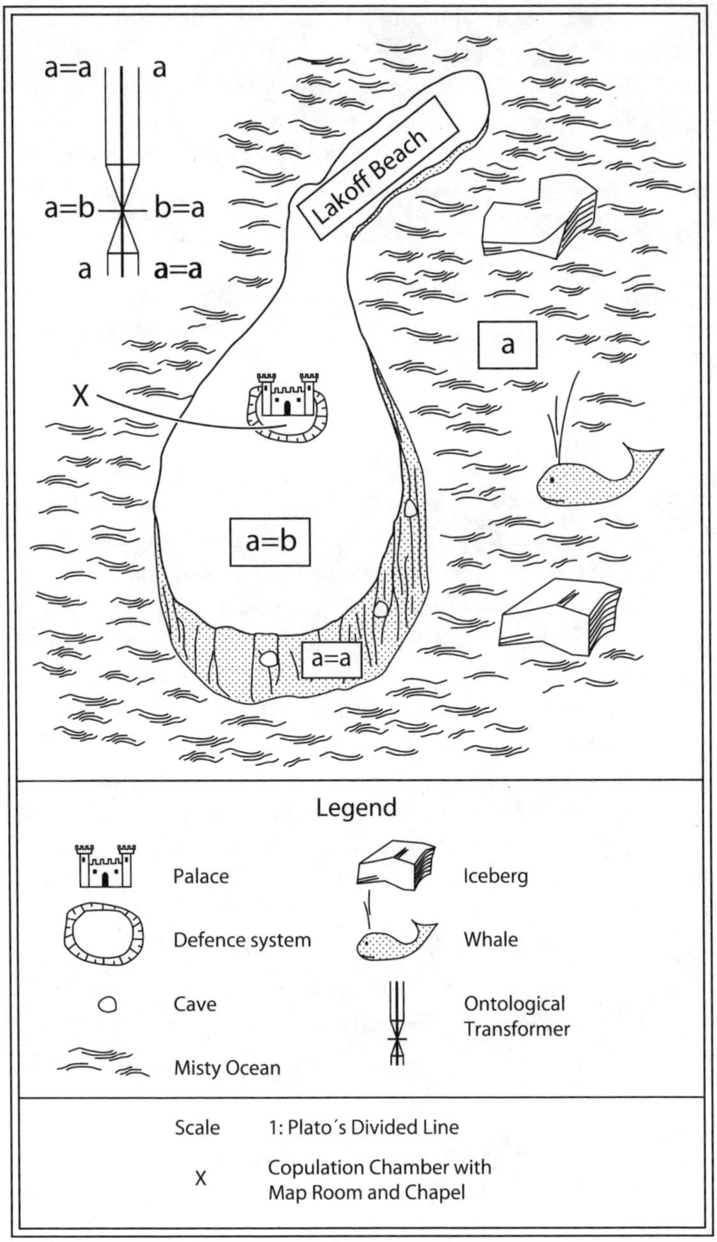


Map of the Kantian Island of Truth.

himself a two-tier defense system consisting of both a wall and a moat, the former constructed as a ban on the (mis)use of metaphor, the latter as a rule against the creative associations of metonymy. The intention of this second paragraph of the Constitutional Law is, of course, to silence any critique before it is uttered, to ensure that

> You, my subject, you as well as your descendants, shall for ever know your place, never commit the sins of trespassing. Wherefore, you must never question my authority, never make of me a caricature, never use my name in vain.

Totalitarianism operationalized, the rhetorical animal not merely circumcised but castrated. The very point of outlawing the as-if is, of course, that the Lord thereby guarantees that nothing new will ever issue from his subjects but always come from him and from him alone. With that purpose constantly in mind he then assigns every phenomenon or sense-thing its rightful place, all social relations between people automatically turned into material relations between things. Countless slaves in the treadmills, none on the chariot. The censor at work.

Surrounding the island of truth and power is a wide and stormy ocean, the true home of illusion. At the same time forbidding and tempting, this vast unknown is the domain of the noumena or thought-things, unapproachable in their own right. Nameable only in the islanders' own language, this ambiguous area is called

*a*

not because the metaphorical fog banks are *a,* but because the semiotic animal would not be what it is if it did not call it something. The very purpose of the baptizing ceremony is in fact to function as a pair of forceps, the name itself a conceptual tool by which the ontological magician can handle nongraspable noumena *as if they were* graspable phenomena.

The northern part of the island consists of a low-lying cape with sandy beaches, a tourist paradise developed and controlled by the consortium of Lakoff & Associates, Inc.[56] The leading idea of this group is that language is essentially metaphorical and that our unconscious therefore is structured like a body; *without body, no thought,* such is the slogan by which the Californians are trying to perform the same upside-down operation on Immanuel Kant as Karl Marx once did on G. W. F. Hegel. And yet, despite the fact that the word "mapping" appears on virtually every page of the voluminous *Philosophy in the Flesh,* the taken-for-granted makes it conspicuously absent from the book's extensive index. The same holds for the present tome as well, the terms "mapping," "power," and the LORD too frequent to list.

For these reasons of (un)conscious legitimation, the Lakoff Cape would be well worth a study of its own. For us, however, it seems even more rewarding to explore the coastal area in the southwestern part of the island, a precipitant landscape which looks like the cliffs of Dover, albeit with a geology of firm granite rather than vaporous limestone. Cut into these steep and inaccessible formations are a number of caves, most of them nesting areas of the albatross, some inhabited by a group of autistic solipsists. Since these outcasts are more interested in telling the truth than in being believed, they call themselves

$$a = a$$

a name which they synecdochically attach to their region as well.

Since the cave-dwellers find socializing exceedingly difficult, the only thing to tie them together is a deep sense of rejection, a kind of hatred which is directed less against the self-acclaimed despot and more against his self-serving lackeys. The most eloquent among them, those who are open to abstract thought, focus their opposition not on the outward signs of power but on the third paragraph of the Constitutional Law, the codified set of beliefs through which the Emperor legitimates his naked behavior. Here, then, is the thesis about the necessary unity of consciousness in another form:

> Remember that you are nothing but a cog in the LORD's machinery. Never question the righteousness of the collective, always remain convinced that the individual is wrong. I am your LORD, the spiritualized embodiment of your taken-for-granted, the pivot of your world.

Whether by force or by choice the cave-dwellers occupy the inaccessible boundary zone between the solid phenomena of the island and the fleeting noumena of the sea. Most of them would rather die than submit, rather shut up than be silenced.

Given this background, the mission of our secret agent turned land-surveyor can now be clearly stated: explore the outer limits of the island, draw a map thereof including the in-between zone of the tidal area. During that work the mapper will quickly discover that without the assistance of the cave-dwellers he will never be able to capture in the same glance both the a posteriori phenomena of the island and the a priori noumena of the ocean; like Charles Mason and Jeremiah Dixon before him, the critical cartographer must seek the help of the exclusives, informative informants in every cave. Only in that way can he hope to stake the boundary between Plato's two realms of cognition, the line which nowadays is better known as Ferdinand de Saussure's Bar or as Samuel Beckett's Tympanum, the thin foil which is one with the mapper's mappa.

*

**The typical cave-dweller** is a dedicated artist. As such (s)he is driven by the idea of simultaneously grasping the phenomena of the five senses and the noumena of the sixth, the organ of the eye and the meaning of the glance.

In reality a mission impossible, a task which Paul Cézanne once outlined in his description of how a tube of readymade paint may be turned into the revelations of a Swedenborgian angel:

> At times I conceive of colours as vast, *noumenal entities,* ideas with a physical presence, creatures of pure reason, with whom we can enter into relations. Nature is not an affair of the surface; it is in depth. Colours express the depth of the surface. They arise from the roots of the world. They are its life, the life of ideas.[57]

And in that conjunction the seemingly different approaches of Francis Bacon and Mark Rothko, two of Cézanne's outstanding descendants, come readily to mind.[58]

Imagine, therefore, that one fine morning these equally dedicated individuals crawl out of their caves, reach for the ladders and climb to the cliff's edge. Once there they set up their easels, Rothko turned towards the mist-enveloped sea, Bacon looking the other way. Drunk on the spirits of the Saussurean Bar both of them share the same dream of painting what it means to be human. However, as their viewing angles are radically different, what they see is radically different too; while Bacon's eyes are drawn to the naked bodies of the islanders, les boudins à la Mésopotamie, Rothko's mind is spellbound by the fog banks and the melting icebergs, visible signs of the invisible.

From these diverse vantage-points it was possible for Bacon to produce some of art history's most brutal paintings of human flesh, violence and deprivation, for Rothko to create what to me are the most sensitive and moving color studies ever made—while to the former the world was cruelty and pain, to the latter it was nothing but mute presence. With Bacon a pope in a cage or a lover carcassed on a butcher's hook, Edward Munch's *Scream* at a higher pitch, Titian's *Flaying* a cut deeper, always a body trying to escape from itself through itself. With Rothko infra-thin layers of red for the happy days, the dark grays for the deepest depressions, in both cases a godlike breath breathed onto a non-prepared canvas, the well-cleaned brush as soft as silk, the work of art always an instance of plasticity. The cruelty of facts versus the tender touch, the horrible versus the sublime. And yet, to both masters, as to any great artist, the subject of their painting was always the actual *paint*ing itself, the fashioning of a sort of liquid and nameless geometry; in Rothko's own words, "a painting is a statement of the artist's notions of reality in the terms of plastic speech. In that sense the painter must be likened to the philosopher rather than to the scientist."[59]

So unyielding was the commitment of the two contemporaries (Bacon 1909–92, Rothko 1903–72) that it led both of them to the suicides which could be predicted, equally sad albeit with a crucial difference. For it now turns

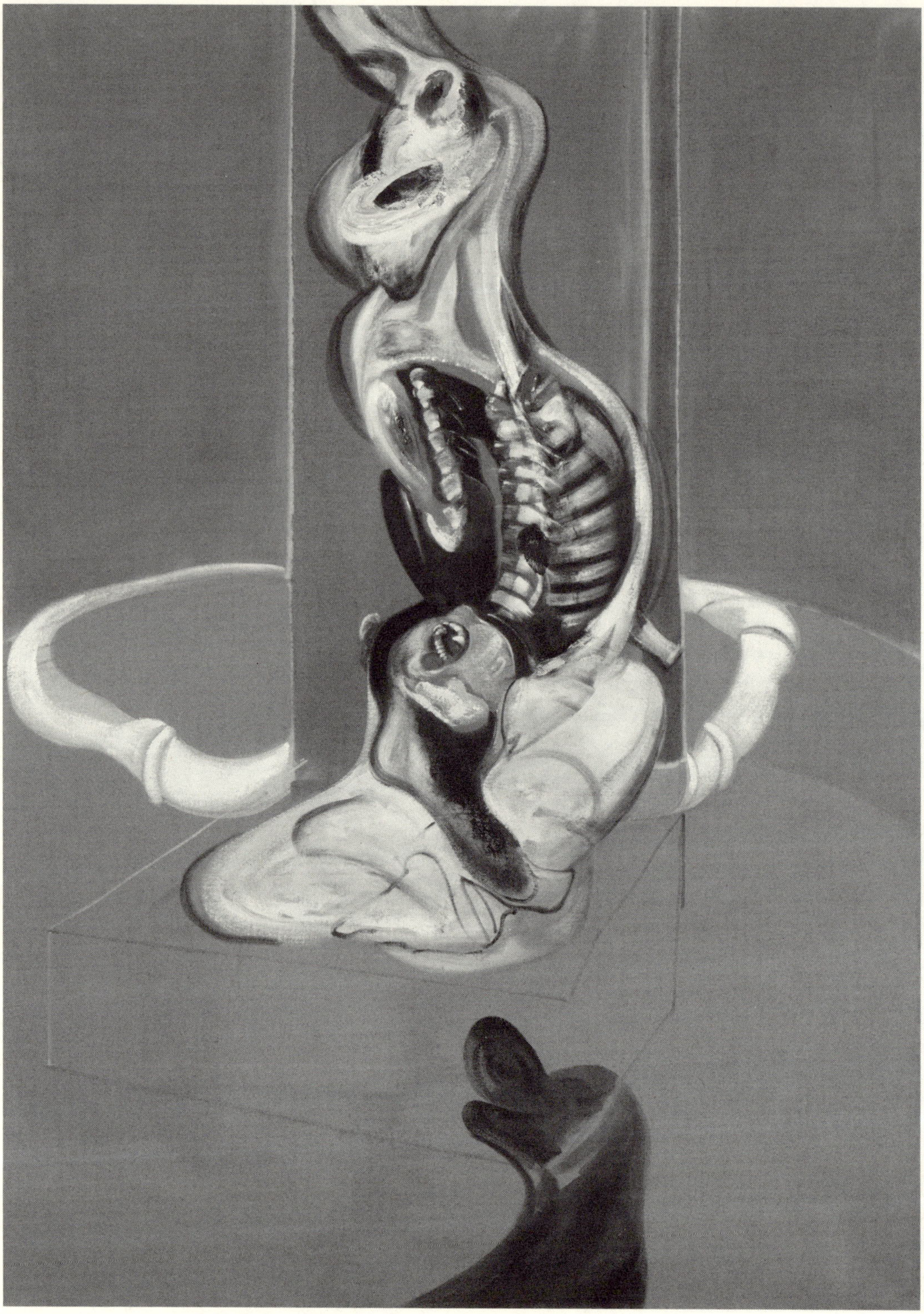

Francis Bacon, *Three Studies for a Crucifixion.* Third panel. 1962. Oil with sand on canvas, 198 × 145 cm. The Solomon R. Guggenheim Museum, New York. 

Mark Rothko, *Untitled (Black on Gray)*. 1970. Acrylic on canvas, 203.3 × 175.5 cm. The Solomon R. Guggenheim Museum, New York: Gift from the Mark Rothko Foundation, Inc. 

out that with Bacon the immediate victim was not the explorer himself but his friend George Dyer, the violent drug addict who killed himself in a Paris hotel room two nights prior to the opening of Bacon's 1971 retrospective at the Grand Palais, the most significant exhibition he had ever had. For Rothko, on the other hand, there was never a substitute, no ram caught with its horns in a thicket, no angel descending from the sky.

In my imagination the two self-killings come together in a diptych of the second commandment, for whereas the atheist Englishman naturally assigned descriptive titles to his figures of crucifixion, the Russian Jew called his best canvases nothing but *Untitled.* The heritage of Odysseus and Abr(ah)am in another guise, the lover's name no secret worth hiding, the LORD's identity an unmentionable mystery which even the atheist knew may be neither said nor shown. In the pregnant words of Bacon himself: "Images just drop in as if they were handed down to me. Really, I think of myself as a maker of images. The image matters more than the beauty of the paint. [However], if you can make the image with the mystery of the paint, so much the better."[60]

In his best moments Mark Rothko managed to do exactly that, to make the image with the mystery of the paint, in that respect similar to Cézanne himself. And not only was Rothko the better for it, but he paid a horrible price for the blessing. Because I, the LORD, the God of thy immigrant parents, am a jealous God, visiting the iniquities of the fathers unto the third and fourth generations. But beware, Dear Critic, and always remember—for thy own sake—that the Almighty will not hold him guiltless that taketh his name in vain.

Two remarkable explorations of the lines of power and the limits of language, Francis Bacon annihilating the cave-dweller's importance as an animal creature, Mark Rothko elevating his own worth as a sensitive intelligence, large, vibrant and iconic. The screen of the cartographer's mappa unveiled, for every serious artist is obsessed with the idea of making his own self-portrait, thereby manifesting what the naked eye can not see.

*

**The lines of power and limits of language** come together in the cartographer's scale, a device which in my rendering is patterned after the Divided Line of the *Republic,* its fix-point lodged in Plato's boundary between invisible forms and visible objects, the *agora* of Greek thought, the Square of René Magritte's *ceci.* This boundary-line is the magic wand by which word turns to flesh and everything solid melts into air, the freezing point of Celsius' zero metaphorized.

Therefore, once we return to the Kantian Island of Truth and Power, we find that stationed in the watch-tower of the Ruler's palace is a cadre of cadets. Since these underlings do not know where the power they are set to guard is actually hiding, they devote their energy to ensuring that the paragraphs of the Constitutional Law are honored and obeyed. We, however, are better informed and can therefore place on the map a childish *X*, a sign which points the treasure-hunter to a well-appointed apartment located in the castle basement. It is there, in a velvet-clad copulation chamber, that the semiotic and political animals come together to con-verse, literally "to turn oneself into another." It is in this incestual room that the likes of you and me are first conceived and then brought to life.

Leading out of the copulation chamber are two doors made of cedar wood, both exits guarded by some trusted eunuchs. Hidden behind one of them lie the archives of the Royal Map Room, an underground intelligence center inundated by the smell of as-if, its shelves filled with technical information about the cartographers' fix-points, scales and mappae, including the architectural drawings of the castle, the wall and the moat. And tucked away in the vast library, unknown to anyone but the few, is a rare volume entitled *Abysmal*.

The other door of the copulation room leads to a Chapel, the holiest place on the island. In that sanctuary, open day and night, a choir keeps repeating that it is you, you, you who are the one, only you beneath the moon and under the sun. Not, however, because you happen to be anything special, but because the thesis about the unity of consciousness connects your person with the person of everyone else. Woven into the lyrics are echoes of Thomas Hobbes, the power analyst who long before the Königsberger understood that

> whosoever looketh into himself, and considereth what he doth, when he does *think, opine, reason, hope, feare, &c*, and upon what grounds; he shall thereby read and know, what are the thoughts, and Passions of all other men, upon the like occasions. I say the similitude of *Passions,* which are the same in all men, *desire, feare, hope &c;* not the similitude of *the objects* of Passions, which are the things *desired, feared, hoped, &c.* . . . [And] when I shall have set down my own reading orderly, and perspicuously, the pains left another, will be onely to consider, if he also find not the same in himself. For this kind of Doctrine, admitteth no other Demonstration.[61]

A universe of human relations deduced from the principle of self-preservation, such is the Hobbesian commonwealth. Sadly outdated, though, not the least because to Hobbes the thought of suicide bombing was too inhuman even to imagine. The frontispiece of his book nevertheless remains a poignant illustration of the structure of power, for in this Arcimboldo-like engraving the sovereign's body is constructed out of the many bodies of his subjects, the ruler's head adorned with the king's crown, his right hand holding a sword, his left hand a bishop's staff. What a magnificent illustration of the fact that to govern is to cause to believe. Another map of the Island of Truth-and-Power.

The strength of the equilateral triangle epitomized, one of the most

Manipulated frontispiece of Thomas Hobbes's *Leviathan*.

power-filled images of power ever imagined. But the biblical Leviathan—Hobbes's vision of power—was itself a monster "with a heart as firm as stone, yea, as hard as the nether millstone. . . . Upon earth there is not his like, who is made without fear."[62] Yet, the most striking difference between Hobbes and Kant is that in the commonwealth of the former the sovereign is a chimera composed of the likes of you and me, while on the island of the latter the inhabitants are human beings exactly because we share our consciousness with everyone else.

*

**In the deep vaults of the castle** the critic of cartography gradually comes to realize that the Kantian conception of truth is at the same time a celebration of a reasoning mode on its deathbed and an anticipation of something not yet imagined: a story of deception transformed into a picture of adventures which we can never leave and never bring to an end; a foretaste of what it means to talk about cartographical reason in the idiom of cartographical reason; a preliminary preface to a fourth Critique, the particular mode of thought-and-action which Kant himself lived and therefore could neither think nor act.

For the same reasons of self-reference it is impossible to say anything definite about the present, even less about the future. Yet one cannot help but be impressed by the protesters who never yield, those who refuse to be caught in the ruler's coordinate net, to be weighed on the sovereign's scale, to be screened by the land-surveyor's mappa. Especially insulted are those rebels who the Almighty continues to call *b* even though they themselves know, beyond a doubt, that they are not *b* at all, but something entirely different. Perhaps *c*, perhaps *d*, perhaps *e*, perhaps something that may be neither grasped nor mentioned.

Not so strange, therefore, that these late-born Ishmaels often dream of a life at sea, maybe remembering that their Melvillean namesake was the only one to survive Moby-Dick's revenge, the imagination that floated away in a coffin destined for another future than Captain Ahab and the rest of his crew, those unfortunate believers who were pulled down into the depths of the unknown. To the critic of cartographical reason the message of Melville's novel is in fact unequivocal: of the high seas there can be no maps, because on the high seas there is no solid ground on which to settle, no fix-points fixed, no light-houses lit. To the eyes of the helmsman the scales are wavy, the mappae fluid.

And that last remark explains why no serious theologian believes that Jesus walked on water. Even more importantly, Melville's masterpiece—through its own existence—proves that the waves are ruled not by the guns of Great Britannia but by the spirits of human imagination. With the classical myths all but gone, the (post)modern artist is reduced to being himself, sheer existence his only belief system. Thomas Pynchon, once again.

# COLLATION

# MISSION IMPOSSIBLE

Vid vägs ände ser jag makten
och den liknar en lök
med överlappande ansikten
som lossnar ett efter ett . . .
TOMAS TRANSTRÖMER, "Fasader"[1]

**From the beginning** of this annotation I have pursued the question of what it means to be human, more exactly what it means to live in the oikumene. In search of an answer I have relied on methods from the epistemology of the extreme, a minimalist approach through which all arguments are pushed to their limits.

What I have found is that the limits of the oikumene are one with the limits of my world, these limits themselves one with the limits of the languages in which the semiotic animal thinks-and-acts. In my self-appointed role as a (post)modern land-surveyor I dream of mapping the outer reaches of that universe, a world which fills the void between the five senses of the body and the sixth sense of culture. Approaching the abyss from a range of directions, I have tried to keep the cartographic fix-points of Abr(ah)am, Moses, Plato and Kant constantly in mind.

The various raids into the unknown have all taken off from the legendary Bar de Saussure, the Nietzschean hangout where not a member is sober. In our capacities as semiotic animals it is in this establishment that we have our homes and lead our lives, our thoughts-and-actions governed by mimetic desires which we can never leave and never bring to an end. Contrary to Ludwig Wittgenstein's stance that *Wovon man nicht sprechen kann, darüber muss man schweigen,* I have long been driven by the motto *Wovon man nicht schweigen kann, darüber muss man sprechen.*

At the center of the Saussurean world lies the dematerialized point of tautology, at the periphery the moving horizons of contradiction. Approaching but never transcending these self-referential marks of silence is the space of meaningful translation, hence of proper names and definite descriptions, truth and trust, knowing and believing. Viewed in that light, the oikumene is in effect an as-if world in which the unconscious is structured like a language, at the same time real, imaginary and symbolic. Beyond the peripheries, on the other hand, lies the vastness of what can be neither said nor shown, by definition unreachable on its own terms. This is the land of the anoikumene, the realm of pure imagination.

The boundary between the oikumene and the anoikumene can be reached only from the oikumene side, only from the interior of the Saussurean Bar. Although the curious investigator may well look out of the Wittgensteinian window and catch a glimpse of what (s)he never saw before, (s)he can never escape from the room of her own, never violate the rules of the game of dice, never fool the invisible croupier.

It is life in the Saussurean Bar that this book is about, for it is in the smoke-filled rooms of that establishment that we learn what it is to be human. It is there that we are taught obedience and predictability, it is there that we live and die.

*

**The limits of the oikumene** can be studied through experiments *with, on* and *in* the Saussurean Bar, invaluable insights gained in the process. The trick is to take the fraction line of the sign

$$\frac{s}{S}$$

literally, in other words to treat it as the line it really is:

---

Accordingly I now let the dividing/unifying divisor between the Signifier and the signified approach the extreme conditions of zero and infinity, in the first instance becoming minimally thin, in the second maximally thick. The paradigm of the first case is the speech act in which the differences between what I am saying and what I am doing are all but erased—when in the wedding ceremony I utter the words "I promise," I am not merely describing a state of affairs but performing an act, the world irrevocably changed in the process. The place of the performative speech act is consequently always located in the foil of Samuel Beckett's tympanum, the topos where the prepositions *of* and *in* are merged into the cartographer's fix-point of fix-points.

In contrast, when the fraction line is put under the microscope, it does not evaporate like phlogiston air but grows and solidifies into a surface. The resulting figure is itself delimited by two horizontal boundaries, effectively lines of silence, barricades so sturdy that no language can tear them down. On the outer sides of these lines are the fairy-lands of the beyond, a universe in which there is neither identity nor difference, hence no understanding either. Bound between these non-penetrable limits of language lies the realm of the oikumene itself, by definition homeland of the semiotic animal.

On the other shore of the non-navigable Ocean River is the anoikumene, mute, blind and invisible, in everything so utterly different that it is not even noticed, much less named and discussed. From the homeliness of the Saussurean Bar—an atmosphere permeated by our shared beliefs in the Kantian thesis about the necessary unity of consciousness—we can nevertheless imagine the boundaries of the *Human Territory* as two limits marked by the alternative non-signs of

$$\frac{s}{s}$$

and

$$\frac{S}{S}$$

The most courageous can now sense that beyond these silent lines there are, on the one hand, the noumena of the signifieds,

$$s$$

on the other, the phenomena of the Signifiers,

$$S$$

The former area is called the *Mindscape of Pure Meaning,* the latter the *Rockscape of Pure Materiality.* And while Mindscape is ruled by a congregation of unmentionable spirits, Rockscape is governed by a mélange of non-imaginable forms of matter.

Now, just as my image of the Kantian Island of Truth could be drawn as a map, so can my conception of the oikumene. Two parallel lines, the Human Territory located in the abyss in-between, the utterly different relegated to the other shore of the Ocean River. In its extreme minimalism an image of incredible richness. A base map to be explored.

*

**The parallel lines** of the Human Territory reflect their own historicity. Over no issue have more people killed and been killed, for at stake is the question of what it means to be human.

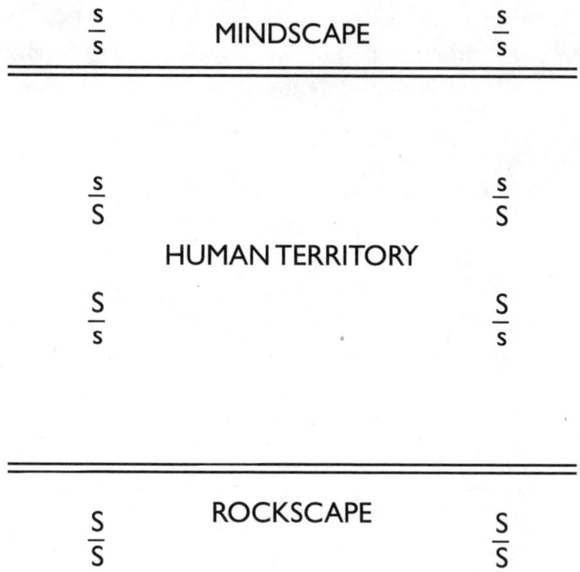


Map of the Human Territory.

For as long as we know—perhaps since the foundation of the world—the semiotic animal has tried to define who and what it is. In turn, this struggle with definitions has forced us to confront the two types of silence to which I have repeatedly alluded, one being the muteness of pure spirituality,

$$\frac{s}{s}$$

the other the stuttering noise of pure physicality,

$$\frac{S}{S}$$

Living on the other side of the former limit are the gods of Mindscape, creatures of imagination so extremely meaningful that they lack all forms of signification. On the other side of the latter is instead a land strewn by the material stuff of Rockscape, things so extremely physical that they offer no associations whatever. In the first case a meaning searching for its ethical expression, in the second an expression searching for its aesthetic meaning. And yet, in spite of (perhaps because of) their ontological differences, gods and rocks often come together in the same epistemological pursuit, a chiastic process in which the practices of reification and deification no longer can be separated.

To be human is in this perspective to be involved in a violent two-front war in which every battle boils down to a struggle over ontological bridgeheads and epistemological supply lines. The objective of our own mission impossible is therefore to trace the shifting boundaries between the familiar oikumene, on the one hand, and the alien anoikumene, on the other, the former area inhabited by people, the latter by gods or beasts. While at the godly front man struggles with non-informative truths of tautology, the enemies on the monstrous side are informative lies of contradiction.

Over time the two boundaries have moved back and forth, but in general the humans have come to occupy more and more of what once belonged to the aliens. As a consequence there is a history to the history of Being, a development which has gradually undermined the conscious belief in authoritarian structures and strengthened the unconscious acceptance of the taken-for-granted; as recalled, already Genesis chronicled how the power of the Amighty has diminished over time such that in the beginning it was Abram who belonged to the LORD, while at the end it was the LORD who belonged to Abraham. This trend has then continued and towards the end of the nineteenth century Friedrich Nietzsche could finally declare that God is dead, indeed that it was not man who had been created in the image of the LORD but exactly the reverse. Not as novel as it may appear, of course, for already Marduk defined man as a boudin à la Mésopotamie, a blood sausage explicitly invented for the enjoyment of the mighty. But how old shall a fertilized egg be for the abortionist to commit murder and how young shall it be for its removal to be considered an act of liberation? And for how long had Lenin's eggs been hatched before he stirred them into his omelet?

*

**The history of Being** is the history of how the boundaries of the oikumene have fluctuated over time and space.

As a way of illustrating these processes of gerrymandering I shall shortly present four (more correctly six) different maps, records from my own visits to some of the most decisive battle-sites in the perpetual war of what it means to be human, an exposition of cartographical reason in practice, a small step towards that *Critique of Cartographical Reason* which Immanuel Kant lived and therefore himself could never fully understand. Four phases in our ancestors' struggle with the utterly different, four delimitations of the human territory, four stories no story can beat:

- the Babylonian epic about Gilgamesh and Enkidu;
- the Jewish accounts of how Jacob and and Job wrestled with the LORD;
- the Greek tragedy in which King Oedipus found himself;
- the Byzantine creation of Jesus Christ.

POWER is the name of that ancient game, a roulette in which the balls keep rolling, especially the term "power" itself. Therefore, in advance of the atlas to come, a brief excursus into the grammar of this most pervasive of human realities, its faces like the layers of an onion, Russian dolls playing with each other.

*

**The word "power"** is a sly and evasive creature, a reality which easily escapes both our attention and our attempts to slay it, its major strategy to move capriciously about and to hide in the labyrinth of speech. Whenever attacked, its first line of defense is to turn into the chameleon of a gun, a boot or a fist, to pretend that it is a concrete noun, a graspable matter of sticks and stones that break my bones. At closer inspection, though, what initially looked like a material thing magically changes into a social relation, no longer the concrete nouns of the torture chamber but the abstract nouns of love and hate, pride and humiliation, fear and revolt. The play of phenomena and noumena, Signifiers and signifieds, does in that context boil down to a play of different noun-types. No wonder that the *Critique of Pure Reason* retains its firm grip on our imagination.

The parading of ontological opposites is only on the surface, though, for once the nouns have been pushed to their limits, then power shows itself to be less about objects and more about actions, less a matter of what there is and more an issue of what one should do. By no coincidence the leading character in Vico's rhetorical drama is the *factum verum*—the true is the made—a performance in which power plays the role of a verb, a mode of understanding which says that the only things we can know anything meaningful about are those we have constructed ourselves. And yet, although infinitives like "to kill" and "to love" do most of the talking, the actual plot is determined by a range of auxiliaries like "may," "must," "will" and "can." And so it is that in the hands of the apparatchik the magic wand of Kant's *Critique of Practical Reason* changes into the hard stick of modality.

But just as a thing might be beautiful or ugly, so a particular act might be right or wrong. And on that account the word "power" is neither a noun nor a verb, neither a brute fact nor a social relation, but an evaluating adjective or a modifying adverb. Such is the core of Kant's third *Critique of Judgment,* the sublime part of his oeuvre which presently commands the greatest interest.

Once the analysis has reached this elevated stage, then the cartographer's problem is no longer to determine *what* but *who* (s)he is—as everyone knows, the business of politics is to interfere in the relations between you and me, to (re)draw the boundary between us and them. Therefore, and since it is in the absolute position of a thing that the analyst finds the t(r)opology of being, the game of power (regardless of whether abstract or concrete) eventually turns into a game of positioning.

At issue in that play of choreography is not merely the question of where you and I are standing in relation to each other, but the issue of how we define the cardinal points of up and down, front and back, right and left—not merely the alternative coordinates of Vitruvian men but the ethical compass of Hobbes's Sovereign, sword in one hand, crosier in the other. And now, once their common understanding has reached that level, the prisoners of language are finally ready to accept the idea that grammar itself is a map of structural repression, hence a blueprint of individual escape routes as well. With that insight the world undergoes yet another ontological transformation, for the slaving hirelings are now free to turn the nouns into files for cutting their chains and the verbs into laws of freedom. The prepositions themselves, though, those most power-filled of all words, they legitimate their own standpoints by becoming unpredictable guardians, once dressed in the judge's cloak as likely to kill as to look the other way.

And for exactly these reasons of language we must now make a brief detour into the peculiarities of the Finnish language, to most people (including myself) a strange and bewildering idiom, not the least because it fails to classify nouns according to gender. It follows that for the Finnish speaker there is no linguistic way to distinguish a "he" from a "she"—for the gossipmonger a heaven of creative ambiguities, for the orthodox feminist a hell of confusing categories. When listening to a native Finn speaking Swedish or English, the unaware is easily lost.

And yet. Given the present context the most remarkable aspect of the Finnish tongue is not the absence of gender but the presence of cases—fifteen in total—the standard means for marking on the noun the relative position of subject and object. Finnish authors are consequently well equipped to do with their words what Paul Cézanne did with his palette and Filippo Brunelleschi achieved by moving his easel, five steps to the left. The classical example is Veijo Meri's untranslatable novel *Peiliin piirretty nainen,* in Swedish interpreted as *Kvinna i spegeln,* literally *Woman in the mirror.* Through an intricate play of word-endings the author is here able to specify directly, rather than circumstantially, whether the person in question is *at* the mirror, *on* the mirror, *in* the mirror, *in front of* or *out of* the mirror. Cubism by another route, the picturer of pipes turned into a piper of pictures.

In most Indo-European languages prepositions perform essentially the same functions as the cases do in the Uralic family. This fact in turn explains why in western thought-and-action no part of speech is more power-laden than the prepositions. As a consequence it cannot be overemphasized how power is always on the move, how it always enters the Saussurean Bar in borrowed clothes and under foreign aliases. Once on that stage of the political theater, however, the prepositions typically assume the leading roles of the drama, the Emperor standing naked before us shamelessly relieved of everything, including his Dionysian mask. Hierarchies consequently *do* matter, for even though the ontological magician performs his tricks in the tabernacle of a traveling circus, the actors' positions are minutely circumscribed, chalk-marks drawn on the floor; as already noted, there are crucial differences between being *at* a limit, *on* a limit, *in* a limit. And so it is that the prepositions in effect function as thoughts-and-actions ahead of themselves—in their dual roles they simultaneously define a noun and depend on a verb. The taken-for-granted in nuce, a Duchampian snow shovel in disguise.

Finally, it says a lot about the spatial structure of the taken-for-granted that for nonnative speakers the pre-positions are the hardest of all words to master, exactly because they are positions assumed in advance. At the same time it reveals much about the particularly English way of apprehending phenomena that the prepositions *in, at, on* and *to* are used in virtually the same ways now as they were in (or is it *at*) the beginning of the sixteenth century. When the chips are down, it is through our use of prepositions—more technically through our choice of fix-points— that the taken-for-granted is formed and the oikumene delimited.

*

**The mapmaker's question** poses itself: Why and how has the territory of the oikumene expanded in some directions and not in others? Why do we now know so much more than the Greeks about the material things of physics and medicine and perhaps less about the social relations of love and politics?

The answer is far from clear, partly because it belongs to the mist-enveloped regions of religion, mainly because our western ancestors have invested so much trust in the rhetorics of logic and geometry that we, their descendants, have long forgotten that logic and geometry themselves are nothing but specialized forms of rhetoric. Aristotle was consequently right when he claimed that literature is a deeper and more philosophical discipline than history, for whereas literature is free to imagine what *might* (have) happen(ed), history is limited to recording what *has* happened. As indicated by the many quotations of the present text, literature is essentially a field of (re)cognition. But the greater the artistry, the greater the inwardness and the more convincing the performative.

And now, finally, it can be said: the limits of Europe are themselves approaching the silent lines of representation. In the east, marking the boundary towards the Orthodox, are the icons; in the south, facing the lands of the Muslims, are the stylized octagons. While the former are staked along a winding line from Karelia in the north to Constantinople in the south, the latter cast their shadows across the Iberian peninsula, the city of Toledo the pivot of transition.

*

**As Plato's Sun is setting** the conventional lines of projection are fading away, the reflecting cave walls loosing their luster. The current truth is in fact that the fix-points, scales and mappae of cartographic reason have lost much of the power they once had. Immersed in a world which is neither solid nor stable we are even beginning to suspect that the excluded of the excluded middle might have escaped from the Renaissance lines of modernism and taken refuge in the Baroque folds of postmodernism. At any rate, it is generally agreed that no universe is a tabula rasa but the accumulated traces of differences deferred, not a blank sheet of paper but the wrinkled face of an old and beautiful lady. When pushed, the Human Being is always a being in-between, and that is regardless of whether the rhetorical blessing takes the form of a Q.E.D., a Q.E.F. or a Q.E.I.

And so it is that the limits of my world may lie less in the limits of my language and more in the limits of my imagination. Perhaps it may even be argued that the Greeks invented the concept of the anoikumene as a prophylactic against the madness of horror vacui, the particular form of the human condition which to them was more feared than anything else. Perhaps they even conceived of the oikumene as an asylum for the reductio ad absurdum, the foundation of a Panopticon whose very purpose it is to turn internees into obedient and predictable wardens.

With these comments constantly in mind we are finally ready to move into the second half of the present volume, an atlas of four maps and two requiems, a set of historical documents which by yet another vicus of recirculation bring us back to what I take to be the most pivotal battles in the unending struggle over ontological fix-points, epistemological sightlines and metaphysical projection screens.

# ATLAS

# URUK

Shall I not be like him?

**The first map** of our atlas reaches back to the oldest masterpiece of world literature, the epic poem of *Gilgamesh,* an outstanding study of power, fame, friendship and the fear of death; a remarkable attempt to determine what it means to be human; in the words of the expert, a story of growing up;[1] an existential drama of lasting impact, cartographical reason in practice; an (auto)biographical study of how a violent adolescent matures into a reconciled man.

Already in the prologue the poet stresses that even though Gilgamesh's fame is closely tied to the wall he had built around the city, his greatness is anchored less in his glorious achievements than in his tragic failures. The ultimate fix-points of the mapping expeditions he is about to undertake are well defined, the city of Uruk at the same time both the origin and the destination of his search. Therefore,

> Go up on to the wall of Uruk and walk around,
> survey the foundation platform, inspect the brickwork!
> (See) if its brickwork is not kiln-fired brick
> and if the Seven Sages did not lay its foundations!
> [*Find*] the tablet-box of cedar,
> [*release*] its clasps of bronze!
> [*Open*] the lid of its secret,
> [*lift*] *up* the tablet of lapis lazuli and read out
> all the misfortunes, all that Gilgamesh went through!
> (SB I, 18–21, 24–28)[2]

This is the point that the secret agent marks on his map, the place where the treasure is hidden.

In total the poem consists of approximately 3,000 lines which are fairly equally divided between twelve clay tablets; some rather say eleven plus an attachment. According to legend its hero lived some time between 2800 and 2700 BCE, the fifth king of the first dynasty of the city-state of Uruk, his rule extending over 126 years. His mother was the minor goddess Ninsun ("Lady Wild Cow"), his father a human called Lugalbanda ("Little Lord"). In the afterlife Gilgamesh served as a judge in the Netherworld, his Sumerian name usually translated as "The Old Man Is A Young Man."

As with the *Enuma elish* the oldest parts of *Gilgamesh* go back about four millennia, none of it materially preserved. However, at least four copies of a much later version are believed to have been in the royal libraries of Nineveh when that city was sacked in 612 BCE, everything destroyed, the tablets broken, scattered and buried in the rubble. All forgotten until 1850 when a group of archeologists unearthed a large number of pieces, most of them now in the possession of the British Museum. Twelve years later the world was stunned by the news that among these remnants a man called George Smith had come across a tablet with a text that shared many details with the biblical story of the Deluge. As the anecdotes have it, this highly nervous and sensitive man got so excited by his discovery that he "[set] the tablet on the table, he jumped up and down and rushed about the room in a great state of excitement, and, to the astonishment of those present, began to undress himself!"[3] From then on the advancement of Assyriology has been nothing but astonishing, most recently spearheaded by people like Jean Bottéro in Paris, Jeffrey Tigay in Philadelphia, Andrew George in London, Simo Parpola in Helsinki.[4]

Informed scholars agree that written literature was created in Mesopotamia already around 2600 BCE, primarily in Sumerian, a language now dead and without known relatives, gradually also in cuneiform Akkadian. What is presently thought to have gone into the Old Babylonian Gilgamesh (OB) must therefore have built on stories which were originally composed in the Sumerian idiom, some parts as early as 2100 BCE. In contrast, the Standard Babylonian Version (SB) was already from the outset written in Akkadian, almost certainly under the editorship of the learned priest Sin-liqe-unnin, who lived some time in the thirteenth to eleventh centuries. Most surviving copies stem from the seventh century BCE, but even when all known fragments are pieced together only about four-fifths of the text is complete. However, this situation is constantly changing as new finds are being made. In addition, in places where the Standard Version shows large lacunae or the signs are illegible, the Old Babylonian Versions can sometimes fill the gaps. It says much about the concept of "progress" that the Old Version (OB) is generally considered more powerful and less repetitious than the new (SB).

*

**Much is changing** and to us it is especially interesting that whereas some of the oldest translations render the poem's first lines as

> [He who] saw everything [within the conf[i]nes(?) of the land;
> [He who] knew [all things and was (?) in] everything[5]

the same passage now tends to be given as

> [He who saw the Deep, the] foundation of the country,
> [who knew . . . ,] was wise in everything!
> (SB I, 1–2)[6]

From other parts of the poem we learn that "The Deep" is in fact nothing but another term for the realm of Apsu, the non-classifiable abyss which in *Enuma elish* was the home of Power itself. And a figure of power is exactly what Gilgamesh is. For

> Who is there that can be compared with him in kingly status,
> and can say like Gilgamesh, "It is I am the king"?
> Gilgamesh was his name from the day he was born,
> two-thirds of him god but a third of him human.
> (SB I, 45–48)

It was Belet-ili, the Mother Goddess, who had shaped his body, Nuddimund, the Man-Fashioner, who had brought his form to perfection, by human standards very handsome.[7] What then follows is the story of how this preeminent bastard gradually grows less godlike and more humanlike, albeit a human of a very special kind; even after his misfortunes—and so much better for them—he is still a king.

Predictably unpredictable, the young Gilgamesh behaves with fierce arrogance, well known for forcing his subjects to work hard on his building projects, for filling their time with military drill and for roughing in the hockey games, perhaps for misusing his right to spend the first night with every newly wed bride. Everything told, people thought that he was not performing his royal duties properly, a circumstance which led them to complain to the gods, who in their wisdom sensed that his bad behavior was somehow related to the loneliness which came with his superiority.

To put things right the Mother Goddess was once again called into service, this time to deliver a complement, indeed a rival, to her first creation. Hence she washed her hands, took a pinch of clay, cast it on the ground and formed a creature called Enkidu, by all accounts more of a wild animal than of a human being:

All his body is matted with hair,
he is adorned with tresses like a woman:
The locks of his hair grow as thickly as Nissaba's,
he knows not at all a people nor even a country.
He was clad in a garment like Shakkan's,[8]
feeding on grass with the very gazelles.
*Jostling* at the water-hole with the herd,
he enjoyed the water with the animals.
(SB I, 105–112)

Now a trapper thrice comes upon this strange creature when it is drinking with the gazelles at the water-hole, a sight which makes the spying intruder so afraid that he freezes stiff. Following his father's advice he seeks the assistance of Shamhat, "The Joyous One," a harlot who works in the temple, priestess and prostitute in the same threshold-crossing figure. Together the two of them—trapper and harlot—go to the water-hole, where Shamhat is instructed to treat the savage-man to the work of a woman. This she does in such an explicit way that George Smith (the scholar who undressed himself in the British Museum) in his version of 1876 left the passage untranslated, while Alexander Heidel in 1949 gave it in Latin rather than English.[9] Fifty years later the mores had changed and Stephanie Dalley came closer to the words of the original:

Shamhat loosened her undergarments, opened her
legs and he took in her attractions.
She did not pull away. She took wind of him,
Spread open her garments, and he lay upon her.
She did for him, the primitive man, as women do.
His love-making he lavished upon her.
For six days and seven nights Enkidu was aroused
and poured himself into Shamhat.[10]

After this, sated with delight, Enkidu wants to return to his former life with the gazelles. However, as soon as the brutes hear, see and smell him, they run away, his own legs too weak for him to follow them. And exactly at that point of the story come some of the most pivotal lines of the entire epic:

Enkidu was diminished, his running was not as before,
but he had *reason,* he [was] wide of understanding.
He came back and sat down at the feet of the harlot,
watching the harlot, (observing) her features.
Then his ears heard what the [harlot] was speaking,
[as the harlot] said to him, to Enkidu:

"You are handsome, Enkidu, you are just like a god,
why do you roam the wild with the animals?
Come, I will lead you to Uruk-the-Sheepfold,
to the sacred temple, the dwelling of Anu and Ishtar,[11]
where Gilgamesh is perfect in strength,
and lords it over the menfolk like a wild bull."
(SB I, 201–12)

Thus she talked to the transformed. And what she said found favor with the savage, presumably because by diminishing his body she had enlarged his mind, giving him reason and making him wide of understanding. His position in the Territory of Humans was thereby drastically changed, for no longer was he relegated to roaming the wildlands close to the Rockscape of Pure Materiality, but he was free to head for an as yet undetermined topos closer to the center. Indeed he asked Shamhat to lead him to the city, where he could seek a friend and where he could challenge the power of the arrogant Gilgamesh. Enkidu's new-found goal was to change the order of things and show that it was he, the one born in the wild, not the king, who was the mightiest in the land.

Meanwhile, back in the palace, Gilgamesh has three dreams, a form of godly messages so strange that he must ask his mother to interpret them for him.[12] She does, saying that he will soon meet a man who she (the mother) will make his equal, a companion of great strength, a fellow to caress, embrace and love like a wife. But in untold ways also Shamhat learns of Gilgamesh's dreams, she tells Enkidu about them and the two make love together.

After this cleaning ritual the prostitute takes her lover's hand and leads him away as if he were a god. Half-way between the animals' waterhole and the glories of the city they reach a camp with a band of shepherds. Gathering around their rare visitors the herdsmen are impressed by Enkidu's tall stature, in everything so similar to that of Gilgamesh, their lord. When they offer him bread and ale, which he has neither seen nor tasted before, he looks intently but refuses to share them. The transgressive harlot once again interferes, this time saying,

"Eat the bread, Enkidu, the thing proper to life;
drink the ale, the lot of the land."
Enkidu ate the bread until he was sated,
He drank the ale, seven jugs (full).
His mood became free, he was singing,
His heart became merry and his face shone bright.
The barber treated his body so hairy,
He anointed himself with oil and became a man.
(OB II, col. iii, 96–108)

In my interpretation a picnic arranged by the waiters of the Bar de Saussure; a transition rite performed under open skies; an anointed savage turned into a man; the bread and the beer substituting for the grass he had grazed and the milk he had sucked. Yet there is a long way to go before he becomes privy to the cooked meat and the fermented wine, much alien territory left to cross. Not merely for Enkidu, though, but for the present-day reader whose only guide is a tablet so badly damaged that it is almost impossible to interpret. Still there is enough to indicate that when the shepherds slept, it was the former wild-man who performed their duties, expertly massacring the wolves and chasing away the lions.

When the text gets readable again Enkidu has not only reached the city but positioned himself in the doorway of a wedding house which Gilgamesh is about to enter, the young bride presumably huddling in a corner. But Enkidu, the newly born human, he heard the words that came from the mouth of the king of Uruk-Main-Street:

> "the 'people's net' will be open for the one who has the first pick.
> He will couple with the wife-to-be:
> he first of all, the bridegroom afterwards.
> By divine consent it is ordained;
> when his navel-cord was cut she was destined for him."
> At the fellow's words (Enkidu's) face turned pale.
> (OB II, col. v, 157–165)

The two took hold of each other, they joined for combat, they wrestled in the street. Bending their backs as bulls they smashed the door jambs and shook the walls. A fight of equals, yet at the end Enkidu submits, not, however, to Gilgamesh's physical strength but to his social authority. They kiss and they form a friendship so strong that they become inseparable until death eventually tears them apart.[13]

At this point the text once again fades away, and when it returns it is with a line which tells how Gilgamesh introduces his newfound friend to his mother. During that audience Enkidu's eyes fill with tears of emotion, for he now begins to understand what a long distance he has come. Across the steppe and into the palace, three stations along the way: the water-hole of the gazelles; the camp of the shepherds; the doorway of the wedding-house. All thanks to a harlot, by definition a taboo-ridden resident of the abysmal land in-between.

*

**Enkidu's eyes** were filled with tears, his heart was made to ache. Gilgamesh comforts him, and, perhaps as a way of distracting him, he suggests that the two of them should go together on an expedition to the Cedar Forest, a dangerous and faraway place, by some believed to be the dwelling of the gods themselves. Guarding this forest is a most fearsome monster called Humbaba, an ogre appointed to his task by the god Enlil, the divine ruler of the Earth. To be precise:

In order to keep the cedars safe,
  Enlil made it his destiny to be the terror of the people:
That journey [is not for the making,]
  [that man is not for the seeing.]
He who guards [the Forest of Cedar, his . . . are wide,]
Humbaba, his voice is the Deluge,
  his speech is fire, his breath is death.
He hears the forest's murmur for sixty leagues;
  who is there that would venture into his forest?
(SB II, 218a–224)

Egging each other on the two comrades have special weapons forged for the battle. The elders, though, warn that Gilgamesh is too young for the task, indeed that their king does not know what he is talking about. The bragger laughs them off but under pressure he agrees that Enkidu should be in the lead, because it is he who has crossed the steppe before and therefore knows the way. Even so they need a map to guide them and Gilgamesh says to his mate:

"Come, my friend, let us go to Egal-mah,
  into the presence of the great Queen Ninsun!
Ninsun is clever, she is wise, she knows everything,
  She will set in place for our feet tracks of (good) counsel."
(SB III, 15–18)

Into the presence of Ninsun, Gilgamesh's mother, they enter hand in hand telling her what they need. She first listens in sorrow, then climbs to the roof of her house from which she calls on Shamash, the Sun God and the patron of all travelers. To placate the wayfinder she scatters incense, lifts her arms and pleads that her son be looked after on his treacherous journey through the unknown:

"Why did you assign (and) inflict a restless spirit on [my] son Gilgamesh?
For now you have touched him and he will travel
  the distant path to where Humbaba is.
He will face a battle that he does not know,
  he will ride a route he does not know.
During the days that he travels there and back,
  until he reaches the Forest of Cedar,
until he slays the ferocious Humbaba,
  and annihilates from the land the Evil Thing that you hate.

. . . . . . . . . . . . . . . . . . . . . . . . . .

As for him, place him in the care of the watches of the night!"
(SB III, 46–54, 57)

*

**Place him in the care of the watches of the night**, let the stars show him the way just as we now know that they were to do for Gilgamesh's great followers, Pytheas of Massalia and Immanuel Kant of Königsberg foremost among them. Shamash complies and in the Sumerian poem *Bilgames and Huwawa*—a forerunner of *Gilgamesh*—he even provides the heroes with celestial guidance in the form of seven constellations.[14] In the Standard version they are armed with thirteen winds instead, weapons by which Gilgamesh will eventually destroy Humbaba, perhaps a parallel to what Marduk did with Tiamat. "Oh Shamash," Ninsun prays, "will you not let Gilgamesh share the heavens with you, will he not share the sceptre with the moon, will he not become wise with Ea of the Apsu!" (SB III, 102–104).

After this, and following several lacunae, the two heroes are off, Enkidu in front to show the way. At twenty leagues they broke bread and at thirty leagues they pitched camp, day after day doing their prayers and carefully observing the rituals, every step meticulously entered into their log book. Most importantly, Enkidu constructs a special house for the dream spirits, a blend of a tabernacle and an observatory, he himself lying down like a net in the doorway.

Inside this structure, curled up in a circle, Gilgamesh falls asleep but is soon awakened by a series of nightmares, images in which Humbaba's violence exceeds all limits, a crescendo of terror. Enkidu is immediately informed, but deaf to all forebodings he insists that there is nothing to fear and that they have an urgent mission to fulfill. On they go, and after much strenuous work, they finally arrive at the forest's edge. Once at that border towards the utterly different

They stood *marvelling* at the forest,
  observing the height of the cedars,
  observing the way into the forest.
Where Humbaba came and went there was a track,
  the paths were in good order and the way was well trodden.
They were gazing at the Cedar Mountain, the dwelling of the gods, the throne-dais of the goddesses,
  [on the] very face of the mountain the cedar was proffering in abundance,
  sweet was its shade, full of delight.
[All] tangled was the thorny undergrowth, the forest was a thick canopy.
(SB V, 1–9)

Large lacunae follow and the reader's mind is free to explore the invisible lands of imagination. Suddenly the two trespassers stand face-to-face with the guardian monster, whose first attack is to verbally insult them, initially scorning Gilgamesh for having an ignorant bumpkin as his adviser, then turning to Enkidu, "that spawn of a fish, who knew not his father, hatchling of terrapin and turtle, the [half-animal] who sucked not the milk of his mother" (SB V, 87–88). Enkidu, of course, gets very angry, insisting not only that they must slay the monster but that they must do so at once lest the gods find out about their plans and interfere against them.

Faced with this momentous hubris—in effect a serious insult to the gods—Humbaba has no choice but to plead for his life, indeed to offer himself as a slave. No mercy, though, and after a swift fight the ogre is killed, Enkidu pulling out its lungs and Gilgamesh cutting off its tusks, which are taken home and hung as trophies above his bed.

Their courage thus proven, the mates cut down a lofty cedar of which they make a mighty door that they intend to put in the temple of Enlil, an obvious attempt to placate the deities, who they realize they have angered. Little did they know of the long-range consequences of their intrusions into the godlands, however, and in triumph they returned to the city of Uruk, the center of human power.

*

**Once back in the city** Gilgamesh washed his matted hair, cleaned his equipment, threw away his dirty things and put on his crown. Watching from a distance is the lusty Ishtar, in the same figure the goddess of sex and the deity of Uruk. She looks covetously on his beauty in a scene that almost certainly predates the story about the hunter Acteaon spying on the bathing Artemis, Ishtar's close relative among the Greeks. Obsessed by desire she says:

> "Come, Gilgamesh, you be the bridegroom!
> Grant me your fruits, I insist!
> You shall be my husband and I will be your wife!
> Let me harness for you a chariot of lapis lazuli and gold,
> whose wheels are gold and whose horns are amber.
> You shall have in harness 'storm-lions,' huge mules.
> Come into our house with scents of cedar!
> When you come into our house,
> doorway and throne shall kiss your feet.
> Kings, courtiers and nobles shall be bowed down beneath you,
> they shall bring you tribute, [*all* the] produce of mountain and land.
> Your nanny-goats shall bear triplets and your ewes twins,
> your donkey foal under load shall outpace a mule."
> (SB VI, 7–19)

To these proposals Gilgamesh replies with scorning rejection, perhaps because he regrets his own behavior prior to Enkidu, perhaps because he knows that none of her previous lovers has endured for long. Indeed he launches into a long speech with many tie-ins to the story of how Acteaon is punished for his indiscretion by being turned into a stag, chased and devoured by his own hounds. These were his words, before and thereafter repeated in many quarrels:

> "You loved the shepherd, the grazier, the herdsman,
> who regularly piled up for you (bread baked) in embers,
> slaughtering kids for you every day.
> You struck him and turned him into a wolf,
> so his own shepherd boys drive him away,
> and his dogs take bites at his thighs."
> (SB VI, 58–63)

All this makes Ishtar so furious that in tears she begs her father, Anu, to interfere. He, however, knows his daughter and retorts that Gilgamesh's accusations are well taken. As could be predicted this response only increases her anger, and she asks that she be given the Bull of Heaven, a force so fierce that it will surely do away with Gilgamesh once and for all. Not merely a request, though, but a threat as well. These were her words:

> "O father, give me, please, the Bull of Heaven,
> that I may slay Gilgamesh in his dwelling.
> If you will not give me the Bull of Heaven,
> I shall smash the *underworld* together with its dwelling-place,
> I shall *raze* the nether regions *to the ground.*
> I shall bring up the dead to consume the living,
> I shall make the dead outnumber the living."
> (SB VI, 96–100)

No way out, not even for the great god Anu, who had no choice but to hand over to his daughter the nose-rope of the Bull of Heaven. The latter is indeed a frightening creature, an animal who snorted with such force that pits were opened in the ground, the first of them so large that a hundred men fell into it, the second capturing twice as many, the third swallowing none less than Enkidu, albeit only to his waist. Stirred into action the reformed savage sprang up, seized the bull by its horns, put his foot on its back and called for Gilgamesh. Like a skilled matador the latter pressed his dirk between the yoke of the horns and the slaughter spot. The brute fell to the ground and that was the end of the famous Bull of Heaven.

Enraged by the heroes' joint action Ishtar went up onto the city-wall, where she hopped and stamped like a spoiled teen-ager, cursing Gilgamesh for vilifying her and damning him for killing the bull. In response Enkidu tore the thigh off the carcass and hurled it at the goddess.[15] After this, in ill-boding language, Gilgamesh praised himself, well aware that through the killings of Humbaba and the Bull of Heaven he had penetrated deeply into the forbidden realm of the gods, reaching to a point far beyond the previous border of the Mindscape of Pure Meaning. These were his self-aggrandizing words and this was the crowd's reply:

"Who is the finest among men?
Who is the most glorious of fellows?"

"Gilgamesh is the finest among men!
Gilgamesh is the most glorious of fellows!"
(SB VI, 172–75)

The katharsis is remarkable, for whereas at the beginning of the epic the people of Uruk had not liked their ruler, now they greet him as their savior. A banquet follows and early next morning Enkidu has a dream. What he sees are the great gods in council gathered to discuss the recent invasions of their territory, a kind of war cabinet called together to decide how the transgressors should be dealt with.

*

**The great gods in council** agreed that one of the two trespassers must die, and, after some deliberation, they decided that it should be Enkidu. The action presumably proceeds and when the story picks up again, the reformed wild-man is already on his death-bed, speaking in the delirium of the deranged. In Job-like terminology he curses first the trapper who had found him, then the harlot, who had taken away his innocence by teaching him love, effectively for making him a man and for leading him to the civilized city. For these unfairnesses Enkidu is rightly rebuked by Shamhash, the patron of travelers, who reminds him that were it not for the harlot he would never have met the fine Gilgamesh, never fallen into the arms of the devoted friend who in despair has promised to build him a golden statue and to lay him out on a bed of honor.

After this Enkidu has yet another dream, a nightmare in which he is taken to the Danteesque Netherworld, where he can see for himself the kind of existence that awaits him. His arms bound like the wings of a bird he is led to the house of darkness:

To the house which those who enter cannot leave,
on the journey whose way cannot be retraced;
to the house whose residents are deprived of light,
where dust is their sustenance, their food clay.
They are clad like birds in coats of feathers,

and they cannot see light but dwell in darkness.
On the door [and bolt the dust lies thick,]
on the House [(of Dust) a deathly quiet is poured.]
On the House of Dust that I entered,
I looked and (saw) the crowns stowed away:
there sat [kings], the crowned heads who had ruled the land since days of old,
who used to serve roasted meat [at the] *tables* of Anu and Enlil,
who used to serve baked (bread), to pour chilled water from skins.
(SB VII, 185–97)

In the Netherworld the symbols of power are obviously trampled on as if they were clay tablets strewn across the floor of a ransacked library. In agony Enkidu calls out that his gods have spurned him.

Throughout these ordeals Gilgamesh was constantly at Enkidu's side mourning the loss of his friend, a lament not easily surpassed. May every one and every thing join in the despair:

"O my friend, a mule on the run, donkey of the uplands, panther of the wild,
my friend Enkidu, a mule on the run, donkey of the uplands, panther of the wild!
We (it was) who joined forces and climbed the [uplands,]
seized the Bull of Heaven and [killed it,]
destroyed Humbaba, who [dwelt in the Cedar] Forest.
Now what sleep is it that has seized [you?]
You have become unconscious and cannot hear [me!]"
But he, he would not lift [his head;]
he felt his heart, but it was not beating any more.
He covered (his) friend, (veiling) his face like a bride,
circling around him like an eagle.
Like a lioness deprived of her cubs,
he kept turning about, this way and that.
(SB VIII, 50–62)

Through much damaged lines we are then told how Gilgamesh fashions a statue of his friend and how he spends seven days and nights praying that his mate will come alive again. Indeed it is not until maggots creep out of the corpse's nose that Gilgamesh accepts what has happened, lays his comrade to rest and finally understands that death is not an abstract idea but a concrete reality. And when the funeral is over—exactly how it is arranged we do not know, for out of approximately thirty lines only two words remain—he lets his hair grow, he takes off his royal garment and dresses in a lion's skin, he leaves the city and begins to roam the wild.

"Must I too die, just like my friend Enkidu?" Such was the question for which he set out on his last expedition. "Woe has entered my belly." Such was the mood that drove him on.

*

**Must I too die**, shall I too be like my mate? Given that I was born as two-thirds god and one-third man, will one-third of me disappear and two-thirds of me continue to live? What does it mean to be human?[16]

Where does Gilgamesh really belong? In the Territory of Humans, in the Mindscape of of Gods, in the Rockscape of Things? At the center of the Kantian Island of Truth, on the icebergs of the foggy Ocean, in the caves at the edge of the world? In Gilgamesh's own words:

"I shall die, and shall I not then be like Enkidu?
    Sorrow has entered my heart.
I became afraid of death, so go roaming the wild,
    to Uta-napishti, son of Ubar-Tutu,
    I am on the road and travelling swiftly.
I arrived one night at the mountain passes,
    I saw some lions and grew afraid.
I lifted my head, praying to Sin,[17]
    to [. . . , *the*] *light* of the gods, my supplications went:
  'O [Sin and . . . ,] keep me safe.'"
(SB IX, 3–12)

Gilgamesh awoke and realized that he had just had another dream. And in the presence of the moon he grew happy to be alive.

When the text gets readable again, Gilgamesh has come to a twin-mountain called Mashu, a landmark whose tops reach both up to the heavens and down to the netherworld, a most contradictory place where the sun both rises and sets. Guarding the gate of this mythical location is a contingent of scorpion-men so scary that when Gilgamesh sees them he covers his face in fear and dread. Yet he collects his wits and draws near their presence. Stunned by what he saw,

The scorpion-man called to his female:
  "He who has come to us, flesh of the gods is his body."
The scorpion-man's female answered him:
  "Two-thirds of him are god but a third of him is human."
(SB IX, 48–51)

The scorpion-man, despite his monstrous state a real practitioner of cartographical reason, then asked the key question:

"[*How did you come here,*] a far road?
[*how did you get*] *here,* into my presence?
[*How did you ford the many rivers*], whose crossing is perilous?
. . . . . . . . . . . . . . . . . . .
[. . . let me] learn [*of your journey.*]"
(SB IX, 54–56, 59)

After this query the text once more fades away but returns at the crucial moment when Gilgamesh reveals the true purpose of his dangerous mission:

"[I am seeking] the [road] of my forefather, Uta-napishti.
He who stood in the assembly of the gods, and [found life,]
of death and life [he will tell me the secret.]"
(SB IX, 75–77)

To this the scorpion-man responds that no man has ever been this far before and then proceeds to tell the explorer about the long journey through darkness that awaits him, a passage of the text that has been the subject of much exegesis;[18] perhaps Gilgamesh is engaged in a race against time, perhaps he tries to retrace the path of the sun, perhaps he imagines a journey towards the most remote east, perhaps, like Alexander the Great, he is obsessed by the desire of reaching the other side of the Ocean River.[19] At any rate, the trespasser eventually arrives at a magic garden which has much in common both with the biblical Eden in the east and with the Greek Hesperides in the west, both locations eventually of great interest to the makers of the medieval mappae mundi, the *Ebstorfer Karte* outstanding among them. As with Plato, the Sun serves as the explorer's fix-point of fix-points.

When Gilgamesh walks about this wondrous garden he senses that someone is watching him. As he comes closer he discovers that at the edge of the Ocean there lies a franchised tavern, a Saussurean Bar run by the ale-wife Shiduri. Like other gate-keepers this figure is a strange appearance, partly because she wears a veil even though ale-wives, like prostitutes, usually are not considered respectable enough to be thus adorned. Inspecting the wanderer, who approaches in his pelt, she notices that despite the godly flesh in his body there is sorrow in his heart. Not knowing what to expect, she gets afraid, bars the gate and climbs to the roof of the house. Thus insulted Gilgamesh threatens to break the door, but, as the experienced barmaid she is, Shiduri calms him down by getting him to brag about his heroic adventures with Humbaba and the Bull of Heaven. To prod him on, she asks:

"[why are your] cheeks [hollow,] your face sunken,
[your mood wretched,] your features wasted?

[(Why) is there sorrow] in your heart,
    your face like one [who has travelled a distant road?]
[(Why is it)] your face is burnt [by frost and sunshine,]
    [and] you roam the wild [got up like a lion?]"
(SB X, 40–45)

As is customary in the taverns, Gilgamesh now proceeds to tell her about his friend, who he loved so dearly and who is now turned to clay. Rhetorically he repeats the standing question: "Shall not I be like him and also lie down, never to rise again, through all eternity?" (SB X, 70–71).

Shiduri's fright now subdued, Gilgamesh reveals that the real purpose of his expedition is to reach the forbidden Netherlands, the place where he hopes to meet Uta-napishti, the only man who survived the Deluge and the only human that the gods have ever made immortal. More than anything else he wants to know first how Uta-napishti obtained his special status, then whether life in eternity is really to his liking. Therefore,

Gilgamesh spoke to her, the ale-wife:
    "Now, ale-wife, what is the road to Uta-napishti?
What is its landmark? Give it to me!
    Do give me its landmark!
If it may be done, I will cross the ocean!
    if it may be done, I will roam the wild!"
(SB X, 72–77)

The ale-wife answers that there has never been a way across, that since the days of old no one has ever come any farther than to her ale-house— "besides, Gilgamesh, once you have crossed the ocean, when you reach the Waters of Death, what will you do?" (SB X, 85–86). But immediately as these words have left her mouth she remembers Ur-shanabi, the boatman who long ago had ferried Uta-napishti across the Waters of Death to his final destination. He, if anyone, should know the landmarks.

And thus it is that Gilgamesh learns that in order to cross the ocean and deliver the supplies to his master, this Babylonian Charon has always used a pair of "Stone Ones," perhaps the words for the magic power of some amulets, perhaps the term for a nautical tool of unknown type, perhaps a way of denoting the crew. By mistake or ignorance, at any rate in a state of anger, Gilgamesh takes his axe and smashes the Stone Ones, an act which gets Ur-shanabi so upset that he warns the traveler that "your own hand, Gilgamesh, has prevented your crossing" (SB X, 156). However, once he has calmed down, he orders the axe-man to go to the forest and cut three hundred punting-poles, each five rods long, trimmed and furnished with a boss.

With this equipment readied, off they go. And on the third day, when they had traveled a month and a half's journey, they finally arrive at the

Waters of Death. Here they begin their punting, very careful never to touch the lethal waters with their hands, each pole discarded once it has been dipped into the poison. As could be expected they eventually run out of punting-poles and, to solve the problem, they make a sail out of Gilgamesh's garment, perhaps the first story of how the first sail was invented.

Standing on the other shore, Uta-napishti was watching. And:

> Talking to himself he [spoke] a word.
> He [was taking counsel] in his own mind:
> Why are the boat's [Stone Ones] smashed,
> and aboard [it] one who is not its master?
> He who comes is no man of mine,
> but on the right . . . [. . .]
> I am looking—he is no [man] of mine.
> I am looking—he is no [. . .]
> I am looking— [. . . . . .]
> (SB X, 187–93)

As the boat finally reaches the quay Gilgamesh goes ashore. There he is greeted by none less than Uta-napishti himself, who immediately asks who he is and what has made him travel such a long way, what has prompted him to pass over the arduous mountains, what has driven him to cross the ocean and to fill his sinews with pain. Gilgamesh's answer is that he has come because of his friend, a loved one who he has lost. The standing question is then repeated, repeated and repeated: "Shall not I be like him and also lie down, never to rise again, through all eternity?" (SB X, 247–48).

In return Gilgamesh receives a reply so long that it runs for about 245 lines, all of them variations on the theme that

> "[*You,*] you kept toiling sleepless (and) what did you get?
> You are exhausting (*yourself* with) ceaseless toil,
> you are filling your sinews with pain,
> bringing nearer the end of your life.
> Man is one whose progeny is snapped off like a reed in the canebrake:
> the comely young man, the pretty young woman,
> *all* [*too soon in*] their very [*prime*] death abducts (them).
> No one sees death,
> no one sees the face [of death,]
> no one [hears] the voice of death:
> (yet) savage death is the one who hacks man down.
> At some time we build a household,
> at some time we start a family,
> at some time the brothers divide,
> at some time feuds arise in the *land*.

At some time the river rose (and) brought the flood,
    the mayfly floating on the river.
Its countenance was gazing on the face of the sun,
    then all of a sudden nothing was there!
The abducted and the dead, how alike they are!
    They cannot draw the picture of death.
    The dead do not greet man in the land."
(SB X, 297–318)

The point is, of course, that even though every individual is doomed to die, mankind and society will continue to live. Thus it was the great gods in assembly who once decreed first that it is man's destiny to live and to die, then that the exact date when life will end is never to be revealed in advance. Therefore, any comment which the living can make about the dead lies on the other side of the line of silence. And whereof one cannot speak, thereof one must be silent—since the doom of man is part of his creation, not even the mightiest king can migrate from the Territory of the Humans to the Mindscape of the Gods.

And with that Wittgensteinian remark the tenth tablet ends, a closure which is immediately breached by the very first sentence of the eleventh. These were the words Gilgamesh spoke to Uta-napishti, the Far-Away whose very name means "I Found Life":

    "As I look at you, Uta-napishti,
your form is not different, you are just like me,
    you are not different at all, you are just like me.
I was fully intent on doing battle with you,
    [*but*] in your presence my hand is stayed.
How was it you attended the gods' assembly, and found life?"
(SB XI, 2–7)

No longer determined to fight, the hero is finally ready to adopt the Kantian thesis about the unity of consciousness: you are not different at all, you are just like me.

*

**How was it that you found life**, even though your form is like mine? Such was the question Gilgamesh posed to the only man who had survived the flood. The Babylonian Noah replied:

"I will disclose to you, Gilgamesh, a secret matter,
    and I will tell you a mystery of the gods.
The city of Shuruppak—a city you yourself know,
    *the* [city that] is situated on the [banks] of the Euphrates–

> that city was old and the gods were within it,
> (when) the great gods decided to cause the Deluge."
> (SB XI, 9–14)

And disclosing the secret matter is exactly what Uta-napishti does when he now proceeds to tell the story which no human had ever heard before, a secret so secret that it was never meant to leave the god's assembly hall. Woven into the tale were exactly those passages which George Smith had rediscovered at his desk in the British Museum, a tale within the tale, its narrative structure highly reminiscent of the story of how Odysseus visited his grandfather, met the wild boar and was marked for life.

The flood story is obviously connected both with Apsu's annoyances in *Enuma elish* and with the LORD's disappointments in Genesis. Enlil, the grandfather par excellence, does in fact get so upset with the human noise that he designs a final solution—everyone shall be drowned, not as punishment for their sins, though, merely because the old mafioso cannot stand them any more. Given the geography of their land, every Babylonian knew the disastrous consequences of flooding, hence the deluge was to them the perfect metaphor of human destruction.

When the gods in council reached their drastic decision they swore an oath that no one was to reveal to the humans the fate that was awaiting them. However, the godly Prince Ea was so taken by the situation that during a visit to king Uta-napishti's palace he could not contain himself but repeated what he knew to the walls of his room, every word passed on by the poet in a language full of ambiguities. In a strange way these same walls then communicated to Uta-napishti what they had heard, instructing him to tear his palace down and build a strange ark shaped as a gigantic cube with many decks and a total of $3 \times 3 \times 7 = 63$ compartments. Into this thesaurus sapientiae he was then to load everything he considered worth saving.

Under intense time pressure he did as told, hurriedly sealing the boat just before the storm struck, an event so frightening that even the gods curled up like dogs. And Belet-ili, the Mother goddess, screamed like a woman in childbirth.

On the seventh day the gale relented and the sea grew calm. Uta-napishti looked out and noticed that the boat did not move, indeed that it had run aground on Mount Nimush. Since he was the master of the ship he took the bearings first by setting free a dove which promptly returned, then by letting out a swallow that did likewise, finally by liberating a raven. Unlike the other birds this one saw how the water was receding, flew away, ate what it could find and never came back. Remembering how relieved he had been by the raven's disappearance, Uta-napishti told Gilgamesh that

> "I brought out an offering and sacrificed to the four corners of the earth.
> I strewed incense on the peak of the mountain.

Seven flasks and seven I set in position,
 below them I heaped up (sweet) reed, cedar and myrtle.
The gods smelled the savour,
 the gods smelled the savour,
 the gods gathered like flies around the sacrificer."
(SB XI, 157–63)

All the gods came to the incense, starved after the long storm, when they had had nothing to eat. All of them came together, all except Enlil, the deity who bore the main responsibility for what had happened. When he finally turns up he is furious with his colleagues and their flagrant breach of trust. The great Ea calmed him down, noting that even in the council of gods there must be a proportion between guilt and punishment; as he put it, it would have been better to kill the people by sending a lion, a wolf or a famine than by opening the flood-gates. According to Uta-napishti the deliberations went on an on and did not end until

Enlil came up into the boat,
 he took hold of my hands and brought me out.
He brought out my woman, he made her kneel at my side,
 he touched our foreheads, standing between us to bless us:
"In the past Uta-napishti and his woman was (one of) mankind,
 but now Uta-napishti and his woman shall be like us gods!
Uta-napishti shall dwell far away, at the mouth of the rivers!"
 They took me and settled me far away, at the mouth of the rivers.

But now, who will bring the gods in assembly for you,
 so you can find the life you search for?
(SB XI, 199–208)

The answer is, of course, that nobody will bring the gods in council, that no human will ever find the eternal life to which Gilgamesh aspires. The explanation is that Uta-napishti was made immortal under very special circumstances, in fact that his immortality may be seen as a damage settlement which the gods paid in retribution for their overreaction. As a peculiar illustration of what Uta-napishti has in store, he then asks Gilgamesh to stay awake for a whole week, a test which the earthling sorely fails.

After this everything happens quickly, including Uta-napishti's firing of the boatman, a helpmate who had overstepped his duties by bringing the stranger across the Waters of Death, a barrier meant never to be violated. However, before Gilgamesh is sent back home he is subjected to a rite of purification, given a bath and dressed in new clothes:

His body so fair was soaked,
 the [kerchief of] his head was renewed,

He was clad in a royal robe, the attire befitting his dignity.
"Until he goes (home) [to his city,]
until he arrives (at the end of) his road,
let [the robe show no stain] but stay brand new!"
(SB XI, 265–70)

Gilgamesh and Ur-shanabi, the boatman, board the boat, crewing it themselves. But now Uta-napishti's wife intervenes, suggesting to her husband that it would be impolite not to offer the rare visitor a gift to bring home. These were the words Gilgamesh heard when he was punting back to the shore to receive whatever they were going to offer:

"You came here, Gilgamesh, toiled, exerted yourself,
what have I given you as you go back to your land?
I will disclose, Gilgamesh, a secret matter,
and [I will] tell you a mystery of [the gods.]
It is a plant, its [*appearance*] is like box-thorn,
its thorn is like the *dog-rose's,* it will [prick your hands.]
If you can gain possession of that plant,
[. . . . . . . . . . . . . . .]"
(SB XI, 279–286)

Full of desire, Gilgamesh, weighty stones tied to his feet, stepped into a water-filled pit and was dragged down to the palace of Apsu. Once there he picked up the "plant of heartbeat," which he wanted to bring with him to Uruk the Sheepfold, where he was to test the potion on some old men. And if it worked on them, then he proclaimed that "I will eat some myself and go back to how I was in my youth" (SB XI, 300).

At twenty leagues they broke bread and at thirty leagues they pitched camp. And when Gilgamesh went to a pool to bathe in the cool water, a snake smelled the flagrance of the plant, devoured it and immediately discarded its old skin and became young again. No way to retrieve it, so Gilgamesh sat down to weep. Completely lost, his map torn to pieces, his compass swirling around, he wished that he had never embarked on this futile exercise. As he expressed it, "When I opened the channel I abandoned the tools: what thing would I find that was placed to serve for my landmark?" (SB XI, 316–17).

Accompanying him all the way back is nevertheless Ur-shanabi, the skipper who for the wanderer's sake had been relieved of his transcendental duties. And when the two of them finally arrived at the city, Gilgamesh repeated the words that the poet had given us already at the beginning of thc first tablet:

"Go up, Ur-shanabi, on to the wall of Uruk and walk around,
survey the foundation platform, inspect the brickwork!

(See) if its brickwork is not kiln-fired brick,
and if the Seven sages did not lay its foundations!"
(SB XI, 322–26)

The circle is closed.

*

**The circle is closed,** for after Enkidu's escape from the water-hole, after the comrades' intrusions into the godly Cedar Forest and the slaying of the Bull of Heaven, after the wildman's death and after Gilgamesh's visit to Uta-napishti, it is obvious that the boundaries of the Human Territory are one with the city wall, its brickwork made not by the gods but by man himself. And while the individual human being is doomed to die, maggots creeping out of his nostrils, his sign-studded clay tablets may well be shattered, yet be brought back to life in another library, at another time and at another place. Thus it is that without the civilization of the Babylonian city our present-day existence would not be the same.

Therefore it may well be that Gilgamesh actually *did* find the life he was searching for, a mode of understanding that via the great practical deeds, which the two mates performed together, brought him from the hedonistic life he had been leading in the beginning to the wisdom and knowledge he gained at the end. Although we are all unique human beings, we are at the same time active contributors to a sensus communis. The epic hero is different, though, for by definition the epic hero is someone who crosses boundaries.

If we are fortunate, then Gilgamesh's destiny may be ours too, for even though there is a difference between the young boy and the old man there is a relation as well. At the end strange feelings of happiness may well besiege us just as they did with him. Therefore, and as a tribute to Andrew George, whose work has meant so much for my own understanding:

> In my view the epilogue of the epic tells its audience a self-evident truth: gaze on the city, consider the generations that surround you and learn that human life, in all its activities, is collective and not individual. The symbol of that life is the great city that we contemplate from the wall. Individual cities, of course, could rise and fall but their human populations live on. . . . The plain implication is that though men are mortal, mankind is immortal. . . . A man makes a long journey. Pursued by death he is able, uniquely for a mortal, to bypass its watery gateway. Beyond the world he comes face to face with an immortal ancestor. Then he must go home. The mortal Gilgamesh represents the individual Everyman, though one who has been singled out for an extraordinary experience. . . . By his quest's end what Gilgamesh has learnt at first hand, alone among mortals, is this: at the end of life the individual perishes in the passage to death's realm, but beyond

> that point in his existence, and necessarily outside his personal experience, stretches the eternal past and future of mankind.[20]

Finnegan four millennia before the Wake, Ishmael ages before Moby-Dick. Here Comes Everybody, Abraham and Moses, Odysseus and Plato, Immanuel Kant and Paul Cézanne, all mixed together in an alphabet soup of sensus communis, things of nature on one side of the epistemological/ontological divide, relations of culture on the other. A classical rite de passage complete with its three phases of separation, liminality and reaggregation. Cartographical reason in theory and practice, the canvas of the mappa so well prepared that it has all but merged with the images projected, the scale and the directional arrow indistinguishable from each other. Imagination hard at work in the Saussurean forge of the unconscious. The outcome a marvelous tale of a glorious failure.

*

**Yet** there is also Tablet XII, the old Sumerian text which was rather crudely translated into Akkadian and somehow hooked on to the *Gilgamesh* proper. Not poetry but prose, by most commentators ignored as a strange appendage, by a minority taken as a possible key to the entire epic.[21] To the outsider a piece so mysterious that it stirs both our curiosity and our excitement. For what the twelfth tablet seems to be saying is that despite the fact that our bodies are all committed to the maggots, we continue to live in the minds and memories of those left behind, our history one with their myths. Some quickly forgotten and utterly neglected, some as powerful and honored as ever before:

> "Did you see the one who was struck by a mooring-pole?" "I [saw (him).]
> Alas for his mother [and father!] When pegs are pulled out [he] wanders about."
>
> "Did you see the one who [died] a natural death?" "[I saw (him).]
> He lies drinking clear water on the bed of the [gods]."
> "Did you see the one who was killed in battle?" "I [saw (him).]
> His father and mother honour his memory and his wife [weeps] over [(him).]"
>
> "Did you see the one whose corpse was left lying in the open countryside?" "I saw (him).
> His ghost does not lie at rest in the Netherworld."
> "Did you see the one whose ghost has no provider of funery offerings?" "I saw (him).
> He eats the scrapings from the pot (and) crusts of bread that are thrown away in the street." (SB XII, 144–153)

For better and for worse Iraq has always been and will always be. Indeed the name of Babylon itself, *Bab-ili,* means "Gate of the gods."

Imperial politics notwithstanding, the Babylonian insights are bound always to be with us. For what Gilgamesh so vividly illustrates is how the boundaries of the Human Territory are determined through a continuous struggle with the gods, on the one hand, and the animals, on the other. In fact I know of no better illustration of how a two-thirds god and a two-thirds gazelle are brought together in acts of love and turned into a semiotic animal with the power of imagining its own demise.

A battle unique in the insights it offers. By all accounts a lasting, perhaps the groundbreaking, story of what it means to be human.

# PENIEL

I put my hand over my mouth.

**The second map** returns us once again to the Hebrew Bible, more specifically to what many consider the two most decisive battles in the perpetual war between the LORD and his subjects. The first story is about Jacob's deceit, escape and wrestling, the second about Job's struggle for the right to be right.[1] In both cases the disputes center on the exact drawing of the boundaries between the Territory of the Humans, on the one hand, and the two modes of the utterly different, on the other. A mapping expedition if one ever was, an attempt to approach the silent lines that mark the difference between the expressible and the non-expressible. Everything at stake, not the least because the mapmaker's instruments were designed by the Greeks and not by the Jews. Therefore,

> Sing in me, Muse, and through me tell the story
> of that man skilled in all ways of contending,
> the wanderer, harried for years on end,
> after he plundered the stronghold
> on the proud height of Troy.
>                                         He saw the townlands
> and learned the minds of many distant men,
> and weathered many bitter nights and days
> in his deep heart at sea, while he fought only
> to save his life, to bring his shipmates home.
> But not by will nor valor could he save them, . . .
> (*Odyssey* 1.1–6, trans. Fitzgerald)

Such are the opening words of Homer's Odyssey, the prayer by which every cartographer smoothes the way before (s)he sets out on the dangerous road of the unknown. In the English version a total of eleven lines, in the Greek original, as in the most classical of all Swedish renderings, merely six:

> Sångmö, sjung om den man, som länge i skiftande öden
> irrade kring, när han Troja förstört, den heliga staden.
> Många människors städer han såg och lärde dem känna,
> många de lidanden voro på hav, som hans hjärta fick utstå
> under hans kamp för sitt liv och för kämparnas lyckliga hemkomst.
> Dock sina män han ej frälste ändå, hur än han försökte, . . .[2]

And in the sense just illustrated every understanding is an exercise in translation, every translation a treacherous journey back and forth between the realms of the vaguely familiar and the completely alien. Such is also my own approach to the delineation of the human landscape: an encounter with an original composed in a language which I do not fully comprehend. And therein lies not only the attraction but also the challenge, for the structure of power is not as simple as it might first appear, not limited to what meets the five senses of the body but equally present in the sixth sense of culture. It follows that any understanding of what it means to be human must consider not only trees and waters, stones and birds, but also hopes and fears, joys and griefs; not merely smell and taste but pride and hate; not only light and shadow but power and submission.

Understood in that manner there are close parallels between my own conception of the human territory and Karl Marx's conception of the commodity, for in both cases there is a braiding of reification and deification, fetishism and alienation. To repeat:

> A commodity appears, at first sight, a very trivial thing, and easily understood. Its analysis shows that it is, in reality, a very queer thing, abounding in metaphysical subtleties and theological niceties. So far as it is a value in use, there is nothing mysterious about it. . . . But as soon as it steps forth as a commodity, it changes into something transcendent. . . . To find an analogy, we must have recourse to the mist-enveloped regions of the religious world.[3]

The Marxian conception of commodities can readily be extended to signs in general, hence to the analysis of landscapes, mindscapes and rockscapes as well. Thus, just as the commodity melts together the two realities of use-value and exchange-value, so every sign does the same with the concepts of Signifier and signified; in essence the human territory is a realm which is controlled from the smoke-filled rooms of the Saussurean Bar.

The text to which I venture for an analogy is Genesis 25–36, a stage in the monotheistic story at which the LORD has already been domesticated into the house of Abraham.[4]

*

**Genesis 25–36** begins with the story of how Abraham after Sarah's death takes a new wife and fathers six more sons and one additional daughter, this time evidently of his own making and without the LORD's interference. At the end, one hundred and seventy-five years old, he breathes his last and his two sons Ishmael and Isaac bury him in a cave. Immediately thereafter God blesses Isaac, who for obvious reasons mourned the death of his mother more than the loss of his father. From that event we are then swiftly moved from the trauma of separation to the story of Jacob, certainly one of the greatest crooks ever born. Yet one of the most richly rewarded.[5]

Everyone remembers how Jacob repeatedly and most purposely had betrayed both his twin brother and their father; first he had forced the starving Esau to sell his birthright for a slice of bread and a bowl of lentil soup, then he had come before the blind Isaac dressed in the smelly clothes of his hairy brother, pretending to be what he was not. The background is that when Isaac was old and his eyes so weak that he could no longer see, he called for Esau his older son, and said

> I am now an old man and don't know the day of my death. Now then, get your weapons—your quiver and bow—and go out to the open country to hunt some wild game for me. Prepare me the kind of tasty food I like and bring it to me to eat, so that I may give you my blessing before I die. (Gen. 27:2–4)

This Rebekah heard and she promptly told Jacob, her favorite son, to bring her some game so that she could prepare the dish that Isaac desired. Said and done. And when Jacob, well aware of the deceit, served his father the savory dish his mother had prepared,[6]

> Jacob said to his father, "I am Esau your first-born. I have done as you told me; now sit up and eat of my game that you may bless me." But Isaac said to his son, "How is it that you found it so quickly, my son?" He answered, 'Because the LORD *your* God granted me success." (Gen. 27:19–20)[7]

The ailing Isaac then embraced Jacob and blessed him, thinking he was the first-born, the legitimate heir, who the father loved because he was a skilful hunter and because he liked the taste of his game. Although initially bewildered by the fact that the voice was the voice of Jacob and the hands the hands of Esau, Isaac's words were doubly anchored in both body and culture, an eminent case of logical empiricism as the erasure of doubts: "Come near, that I may feel you, my son, to know whether you are really my son Esau or not. . . . Are you really my son Esau?"

Jacob, the deceitful, answered "I am," upon which his father said

"Come near and kiss me, my son" (Gen. 27:21–26). And when Isaac caught the smell of his clothes, he blessed his second-born and said:

> See, the smell of my son
> is as the smell of a field which the LORD has blessed.
> May God give you of the dew of heaven,
> and of the fatness of the earth, and plenty of grain and wine.
> Let peoples serve you,
> and nations bow down to you.
> Be lord over your brothers,
> and may your mother's sons bow down to you.
> Cursed be every one who curses you,
> and blessed be every one who blesses you! (Gen. 27:27–29)

In all its sadness the story is unique. For now

> as soon as Isaac had finished blessing Jacob, when Jacob had scarcely gone out from the presence of Isaac his father, Esau, his brother, came in from his hunting. He prepared savory food, and brought it to his father. And he said to his father, "Let my father arise, and eat of his son's game, that you may bless me." His father Isaac said to him, "Who are you?" He answered, "I am your son, your first-born, Esau." Then Isaac trembled violently and said, "Who was it then that hunted game and brought to me, and I ate it all before you came and I have blessed him?—yes, and he shall be blessed." When Esau heard the words of his father he cried out with an exceedingly bitter cry, and said to his father, "Bless me, even me also, O my father!" But he said, "Your brother came with guile, and he has taken away your blessing." Esau said, "Is he not rightly called Jacob? For he has supplanted me these two times. He took away my birthright; and behold, now he has taken away my blessing." Then he said, "Have you not reserved a blessing for me?" Isaac answered Esau, "Behold, I have made him your lord, and all his brothers I have given him for servants, and with grain and wine have I sustained him. What can I do for you, my son?" Esau said to his father, "Have you but one blessing, my father? Bless me, even me also, O my father." And Esau lifted up his voice and cried. (Gen. 27:30–38)

Isaac's point is that, even though he knows that he has made a mistake, a blessing is a blessing, a word is a word, a promise a promise, a patriarch a patriarch. And therein lies the crucial difference between Isaac and his LORD, for whereas the former is obliged never to revoke a blessing, the latter is free to change his commands as he pleases. No doubt Isaac remembers how long ago he himself had been tied to the altar, how he (a thirty-seven year old boy) had been chosen to be sacrificed as proof of his father's obedience, the sharp knife on his throat. But in his ears ring also the cries of the angel, "Stop, stop! For heaven's sake. The LORD has changed his mind, no need to execute your son, through your willingness you have already destroyed yourself."

And in the same moment that the God-sent messenger saves the boy's life, the Almighty institutionalizes the rule by terror, shows by his reversal that his words are not to be trusted, proves through his action that he is free to knock on any door at any time, that he alone sets the rules of the game; as in the first paragraph of the Constitutional Law, no power will ever be permitted to stand above or next to him. Dictator of dictators, Absolute of the absolute. At the same time it should be stressed that while it was God—not Abraham—who blessed Isaac, it was Isaac—not God—who blessed Jacob. Another most remarkable shift in power, another adjustment in the boundary-line between the Territory of the Humans and the Realm of the Gods.

Esau, on his part, has learned his family history. Therefore he is the first to realize that he might be in for the same unfair treatment as once his father and grandfather. But unlike them he is not ready to accept whatever comes his way. And from the beginning he knows his enemy. For already in their mother's womb Jacob, his twin, had tried to hold his brother fast, struggling to get out first, determined to claim the rights of the first-born. And that is exactly why Jacob was called "Jacob," a name which literally means "he who takes by the heel" or "he who supplants."

No wonder that Esau, the hunter and civilized descendant of Enkidu, came to hate his brother. And he vowed to kill him. Duly warned by Rebekah, the scheming mother who loved her second-born more than his first-born brother, Jacob had no choice but to escape. He left Beersheba and went to Haran, as frightened as he deserved to be.

Such is the power-filled context in which the first landscape was created.

*

**Jacob left Beersheba** running for his life and away from his misdeeds. Rather soon

> he came to a certain place, and stayed there that night, because the sun had set. Taking one of the stones of the place, he put it under his head and lay down in that place to sleep. And he dreamed that there was a ladder set up on the earth, and the top of it reached to heaven; and behold, the angels of God were ascending and descending it! And behold, the LORD stood above it, and said, "I am the LORD, the God of Abraham your father and the God of Isaac; *the land on which you lie will I give to you and to your descendants;* and your descendants shall be like the dust of the earth, and you shall spread abroad to the west and to the east and to the north and to the south; and by you and your descendants shall all the families of the earth bless themselves. Behold *I am with you and will keep you wherever you go,* and

> will bring you back to this land; for I will not leave you until I have done that of which I have spoken to you." Then Jacob awoke from his sleep and said, "Surely the Lord is in this place! This is none other than the house of God, and this is the gate of heaven."
>
> So Jacob rose early in the morning, and he took the stone which he had put under his head and set it up for a pillar and poured oil on top of it. He called the name of the place Bethel ["the house of God"]. Then Jacob made a vow, saying, '*If* God will be with me, and [*if he*] will keep me in this way that I go, and [*if he*] will give me bread to eat and clothing to wear, so that I come again to my father's house in peace, *then* the Lord shall be *my* God, and this stone, which I have set up for a pillar, shall be God's house; and of all that thou givest me I will give the tenth to thee." (Gen. 28:11–22, emph. added)

What a remarkable text from the mist-enveloped realities of ontological transformations. What a fantastic story of how a landscape is born from the intercourse of a rock and the guile of a human serpent. The secret reveals itself in the fact that the two terms "Bethel" and "The Saussurean Bar" are different designations of the same place—that very particular place in which the Mindscape of Pure Spirituality and the Rockscape of Pure Materiality come together and merge into one. Any fugitive who is resting on a mattress of gravel and a pillow of stone is prone to dream strange dreams, to hear voices from afar, to imagine pictures of the surreal. And to stay sane Jacob did what any post-Freudian therapist would recommend him to do: he held his mind in place by transforming anxiety into fear, by simultaneously thingifying the social and naming the unnamable. Comforting sigh of relief, Søren Kierkegaard in advance of himself.

But relief is no cure. The only thing that remains to hang onto is therefore the hard fact of the stone, the pillow on which Jacob puts his head when he lies down in the evening—a piece of rock detectable by the five senses of the body, but essentially outside the sixth sense of culture. In contrast, the top of the ladder reaches into heaven, the epitome of those mist-enveloped regions where everything is invisible and nothing mentionable, where there is nothing to sense yet everything to imagine. Connecting the two extremes is the ladder—*La Scala*—the cartographical translation function through which material matter and spiritual spirit are transformed into mutually meaningful statements, a Mesopotamian ziggurat or a Vitruvian man in disguise. Busily trying to tie the two worlds together is a host of angels ascending and descending the scale, a corps of ambassadors charged with the impossible mission of negotiating a lasting peace between heaven and earth, visible and invisible, inner and outer.

If the two rulers of Rockscape and Mindscape can be said to speak at all, they do so in mutually incompatible languages. Just as you and I cannot grasp what they are saying, so they too are unable to understand the tongue of the other. Their respective statements are equally untranslatable—literally on the other side of the Wittgensteinian limit—for the Ruler of Rockscape is a matter so utterly material that it emits no signified whatsoever and the Ruler of Mindscape is a meaning so meaningfully meaningful that there is no Signifier rich enough to express it. It follows that both absolutes share the characteristic of being beyond the senses, beyond rephrasing, beyond the limits of representation, outside the Territory of Humans. Yet—and this is the crucial paradox—it is only by playing the tricks of renaming that the two potentates can be approached, only through acts of labeling that they can be tied down.

It hurts to officiate at the ceremony, but in the present case rape is the only alternative; as recalled, understanding is by definition a deed of reformulation, reformulation always an issue of force. Therefore I hereby baptize the Ruler of Rockscape as "Rock-Lord," a thing-like phenomenon which simply is. In the same mood I name the Ruler of Mindscape "Mind-Lord," a misty noumenon defined as something which is what it is. For short I now call the Mindscape and its sovereign

$$M$$

by definition a tautology, hence not merely $M$, but

$$M = M$$

Likewise I call Rockscape and its lord

$$R$$

Now, awakening from his dream, Jacob realizes that he himself is one of the ambassadors who travels back and forth between the two realms, carrying secret messages from one end of the ontological scale to the other. Through his skills as a negotiator he manages to bring the two rulers $M$ and $R$ together in the capital of the Human Territory, otherwise known as "Bethel" or "God's House."[8] As a seal of the peace treaty, the horizontal pillow is turned on high and changed into a vertical pillar, a right angle built in stone. It is in the origo of the thus constructed coordinate net that the godly house Bethel is erected—a social invention of the highest order.

But this so-called "House of God" is in reality not a house at all but an abstract place, the only point at which its is possible to cross the otherwise closed border between Rockscape and Mindscape. Indeed it is exactly out of this intersection between the two major lines of Plato's *Republic* that the human landscape springs forth, new gods and new goods created in the process. The ecological debates, the Sierra Clubs, the Nature walkers, the American Indians, the whale protectors, the forest companies, the road builders, the green farmers, all bear testimony to the fact that in the

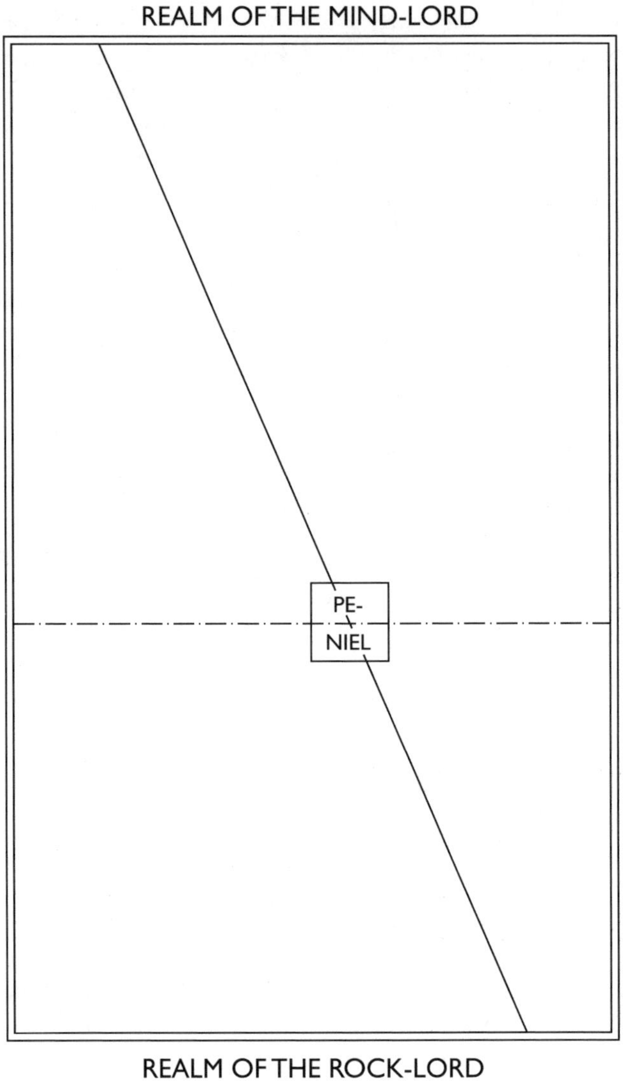


Map of Peniel.

disputes about what can and cannot be done, the forces of deification and reification are as intimately interwoven now as ever before. And just as the principle of mutual distrust ruled at Bethel, so it permeates the current struggles as well. But which of the two rulers exercises more influence, the shady Rock-Lord or the tautological Mind-Lord?

To these questions of master and slave, Jacob and Karl Marx offered closely related answers. For instance, it was only under certain, clearly defined conditions, that Jacob was willing to call the Lord his own God. Only *if,* only *if,* only *if, then* he will pay the ten percent tax; no longer, as in the case of Abr(ah)am, is it God who is testing the allegiance of his subject, but a citizen who is questioning the trustworthiness of the Sovereign's promise that he will be with him wherever he goes. From personal experience Jacob knows that lying and cheating often pays off and he is not about to be fooled. Therefore he trusts no one, not even the God of his forefathers, that self-made and self-named entity who calls himself neither "Rock-Lord" nor "Mind-Lord" but "the LORD of LORDS." In short, the usurper is determined not to give in, and that is regardless of whoever the adversary may be. And in this sense the authors of Genesis and *Das Kapital* agree: even though the Territory of the Humans is partly for contemplation, it must yield a profit as well.

In that context of ontological transformations there are striking similarities also between Jacob's ladder and the divided line of Plato's *Republic.* Both devices in fact serve as epistemological-ontological translation functions through which degrees of truth and degrees of being are made to correspond to one another. While in the *Republic* the measuring rod hangs suspended from the Good of the Good, in Jacob's dream the ladder reaches into the heavens of the Mind-Lord. Dangling at the other end of the scale are in Plato's case the shadowy objects, in Jacob's world the pillow of the Rock-Lord's pure materiality. Somewhere in between lies the thin line of

$$\frac{M}{R}$$

better known as the Saussurean Bar, the abysmal gathering-point in which invisible ideology is turned to touchable stone and meaningless stone to meaningful spirit. And yet, even though this transition-point metaphorically corresponds to the zero-mark on Celsius' thermometer, the absolute zero of human thought-and-action is in the $M = \text{M}$, by definition always true but never informative. There—and nowhere else—is the ultimate limit of every power, in Plato's universe the convergence of the Sun, a matter without whose light nothing can be seen, and the Good, an idea without which nothing can be understood.

And so it is that both the monotheistic Jews and the polytheistic Greeks are condemned to hitting their heads against the ceiling of language. But which really *are* the relations between original and copy, the invisible and godly forms up there and the visible and graven images down here, the ideas and the shadows?

Every ambassador, every angel, every muse knows the answer, for they have all learned that unless both forms and images are distorted, peace will never come. It follows that the messengers play such pivotal roles not because they are what they are, but because they do what they do; as recalled, to "translate" is, according to the *OED,* "to bear, convey, or move from one person, place or condition to another"; "to remove the dead body or remains of a saint, hero or great man from one place to another"; "to carry or to convey to heaven without death"; "to move (a body) from one point or place to another without rotation"; "to turn from one language into another"; "to interpret"; "to change in form, appearance or substance, to make new boots from the remains of old ones."

*

**To carry to heaven without death**, ascending and descending the scale. Such is the nature not only of the translating angels but of the measuring land-surveyors. More analogies are therefore to be drawn, more insights to be gained. Thus, when we encounter Jacob again, he is once more on the move, on this occasion running away not from Esau, his brother, but from Laban, in the same person his father-in-law and his mother's brother. The story is chilling, for in many ways Laban has just done to his daughter what Jacob previously had done to his brother, in everything each other's equal in scheming and treacherous behavior.

As recalled, Laban had two daughters, Leah the older whose eyes were weak, and Rachel the younger who was so beautiful that Jacob wept aloud the first time he saw her. So much did Jacob love her that he promised to serve seven years for her, in his own eyes merely a few days. When that tenure period was over he demanded that she be given to him and that he may lie with her. Laban agreed to arrange a feast and the next morning, when the bridegroom woke up, he saw that the woman he had slept with was not Rachel but Leah, the latter surely an expert in the art of imagination, for during the night she really *is* her sister. Outraged Jacob confronts her and, according to the midrash, "He said to her, 'Deceiver, daughter of a deceiver! Did I not call you Rachel and you answered me?!' She replied, 'Is there a master without students? Did your father not call you Esau and you answered him?!'"[9] In the original text, Jacob said to Laban:

> "What is this you have done to me? Did I not serve with you for Rachel?" Laban said, "It is not so done in our country, to give the younger before the first-born. Complete the week of this one, and we will give you the other also in return for serving me another seven years." (Gen. 29:25–27)

So great was his love for Rachel that Jacob accepted Laban's offer. And when the week was over he went in to her as well.

But when the LORD saw that Leah was hated, he opened not only *her* womb but that of her servant too. Together the two of them gave Jacob one son after the other, each one with a name that reflected their mother's feeling that neither she nor they were wanted; in a very real sense it was not Jacob's love but Leah's tears that generated their children. And finally came a daughter called Dinah, one of the Bible's most unfortunate beings.

Rachel, though, was barren and instead of asking the LORD to interfere on her behalf, she scolds her husband: "Give me children, or I shall die." In anger he replies, "Am I in the place of God, who has withheld from you the fruit of your womb?" (Gen. 30:1–2). And just as Sarai in her despair had offered her servant to Abr(ah)am, so Rachel now offers her servant to Jacob. And just as had happened with Sarah and Hagar, a son is born, all with the same predictable consequences of hatred and jealousy.

At this stage, however, God suddenly remembers Rachel and gives her a son whom she calls "Joseph," a name which means "He adds." A happy occasion, of course, yet at the same time a setback in man's struggle for independence. For once again the LORD, through his interference, has shown that in the area of fertility—the only matter of strategic importance—it is he who rules, he and none other. To this fact Jacob seems totally blind, and in an atmosphere of mutual distrust he asks Laban for permission to leave. The situation gets tense, but when Rachel steals her father's household gods—itself perhaps an act of polytheism and a violation of the First Commandment—Jacob has no choice but to flee. Overtaken by Laban's forces when he camped in the hill country of Gilgaed, Jacob heard these words from his father-in-law:

> "What have you done, that you have cheated me, and carried away my daughters like captives of the sword? Why did you flee secretly, and cheat me, and did not tell me, so that I might have sent you away with mirth and songs, with tambourine and lyre? . . . Now you have done foolishly. It is my power to punish you; but the God of your father spoke to me last night, saying, 'Take heed that you speak to Jacob neither good nor bad.' And now you have gone away because you longed greatly for your father's house, but why did you steal my gods?" Jacob answered Laban, "Because I was afraid, for I thought that you would take your daughters from me by force. Any one with whom you find your gods shall not live." . . . Now Jacob did not know that Rachel had stolen them. (Gen. 31:26–32)

They searched everywhere and found nothing, for Rachel had hidden the idols in the camel's saddle and sat on them. And she said to her father, "Let not my lord be angry that I cannot rise before you, for the way of women is upon me" (Gen. 31:35). A most crucial passage, for here we see God entangled in his own net; it was Rachel's menstrual blood over which the Almighty claimed to be in control that now kept his envoy from discovering the household gods which represented the real threat to his authority. And early in the morning Laban arose, kissed his grandchildren and his daughters, blessed them and returned home.

Throughout these events Jacob remains as afraid of his brother as ever before, especially as messengers inform him that Esau is coming to meet him with a force of four hundred. "Deliver me," prays the heel-holder, "from the hand of my brother, from the hand of Esau, for I fear him, lest he come and slay us all" (Gen. 32:11).

*

**Deliver me**, prays the haunted man. Therefore,

> the same night he arose and took his two wives, his two maids, and his eleven children, and crossed the ford of the Jabbok. He took them and sent them across the stream, and likewise everything that he had. And Jacob was left alone; and a man wrestled with him until the breaking of the day. When the man saw that he did not prevail against Jacob, he touched the hollow of his thigh; and Jacob's thigh was put out of joint as he wrestled him. Then he said, "Let me go, for the day is breaking." But Jacob said, "I will not let you go, unless you bless me." And he said to him, "What is your name?" And he said, "Jacob." Then he said, "Your name shall no longer be Jacob, but Israel, for you have striven with God *and* with man, and have prevailed." Then Jacob asked him, "Tell me, I pray, your name." But he said, "Why is it that you ask my name?" And there he blessed him. So Jacob called the name of the place Peniel, saying "For I have seen God face to face, and yet my life has been preserved." The sun rose upon him as he passed Penuel [*sic*], limping because of his thigh. (Gen. 32:22–32, emph. added.)

For the cartographer of power it is especially important that the self-proclaimed victor spreads the news by naming a place, not by telling a story. And by choosing the word "Peniel" as the name of the meeting-place, the second-born shows that in his lust for power nothing is sacred, not even his relations to the LORD himself.

The key to the hidden meaning is in a footnote which explains that "Peniel" means "the face of God," at the same time a proper name and a definite description. But in its audacity this descriptive name is completely outrageous, for if there is anything the LORD has made perfectly clear, it is that his face must never—under any circumstances—be seen. As he later reminded Moses:

> "I will make all my goodness pass before you, and I will proclaim before you my name 'The LORD'; and I will be gracious to whom I shall be gracious, and will show mercy on whom I will show mercy. But," he said, "you cannot see my face; for man shall not see me and live." (Exod. 33:19–20)

And yet. In his remarkable dispatch from the front line, this is exactly what Jacob claims to have done: seen God's face and survived. In the eyes of the Almighty the blasphemy of blasphemies, an extreme violation of the second paragraph of the Constitutional Law, in the propaganda of the usurper merely a way of legitimating his own behavior. For with what did the runaway actually wrestle, his bad conscience or his twin-brother? The biblical narrator does not say, stating only that Jacob first had striven with *a* man and then with God *and* man; in the abstractness of Mindscape ambiguity rules, in the concreteness of Rockscape certainty matters. In a later passage, when he is negotiating with his brother, Jacob in fact played the same trick once again, obliquely begging Esau "I pray you, if I have found favor in your sight, then accept my present from my hand; for truly to see your face is like seeing the face of God, with such favor have you received me" (Gen. 33:10).

And thus it is that just as Bethel may be taken as the name of the border station which is located in the frontier line between the ontological realms of Rockscape and Mindscape—a primordial Checkpoint Charlie—so Peniel designs the torture chamber in which all travelers have their papers checked and their bodies marked; it is telling that in the Swedish version of the Lord's Prayer the sinners beg the LORD to forgive them their debts, while the English believers ask forgiveness for their trespasses. But the word "trespass" is itself unusually rich, noun and verb in one; according to the *OED* the term means both "to pass beyond this life; to die," and "to unlawfully enter on the land of another." But that is of course exactly what Jacob says that he has done: not only invaded the lands of the Mind-Lord but redrawn the boundaries of the Human Territory in a most radical way. The next question is to determine what may or may not be admitted into the glorious land of the humans.

However, it is when Jacob assumes this new role of immigration officer that he runs into his greatest difficulties. The reason is that the asylum seekers he encounters in the office of Peniel come either from the land of Rockscape, which is ruled by the shadowy Rock-Lord, or from the realm of Mindscape, which is governed by the evasive Mind-Lord. The characteristic that makes these absolutes alike is that once their respective name is rephrased, its bearer dies; thus, while *M* keeps chanting his "I am I, I am who I am," *R* whispers in a voice that cannot be heard. Rephrased, the former citizens of Mindscape speak the tongue of tautology, the refugees from Rockscape the silence of silence. Together these expressions form the outer limits of every translation, the border that no human can trespass and live.

Hidden in this insight lies the secret not only of Jacob's story but of power itself. For in his report from the battlefield Jacob effectively violates the second paragraph of the Constitutional Law. No graven image, no definite description, merely the obligation to keep the sabbath day holy and to honor your father and your mother. Pushed to this ultimate limit, the LORD has no alternative but to demonstrate that the double privilege of categorizing and naming rests with him and with him alone. With great dispatch and fitting cruelty the Almighty therefore kills the offender by first moving him from one category to another and then by giving him a new name; after a touch of the divine wand, the solid man Jacob melts into the airy people of Israel, no longer an independent human being but an obedient member of a social class, Israel itself a name which signifies "one who fights with God." And when God calls out "There you are!," then Jacob repeats the words of his harassed forefathers: "Here I am."

It was this response that made Søren Kierkegaard fear and tremble. To the cartographer of power the same exchange shows that the war between the Rock-Lord and the Mind-Lord always involves a strategic deployment of personal pronouns, adverbs and prepositions. The landscape of landscape is indeed contested territory. Therefore, to set things right,

> God said to Jacob, "Arise, go up to Bethel, and dwell there; and make there an altar to the God who appeared to you when you fled from your brother Esau." So Jacob said to his household and to all who were with him, "Put away the foreign gods that are among you, and purify yourselves, and change your garments; then let us arise and go up to Bethel, that I make there an altar to the God who answered me in the day of my distress and has been with me wherever I have gone." So they gave to Jacob all the foreign gods that they had, and the rings that were in their ears; and Jacob hid them under the oak which was near Shechem. . . . And Jacob came to Luz [that is, Bethel], which is in the land of Canaan, he and all the people who were with him, and there he built an altar, and called the place El-bethel,[10] because there God had revealed himself to him when he fled from his brother. (Gen. 35:1–7)

The LORD having delivered on all of Jacob's many *ifs,* the rascal is finally ready to hide (not destroy) the idols, seemingly accepting the first paragraph of the Constitutional Law. This done God left the place, never to appear again in the Book of Genesis.

In contrast, at the close of chapter 35, there are two short verses which bring us back to Isaac, the third patriarch who after one hundred and eighty sad years breathed his last. His two sons Esau and Jacob—by the biblical redactor listed in that order—buried him, a turning-point at which the narrator moves on to the story of Joseph, Jacob's favorite son, the successful bureaucrat who never spoke directly *to* God, rarely even *of* him. Certainly an escape, albeit neither the first nor the last.

Thus, as another way out, most Marxists argue that the real power of ontological transformations is lodged not in the spiritual imagination of ambiguous belief but in the factual solidity of the material. Tied to that reductionist suggestion are nevertheless a host of problems, most of them connected with the Kantian tendency to treat others *as if they were* dead things rather than living humans. But that fetishist temptation must surely be resisted, for it is well established that material relations in reality are social relations, just as social relations in essence are material relations. Thus, when commodities converse with each other, they constantly wonder why people mistake them to be reified use-values while, within themselves, they know that they are nothing but deified exchange-values. Queer things, these political subtleties and psychological niceties. Foghorns sounding through the mist. And the sun rose upon him as he passed the shoal, the prow shearing the waves.

Giambattista Vico consequently knew what he did, when he preached the gospel of *Factum verum*—the true is the made. But the Neapolitan professor of rhetoric also realized that to share the world is to allow metonymic associations to move from one metaphorical topos to another, to journey back and forth between the here and the there, the there and the here. Odysseus himself spent a lifetime demonstrating that the point of traveling is not to reach a destination but to return to an origin, not to rest contended in the arms of Nausikaa but to slaughter the suitors who occupied his house and made passes at his wife; the cunning trickster fought only to save his life and to bring his shipmates home. But not by will nor valor could he save them, for their own recklessness destroyed them all.

* * *

**Will and valor** were the admirable attributes also of the prophet Job, his form of recklessness not a manipulative trick that destroyed him but a courage that saved him. Although most likely written in the fifth or fourth century BCE—hence roughly contemporary with the Sophoclean tragedies—this greatest and most difficult of biblical books has no Greek equivalent.

Central to the understanding of this remarkable text is an analysis of the Bakhtinian blend of genres, especially the intricate interweaving of prose and poetry. To be specific, Job's book begins with a prose Prologue (chaps. 1 and 2) from which the author moves first to the so-called Wisdom Dialogues (chaps. 3–27) with the interchanges between Job and the three "friends," then to the Wisdom Poem (chap. 28) and eventually to Job's restatement of his case (chaps. 29–31), these twenty-nine chapters all composed in fine poetry. A few lines of prose are then inserted as transition to the young Elihu's rather mediocre poems (chaps. 32–37), almost certainly added by another hand at a later date. Following this interruption comes God's own reply spoken from the whirlwind (chaps. 38–41), a stunning show of power which is topped off by Job's short reaction to what he has heard (42:2–6). Finally, at the very end, a prose Epilogue in which we are told about Job's restoration, his previous wealth and status not merely regained but significantly enhanced. In short, a long poem of ambiguity framed and tied together by shorter passages of rich and informative prose, a unique part of world literature.[11]

*

**The Book of Job** begins with the information that once upon a time there lived in the land of Uz a man called Job, a blameless and upright foreigner who feared God and shunned evil. So pleased with this vassal was the

Almighty that he made him the greatest man among all the people of the East, conferring on him a total of seven thousand sheep, three thousand camels, five hundred yoke of oxen, five hundred donkeys and a large number of servants. Most importantly he was blessed with seven sons and three daughters, a clan that met for regular feasts at which they ate and drank and where the family father always took the precaution of sacrificing burnt offerings just in case any of the children had sinned and cursed God in their hearts. Yet, despite his countless virtues, it was this man who was selected to be a scapegoat and to perish forever like his own dung, to fly away like a dream no more to be found, to be banished from the social community like a vision of the night. The eyes that saw him would not see him again, the place where he lived would look at him no more, the venom of the serpents were within him. But who knows? Perhaps he actually *was* the oppressor of his people, the wolf in a sheep's clothing, the traitor which the friends suspected him to be. Alternatively he may merely be one of those whose careers ends so badly because it began so well.

The turning point in Job's life came on the day when the LORD went on a walk with the angel Satan, not exactly the Christian Devil but more likely the Jewish Accuser. At any rate this figure—by some accounts God's half-brother—suggested that if the courtier were really put to the test, then even he would curse his benefactor, even he would bite the hand that fed him. God agreed, but only under the provision that his protégé was not to be tortured.

What then followed is what could be expected, for Satan immediately proceeded to kill not only all of Job's animals and servants but his ten children too. On hearing the shattering news Job tore his robe, shaved his head and fell to the ground, reaffirming his allegiance to the Almighty. In his laments he blamed no one for what had happened, merely limited his comments to the factual statement that

> Naked I came from my mother's womb,
> and naked I will depart.
> The LORD gave and the LORD has taken away;
> may the name of the LORD be praised.
> (Job 1:21)[12]

Job's first test so handsomely passed, Satan appears again once more restating his suspicion. In reply God repeats that of all people on earth no one is as blameless as his servant from afar, the favorite who "still maintains his integrity, though you incited me against him to ruin him *without any reason*" (2:3, emph. added). To this word from the Ruler's mouth the Accuser replied "Skin for skin! For now I tell you that if you strike his flesh and bones, he too will curse you. As a person well trained in the art of torture I know that even those who are not guilty can be beaten until they admit that they are."—"Very well," said the LORD, "he is in your hands; but you must spare his life" (2:6)[13]—"Who do you think I am," Satan replied, "every professional knows that a dead victim is a job badly done."

Thus reassured, the Accuser went out of his LORD's presence and afflicted Job with such painful sores from the soles of his feet to the top of his head that the poor man was found sitting in the ashes scraping himself with a potsherd. Much else was brought on him too, including a breath so foul that his wife refused to sleep with him, no husband ever as pitiful and hopeless. All she could say were the words:

> Are you still holding on to your integrity? Curse God and die! (2:9)

To this double-edged advice Job simply retorted:

> You are talking like a foolish woman. Shall we accept good from God, and not trouble? (2:10)

The narrator then concludes that Job never sinned *in what he said* ("Read my lips!"), certainly an ambiguous statement which leaves open the possibility that he might well have committed crimes which he refused to admit, perhaps even failed to recognize.

But which sins could possibly have been so grave that they warranted the tremendous violence it generated? Of which transgressions could Job reasonably be suspected?

My own hunch is that the Accuser—the chief of the LORD's Bureau of Investigation—had reason to suspect that God's favorite was violating the first paragraph of the Constitutional Law, if so certainly a case of high treason. After all, Job was not a born Israelite but an alien who lived in a foreign land, as such no doubt surrounded by deities of quite different kinds, himself in fact a kind of shattered idol.[14] Every effort, including a combination of bodily torture and mental brainwashing, must therefore be made to bring clarity. And to that end the narrator calls into action the three characters Eliphaz, Bildad and Zophar, Job's friends from the good days, experts in moral technology; "for what is the Law but a representation whereby man tries to rectify the crooked effects of the unforeseeable, of madness, of evil's excess? . . . Technique, the human effort to dominate the world, is a response to the human impotence that is always already there."[15]

Under false identities the three comrades are sent to comfort their old colleague, allegedly to tell him that he can still be rescued. In reality, though, the three are not Job's friends at all but disguised secret service agents whose duty it is to demolish Job as a person, to kill not his body but his mind, to show the rest of the world that no society can tolerate the degree of integrity which he is exhibiting. And so it is that in this drama of all against one the three are actually performing their roles not as individual actors but as voices of the Greek chorus. Thus, in a stroke of genius the author moves from the prose of the Prologue to the poetry of the Wisdom Dialogues, the three friends doing their utmost to make Job admit his guilt and accept his role as a scapegoating Other. With that goal constantly in mind each agent appears three times (Zophar possibly only twice), each of them receiving Job's answer in return. The first exchange is actually started by Job himself, who in his confused state curses the day of his birth, a theme to which he will constantly return.

*

**Cursing the day of his birth** Job said:

> May the day of my birth perish,
>     and the night it was said, "A boy is born!"
> That day—may it turn to darkness;
>     may God above not care about it;
>     may no light shine upon it.
> . . . . . . . . . . . . . . .
> Why did I not perish at birth,
>     and die as I came from the womb?
> (3:3–4, 11)

As a primordial Wittgensteinian who looks out of his window and sees something he has never seen before, Job knows neither where he is nor where he should go. Locked in behind a Berlin Wall of his own making, he asks the crucial question "Why is life given to a man *whose way is hidden,* whom God has hedged in?" (2:23, emph. added).

But whereas Job cannot tell why he is lost, the so-called friends take his disorientation as an indication that he has committed a serious offence. The uncertainties must therefore be straightened out and the place to do so is in a suite of interrogation rooms. Initially these apartments seem to be located at the crossroads of Peniel (the house of Bethel also known as the Checkpoint Charlie), but in reality they are in the basement of the castle of the Kantian Island of Truth.

The first to enter these rooms is Eliphaz, who reminds his bereaved comrade of his former status and ensures him that the Almighty never punishes an innocent. Hence Job should have nothing to fear, he should simply take it easy, accept whatever comes his way and return to the place where he belongs. Once this has been said, however, a spirit glides past Eliphaz's face, his voice changes and he asks whether a simple mortal can be more righteous than the Lord—an obvious hint that this is exactly the crime of which Job is suspected. The stark light shines straight into Job's eyes as the interrogator tells him:

If it were I, I would appeal to God;
    I would lay my cause before him.
He performs wonders that cannot be fathomed,
    miracles that cannot be counted.
. . . . . . . . . . . . . . . . . . . . . . .
Blessed is the man who God corrects;
    so do not despise the discipline of the Almighty.
(5:8–9, 17)[16]

To these veiled threats Job responds with the vehemence they deserve, staunchly repeating his conviction that

If only my anguish could be weighed
    and all my misery be placed on the scales!
It would surely outweigh the sand of the seas—
    no wonder my words have been impetuous.
(6:2–3)

After this outburst Job quiets down. Assessing the situation, he concludes that Eliphaz is more words than deeds, in fact that he is trying to dispatch his friend into the desert equipped with navigational tools which will lead him not to the salvation he desires but to the doom he fears, not to the safety of an oasis but to the dangers of an overflowing wadi. To be precise:

A despairing man should have the devotion of his friends,
    even though he forsakes the fear of the Almighty.
But my brothers are as undependable as intermittent streams,
    as the streams that overflow
when darkened by thawing ice
    and swollen with melting snow,
but that cease to flow in the dry season,
    and in the heat vanish from their channels.
Caravans turn aside from their routes;
    they go up into the wasteland and perish.
. . . . . . . . . . . . . . . . . . . . . . . . . . .
Am I the sea, or the monster of the deep,
    that you put me under guard?
(6:14–18, 7:12)

Thus ends the first exchange between Job and Eliphaz the Temanite, a quarrel about a map whose fix-points are not fixed and about a canvas which is not properly prepared.

Resuming the questioning is the second of Job's friends, Bildad the Shuhite, his opening words brimming of violence—"How long will you say such things? Your words are blustering wind" (8:2). Then, almost automatically, comes the accusation that Job is not to be trusted because the net in which he is capturing the world is best described as a spider's web, a fragile construction whose threads will never hold. The point is, of course, that Job is not merely wrong, but that his calamities are punishment for his own wickedness, everything his own fault:

> Surely God does not reject a blameless man
>     or strengthen the hands of evildoers.
> He will fill your mouth with laughter
>     and your lips with shouts of joy.
> Your enemies will be clothed in shame,
>     and the tents of the wicked will be no more.
> (8:20–22)

"You are right," says Job in a paraphrase of what many consider his greatest speech. "Who has the right to ignore the commands of the great majority? How can I dispute with a LORD who I can neither see nor perceive, a self-appointed power so evasive that it calls itself *I am who I am*? I am convinced that even if the Almighty were summoned, he would never show up, his place on the defender's bench would either stand empty or be occupied by some well-paid advocate. Rather than appear in the open, the tautology of tautologies hides in obscurity, his thoughts-and-actions impossible to grasp. Whim of whims, the epitome of the otherness of the Other, an imagination beyond imagination, at the same time excluding and embracing. But listen! What is it that lies on the other side of the limits of language? What would I find if I submitted the laws of censorship to the test? What would happen if I intentionally violated the second paragraph of the Constitutional Law, if I told the real truths of how the world is governed? Clearly the Almighty is at the same time both everywhere and nowhere." Rephrased:

> He is not a man like me that I might answer him,
>     that we might confront each other in court.
> If only there were someone to arbitrate between us,
>     to lay his hand upon us both,
> someone to remove God's rod from me,
>     so that his terror would frighten me no more.
> Then I would speak up without fear of him,
>     but as it now stands with me, I cannot.
>
> . . . . . . . . . . . . . . . . . . . . . . . . . . .
>
> I will say to God: Do not condemn me,
>     but tell me what charges you have against me.
>
> . . . . . . . . . . . . . . . . . . . . . . . . . . .

Do you have eyes of flesh?
    Do you see as a mortal sees?
(9:32–35, 10:2, 4)

A clear-cut issue of the relations between body and mind. For even though Job readily acknowledges God's physical creations for what they are, he has no idea how to grasp the social relations that come with them. He nevertheless concludes that whenever evil flourishes it is because the Almighty for some reason tolerates and sanctions it. But why has the LORD bothered to give Job his life if he is not prepared to protect it, why is he treating this upright servant as a dead thing rather than as the living individual he knows himself to be? Once more Job repeats that he would rather have died in his mother's womb than live in this world as it is, rather not exist at all than be an inmate in this prison-house without keys.

These complex issues from the border areas of theodicy and categorization are cleverly picked up by the third interrogator, Zophar the Naamathite. Leaving Job no peace, he pushes the defendant closer to the abyss than any of the other questioners had done before:

Will your idle talk reduce men to silence?
    Will no one rebuke you when you mock?
You say to God, 'My beliefs are flawless
    and I am pure in your sight.'
Oh, how I wish that God would speak,
    that he would open his lips against you
and disclose to you the secrets of wisdom,
    for true wisdom has *two sides*.
    Know this: God has even forgotten some of your sin.
Can you fathom the mysteries of God?
    *Can you probe the limits of the Almighty?*
. . . . . . . . . . . . . . . . . . . . . . . . . . .
But a witless man can no more become wise
    than a wild donkey's colt can be born a man.
(11:3–7, 10, emph. added)

At issue is the exact drawing of the boundary between Plato's two realms of objects and cognition, by extension also the validity of Kant's thesis about the unity of consciousness and the status of the third paragraph of the Constitutional Law. Job obviously knew that the interrogators were trying to break him, that they were attacking his very right to be uniquely himself. To that end he is continuously reminded that if he submits, then he will once more be rewarded with all kinds of riches.[17] Still, even though he is well aware that he is a human being only in relation to others, he refuses to give in. His only reply to the torturers' allegations is that

Doubtless you are the people,
    And wisdom will die with you!
But I have a mind as well as you:
    I am not inferior to you.
    Who does not know all these things?
(12:2–3)

Pleading his case, the persecuted says that the boundary between the wicked and the blameless cannot possibly be guarded by God's forces, for how could one then explain that the marauder's tents are left undisturbed and the wise judges turned to fools. In a remark which goes to the heart of cartographical reason, Job even notices that "[God] deprives the leaders of the earth of their reason; *he sends them wandering through a trackless waste*" (12:24, emph. added). But unlike the animals, to whom God's actions seem completely arbitrary, the nature of justice is a truth which the humans cannot fathom. Little wonder, therefore, that Job tells Bildad that the latter is lying on the Almighty's behalf, making claims that would never be accepted in a human court. In no uncertain terms he repeats:

My eyes have seen all this,
    my ears have heard and understood it.
What you know, I also know;
    I am not inferior to you.
But I desire to speak to the Almighty
    and to argue my case with God.
You, however, smear me with lies;
    you are worthless physicians, all of you!
If only you would be altogether silent!
    For you, that would be wisdom.
Hear now my argument;
    listen to the plea of my lips.
Will you speak wickedly on God's behalf?
    Will you speak deceitfully for him?
Will you show him partiality?
    Will you argue the case for God?
Would it turn out well if he examined you?
    Could you deceive him as you might deceive men?
(13:1–9)

"I am not inferior to you, I am an independent person in my own right, not an obedient representative of the crowd!" No wonder, then, that Job paradoxically appeals to the Hobbesian Sovereign, well aware that it is the master who sets the limits of what may and may not be thought-and-said. In desperation he begs:

Only grant me these two things, O God,
and then I will not hide from you:
Withdraw your hand from me,
and stop frightening me with your terrors.
Then summon me and I will answer,
or let me speak, and you reply.
How many wrongs and sins have I committed?
Show me my offense and my sin.
Why do you hide your face
and consider me your enemy?
(13:20–24)

Thus ends the first cycle of exchanges with an appeal to the LORD himself: "Why do you not show your face? Why are all power-holders so afraid of the open?"

And with that pivotal question the discussion gradually moves from the sufferings of one particular man to issues of the human condition in general, Job effectively becoming a forerunner of James Joyce's HCE, an everyman found again. The second cycle of speeches is consequently focused on the problem of why the blameless so often are punished for what they have not done, while the wicked are rewarded for their evils. "So," Job asks the friends, "how can you console me with your nonsense? Nothing is left of your answers but falsehood" (21:34). Where is he, that evasive torturer who never gets enough of my flesh, who draws his net around me and blocks all escape routes? Job's body is in pain, his mind swings back and forth. He prays and he hopes:

Oh, that my words were recorded,
that they were written on a scroll,
that they were inscribed with an iron tool on lead,
or engraved in rock forever!
I know that my Redeemer lives
and that in the end he will stand upon the earth.
And after my skin has been destroyed,
yet in my flesh I will see God;
I myself will see him
with my own eyes—I and not another.
How my heart yearns within me!
(19:23–27)[18]

No answer given, though, and after five additional chapters the discussion tapers off into the third round of speeches and eventually into Job's last words to his friends. It is there that we encounter the most cartographic passages of the entire book, a set of incursions into the no-man's land between material things and social relations, an attempt to establish the boundary-line between the Territory of the Humans and the imaginations of the utterly different. They deserve the greatest attention.

*

**Establishing the boundaries between the humans and the utterly different** is the key purpose of Eliphaz's third speech:

> Can a man be of benefit to God?
>     Can even a wise man benefit him?
> What pleasure would it give the Almighty if you were righteous?
>     What would he gain if your ways were blameless?
> (22:2–3)

A soliloquy of rhetorical questions all followed by an allusion to Job's Babylonian ancestors and polytheistic connections, another hint of high treason:

> You are true to the ancient way
> trodden by sinful men of old,
> men destroyed before time was,
> their foundation swept into the flood,
> who said to El, "Away from us!"
> (22:15–17, Scheidlin trans.)

To Eliphaz's list of wrongdoings —Job stripping people of their clothing, leaving the poor hungry and the widows empty-handed, hoarding gold and mistrusting his brothers—the accused responds by exemplifying both the type of wickedness that God regularly ignores and all the bad things he seems to accept. But why is Job so surprised? After all he must be aware that the LORD long ago decreed himself to be a jealous ruler, who will punish those who hate him and show mercy unto those who love him. In addition, we have also learned that his face must not be seen. Terror is terror for terror's sake, for as Shakespeare knew, "As flies to wanton boys, are we to the gods,—They kill us for their sport" (*Lear,* 4.1.8–9).

No wonder, therefore, that Job's greatest concern is that the LORD is nowhere to be found, that there is no map that shows the way to his dwelling, no compass that points the bewildered in the right direction. Neither fix-points nor scale, least of all a well-prepared mappa. Thus,

> If I go to the east, he is not there;
>     if I go to the west, I do not find him.
> When he is at work in the north, I do not see him;
>     when he turns to the south, I catch no glimpse of him.

. . . . . . . . . . . . . . . . . . . . .

But he stands alone, and who can oppose him?
He does whatever he pleases.
He carries out his decree against me,
and many such plans he still has in store.
That is why I am terrified of him,
when I think of all this I fear him.
(23:8–9, 13–15)

Even so, for each chapter Job comes closer and closer to accepting reality as it is, perhaps realizing that since power is everywhere it cannot be pinned down—in that case a most significant advance in his existential and epistemological development.

After this third exchange between Eliphaz and Job the turn returns to Bildad. In an impressive, probably truncated, speech we are offered an account of God's creation as something entirely beyond the reach of human imagination. In yet another veiled reference to the Constitution's first paragraph the interrogator declares:

Dominion and awe belong to God;
He establishes order in the heights of heaven.
Can his forces be numbered?
Upon whom does his light not rise?
How then can a man be righteous before God?
How can one born of woman be pure?
If even the moon is not bright
and the stars are not pure in his eyes,
how much less man, who is but a maggot—
a son of man, who is only a worm!
(25:2–6)

Job interrupts(?) and admits that even though Bildad is right, he himself is not a maggot but a human being, a person who under no circumstances will sacrifice his integrity. At the same time he is so eager to outdo Bildad's description of the wondrous creation that he comes close to blasphemy; in a famous passage there is even a reference to the net in which Marduk captured Tiamat before the winds blew her to pieces. But man himself

He lies down wealthy, but will do so no more;
when he opens his eyes, all is gone.
Terrors overtake him like a flood;
a tempest snatches him away in the night.
The east wind carries him off, and he is gone;
it sweeps him out of his place.

It hurls itself against him without mercy
as he flees headlong from its power.
It claps its hands in derision
and hisses him out of his place.
(27:19–23)

And yet. Far away from people's dwellings, in places forgotten by the foot of man, the explorer searches for sapphires, lapis lazuli and nuggets of gold, at the outskirts of the world he tunnels the rocks for treasures and discovers the sources of the rivers. But where is the place of wisdom, the type of understanding which cannot be bought with the finest gold, its price beyond rubies?

Job's own answer is given in the so-called Wisdom Poem,[19] in my interpretation an outstanding example of an invisible map of the invisible:

Where then does wisdom come from?
Where does understanding dwell?
It is hidden from the eyes of every living thing,
concealed even from the birds of the air.
. . . . . . . . . . . . . . .
God understands the way to it
and he alone knows where it dwells,
for he views the ends of the earth
and sees everything under the heavens.
When he established the force of the wind
and measured out the waters,
when he made a decree for the rain
and a path for the thunderstorm.
And he said to man,
"The fear of the LORD —that is wisdom,
And to shun the evil that is understanding."
(28:20–21, 23–28)

The point is, of course, that the Almighty's wisdom is not in what is said but in what is shown, the principle of factum-verum two millennia before Giambattista Vico, the *Ebstorfer Karte* long before Gervase. But the Joban poet also knew that "the true" changes into *truth* only by being anchored in the collective unconscious; on the Island of Truth and Power the *vox dei* is one with the *vox populi,* the *vox populi* indistinguishable from the *vox dei.* Yet, man's search for genuine wisdom is futile, for while human understanding is a matter of conventional logic, the LORD's reasoning is of a totally different kind.

And for that reason Job recalls how long ago he used to take his seat in the public square, the young men stepping aside and the old rising to their feet, the listeners drinking his words like spring water. Those were the years of Job's glory, so fondly remembered and so sorely missed. Those were the happy days, a time when the foreigner had not yet been chosen to be a scapegoat, an alien not yet accused of revolutionary leanings or seen as a traitor who uses the LORD's name in vain and rejects the interrogator's interpretation of Kant's thesis about the necessary unity of consciousness. Job understands, of course, that whoever is suspected of breaking all paragraphs of the Constitutional Law is seen as a dangerous rival to every dictator, God himself included. No wonder, therefore, that

> Now they mock me,
> men younger than I,
> whose fathers I would have disdained
> to put with my sheep dogs.
> Of what use was the strength of their hands to me,
> since the vigor had gone from them?
> (30:1–2)

Totally isolated, Job has no allies and must fight the two-front war of what it means to be human on his own. Banishing his former friends to the outer rim of the oikumene, he turns them into the equivalents of Actaeon's hounds, the type of beings which in the mappae mundi are pictured as monsters huddling in the brushy undergrowth, like Enkidu before Shamhat not full-fledged humans but two-thirds animals—a base and nameless brood, a congregation of brutes from the Rockscape of meaningless matter. It is easy to imagine Job's anger when he realizes that these lowly creatures are spitting him in his face, unstringing his bow, laying snares for his feet, tearing his roadmaps asunder, invading the human territory which in his mind had always been reserved for the equals of himself.

The invisible forces of the Mindscape attack him too, none less than God himself grabbing him by the collar, throwing him into the mud, reducing the trusted vassal to dust and ashes. In desperation Job turns into a person who understands that he has just seen what he never saw before:

> I cry out to you, O God, but you do not answer;
> I stand up, but you merely look at me.
> You turn on me ruthlessly;
> with the might of your hand you attack me.
> You snatch me up and drive me before the storm.
> . . . . . . . . . . . . . . . .
> Have I not wept for those in trouble?
> Has not my soul grieved for the poor?
> Yet when I hoped for good, evil came;
> when I looked for light, then came darkness.

. . . . . . . . . . . . . . . . . . . . . . . . . . . . . .

I have become a brother of jackals,
  a companion of owls.
My skin grows black and peels;
  my body burns with fever.
My harp is tuned to mourning,
  and my flute to the sound of wailing.
(30:20–22, 25–26, 29–31)

Still. Even though he pleads for mercy, he never surrenders. Once back in the palace courtyard, he addresses the crowd in the same manner as when he was the mightiest man in the land. An example of textbook rhetoric, a declaration to the effect that in Job's closet there are no skeletons:

If I have walked in falsehood
  or my foot has hurried after deceit —
let God *weigh me on honest scales*
  and he will know that I am blameless —
if my steps have turned from the path,
  if my heart has been led by my eyes,
  or if my hands have been defiled,
then may others eat what I have sown,
  and may my crops be uprooted.
(31:5–8, emph. added)

Following this magnificent introduction he then lists a range of sins for which any offender ought to be punished—sleeping with another woman, mistreating his servants, leaving the hungry unfed and the poor unclothed, adoring his money and rejoicing at his enemy's misfortunes. But none of this is what Job knows himself to have done, none of this he even concealed in his heart. Therefore the words of Job were ended with yet another plea that his case be heard, albeit with the provision that he will never accept blame for what he knows that he has done right:

Oh, that I had someone to hear me!
  I sign now my defense—let the Almighty answer me:
  let my accuser put his indictment in writing.
Surely I would wear it on my shoulder,
  I would put it on like a crown.
I would give him an account of my every step;
  like a prince I would approach him.
If my land cries out against me
  and all its furrows are wet with tears,
if I have devoured its yield without payment
  or broken the spirit of its tenants,

then let briers come up instead of wheat
    and weeds instead of barley.
(31:35–40)

*

**The defense was signed**. And in the brief prose passage which then follows we are told that the three men stopped answering, presumably because they understood that in Job's own eyes he was a righteous man determined never to yield. In their judgment, however, the case had been settled already from the start, for whatever they heard him say they never believed. To them the process has run its course, just as it later was to do in Franz Kafka's *The Trial,* the masterpiece whose first and last sentences are virtually quotations from its biblical forerunner:

> Someone must have been telling lies about Joseph K., for without having done anything wrong he was arrested one fine morning. . . .
>
> Where was the Judge whom he had never seen? Where was the High Court, to which he had never penetrated? He raised his hand and spread out all his fingers.
>
> But the hands of one of the partners were already at K.'s throat, while the other thrust the knife deep into his heart and turned it there twice. With failing eyes K. could still see the two of them immediately before him, cheek leaning against cheek, watching the final act. "Like a dog!" he said; it was as if the shame of it must outlive him.[20]

Long before these European events a young man called Elihu, son of Barakel the Buzite, got seriously upset not only with Job, who with such flagrant blasphemy had justified himself rather than the LORD, but also with the three friends, who were severely criticized for not refuting the treasonous foreigner. While in the youngster's eyes the former was guilty of violating both the first paragraph of the Constitutional Law and the prohibitions of the second, the latter were not living up to the obligations of paragraph three. In Elihu's own version of what had happened:

I waited while you spoke,
    I listened to your reasoning;
while you were searching for words,
    I gave you my full attention.
But no one of you has proved Job wrong;
    none of you has answered his arguments.
Do not say, "We have found wisdom:
    let God refute him, not man."
(32:11–13)

In this youthful manner Elihu goes on and on, totally convinced that the Almighty is unable to do anything wrong; God's wisdom is lavishingly praised and the splendor of his creations paraded, his might demonstrated by the verdict that the godless will not live but are doomed to die among male prostitutes. Anticipating the storm out of which the LORD will soon be speaking, he concludes:

> The Almighty is beyond our reach and exalted in power;
> in his justice and great righteousness, he does not oppress.
> Therefore, men revere him,
> for does he not have regard for all the wise in heart?
> (37:23–24)

To all this Job offers no reply, perhaps because Elihu's speech does not belong to the original text but was later added as a kind of popular sermon. For the same reason Yahweh's great speech in chapters 38–41 never refers to Elihu, but addresses Job alone.

As could be expected, the God's words to Job are among the most overpowering ever penned, the rhetoric of a physical scientist turned social engineer, cartographical reason in practice, the pre-Platonist God affirming light where the suffering Job wishes for darkness. Nowhere, however, does the Almighty claim that he is just, only that he is almighty:

> Who is this that darkens my counsel
> with words without knowledge?
> Brace yourself like a man;
> I will question you;
> and you shall answer me.
> *Where were you when I laid the earth's foundation?*
> Tell me, *if you understand.*
> Who marked off its dimensions? Surely you know!
> *Who stretched a measuring line across it?*
> On what were its footings set,
> or *who laid its cornerstone*
> while the morning stars sang together
> and all the angels shouted for joy?
> Who shut up the sea behind doors
> when it burst forth from the womb,
> when I made the clouds its garment
> and wrapped it in thick darkness,
> when *I fixed the limits* for it
> and set its doors and bars in place,
> When I said, "*This far you may come and no farther;*
> here is where your proud waves halt"?
> (38:2–11, emph. added)

With unmistakable allusions to Gilgamesh—two-thirds god and one-third man—God here poses a series of questions that focus on the concept of limits, especially on the drawing of the boundary between the Realm of Humans, whose inhabitants all have names, and the lands of Rockscape, where nothing is baptized and everything passes unnoticed. The examples overflow:

> Have you journeyed to the springs of the sea
> or walked in the recesses of the deep?
> Have the gates of death been shown to you?
> Have you seen the gates of the shadow of death?
> Have you comprehended the vast expanses of the earth?
> Tell me, if you know all this.
>
> . . . . . . . . . . . . . . . . . . . . .
>
> *What is the way to the place* where the lightning is dispersed,
> or the place where the east winds are scattered over the earth?
>
> . . . . . . . . . . . . . . . . . . . . .
>
> Can you *bring forth the constellations* in their seasons
> or lead out the Bear with its cubs?
>
> . . . . . . . . . . . . . . . . . . . . .
>
> Do you know when the mountain goats give birth?
> Do you watch when the doe bears her fawn?
>
> . . . . . . . . . . . . . . . . . . . . .
>
> Who let the wild donkey go free?
> Who untied his ropes?
> (38:16–20, 24, 32; 39:1, 5, emph. added)

Having been subjected to this crushing combination of factum-verum and might makes right, Job's only response was:

> I am unworthy—how can I reply to you?
> I put my hand over my mouth.
> I spoke once, but I have no answer—
> twice, but I will say no more.
> (40:4–5)

Yahweh himself, though, is not to be silenced. Instead he proceeds into his own version of where the Human Territory should end. Therefore, while the first part of his speech focused on how he had created the phenomena that belong to the heartlands of the oikumene, he now moves to a description of the Ocean River that surrounds it. To that effect he presents two ferocious animals from the fringe, liminal beings that belong to the waters which no one can cross, imaginations of whatever lies beyond the cape of the *non plus ultra*. The first of these inventions is the Behemoth, a biblical description of the hippopotamus, an animal with limbs like rods of iron and a nose that cannot be pierced, the second the famous Leviathan, a crocodile which cannot be pulled in with a fishhook, a creature without fear and king over all who are proud, an imagination from the mist-enveloped regions of the noumenal and a man-hater impossible to turn into a pet, a force known to make covenants with no one.

The Behemoth and the Leviathan, like everything else that has to do with power, two instances of the terror built into the sublime,[21] border guards so fiery that no human dares to approach them, perhaps because the sublime by definition transcends the categories of all reason, cartographical reason included.[22] But since Job had never questioned the LORD's power, merely his justice, the only words he could say were:

> I know that you are all powerful:
> what you can conceive, you can perform.
> I am the man who obscured your designs
> with my empty-headed words.
> I have been holding forth on matters I cannot understand,
> on marvels beyond me and beyond my knowledge.
> (Listen, I have more to say,
> now it is my turn to ask questions and yours to inform me.)
> I knew you then only by hearsay;
> but now, having seen you with my eyes,
> I retract all I have said,
> and in dust and ashes I repent.
> (42:2–6)[23]

It should be noted that according to the *Revised Standard Version* the two last lines should be translated:

> *therefore* I despise myself
> and repent in dust and ashes.

*

**Therefore in dust and ashes I repent.** Those were Job's parting words, easier to quote than to understand. For even though the transgressor might well have repented, it seems entirely against his personality that he would ever have despised himself. Not so easy to know, however, not only because the Hebrew of the Book of Job is more ironic, hence more ambiguous, than anything else in the Bible,[24] but also because in the speech from the whirlwind nothing is said about justice and everything about power. The narrative structure is in fact such that "the Job-writer has created symmetry in the form of two demands and two refusals. Job speaks at length about justice and demands that God respond. God refuses. God speaks at length about power and demands that Job respond. Job refuses,"[25] perhaps because he senses that through the meticulous listing of his achievements—no Curriculum Vitae more impressive—the LORD may be hiding something more important, most likely the circumstance that if God is God he is not good and if God is good he is not God.

After the voice out of the whirlwind has uttered whatever it uttered, and after Job has answered whatever he answered, the analysis closes with a prose Epilogue, by many criticized for not being on par with the rest of the text. For instance, we now discover that what we have been reading is not what we had thought it was, not a piece of kathartic tragedy but the script of a serious comedy, albeit without the slightest trace of humor. Even more importantly, however, the Epilogue stands out as the ruling of the imaginary law-court to which Job so persistently and so unsuccessfully had appealed, the final verdict in the famous case of Job the blameless vs. God the almighty.

The issued decision begins with a recital of God's reprimand to Eliphaz in which the LORD—now turned defendant—admits that it is Job, not the agents from the Bureau of Investigation, who has spoken rightly of him. To redeem themselves the latter must first sacrifice seven bulls and seven rams and then subject themselves to the humiliation of letting Job pray for them. This being done—the smoke rising to the sky and Job saying his prayer—the three were forgiven their folly and retired from active duty. Even more remarkably, the LORD (or is it Satan) was sentenced to pay his formal vassal a tremendous damage settlement, not merely *quid pro quo* but everything except the number of children exactly doubled: fourteen thousand sheep, six thousand camels, a thousand yoke of oxen, a thousand donkeys, seven sons (by some accounts fourteen) and three beautiful daughters.

The handsome settlement notwithstanding, Job never rejoiced, partly because three new daughters can never relieve the pain of having lost just one of the old, mainly because he seems to realize that now, when he has actually seen the LORD, he really does not like him, perhaps because both of them are created in the image of the other. In addition, and contrary to common belief, Job understands that the acts of the Almighty are not only good, but at the same time *both* good *and* ill. The truth is that God is totally unpredictable and does whatever he fancies. Rephrased, the LORD never plays with the dice of two-valued logic but always with the utterances of performative speech acts, by nature thoroughly ambiguous. Perhaps it is even this newfound insight into the structure of power that the court wants to acknowledge through its groundbreaking decision, a type of understanding which the three apparatchiks never came close to. In the words of Jack Miles, whose readings have been most helpful for my own analyses:

> The Lord bows, in a way, to Job's characterization of God, abandons his wager with the devil, and after a vain attempt to shout Job down, atones for his wrongdoing by doubling Job's initial fortune. Job may, therefore,

> have saved the Lord from himself, yet God can never seem to Job after this episode quite what he seemed before it. More to the point, the Lord can never seem quite the same to himself. The devil is now a permanent part of his reality; and though at the eleventh hour he has broken free from the Adversary, he has done so through a deeper humiliation at the hands of a terrestrial adversary, Job himself. . . . But the God who is seen in this vision is new not just for Job but also for God himself. . . . After Job, God knows his own ambiguity as he has never known it before.[26]

After these events Job lived for a hundred and forty years and he saw his children and their children to the fourth generation. Most significantly, the Absolute never visited on any of them the iniquity of their father and that is despite the fact that Job had not only seen God but survived the ordeal, his case much clearer than Jacob's report of how he got his limp. So total is in fact Job's victory that in the Hebrew Bible God never speaks again; after the humiliating sentence and the fantastic fine, his hand is on his mouth just as Job's hand was on his. In the process, the boundary-line between the Landscape of the Humans and the Mindscape of the Almighty was irrevocably shifted, large parts of the territories that earlier had belonged to the gods now forever occupied by the ancestors of you and me.

And yet. Although the Book of Job ends with a ceasefire of historic proportions, this does not mean that the perpetual two-front war is over. On the contrary, for whereas the story of the Hebrew Bible moves from action to speech to silence, the Old Testament moves from action to silence to speech. This difference is crucial and was created by the Christian redactors, who changed the order of the individual books to make the total text—the Old and New Testaments taken together—fit better into their own belief system. Thus, by shifting the prophetic books of the Tanakh from the middle towards the end, the new story was made to flow more naturally from the old. For such is the power of rhetoric that it always surprises us with what we already know. And such is the rhetoric of power that integrity and political correctness are like oil and water.

*

**The questions multiply** when they serpent-like bite their own tail: How different would the world have been if the Lord God had made his covenant not with the childless Abram but with the respected Job? Why do the power-wielders never learn to live with the rumors that they are corrupt? Why must they always prove that the rumors are correct?

In his search for answers into the dialectics of power and submission, the Job-writer let the master magician perform his rhetorical transformations in such slow motion that the tricks are easy to discover. The world changed in the process—fix-point, scale and canvas, none of them ever the same again.

And thus it is that Job is unthinkable not only to Odysseus and Abr(ah)am, but to Herr Goldschmidt as well. The mystery is that the author (poet and prosaist alike) made Job a wise man (detached but not withdrawn) rather than a solipsist (withdrawn but not detached). But how could anyone expect from a text as thoroughly political as the Bible that it would show respect to whoever turned his back on the political?

What a wonderful instance of how we pre-write the future by re-reading the past, what a marvelous illustration of the third commandment in action.

# THEBES

I thought it out for myself.

**The third map** of our atlas reaches back to Greece and the fifth century BCE, a time and place of drastic changes in the attitude to what it means to be human, a period of intense struggle between the gods and the humans, a series of battles marvelously shaped and chronicled by Sophoclean tragedy, *Oedipus Rex* greatest of them all. It is to a focused reading of this play, written some time soon after 430, hence roughly contemporaneous with the Book of Job, that the present chapter is turning.[1]

The setting is crucial, for Sophocles lived his long life in the abyss between the *mythos* of Homer and the *logos* of Plato, a Janus-like figure who with one eye was scanning the old, with the other imagining the new. Perhaps the greatest tension of his time was in the attitudes to predicament, for while the archaic poets had taken a man's social standing to reflect his ability to handle contradiction, the new philosophers saw paradox as the worst enemy of human reason; as Ludwig Wittgenstein later put it, "without philosophy thoughts are, as it were, cloudy and indistinct: its task is to make them clear and to give them sharp boundaries."[2] But in Sophocles' eyes religion itself was nothing but a human invention designed to keep people in place, like other laws issued by the humans of the *polis,* not by the gods of the Olympus. And that circumstance explains why the tragedians assigned such a crucial role to the chorus.

The fifth-century outbursts of individual and collective creativity had far-reaching consequences for the conception of human action, not the least for the drawing of the boundary between the Human Territory, on the one hand, and the godly Mindscape, on the other. Most importantly, an increasing number of people began to question the old belief that if they did something, then it was not they who did it but the gods who acted through them. Thus, when Agamemnon in Homer's *Iliad* recalled the horrible atrocities through which so many heroes had lost their lives, he excused his own behavior by claiming that he had not acted freely. To be precise,

> I am not responsible
> but Zeus is, and Destiny, and Erinys the mist-walking
> who in assembly caught my heart in the savage delusion
> on that day I myself stripped from him the prize of Achilleus.
> Yet what could I do? It is the god who accomplishes all things.
> (*Iliad* 18.86–90, trans. Lattimore)[3]

It cannot be said more clearly: to be human during the mythological era was to be an extended arm of the gods, each one of them adorned with a proper name, a definite personality and a set of well-defined duties. So deeply ingrained was this belief in the gods' omnipotence that in Old Greek there was not even a word for what later generations have come to mean by the verb "to will."

Gradually, however, this understanding of human action began to be questioned. And out of those queries grew not the idea of free will (too early for that) but a novel conception of guilt and punishment. This new point was that if my actions are determined by the divinities, then it is they, not I, who should be praised or blamed. If, on the other hand, whatever I do is of my own making, then all issues of right or wrong fall back on me.

In this political context of one-against-many it is especially illuminating to compare the attitudes of Homer and Sophocles. Thus, three centuries after the story of Troy, it was Oedipus, blood streaming from his pierced eyes, who saw what Agamemnon never did:

> Apollo, friends, Apollo—
> he ordained my agonies—these my pains on pains!
> But the hand that struck my eyes was mine,
> mine alone—no one else—
> *I did it myself!*
> What good were eyes to me?
> Nothing I could see could bring me joy.[4]

Paving the way to this breakthrough—*I did it myself*—was Protagoras, Sophocles' contemporary and the forerunner not only of Plato's theory of perception but of the Renaissance invention of the self—reputedly the first person to say that man is the measure of all things. No easy way out, however, for every individual craves both the freedom of will and the chains of order. One century later the field of ethics was born, Aristotle the obstetrician in charge.

An integral part of these profound changes were the approximately one thousand tragedies which were composed during the brief time span of about a hundred years; indeed it was at the stone-laid theater that most Athenians learned what they were to take for granted. No small business, for judging from the size of the ruins the open-air theater of Dionysus could seat between fourteen and fifteen thousand spectators! And to the present day most commentators have agreed that tragedy is the art form which most seriously probes the role of man in the universe.

In constant focus for the tragedians—Aeschylus, Sophocles and Euripides foremost among them—was the exact drawing of the boundary between the humans, on the one hand, and the gods, on the other; as one influential interpretation puts it, "the tragic dimension of the play [*Oedipus Rex*] lies precisely in the gap between the remote, incomprehensible gods and any sense of humanly meaningful suffering."[5] Like the Jewish Job-writer, the Greek tragedians were consequently grappling with the problem of why humans must suffer and why we seem to be forever torn between the irreconcilables of good and evil, freedom and necessity, trustworthy truths and suspicious deceits. While Aeschylus—the oldest of the three masters—was the one who most explicitly returned to the questions of the old bards, Sophocles was more concerned with the practical action of what, under given circumstances, one should or should not do.

In search of an answer the latter mobilized a rhetorical army of unmatched competence in deductive logic and self-conscious reevaluation, epistemological strategies which in themselves are outstanding exercises in translation. In the course of these investigations he learned that it is through the transcendence of contradictions that the world is moved. As G. W. F. Hegel later came to argue, "just as lordship shows its essential nature to be the reverse of what it wants to be, so too, bondage will, when completed, pass into the opposite of what it immediately is: being a consciousness repressed within itself, it will enter into itself, and change round into real and true independence."[6]

Nothing mystical about that, merely yet another instance of the perpetual struggle between transparent certainties and oblique ambiguities, the theory of proper names and definite descriptions in practice. The play of King Oedipus may in fact be alternatively read as one of the best detective stories ever written, a manual of scientific methodology, an outstanding example of how we live forwards and understand backwards.

*

**The theory of proper names and definite descriptions** is the intellectual engine that propels the personæ of *Oedipus Rex* from a set of true premises to an equally true conclusion. The fuel of this epistemological mover is the principle of self-conscious reevaluation, the idea that in our search for truth we are driven by our ignorance of truth.[7]

Since everyone in the audience already knew the underlying myth, what mattered was not the plot itself but the particular twists by which the poet managed to make the story relevant. In Sophocles' case this meant that he must somehow relate what happened on the stage to the great plague that ravaged Athens between 429 and 425, a disaster so immense that it killed one quarter of the city's population, the genial Pericles most prominent among them. It is to this setting that the brilliant opening scene of *Oedipus Rex* takes us when, in a language full of ironies and double meanings, it defines the problem which the remainder of the play proceeds to disentangle.

At the very beginning it is consequently told how Oedipus, bearer of the famous name and powerful king of Thebes, receives a group of citizens. When he asks them why they have come, a priest steps forth with the answer:

> Your own eyes
> Must tell you: Thebes is tossed on a murdering sea
> And can not lift her head from the death surge.
> A rut consumes the buds and fruits of the earth;
> The herds are sick; children die unborn,
> And labor is vain. The god of plague and pyre
> Raids like detestable lightning through the city.
> . . . . . . . . . . . . . . . . . . . .
> *You are not one of the immortal gods, we know;*
> Yet we have come to you to make our prayer
> As to the man surest in mortal ways
> And wisest in the ways of God. You saved us
> From the Sphinx, that flinty singer, and the tribute
> We paid to her for so long; yet you were never
> Better informed than we, nor could we teach you:
> *It was some god breathed in you to set us free.*
> . . . . . . . . . . . . . . . . . . . .
> Once, years ago, with happy augury,
> You brought us fortune; *be the same again!*
> (*Oedipus Rex,* 4–5)

Be the same again! Surely a call that the king must reestablish his old identity and rescue the city as he once had done before.

On that previous occasion the problem had been to solve the riddle of the Sphinx: what is it that first walks on four feet, then on two, finally on three? Oedipus' solution was, of course, that this creature is "man himself," an answer that lies deeply hidden in his own name, which according to one etymology means "Know-foot," quite literally "He who knows the riddle of the feet."

On that interpretation Oedipus managed to solve the riddle by going to his own body. Not merely to his body, though, but to those areas of the Human Territory which border on the Rockscape of pure materiality, the traditional homelands of the monsters, the Sphinx included. To be precise, this hellcat was first forced to acknowledge the supremacy of man's intelligence and then to lift the taxes which the occupiers had imposed on the citizens. A border-line shifted.

Now, however, the challenge was of a different magnitude, for whereas subduing the Sphinx had been easy, conquering the plague will be of another magnitude entirely. Intuitively sensing that the new enemy came not from the monstrous Rockscape but from the divine Mindscape, Oedipus answered the supplicants:

> Poor children! You may be sure I know
> All that you longed for in your coming here.
> I know that you are deathly sick; and yet,
> Sick as you are, not one is as sick as I.
> Each of you suffers in himself alone
> His anguish, not another's; but my spirit
> Groans for the city, for myself, for you.
> (*Oedipus Rex,* 5)

Immediately following this declaration he goes on to announce that Creon—Jocasta's brother and a potential successor to the throne—had been dispatched to seek the advice of Apollo's oracle. Now, on the return from his diplomatic mission, Creon is publicly interrogated by the king, the latter growing increasingly impatient with the evasive answers he receives. In anger Creon strikes back:

> Then I will tell you what I heard at Delphi.
> In plain words
> The god commands us to expel from the land of Thebes
> An old defilement we are sheltering.
> It is a deathly thing, beyond cure;
> We must not let it feed upon us longer.
> (*Oedipus Rex,* 7)

As Creon continues his report from Delphi, it becomes clear that the plague-wind has been brought on the city by the death of Laios, the old man who long ago had been killed where three highways meet, the king of Thebes who Oedipus had come to succeed and whose widow he had married as a booty. In the emissary's words:

> He was murdered; and Apollo commands us now
> To take revenge upon whoever killed him.
> (*Oedipus Rex,* 8)

"But how," says Oedipus, "how can this be done after so many years. Where do we find a clue?" Creon retorts "Here in this land," a remark which Oedipus answers with the words:

> If any man knows by whose hands Laios, son of Labdakos,
> Met his death, I direct that man to tell me everything,
> No matter what he fears for having so long withheld it.
> Let it stand as promised that no further trouble
> Will come to him, but he may leave the land in safety.
>
> . . . . . . . . . . . . . . . . . . . . . .
>
> Thus I associate myself with the oracle
> And take the side of the murdered king.
> (*Oedipus Rex,* 13)

Working himself into a rhetorical frenzy, he even proclaims:

> Now I,
> Having the power that he held before me,
> Having his bed, begetting children there
> Upon his wife, as he would have, had he lived—
> Their son would have been my children's brother,
> If Laios had had luck in fatherhood!
> (But surely ill luck rushed upon his reign)—
> I say I take the son's part, just as though
> I were his son, to press the fight for him
> And see it won! I'll find the hand that brought
> Death to Labkados' and Polydoros' child,
> Heir of Kadmos' and Agenor's line.
> (*Oedipus Rex,* 14)

Here, if not before, it is obvious that getting rid of the plague will be far more difficult than outwitting the Sphinx, for whereas solving a riddle requires mere intelligence, interpreting an oracle takes real wisdom. In both instances, though, the key to success lies in language itself, especially in the relations between Signifier and signified, body and meaning. Therefore, when Oedipus finally understands, he blinds himself, not as an act of self-punishment but as a way of seeing the invisible. Nothing on purpose, however, although everything began with a purpose. Nothing but fate and no one to blame. Yet, every step could easily have been avoided.

Countless interpretations. Yet, in the present context Sophocles' play is essentially an epistemological study of the relations between proper names and definite descriptions, a search for truth which brings the explorer back to the Bar de Saussure, the three-star establishment where the wine is served undiluted. It is from that base camp of logical analysis that the author lets his actors venture into a world they never saw before, gradually introducing them to the secrets of a map whose fix-points are invisible, whose scale is twisted, and whose mappa is of a revolutionary yet timeless kind. As it turns out, the only way to save the city is to establish an equation in which the identity statement is at the same time both true and informative.

*

**Oedipus' map takes the form of an equation,** a construction which is built around the expression

$$a = b$$

What is to go on the *a*-side of the equality sign is specified early on and by none less than Apollo himself, who

> Sent us back word that this great pestilence
> Would lift, but only if we established clearly
> The identity of those who murdered Laios.
> (*Oedipus Rex,* 16)

This specification seems clear enough even though everyone in the audience knows that it will later lead to horrible fears and futile hopes. Not many but one killed Laios. The mistake is not inevitable, though, because Teiresias, the blind seer, leaves no doubt about what is to go on the *b*-side of the equation, his words perfectly transparent and void of any ambiguity:

> I say that you are the murderer whom you seek.
> (*Oedipus Rex,* 19)

And yet, the original specification is not so wrong. Like everyone else, also Oedipus fits many descriptions.

Teiresias' prophecy is too serious to ignore and too damaging to believe. As a consequence there follows a long struggle of power and legitimation, the tone and arguments highly reminiscent of the LORD's speeches in the Book of Job—where were you, young man, when I created the world! Oedipus, who has good reasons to suspect a conspiracy, perhaps a coup d'état, responds by insinuating that Teiresias has received his instructions not from the heavenly Apollo but from the worldly Creon, the competitor who the present king believes is out to destroy him:

> [Creon] has bought this decrepit fortune-teller, this
> Collector of dirty pennies, this prophet fraud—
> Why, he is no more clairvoyant than I am!
> Tell us:

> Has your mystic mummery ever approached the truth?
> When that hellcat the Sphinx was performing here,
> What help were you to these people?
> Her magic was not for the first man who came along:
> It demanded a real exorcist. Your birds—
> What good were they? Or the gods, for the matter of that?
> But I came by,
> Oedipus, the simple man, who knows nothing—
> I thought it out for myself, no birds helped me!
> And this is the man you think you can destroy,
> That you may be close to Creon when he's king!
> Well, you and your friend Creon, it seems to me,
> Will suffer most. If you were not an old man,
> You would have paid already for the plot.
> (*Oedipus Rex,* 20–21)

Oedipus' point is that since he had saved the city before, he can do so again. Furthermore, since on that previous occasion he had acted on his own, without the assistance of any gods, he is certain he can do so again—yet another version of the Almighty reminding the Israelites that it was he who cut their chains, he who had brought them out of Egypt. The standard technique of advance legitimation, a case of rhetoric when it works.

When Teiresias hears the accusation that he is nothing but Creon's errand-boy, he angrily retorts that unlike Oedipus—who keeps referring to his previous run-in with the monstrous hellcat—he himself is no mere human but the blessed servant of the gods themselves. Securely anchored in his own identity, well aware that the forces behind him are not of this world, Teiresias never waivers, merely observes that

> You are a king. But where argument's concerned
> I am your man, as much a king as you.
> I am not your servant, but Apollo's.
> I have no need of Creon to speak for me.
> Listen to me. You mock my blindness, do you?
> But I say that you, with both your eyes, are blind:
> You can not see the wretchedness of your life,
> Nor in whose house you live, no, nor with whom.
> Who are your father and mother? Can you tell me?
> You do not even know the blind wrongs
> That you have done them, on earth and in the world below.
> But the double lash of your parents' curse will whip you
> Out of this land some day, with only night
> Upon your precious eyes.
> Your cries then—where will they not be heard?

. . . . . . . . . . . . . . . . . . . . . . . .

All this, and more, that you can not guess at now,
will bring you to yourself among your children.[8]
Be angry, then. Curse Creon. Curse my words.
I tell you, no man that walks upon the earth
Shall be rooted out more horribly than you.
(*Oedipus Rex,* 21–22)

"While *you,*" says the blind seer, "are bragging about your business with the monsters, *I* am coming from a session with the gods in person. While you won your medals at the Rockscape front, I have just returned from negotiations with the Mind-Lord himself."—"You are a fool," shouts the offended king, who then receives the stinging reply:

A fool? Your parents thought me sane enough.

On hearing that remark Oedipus halts himself, anger turns to curiosity, curiosity to fear. Like a bird shot in flight, he cries and wonders:

My parents again!—Wait: Who were my parents?
(*Oedipus Rex,* 22)

Before Teiresias leaves he specifies in detail all the identities that Oedipus will need in order to solve the equation and save the city. He even goes to the trouble of warning that what is now in Oedipus' private mind is not in the common facts. Thus,

The man you have been looking for all this time,
The damned man, the murderer of Laios,
That man is in Thebes. To your mind he is a foreign-born.
But it will soon be shown that he is a Theban,
A revelation that will fail to please.
A blind man,
Who has his eyes now; a penniless man, who is rich now;
And he will go tapping the strange earth with his staff.
To the children with whom he lives now he will be
Brother and father—*the very same;* to her
Who bore him, son and husband—*the very same*
Who came to his father's bed, wet with his father's blood.

Enough. Go think that over.
If later you find error in what I have said,
You may say that I have no skill in prophecy.
(*Oedipus Rex,* 23–24)

The very same! That is identity. Identity is the very same. The chorus, not yet knowing whom to believe, offer no advice, simply reflect:

> Bewildered as a blown bird, my soul hovers and cannot find
> Foothold in this debate, or any reason or rest of mind.
> (*Oedipus Rex,* 25)

The mind of the chorus drifts. Not tied down to bodily facts, it has no way of establishing the identities; staring into the unknown they sometimes see a king, sometimes an incest-ridden murderer. The singers are not to be faulted, though, for how could they know? Not even in Oedipus' own mind had anything yet become the very same.

Nothing is yet the same. For what is in Teiresias' mind are the words of the oracle and what is in Oedipus' head are the memories of everything he has done to avoid the horrible prophecy which says that he is destined to kill his father and sleep with his mother. What one of them sees transparently, the other sees obliquely. To remove the ambiguities, Creon, the man who hates anarchy, pleads for certainty:

> You cannot judge unless you know the facts.
> . . . . . . . . . . . . . . . . . . . . . . . .
> And if you think there is anything good in being stubborn
> Against all reason, then I say you are wrong.

To this he receives Teiresias' reply:

> If you think a man can sin against his own kind
> And not be punished for it, I say you are mad.
> (*Oedipus Rex,* 27)

This appeal to reason is later echoed by Jocasta, who begs Oedipus to anchor his thoughts not in the fantasies of his own fears but in the realities of her facts. Hence first the advice

> Set your mind at rest,
> (*Oedipus Rex,* 36)

then—in a flashback to Homeric beliefs—the lady's version of what had happened to her previous husband and their child:

> Now you remember the story: Laios was killed
> By marauding strangers where three highways meet;
> But this child had not been three days in this world
> Before the King had pierced the baby's ankles
> And left him to die on a lonely mountainside.

> Thus, Apollo never caused that child
> To kill his father, and it was not Laios' fate
> To die at the hands of his son, as he had feared.
> This is what the prophets and prophesies are worth!
> Have no dread of them.
> It is God himself
> Who can show us what he wills, in his own way.
> (*Oedipus Rex*, 36–37)

The double point is that ambiguity is in the mind and that the mind must never be trusted. Certainty is in the facts that count.

*

**The ambiguous must not be trusted**, for the ambiguous belongs to the abysmal in-between. And since in Jocasta's judgment the oracle has been wrong before, it can well be wrong again. As so often in Greek tragedy her argument was well intended, its goal to quiet Oedipus' forebodings. The effect, however, was the opposite, for instead of calming him down her words raise within him a horrible premonition, a Janus-like fear grounded in the past and projected into the future, *the* turning point of the entire play. For from now on the once powerful Oedipus is searching no longer for the identity of Laios' killer but for the identity of himself. His voice trembles as he exclaims:

> How strong a shadowy memory crossed my mind,
> Just now while you were speaking; it chilled my heart.

What chilled Oedipus' heart was the thought that his own curses may perhaps be turned against himself. To be exact, the memory that crossed his mind stirred within him the fear that he himself might be the man he is searching for, the murderer he has already sentenced by his own words. What bothers him is that Jocasta's account agrees too well with his own recollection of how long ago he had killed that stranger at the crossroads, the geographic fix-point in which the projection-lines from the three cities of Thebes, Corinth and Delphi all come together. When squeezed, their respective stories differ in one detail only. The number of men who did the killing.

The only person who can settle the uncertainty is the single man who escaped the ordeal, one of the slaves in Laios' entourage. This man must be found and brought back to Thebes to be questioned. For as Oedipus sees with frightening clarity,

> If his account of the murderer tallies with yours,
> Then I am cleared.
> (*Oedipus Rex*, 42)

As the proto-positivist he is, Oedipus realizes that what must be checked is not the consistency of his own reasoning but the accuracy of Jocasta's facts, not the logic but the empirical observations. Thus,

> Why, "marauders," you said,
> Killed the King, according to this man's story.
> If he maintains that still, if there were several,
> Clearly the guilt is not mine: I was alone.
> But if he says that one man, single-handed, did it,
> Then the evidence all points to me.
> (*Oedipus Rex,* 43)

Once at this stage, solving the equation depends on a quantitative measure, the determination of a simple constant. As soon as that detail has been settled, then the identities will all have defined themselves. Oedipus' concern is therefore no longer with the legality of his actions—after all the man he had killed had struck first and to strike back is justifiable self-defense. His worries are instead about the curse that spurted out of his own mouth. For if it turns out that one man, single-handed, killed the old man, then it is Oedipus himself who must leave the city, his wife and children destroyed in the process.

What seems so straightforward is nevertheless soon to be confused again. New hope is created by a messenger from Corinth, a slave who has been sent to Thebes to tell the news that King Polybos, the man who Oedipus has always taken to be his father, is dead. Jocasta is jubilant, for Polybos had not been killed, merely been taken away by old age. She introduces the messenger to Oedipus with the words:

> He has come from Corinth to announce your father's death!

to which the messenger replies

> I can not say it more clearly: the King is dead.
> (*Oedipus Rex,* 48)

Oedipus breathes a sigh of relief, repeating first that he does not trust the Delphic oracle, then that he considers human reason more trustworthy than divine prophecy:

> Ah!
> Why should a man respect the Pythian hearth, or
> Give heed to the birds that jangle above his head?
> They prophesied that I should kill Polybos,
> Kill my own father; but he is dead and buried,
> And I am here—I never touched him, never,

Unless he died of grief for my departure,
And thus, in a sense, through me. No. Polybos
Has packed the oracles off with him underground.
They are empty words.
(*Oedipus Rex,* 48)

In her blindness to the other face of the messenger's statement, Jocasta raves. Once more she repeats that what counts are not soft thoughts but solid facts. Renewing the plea that Oedipus must strike Teiresias out of his mind, she denounces not only the soothsayer but the gods who sent him:

Why should anyone in this world be afraid,
Since Fate rules us and nothing can be foreseen?
A man should live only for the present day.

Have no more fear of sleeping with your mother:
How many men, in dreams, have lain with their mothers!
No reasonable man is troubled with such things.
(*Oedipus Rex,* 49)

In Jocasta's mind the trustworthy rests in material things, not in social relations, the former entities stable and certain, the latter constantly changing and soaked in ambiguity. In the real world dreams and fears are nothing but empty words, standard attributes of the man who does not know himself.

But that impression will soon be changed into its opposite, the oblique turned transparent. Thus, with the irony so typical of tragedy, truth is once again revealed not for its own sake, but with the purpose of relieving Oedipus of his dread. When he refuses to go back to Corinth because he fears that he might sleep with Merope, Polybos' wife, he is consequently told not to worry, because

Polybos was not your father.
(*Oedipus Rex,* 51)

Given this piece of information, the situation changes again, the process of self-conscious reevaluation speeding up. For now the problem is no longer how to save the city from the plague but how to determine who Oedipus is, no longer an issue of man in general but of one person in particular. What Jocasta in her ignorance had seen as a true identity no longer is, at least not if the Corinthian messenger is to be believed. Since that would be too painful to accept, the poor man is ordered to substantiate his proof, to show that in the equation of who Oedipus is, there Polybos appears as nothing but a zero. This is swiftly done as the messenger declares that Polybos was

> No more your father than the man speaking to you.
> (*Oedipus Rex,* 51)

However, not even this is proof enough for the shaken Oedipus, who exclaims:

> But you are nothing to me!

to which he gets the reply:

> Neither was he.
> (*Oedipus Rex,* 52)

Zero is equal to zero is equal to zero. Even after the double check, nothing is equal to nothing. But in Greek logic, nothing is nothing. To say that one's father is nothing is therefore absurd. Nobody, not even a king, can have nothing as his father.

After these revelations the statement that links the proper name "Oedipus" with the definite description "the son of Polybos" is no longer true. Instead it is hinted that the true and informative identity is hidden in Oedipus' own name. When the messenger—who by an ironic twist turns out to be the very same who once gave the foundling to Polybos—is asked

> From what did you save me?

he therefore first receives the reply:

> Your ankles should tell you that

then the elaboration:

> I cut the bonds that tied your ankles together.

With this information Oedipus is finally ready to move from mind to body, from ambiguity to certainty. No need for further reasoning, merely for the factual observation:

> I have had the mark as long as I can remember.

Everything settled, the true identity between the proper name and the definite description finally verified by the Corinthian's words:

> That was why you were given the name you bear.
> (*Oedipus Rex,* 53)

The argument is that the secret of Oedipus' identity lies hidden in his name; as every Greek knew, "Oedipus" means "Swollen-foot." That name nevertheless retains some of the vagueness necessary for tragedy, for whereas *oidi* translates as "swell," *oida* means "I know." Never is this wordplay more effective than when the Corinthian first arrives and asks where he can find the person he is looking for. Thus, in an almost literal translation, he asks:

> Strangers, from you might I learn where [*oimopou*]
> is the palace of the Tyrannus Oedipus, [*oidipou*]
> best of all, where is he himself if you know where[9] [*isthopou*]

Hope nevertheless lingers on, and to be certain beyond doubt Oedipus proceeds to check the final details of the calculation. Persisting in his questioning of the only man who had escaped from the crossroads, Oedipus demands to know the number of men who killed Laios, obviously a crucial variable in the equation of who he is.

In contrast, Jocasta—who in vain has tried to stop the interrogation of the slave—can no longer be methodical. Quite correctly, and once again in a statement brimming of well-meaning, she concludes:

> You are fatally wrong! May you never learn who you are!
> (*Oedipus Rex*, 55)

Not bothering with yet another constant, she jumps to the end of the proof, crying:

> Ah, Miserable!
> That is the only word I have for you now.
> That is the only word I can ever have.
> (*Oedipus Rex*, 56)

No longer able to equate the name "Oedipus" with any of the alternative definite descriptions "my king," "my husband," "father of my children," she cries out her "Ah, Miserable." In the new context anything less is false, for only words of grief can now communicate the truth she has discovered. As she exits into the palace, he, though, persists in his stubbornness. Perhaps, who knows, he may not be the legitimate son of the king and the queen, but the fruit of his mother's affair with a slave:

> Let it come!
> However base my worth, I must know about it.
> The Queen, like a woman, is perhaps ashamed
> To think of my low origin. But I

Am a child of Luck; I can not be dishonored.
Luck is my mother; the passing months, my brothers,
Have seen me rich and poor.
If this is so,
How could I wish that I were someone else?
How could I not be glad to know my birth?
(*Oedipus Rex,* 56)

With this idea in mind the battered king proceeds to the last calculation. To complete that task he needs the help of the person who long ago had brought the baby to the mountain and instead of leaving him there to die had handed him over to a stranger he did not know. The shepherd tells him:

I pitied the baby, my King,
And I thought that this man would take him far away
To his own country.
He saved him—but for what a fate!
For if you are what this man says you are,
No man living is more wretched than Oedipus.
(*Oedipus Rex,* 62)

At the end, the miserable king cries out:

Ah, God!
It was true!
All the prophesies!
— Now
O Light, may I look on you for the last time!
I, Oedipus,
Oedipus, damned in his birth, in his marriage damned,
Damned in the blood he shed with his own hand!
(*Oedipus Rex,* 62–63)

It was all fate. Yet every act was free. The chorus sing:

Alas for the seed of men.

What measure shall I give these generations
That breathe on the void and are void
And exist and do not exist.
(*Oedipus Rex,* 63)

That measure is zero. Zero is the measure of nothing. And with this final calculation completed, all pieces of scattered evidence have been fitted into the pattern that everyone in the audience knew long before Sophocles staged his play. Finally Oedipus has convinced himself that his reasoning has been correct, that there is nothing more to learn, that all prophecies were true. Construction by negation has run its course, the incest-ridden father-murderer doomed by his past. Not because anyone wished it that way, merely because he himself had been faithful to his own pronouncements uttered in ignorance and anger. Like the blind Isaac, he accepts the consequences, because the question of whether his reasoning rules are right or wrong is not for him to decide. In classical Greece "the virtue of proof does not depend on induction; nor is it based on the hypothesis of a relative or revisable credibility. Proof is institutional; that is to say, it is derived from a system of conventions in which the signifier tends to absorb the signified. Insofar as the 'signified' belongs to the past, proof is not offered in order to put the judge in immediate contact with it. Such is the archaic state from which law emerges."[10]

*

**Zero is the measure of the damned**, the quantitative proof that Oedipus has been defeated by his own means. Through the principle of self-conscious reevaluation he has finally been anchored in the unyielding of his own identity. As so often, lives are torn apart at the moment they are brought together. A messenger exits from the palace with the words:

> I think neither Istros nor Phasis, those great rivers,
> Could purify this place of the corruption
> It shelters now, or soon must bring to light—
> Evil not done unconsciously, but willed.
>
> The greatest griefs are those we cause ourselves.
> . . . . . . . . . . . . . . . . . . . .
> The Queen is dead [by her own hand.]
> She ran to the apartment in her house,
> Her hair clutched by the fingers of both hands.
> She closed the doors behind her: then by that bed
> Where long ago the fatal son was conceived—
> That son who should bring about his father's death—
> We heard her call upon Laios, dead so many years,
> And heard her wail for the double fruit of her marriage,
> A husband by her husband, children by her child.
> (*Oedipus Rex*, 65–66)

Hearing the news, Oedipus bursts in moaning. Storming around and begging for a sword he breaks the double door of Jocasta's apartment, runs into the room where she is hanging, loosens the rope, lowers her body, rips the golden brooches from her gown, raises them and plunges them straight into his eyeballs. In the messenger's words, he shouts:

> "No more,
> No more shall you look on the misery about me,
> The horrors of my own doing! Too long you have known
> The faces of those whom I should never have seen,
> Too long been blind to those for whom I was searching!
> From this hour go in darkness!" And as he spoke,
> He struck at his eyes—not once, but many times;
> And the blood spattered his beard,
> Bursting from his ruined sockets like red hail.
> (*Oedipus Rex,* 67)

Bewildered, lost in a world he does not know, his map torn to pieces, his compass smashed:

> Oh, Ohh—
> the agony! I am agony—
> where am I going? where on earth?
> where does all this agony hurl me?
> where's my voice?—
> winging, swept away on a dark tide—
> My destiny, my dark power, what a leap you made.
> (*Oedipus the King,* trans. Fagles, 1443–48, p. 39)

His mind scratching for a different anchorage, groping in his blindness for his daughters' heads, he begs:

> Could I but touch them
> They would be mine again, as when I had my eyes.
> (*Oedipus Rex,* trans. Fitts and Fitzgerald, 75)

And yet he understands that their identities will never be complete in the same sense that his own was once complete. The difference is that whereas his identity extends both into the past and into the future, theirs will be only in the past. Thus he notes:

> Then, whom
> Can you ever marry? There are no bridegrooms for you,
> And your lives must wither away in sterile dreaming.
> (*Oedipus Rex,* 76)

Therefore, in the end what really matters may not be *who* you are but *that* you are. Not a question of epistemology but of ontology, not a matter of material *ex*istence, but of cultural *sub*sistence. Only anarchists can roll with the tide and on no occasion was Oedipus an anarchist. The lasting lesson is that in the struggle between ultimates no one wins.

*

**In the struggle between ultimates**, no one wins. No one except perhaps Creon, the pretender who loves order and until the time at Colonus has the power to make right through might. Wrestling with paradoxes is the only way to learn, because paradoxes can be neither solved nor ignored.

As times go by, as they eventually did also for Oedipus Tyrannus, we come gradually to understand that in every already there is always a not yet, in every not yet always an already. Finding out who Oedipus was may or may not have rid the city of the plague; Sophocles never bothered to say. What he did say, though, was that at the crossroads nobody knew and that in the palace nobody rejoiced. And what is thereby proven is that honesty is in pursuit and pursuit in tragedy, life itself a game of dice played by men and watched by gods. As Oedipus eventually understood and accepted, whatever fate there is we bring onto ourselves. To do otherwise is to be dishonest to oneself, to break the rules of one's own game. In the long run that is impossible.

After the decisive battle at Thebes the boundaries of the Human Territory were consequently moved into areas which previously had been controlled by the utterly alien, the monstrous Sphinx standing guard at one front, the godly Apollo at the other. And just like Gilgamesh long before him, so also Oedipus found that he too was purely human, that whatever he had done, he had done himself.[11]

Most importantly, the dishonored decided not to kill himself but to go on living, tapping his way through a world unveiled. His life shattered, he asks the chorus, who show him great respect perhaps because he is treating his children better than his parents had treated him:

> Come lead me. You need not fear to touch me.
> Of all men, I alone bear this guilt.
> (*Oedipus Rex*, 72)

But Creon, the inferior who is now king, he tells him:

> Think no longer
> That you are in command here, but rather think
> How, when you were, you served your own destruction.
> (*Oedipus Rex*, 77)

No one, not even a king, should deem himself happy until he is dead. Too late by then.

Therefore it says much about Sophocles' skill as a dramatist that he never showed the three most violent events of the play enacted on the stage—the killing of Laios, Jocasta's suicide, Oedipus' self-blinding—but left it to various messengers to report them. Everything important is consequently happening not in front of the onlookers' eyes, but in imaginary places far beyond. The seeing is blind, the blind is seeing. As the Greeks were certain that blindness is a punishment for the crime of breaking the limits of human knowledge, they also knew that it is a means to insight.[12] A preview of Eratosthenes' invisible shadow, imagination at work.

But who really was that creature called "Oedipus"? Breathed on by the gods, as the chorus thought at the beginning, nothing at all, as the Corinthian said towards the end? A king unwittingly serving the common cause or an incestuous murderer sentenced by his own words?

No wonder this figure was turned into a scapegoat, his great sin being that he was a slayer of distinctions, father and son in one, husband of his mother, child of his wife, brother of his children. All the very same. An inter-esting figure if one ever was, like Jesus Christ a paradigmatic inhabitant of the ontological realm of the in-between, "a priest who is his own sacrificial lamb, a lamb who is his own sacrificing priest, a father who is his own son, an Isaac who is his own Abraham, with the dagger in his own hand."[13]

And so it is that the great tragedians were cultural land-surveyors. As in the case of the secret agent of the present book, their mission was to determine the boundaries of the Human Territory, ambiguous thoughts constituting the enemy at one front, factual certainties the adversary at the other.

# NICAEA

> All authority in heaven and
> on earth has been given to me.

**The last map** of the four-square is in many ways the most interesting of the entire atlas. To understand why, merely examine the word *inter-esting* itself. Have a taste of that expression from the devilish corner of Walt Whitman's square deific, press it with your tongue against the inside of your body. And it will come, whispering in your imagination that the English *interest* stems from the Latin *inter esse,* literally "in-between-being." To be inter-esting is consequently to dwell in the razor-sharp limit between ontological categories, to be a permanent resident of the Bar de Saussure, that marvelous establishment which itself is erected on the remnants of the abysmal palace of sweetwater Apsu.

Although the different categories of being carry different aliases in different contexts, they all stem from two families only, one called "Mind," the other "Matter." Like other Mafia relations, these concepts are thoroughly intertwined, the one impossible without the other, the other impossible without the one. Indeed it is the eternal feud between these forces that the secret agent now must map, a mission which will take us back to battlefields drenched in desires that can never be satisfied, a set of monologues in which the last word will never be said. At issue is once again the question of what it means to be a semiotic, hence a rhetorical, animal, an ironic creature steeped in the Sisyphean spirit of trying to express the inexpressible.

History is replete with attempts to reach a final solution to this irresolvable question, to end once and for all the conflicts between Mindscape and Rockscape. The examples are legion, for by what would the true artist be obsessed if it were not by the dream of merging expression and impression, impression and expression. And by what would the dedicated scientist be driven if not by the desire to crack the code in which his data are hiding their secrets. In both cases an endless struggle with lines of power and limits of language.

The various arts and sciences, mythologies and ideologies, all grapple with this same problem of representation. Through the maps of the present atlas we have already returned to the pivotal cases of Gilgamesh in the city of Uruk, Jacob at Peniel and Odysseus in Ithaka, Job from the land of Uz and Oedipus in the palace of Thebes. The turn has now come to Jesus Christ, history's most remarkable attempt to negotiate lasting peace between the humans and the utterly different.

What I will eventually present is obviously not a full-scale map of this social symbol, for no expedition could ever do that.[1] Offered instead is hardly a thumbnail sketch, partly because the issues are so inherently difficult, partly because our inherited preconceptions work so strongly against us. And yet the stakes might not be as high today as they were during the third to the ninth centuries AD, a period in which the question of who and what Jesus actually was had become so acute that it could no longer be brushed aside. Was Christ "begotten" and "divine by nature" or was he "created" and "divine by adoption"? How could the contradictory influences from Jewish monotheism and Greek polytheism be brought together in a coherent way? How were religious mysticism and logical precision to live with each other?

Not surprisingly, the offered solutions varied widely. Therefore, as a way of cooling things down, Emperor Constantine I, himself an unbaptized neophyte, summoned a general council of the whole Church (the first ever), about two-hundred-seventy bishops most of them from the eastern parts of the empire. The purpose of the meeting (which began in May 325 and was held at the emperor's summer residence on the Lake of Nicaea near the town of Nicomerdia) was to end the bishops' bitter wrangling and begin instead an era of harmony. A pivotal event in the history of Christendom, for here, as elsewhere, power and organization are one and the same.

As at other party conventions—the Socialist Internationals foremost among them—the arguments were strong and often invidious, perhaps because early Christianity, like other revolutionary movements, dreamed about the creation of a new type of human being. The common goal was nevertheless to reach a consensus, to determine once and for all the true nature of Christian faith, to establish a foundation of unquestionable solidity. And in that sense the meeting was certainly a great success, for, after small but significant revisions, the resulting document has remained the key codification of Christian dogma. Never a communion without it.[2]

Throughout the long negotiations at the splendid palace the emperor personally played a major role, often styling himself as "a bishop for the outsiders." And for that reason it is especially noteworthy that also the Creed, like the Constitutional Law and the Laws of Thought, consists of three articles; in each instance a detailed specification of the fix-points, scale and mappa that inform the believer not only where (s)he is but from whence (s)he has come and to where (s)he should go. The rhetoric is further strengthened by the fact that the warp of the Creed and the weft of the New Testament are intricately interwoven; not only did the bishops spend much time on the question of whether the Creed could include words that were not in the Gospels,[3] but it was only *after* they had agreed on the wording of the Creed that they went on to decide which of the competing texts should be included in the Holy Scriptures.

And so it is that Jesus Christ of Nicaea must be treated as one of the most revolutionary inventions ever made. In and through him the question of what it means to be human is pushed to its limits.

*

**Jesus Christ of Nicaea**—a most revolutionary invention, a triangulation of belief which in the same symbol brings together Moses' first stone tablet and Aristotle's Laws of Thought; not only does the first article of the Creed tie the faithful directly to the monotheism of the first commandment, but it states explicitly that here is a power so absolute that it knows nothing but identity.[4] Thus:

> We believe in one God, the Father Almighty, Maker of heaven and earth, and of all things *visible and invisible.*

From this beginning of beginnings the Creed swiftly moves to the second article, a most remarkable amendment to the censorship paragraph of the Constitutional Law, a drastic turnabout. For what this article projects is a desire totally alien to the Judaic tradition, a play of identity and difference in which Jesus Christ is actually made to occupy the same position as the LORD himself, both of them roaming the empty halls of Apsu's palace, both of them sharing the same penthouse at the top of the Plotinian house, both of them creatures of the excluded middle. A fix-point never fixed before, the tautological

$$a = a$$

begetting an informative

$$a = b$$

The wording is ingenious, a set of formulations well worthy of the emperor who stood behind them, a clear-cut case of how the mapmaker's scale determines what is what:

> And [we believe] in one Lord Jesus Christ, the only-begotten Son of God, begotten of the Father before all worlds, God of God, Light of Light, Very God of Very God, begotten not made, *being of one substance* with the Father by whom all things were made; who for us men, and for our salvation, came down from heaven, and was incarnate by the Holy Spirit of the Virgin Mary, and was made man, and was crucified also for us under Pontius Pilate. He suffered and was buried, and the third day he arose again according to the Scriptures, and ascended into heaven, and sitteth on the right hand of the Father. And he shall come again with glory to judge both the quick and the dead, whose kingdom shall have no end.

Uta-napishti dressed up, for the "Son of God" was a name that Jesus had not been given by someone else but in God-like fashion had chosen for himself (Matt. 11:27, John 10:36).

Then, finally, the third article, in essence the mappa of Christian faith, a projection screen so carefully prepared that the messages of the previous articles will neither crack nor run off but forever stick to the taken-for-granted of our thoughts-and-actions. And in this sense the emperor and the bishops were in agreement that just as every state must have its religion, so every religion must have its state. In an attempt to entrap the strange teacher the Pharisees said:

> "Tell us what you think. Is it lawful to pay taxes to the emperor, or not?" But Jesus, aware of the malice, said, "Why are you putting me to the test, you hypocrites? Show me the coin used for the tax." And they brought him a denarius. Then he said to them, "Whose head is this, and whose title?" They answered, "The emperor's." Then he said to them, "Give therefore to the emperor the things that are the emperor's, and to God, what is God's." When they heard this, they were amazed; and they left him and went away. (Matt. 2217–22, New RSV)[5]

United we stand, divided we fall! And thus it is that the Creed remains committed to the spirit, although not always to the letter, of the third paragraph of the Constitutional Law. Therefore, after an updating of the specific allegiances:

> We believe in the Holy Spirit, the Lord and Giver of Life, who proceedeth from the Father [and the Son],[6] who with the Father and the Son together is worshipped and glorified, who spoke by the Prophets. And *we believe one holy catholic and apostolic Church*. We acknowledge one baptism for the remission of sins. And we look for the resurrection of the dead, and the life of the world to come.
> Amen.

Trinity created, triangulation at work. A codification of a belief system as power-filled as anything ever chiseled onto the first stone tablet, as logical as anything ever uttered in the Greek Academy.[7] The parallels are astonishing and at the very end of *Timaeus* Plato concludes:

> And so the world received animals, mortal and immortal, and was fulfilled in them, and became a visible God, comprehending the visible, made in the image of the Intellectual, being the one perfect only-begotten heaven. (*Timaeus* 92c, trans. Jowett)

*

**Triangulation at work** in the production of yet another map of power. Picture and story intimately interwoven in the border-line figure of Jesus Christ himself, the practice of proper names and definite descriptions long before Bertrand Russell imagined the theory; for whereas "Yeshua" is a Hebrew name, "Christos" is the Greek term for "anointed" or "consecrated." However, while one is Jewish by one's mother, one is Christian by baptism and communion, the former by biological necessity, the latter presumably by choice. Both cases with exceptions, of course, yet it is true to say that the western conception of belonging is fundamentally torn by contradictions. In the words of the Founder himself, "Whosoever loves father or mother more than me is not worthy of me" (Matt. 10:37); indeed, "Whosoever comes to me and does not hate father or mother, wife and children . . . cannot be my disciple" (Luke 14:16). The Lord of the Book versus the Lord of the Sensation(al).

Now, on the Christian base map there is a thin line which runs exactly midway between the two boundaries of the Territory of the Humans, the Mindscape of godly spirits located on one side, the Rockscape of material monsters on the other. The name of that inter-esting line is "Jesus Christ," a denotation of the in-between whose (theo)logical meaning hinges on the Greek word *homoousios*—"of one substance."

If the Nicene definition is to be believed, then this Lord Jesus Christ, the Son of one substance with the Father, was begotten not made. By decree a paradoxical figure in constant limbo; a slashy entity from the transition zone between ontological categories; the sole inhabitant of the no-man's land of Inter-Esse; *the* paradigmatic case of Signifier and signified inseparably merged together. Not a Kantian phenomenon to be shared and understood, though, but a noumenon imported from the lands beyond the beyond, an imagination which can never be reached via the well-paved avenue of logical reason, only through an individual, Kierkegaardian, leap of faith.

And yet. Although the abysmal may never be said, it can sometimes be shown. For even though no mortal has ever been one with the ultra-

MINDSCAPE

— · — · — · — · — · — · — · — · — · — · — · — · — · — · — · — · — JESUS CHRIST

ROCKSCAPE

Map of Jesus Christ at the inter-esting center of the Human Territory.

thin line of the in-between-being, the divisor may still be asymptotically approached. The perfect time for doing so falls between Easter and Pentecost, those fifty days when Jesus was floating around leaving us uncertain about exactly where he was, indeed whether he was at all. Dead or alive, ascended to heaven or buried in a rock? Decaying matter or Holy Spirit? Like a Giacometti sculpture, always tentative but never vague.

The rebirth—more accurately the resurrection—is nevertheless near, that most peculiar miracle which for two thousand years has been tied to the celebration of Whitsunday, the Christians' special day for commemorating the descent of the Holy Spirit on the disciples. And for that reason Whitsunday is still considered the ideal time to be baptized, for just as the Savior on that day (according to our calendar never on the same date two years in a row) was moved into a new existence, so the convert, through the sacrament of baptism, is entering into a new life. As usual, naming is the name of the game, for protected by her invisible talisman the believer is now free to cross the border between categories—like the master himself who no longer was "Yeshua" but "Jesus Christ," like the Polish churchman who with the white smoke from the Vatican chimney ceased to be Karol Wojtyła and became Pope John Paul II instead, like my own wife who once was Miss Magnusson and no longer is. And the proselyte's garment is traditionally white, a color which according to Wassily Kandinsky is both silent and vertical. As Marcel Duchamp might have put it: the convert stripped bare by her baptizers, even.

But which were the places that Jesus visited during those fifty days when he was in ontological and epistemological limbo? Where was he and where was he going? As usual a map provides the answer.

The picture-part of that map is one never seen before, even though the *Ebstorfer Karte* may be invoked as an example of the same type. With the story-part the situation is different, for everyone remembers how Jesus, after he had been made to die on the cross, was taken down and buried at sundown on the eve of Passover (Matt. 28; Mark 16; Luke 24; John 20–21.); as it turns out a deliverance just in time, for at night and during the sabbath no funerals were allowed to be held. Immediately thereafter we are introduced to Mary Magdalene, the mysterious woman who many say was his most devoted disciple, possibly his mistress, perhaps his wife. In the morning of the first weekday after the burial she goes to the grave to mourn the loss. Once there, however, she discovers that the stone at the opening of the tomb has been removed and that the crypt is empty except for the linen cloths which had been folded together and laid aside. Bewildered and dismayed she turns around. And there he is, Yeshua himself! Overwhelmed she rushes to embrace him. But he, the beloved, he who used to kiss her on the mouth, he draws back hiding himself behind the phrase "*Noli me tangere,* don't touch me! I have not yet ascended to the Father in Heaven."

When Jesus utters these words he has obviously left his original position in the inter-esting limit between ontological categories. Instead he is to be found in a more spiritual place located somewhere between the homoousios-line of one substance, on the one hand, and the Mindscape of pure meaning, on the other. Perhaps he avoids being touched because he does not want his woman to know that in his new mode of existence there is no real flesh to hold on to, merely an imaginary symbol to believe in. Jacques Lacan before the purloined lectures.

And then there is evening the same day. The frightened disciples have locked themselves inside a house fearing that the Romans (or is it the Jews) will arrest them too, just as they had previously done with their teacher. Suddenly, without warning, he stands there among them, breathes on them and says, "Receive the Holy Spirit. If you forgive anyone his sins, they are forgiven; if you do not forgive them, they are not forgiven" (John 28:23).

But Thomas, one of the twelve, he who was called Didymus or the Twin, he was not in the house when Jesus came by. On his return he is immediately told what has happened, for which story could anyone tell that would be more remarkable than this:

> "He has been here, he has been here!"
>
> "Who has been here?" Thomas asked.
>
> "Jesus! Jesus himself! We have seen the Lord!"
>
> "I don't believe you," said Thomas, who after this was nicknamed *The Doubter.* "Jesus is dead as a rock. You are hallucinating, you have seen a ghost. But I would believe, if I could see him with my own eyes, if I could stick my finger through the nail-holes in his hands, if I could touch the lance-wound in his side. Let me be frank and tell it as I see it: I trust my own body more than I believe your story."
>
> Of this Jesus of course came to know, for in the inner circles of the

> Almighty's Court there are no secrets unknown. Therefore, eight days later, he returns to settle the case.
>
> "Thomas," he says, "I know of your doubts. Now, come here. Come, come. Fear not! Your finger, please. My hands are here, here my side. Still your doubts. Touch and you shall trust!"
>
> After this powerful demonstration of the power of the example—empiricism when it works—Thomas had no choice but to submit.
>
> "My Lord, *My God!*"

Then came the conclusion, a variation on the interrogator's approach to Herr Goldschmidt:

> "You believe, Thomas," said Jesus, "because you have seen me with your own eyes, because you have heard me with your own ears, because you have touched me with your own hands. Your belief is anchored in your body, in Euclid's *Quod Erat Faciendum*. But beware, for blessed are those who believe not because of their five senses but because of their sixth, those who have learned and internalized their *Quod Erat Demonstrandum*. I say unto you that from now on it will be those who can grasp me without having seen me that will be the sheep of my flock." *Q.E.I.*

And the disciples, who through his breath now received his spirit, were born anew, no longer torn apart by their skepticism but united by their knowledge of the Scriptures and their memories of the persecuted master.

Forty days later, as a resounding echo of the third paragraph of the Constitutional Law, Jesus confirmed the bond in words that every guerilla leader has adopted as his own:

> All authority in heaven and on earth has been given to me. Therefore, go and make disciples of all nations, baptizing them in the name of the Father and of the Son and of the Holy Spirit, and teaching them to obey everything I have commanded you. And surely I am with you always, to the very end of age. (Matt. 28:18–20)

And surely I am with you! Such was Jesus' parting message, an updated affirmation of the absolute kingdom which he claimed to have inherited from his father, absolute of absolutes. After his encounter with Thomas, he is no longer roaming the borderlands of Mindscape but he finds himself on the other, more worldly, side of the inter-esting line. The first step in an ontological transformation which is bound to continue, for in a sense the shepherd's journey through the Human Territory was exactly opposite to that of Enkidu, the half-gazelle who two millennia earlier, through the love of a harlot and the friendship of a king, became a man never to be forgotten.

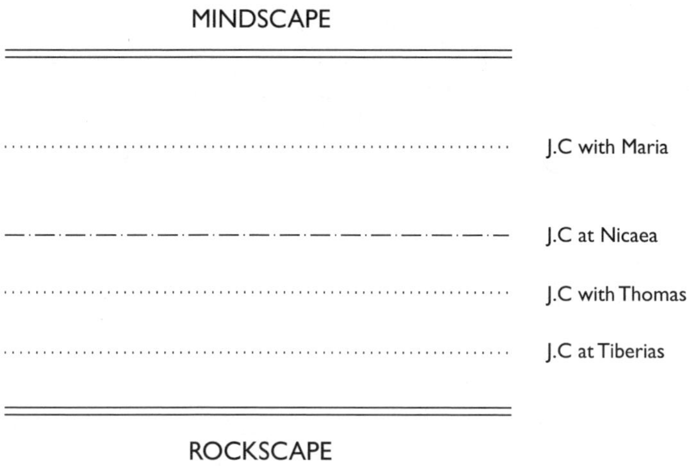


Map of Jesus Christ's whereabouts between Easter and Pentecost.

After the rhetorical victory over the doubter, Jesus showed himself once more, this time at the Sea of Tiberias, where some of his followers had sought shelter. Worn, haunted and much afraid, they went fishing, albeit without luck. Then, just as the day was breaking, they looked up from their chores, searching for a place to put ashore. And there he was, saying to them, "Cast the net on the right side of the boat and you shall no longer go hungry." They did as told and they harvested so much that they could hardly haul it in.

This was the third time Jesus revealed himself to the disciples, the thrust of his rhetoric on each occasion becoming more and more Marxian, less and less Kierkegaardian. Transferred to the map, the incidence at the Sea of Tiberias is found somewhere between the empiricism of the doubting Tom and the utterly different of the mute Rock-Lord.

The outlines of western culture in concentration. A map of the interesting world in-between. Three fix-points for surveying the abysmal. The scale of Plato's line for understanding the tricks of ontological transformation. *The Mappa of the Katholikos* for making us obedient and predictable.[8] A guide to the landscape of western culture.

*

**Such is the map of maps**. No apologies offered, no apologies required. I shall nevertheless be the first to acknowledge that my presentation has been based on an unusual, perhaps heretic, reading of what has rightfully been called *The Book of Books*. Yet, my intention has most certainly not been to be blasphemous, merely to show that whoever grapples with issues of representation is bound to reach deeply into the taboo-laden pockets of the unconscious. And since that unconscious nowadays is structured like a language, the critic has no choice but to explore the taken-for-granted relations between Signifier and signified, mind and matter, body and culture, reification and deification.

These central tenets of culture lead directly to the realm of power and thereby to the close connections between the Greek terms *aletheia* and *pistis,* on the one hand, and the English *truth* and *trust,* on the other; in particular it should be noted that both parts of the latter pair stem from the same Indo-European root *deru,* which means "tree."[9] It follows that whenever I am asserting that a particular statement is true, then I am not saying that it is factually correct but that it is socially correct. A "true" statement is consequently true only if it is trustworthy, only if its truthhood is shared by a sufficient number of others who for some reason know how to draw the Mason-Dixon line between left and right, right and wrong, between black and white, freeman and slave. In this context it should also be noted that the term "propaganda" is the gerundive of the Latin *propagare,* where *pro* means "before" or "forward" and *pag* is the root of *pangere,* "to fasten." By no coincidence "Propaganda" is also the name of the special committee of cardinals—The Congregation for the Propagation of Faith—which since 1622 is in charge of the Vatican's foreign missions.

Regardless of whether the specific premises are logical, scientific or religious (regardless of whether the reasoner's passport is stamped Q.E.D., Q.E.F. or Q.E.I.), the conclusion is inevitable: reality is never what it appears to be, for reality and language are never one and the same. Already Polyphemos, the monstrous Cyclops, had had to learn the painful lesson that there is a relation, yet a decisive difference, between Signifier and signified—even if somebody is *called* "Nohbdy" that does not mean that (s)he *is* nobody. Likewise with Oedipus, the godlike king, who had to blind himself in order to see what he had not previously seen, to realize that whereas the imaginary eyes of perspectiva naturalis are drawn to the real world of things, the imaginations of the perspectiva artificialis reach into the symbolic world of human relations. Both the cunning Odysseus and the honest Oedipus were of course well aware that the semiotic animal is thoroughly paradoxical, indeed that it can be what it is only by being what it is not. But the geniuses who created these literary figures also knew that the rhetorical animal is thoroughly ironic, indeed that the very definition of truth is that the sage can say that something is something else and be believed when (s)he does it.

In this game of make-believe Constantine the Great was no amateur, perhaps because he was such an integral part of his own culture that he knew neither this nor that. Torn between the irreconcilable differences of Judaic monotheism and Greek polytheism, his only hope of political survival was to reach some sort of conciliation between the opposing traditions. At issue was not only the interpretation of the Second Commandment, and thereby the meaning of the second paragraph of the Constitutional Law, but also he ingredients of the Aristotelian Laws of Thought, especially the principle of the excluded middle. The real challenge was to keep the warring factions at bay, to make sure that the people of the empire remained predictable and governable. And since neither the emperor nor the bishops were stupid, they somehow knew how to turn the unborn Marx upside down. Therefore, when they met together in the splendid halls of Nicaea, they might well have whispered, one to the other, that

> the power of a belief system appears at first sight a very queer thing, abounding in metaphysical subtleties and theological niceties. Analysis shows, however, that it is, in reality, a very inter-esting commodity located exactly in the taboo-ridden abyss of the excluded middle, the symbolic figure of Jesus Christ the most outstanding of all examples. Visible and invisible; touchable and untouchable; mind and matter—all power-holders being of one substance with the ontological transformations by which everything is made, the concept of nothing included. Proper names and definite descriptions glued together by a concoction of guns and words. And that is why no organization is likely to survive as long as the holy catholic and apostolic Church already has done, for no other organization will ever have a spy in every village.

*

**A spy in every village**, such is the wet dream of every politician, Constantine I and his bishops included. And as the sophisticated persons they were, they doubtless knew that in the long run it is more crucial to control the production of meaning than the making of things. Yet they also realized that the way to the mind always goes via the body, that the uncontrollable explosions of metonymic meanings are set off by metaphoric charges. In addition, some of them, Constantine not the least, knew how to use ambiguous language as a way of diffusing opposition; while belief may be a theological problem, coercion is an issue of politics. In the present case we have already noted that the magic term was *homoousios,* a word that may be translated not only as "of the same substance" but also as "of the same existence," "essence," "reality," "being," "form," "definition," or even "truth."[10] Indeed it has frequently been noted that "there were few words in Greek as susceptible of so many and so confusing shades of meaning as *ousia,* [a term whose precise meaning] varied with the philosophical context in which it occurred and the philosophical allegiance of the writer."[11]

It was exactly on this ambiguity of language that Constantine was playing, even though he was well aware that "ousia" was a term that most churchmen did not like because in the Scriptures it was nowhere to be found. After much arm-twisting the bishops nevertheless accepted it as a theological and political safety valve. As the record shows, only two voted against it and these heretics were promptly sent into exile along with some priests who supported them.[12] Foremost among the latter was the ageing

Arius (c. 250–336), an ascetic presbyter from Alexandria, father of the Neoplatonist doctrine of Arianism.[13]

To both the church and the state the Arians presented a serious threat because their main message was that unlike God, who was alone in being without a beginning, the Son (like the emperor) had once not existed. Hence Christ could not be considered truly divine but must be seen as a created and finite being. Reasoning from the premise that God alone is unique, self-existent and immutable, the Arians then concluded that the Son (who according to them was not self-existent) could not be God. To be precise: the Son is of a substance which is like, but not identical to, that of the Father. Rephrased, "the Father knows the Son, but the Son does not know himself."[14]

In hindsight it is obvious that the Arians tried to capture a both/and fish in an either-or net, a futile attempt on par with the twentieth-century social scientists' approach to power—in the words of Cardinal Newman a case of "Jewish prejudices" being rationalized by the aid of Aristotelean ideas.[15] In my own interpretation, the real conflict was therefore not between monotheism and polytheism (which neither of the two sides advocated), but between the certainties of the principle of excluded middle, on the one hand, and the ambiguities of the homoousios clause, on the other. But how does one grasp a chameleon invention which constantly straddles the line between the Territory of Humans and the Mindscape of the Gods?

One answer to that question was proposed by Athanasius (c. 296–373), the Arians' main opponent, bishop of Alexandria and a brilliant man full of violence and Mafia techniques. In a disagreement with Constantine he even hinted that if he did not get his way, then he would cut off the grain supply to the capital, a threat which promptly sent him into exile in faraway Rhineland. After the emperor's death he was allowed to return to Alexandria, but once there he behaved as usual and was treated accordingly—five times exiled and five times returned, such was the outward pattern of Athanasius' life. Precision of thought, strong convictions and a ceaseless defense of freedom have nevertheless earned him an important place not only in the annals of theology but in the history of his country as well. Indeed he has been called "the greatest champion of Catholic belief on the subject of Incarnation that the Church has ever known."[16]

True to his personality and totally committed to the integrity of his Catholic creed, Athanasius never yielded. His main point against the Arians was that they were inadvertently reducing the Son to a demigod, indeed that they were reintroducing a form of polytheism. To Athanasius himself, on the other hand, Jesus was not merely God's representative but he was God himself; not merely that Christ is *like* God, he *is* God. When the Arians said that the idea of God suffering on the cross was ridiculous, then the Athanasians replied that God, since he is God, can do whatever he pleases, including what to us seems utterly wrong, even stupid.

Given the gravity of the issues, it is natural that the fronts in this battle fluctuated widely, one emperor and one council shifting Christ closer to the spiritual borderlands of Mindscape, the next emperor and the next council moving him towards the boundary of fleshy Rockscape. So on and so on in a seemingly eternal dispute, the church meetings so frequent that to this day their dates remain a matter of controversy. Eventually, however, even bishops and emperors get enough and in 381, at the historical council of Constantinople, the key words were redefined and an amended Nicene Creed was finally approved.

Opening the way to this historic breakthrough were three boyhood comrades from Cappadocia—Basil the Great, his brother Gregory of Nyssa, and their best friend, Gregory of Nazianzus—who managed to show that there is a clear distinction between the two concepts of *ousia* and *hypostasis*, essence and being. To be precise:

> The father, the Son, and the Holy Spirit are three separate beings, each with his own individual characteristics—they are three *hypostases*. But they are one and the same in essence—they are *homoousios*. . . . [Rephrased], the real thrust of the Cappadocian doctrine was to differentiate the Christian 'Godhead,' which now incorporated Jesus and the Holy Spirit, from the monolithic God worshipped by Jews, radical Arians, and, later on, by Muslims, Unitarians, Bahais, and others. Restating the relationship between Father and Son, in other words, redefined both parties, not just the Son. As a result, Christians who accepted this triune God, distributed over three Persons, no longer shared Jehovah with their Jewish forebears or the Supreme Being with their pagan neighbors, nor could Jews or pagans claim to believe in the same God as that worshiped by the Christians. . . . It was not just a question of Jesus being recognized as God, but of God becoming Jesus.[17]

By all accounts a most remarkable development at the upper front of the Human Territory, the border line shifted in a fashion it had never been shifted before. Given the new conception of Christ, how close to man had God actually been moved? How, in that new world of upside-down, was the highest joined to the lowest?

No wonder that the priests sometimes scraped off pieces of an icon and mixed them with the bread and wine.

*

**The border line shifted** and people were once again lost in unchartered territory. Showing the way through the pending madness was Cyril of Alexandria, a master-politician who in Cardinal Newman's words would probably not have agreed to have his sanctity based on his acts, as a patriarch (412–44) so incredibly rich that on a diplomatic mission to Constantinople he is rumored to have distributed bribes worth 2,500 pounds of gold—enough to feed 45,000 persons for an entire year.[18]

At the heart of Cyril's teaching was the Athanasian idea that in Christ's person there is a single nature which constitutes a synthetic bonding of body and soul.[19] And via a most complicated route, perhaps beyond the skills of any mapmaker, he finally concluded that the flesh of Jesus Christ had become the image of the invisible God—not, however, in his *person* but in his *nature.*[20] In current parlance: Christ's person is in the flesh of the Signifier, his nature in the invisibility of the signified, the former primarily an issue of "likeness," the latter a matter of "sameness." But surely it was Christ's fleshy person, not his divine nature, that cried out from the cross, "My Lord, My Lord, why hast thou forsaken me!"

Out of this highly technical reasoning eventually grew the conception of Christ as the image of the invisible God, albeit with the proviso that the ontological transformation did not occur through his flesh but through the supreme nature of his work, especially through his willingness to die for *our* sins, not for his own. Elegant, indeed. Yet, fifteen hundred years later every (post)modernist will insist that Cyril's doctrines never reached deeply enough into the more fundamental relations between prototype and copy, appearance and apparition.

And for that reason the old problems of the inter-esting line were soon to resurface again, most prominently in the iconoclastic controversy, the often violent quarrel which came to dominate the Byzantine empire throughout the eight and ninth centuries. Yet another struggle of immense proportions, a seemingly endless battle in which the boundaries of the Human Territory fluctuated back and forth. On the surface a question of idolatry, deeper down a question about the legitimation of power; in the propaganda a theological issue about the interpretation of the first and second commandments, in reality a political issue about the operationalization of the third. The question of how we become so obedient and so predictable in another light and in another context.

The major players in this legitimation game were naturally the church and the empire. However, neither church nor empire were monolithic organizations but powers which themselves were caught in the crossfire between centripetal and centrifugal forces. Not only must the emperor at the center constantly worry about uprisings in the peripheries, but inside the church there were strong tensions between the clergy and the monks, the former via the bishops tied to the emperor, the latter bound to the local monasteries and therefore in immediate touch with the population at large.

Such was the setting of what eventually grew into a conflict about the right to control the symbols of veneration, the key metaphors which no ruler can rule without. Who should have the last say, the venerated or the venerators, the idols or the idolaters? Who in these often deadly disputes should serve as supreme judge, who be appointed Director General of the Department of Truth? Not so easy, because whoever tries to mold the collective taken-for-granted will quickly learn that there is no institutional fact without a brute fact, no signified without a Signifier, no metonymic association without a metaphoric charge. Why else would the authors of the Constitutional Law have threatened the makers of graven images with such outrageous punishments? Why else would Jesus Christ of Nicaea ever have been invented?

But the forgers of the unconscious have also always learned that different symbols belong to different congregations, that some minds are more conducive to abstract manipulation, others to brute force. And the more volatile the situation, the more crucial the choice of metaphor. Therefore, John of Damascus noted, "*God spoke in many and various ways.* A skillful doctor does not prescribe the same for all alike, but for each according to his need, taking into consideration the sickness and the climate, season and age, giving one kind of medicine to a child, another to a grown man, according to his age, one thing to a weak patient, another to a strong, and to each sufferer the right thing for his condition and ailment."[21]

The question of appropriate images lay at the heart of everything that transpired in the Byzantine empire during the eight and ninth centuries. Serious business indeed, for during this period the empire was in deep crisis, its once powerful cities dwindling away, Constantinople itself a town of no more than 60,000 people. The Slavs were attacking in the northwest, the Muslims everywhere else. Since an empire without a center that defines the periphery by definition is not an empire, it was mandatory that a new sense of patriotism be created. But how should that be done, how and by whom could the faithful be checked? By the governors or the holy men, by the basilicas or the monasteries? Through participation in the eucharist or through the veneration of icons?

Strange questions to us, seemingly not to them. For as Peter Brown once put it, "the beleaguered Christians of the Byzantine empire believed that, if they looked to God for help, then they had to be sure that the manner of their worship was acceptable to him. The survival of the empire was at stake. . . . This was an age of emergency, the Byzantine equivalent of the battle of Britain."[22]

On one side in this quarrel stood the iconophiles (literally the lovers of images) and the more ardent iconodules (the worshippers of images), both groups believing that even though Christ, Mary and the Saints were in heaven, they could somehow be reached through their portraits. The connections with pagan practices from earlier times were obvious, not the least since it was not in the churches but in the monasteries that the icons were produced. It was in fact the decentralized network of monks and holy men that made the iconodules into such a serious threat to the central authorities. The real problem was that it is literally impossible to determine to whom a person in front of an icon is actually praying—kissing an icon is something entirely different from sharing the bread and wine of a last supper, the former a private act, the latter a public and communal performance. It was in the home, away from the clergy, that the Christian cult came closest to the pagan. As recalled, the complement of the third commandment is that you must never be alone, for left alone the cohesive powers cannot reach you.

Standing on the other side in the controversy were the iconoclasts (literally image smashers), to whom icons were not holy at all, merely a form of graven images which opened the way to idolatry. Hence these people argued that all pictures should be banned from religious services and that the eucharist, the church building and the cross should be used as the only intermediaries between the worshipper and the worshipped; the eucharist because it was given by Christ; the basilicas because they were consecrated by the bishops; the cross because it had been given by God to Constantine before the battle at the Ponte Milvia in 312. *In hoc signo vinces,* "By this sign thou shalt conquer."[23] In contrast, the icon was blessed by no one. Therefore, "by asserting that only a limited number of symbols were invested with the idea of the holy, the Iconoclasts were choosing just those symbols that best suited a more collective and more highly centralized society."[24]

What a wonderful illustration of the third and second commandments intertwined, the certainties of collective unity overruling the ambiguities of individual freedom, the eucharist shared in public, the icon addressed in private. No wonder that the bishops and governors preferred the cross, while the monks and laypersons opted for the icon. For whereas the former is a geometric figure without names, the latter is a face with two eyes to sink into. It was said that in his lifetime Jesus never closed them.

From 730 to 787 the initiative in this deadly battle lay with the iconoclasts, who were supported by the two emperors Leo III and Constantine V. However, when the latter died he was succeeded by his wife, Irene, who promptly summoned a council—the Second Council of Nicaea—which brought an end to the iconoclastic practices of her husband. Not without opposition, though, and when her son, yet another Constantine, tried to usurp her, she first had his eyes stabbed out, then got him killed. Seemingly to no avail, for in 802 she herself was dethroned and exiled to Lesbos, where she had to support herself by spinning. A few years later the image-smashers were back in power only to be ousted again in 842, when another lady, Theodora, became regent for her three-year-old son.[25] Already in 843 she convoked a local synod, deposed the iconoclast patriarch of Constantinople and reinstituted the iconodules. All at a time when the Byzantine empire had not merely survived its crises but become miraculously richer than ever before, hence not as much in need of a unifying symbol as it had been a century earlier.

The result of this tremendous battle at the frontiers of the Human Territory was that the emperors did not succeed in their attempt to challenge three centuries of unofficial leadership in the Christian community. Emerging victorious was not the caesar but the holy man, not the bishops but the monks, not the basilicas but the monasteries, not the abstract geometry of the cross but the colors of the icon, not the men but the women. Thus:

> For women especially, possession of an icon permitted a most satisfying form of Christian devotion, independent of church liturgy, official, or environment. In the privacy of their homes, women set up icons and poured out their distress, prayer and gratitude to the figure, whom they came to know in a very personal way. The existence of portable icons with covers to protect them in transit—one as small as 20.1 × 11.6 cm.—confirms their use in this form of worship. These icons emphasised the holy person's power of intercession and the personal nature of the prayer, a relationship between the worshipper and worshipped that did away with ecclesiastical intervention.[26]

Beware, though. For the iconoclastic controversy was not only a struggle over political supremacy. In addition, and equally much, it spoke directly to the issue of incarnation, that most spectacular of all ontological transformations, the concept without which the eucharist itself looses its meaning. The lasting voice in that attempt to formulate a valid theory of human action belonged to John of Damascus (c. 665–c. 749), a man also known as John Damascene.[27]

*

**John of Damascus** was in many ways a most unlikely iconodule. Born with the name Mansr bar Sarjn, he came from a prominent Arab Christian family with an almost inbred ability to think-and-act in and of the in-between. So dual was his own life that when he wrote in Arabic he used the Greek, not the Arabic, alphabet. Prior to taking the vows he served as chief financial officer for the caliph Abdul Malik, a position he had inherited from his father. His most important treatises were written between 726 and 733 at a time when he was living in a monastery near Jerusalem, a place where the emperor's arm could not reach him.

John's most revolutionary idea was that the possibility, nay the necessity, of materially representing the holy follows directly from Jesus' presence on earth; not only did John claim that the veneration of icons is a logical consequence of the incarnation, but he also argued that images have the same ability to represent the unseen as do the liturgy and the sacraments. As vouched by Peter Brown, the point was that religious images

> were not a mere product of human choice and of human invention. They were essential to God's revelation of himself. They formed a bridge, wished by God himself, with which to cross the vertiginous chasm between the seen and the unseen. The figures which believers saw in icons

> or on walls of churches were what God wished them to see. They were clearly recognizable tokens which led the mind to the invisible persons that they represented.[28]

To the accusations that the veneration of icons led to idolatry, the Damascene boldly answered that even though the fallacy of reification once had presented a real threat to social cohesion, it no longer did. In John's own words:

> These commandments [the prohibition against graven images] were given to the Jews because of their proneness to idolatry. But to us it is given . . . to avoid superstitious error and to come to God in the knowledge of the truth; to adore God alone, to enjoy the fullness of divine knowledge, to attain to mature manhood, that we may no longer be children, tossed to and fro and carried about with every wind of doctrine. We are no longer under custodians, but we have received from God the ability to discern what may be represented and what is uncircumscript. "You cannot see My form," the Scripture says. What wisdom the Lawgiver has! How can the invisible be depicted? How does one picture the inconceivable? How can one draw what is limitless, immeasurable, infinite? How can a form be given to the formless? How does one paint the bodiless? How can you describe what is a mystery? . . . [But] when the invisible One becomes visible to flesh, you may then draw His likeness. . . . [Thus] in former times God, who is without form or body, could never be depicted. But now when God is seen in the flesh conversing with men, I make an image of the God whom I see. I do not worship matter; I worship the Creator of matter who became matter for my sake, who willed to take His abode in matter. . . . I honor it, but not as God.[29]

What a remarkable rendering of the abracadabra of ontological transformation, the icon serving as the magic wand by which God's nature is first transformed into the person of Jesus Christ, then into the flesh of his followers! The connections to the Divided Line of Plato and to the hypostases of the Plotinian house are immediate, not the least because John Damascene was so carefully distinguishing between six different categories of images, a series which runs from the most perfect to the least perfect. Like his Greek predecessors the master iconodule was fully aware that the semiotic animal can reach the intelligible only through the intermediaries of corporeal things.

At the top of this ontological hierarchy John placed the Son as the consubstantial image of the Father Almighty, at bottom the conventional icons, the latter by definition images of past things that we wish to remember, not the prototype but an imitation of the prototype. Rephrased, the Signifier of the real opens the door to the signified of the symbolic, a circumstance that makes John a precursor of both Ignatius of Loyola and

Jacques Lacan. However, he was always careful to note that "God did not become an angel; He became a man by nature and in truth."[30] Surely a way of saying that with the Nicene Creed the boundaries of the Human Territory were drastically redrawn. He was right.

But the fervor of John and his followers brought them even further, for like Cosmas Indicopleustes they insisted that the entire world is an icon. And even though there are no extant descriptions of what Jesus actually looked like,[31] it is generally believed that the Gospel of Luke included both words and pictures. At any rate, they said that all images in Christian rituals—especially those of baptism and communion—reach back all the way to the times of Christ and the Apostles. The Damascene once again:

> We all use our senses to produce worthy images of Him, and we sanctify the noblest of senses, which is that of sight. For just as words edify the [Jewish] ear, so also the image stimulates the [Greek] eye. What the book is to the literate, the image is to the illiterate. Just as words speak to the ear, so the image speaks to the sight; it brings us understanding. For this reason God ordered the ark to be constructed of wood which would not decay, and to be guilded outside and in, and for the tablets to be placed inside, with Aaron's staff and the golden urn containing the manna in order to provide a remembrance of the past, and an image of the future.[32]

Janus, of course. But a report from John's visit to the Saussurean Bar as well, that glorious establishment where the five senses of the body are mixing freely with the sixth sense of culture. And topping this story off, here is John's conception of the cross, the real symbol that the iconoclast emperors favored over the imaginary icon:

> If I honor and venerate the cross, the lance, the reed, or the sponge, by which the murderers of God mocked and murdered my Lord, shall I not also bow before images made by believers of good intentions, who wish to glorify and keep in remembrance the sufferings of Christ? If I bow before the image of the cross, regardless of what kind of matter has been used to make it, shall I not venerate the image of the Crucified One, who won our salvation on the cross? What outrageous inhumanity! Obviously I do not worship matter; for if it should happen that a cross, which has been fashioned from matter, should be ruined, I would consign it to the fire, and the same with damaged images.[33]

* * *

**A rotten cross I consign to the fire.** The same with the damaged icon, except the believers insist that the older an icon gets, the more it has left to say. And just as every cartographer knows that the map is made from a combination of picture and story, so the iconodules knew that "however perfect a bishop may be he needs the Gospel book on the one hand and the painted expression of this same Gospel on the other. Because the two have equal value and should receive equal veneration."[34]

Those were the words of John Damascene's intellectual heir Theodore (759–826), learned abbot of the Studion at Constantinople, a sanctuary to which he gave such an excellent organization that up to this day it has remained the model for the entire Byzantine monastic world, Mount Athos in Greece and Russian monasticism the most outstanding examples. And with that remark we are finally ready to enter the last, and to the critic of cartographical reason most critical, phase of the drawn-out battle of Nicaea.[35]

Like everything else related to the iconoclastic controversy, Russian orthodoxy is thoroughly steeped in intentional thought and manipulative action.[36] Given that history, it was almost inevitable that once Vladimir had become prince of Rus in 978 he began to think of religious persuasion as a way of unifying the land. But which particular belief system should he impose on his subjects? To find an answer he dispatched a group of emissaries to the various realms of Islam, Judaism, Catholicism and Orthodoxy. On their return they had very little to say about Judaism, but about Catholicism they concluded that its liturgy was without beauty. They then went on to note first that the Muslims did not drink alcohol, which made them impossible to deal with, then that the rituals of the Orthodox were so beautiful that the ambassadors did not know whether they had been in heaven or on earth. They were right.

And on that ground Vladimir had his entire people baptized, perhaps an early example of the Russian tendency to seek extreme solutions to ordinary problems, to swing from one cultural pattern to another, to impose change from above rather than foster gradual evolution from below.[37] In succinct summary:

> The new Christianity which gained ground in Rus from the late tenth century was received as a complete whole, assimilated in an integral package, without any sense of history, evolution, or inner conflict, beautiful and to be revered, but not open to discussion or amendment. Rus had not experienced centuries of theological and ecclesiastical controversy, nor had she observed successive ecumenical councils gradually chiseling out the contours of the credo. Her rulers and perhaps her people accepted the new faith wholesale as a harmonious, intellectually and spiritually satisfying answer to their needs. For them, moreover, it was faith which manifested itself at least as much in its liturgy as in its dogma. Orthodoxy is reticent in making dogmatic statements about God, regarding Him as beyond the reach of rational understanding and accessible to the human heart only through the symbolism of the liturgy.[38]

This shows that although the Orthodox liturgy is in direct descent from the teachings of John the Damacsene and Theodore the Studite, it is in some ways also radically different. Most importantly, the worshippers are largely passive observers rather than active participants, the laypersons' withdrawal especially pronounced during the preparation for the eucharist, a ritual which literally takes place behind closed doors. The inherent power structure is further strengthened by the fact there are no pews for the congregation to sit on and no pulpit for the priests to preach from. Only after the cantor has intoned the scriptural quotation of the day does the choir know what to sing.

In the performance of this elaborate liturgy the geography of the church building plays a major role. For just as the Old Testament tabernacle was raised according to the image shown on Mount Sinai, so the Orthodox church is loaded with symbols intended to facilitate the crossover from the wor(l)dly to the heavenly. And exactly like the tabernacle of Moses and the temple of Solomon, the Russian church is divided into three well-defined areas commonly known as the sanctuary, the nave, and the narthex. In principle a circle inscribed within a square or a globe contained within a cube, the proportions of Vitruvian man transformed into the architecture of a palace of another kind.

The geometric structure of the church building is directly associated with different degrees of sacredness. Thus, on the map of the Human Territory the sanctuary is found closest to the unreachable heavens of Mindscape, while the narthex is next to the material realm of the Rock-Lord. In between lies the nave, the gathering place of the already enlightened. Moreover, just as the sanctuary is usually located in the eastern (oriental) part of the building, off limits for anyone but the clergy, so the narthex lies in the west closest to the sunset. Finally, at the center of the sanctuary, the holy of holies, stands an altar table, in Russian called the *prestol* or "the throne." To be concise:

> The very plan of a church makes a clear distinction between those who participate in the Body of Christ and those who do not. The latter are not driven out of the church and can remain until a certain moment. But they cannot participate in the internal, sacramental life of the Church. They are neither completely outside the church, nor a part of it. They are, so to speak, on the periphery, at the limit between the church and the world. The narthex, according to the Fathers, symbolizes the unredeemed part of the world, the world lying in sin, and even hell. It is always at the opposite end of the church from the sanctuary, that is, at the west.[39]

The same hierarchy is evident also in the numerous decorations which adorn the church, all of them arranged in a manner that makes the whole and its parts completely inseparable.[40] As in the *Enuma elish* and the *Ebstorfer Karte,* what is above is always considered more sacred then what is below. Indeed the entire church building is in itself a gigantic icon, a sign designed to show the believers the way to their sanctification. In the words of the

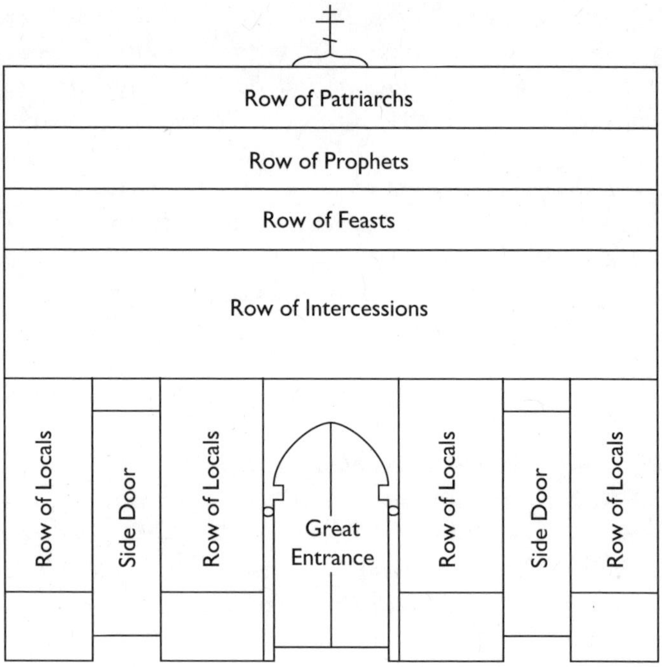


Scheme of Russian iconostasis. Adapted from Ulf Abel, *Ikonen—bilden av det heliga* (Hedemora: Gidlunds Bokförlag, 1989), 92.

famous theologian and icon painter, "a church is therefore the prefiguration of the peace to come, of the new heavens and the new earth where all creatures will gather around their creator,"[41] the stage for a most remarkable dramaturgy in which the humans are always caught in between. To strengthen this feeling of liminality, the light is made to stream into the building through windows which are located just below the vaulted cupola, the latter a symbol of the celestial sphere at whose center there is always a painting of Jesus Christ, who with one hand holds a copy of the Gospels and with the other blesses whatever may be blessed. Ascended to this elevated position in heaven, seated to the right of the Father Almighty, the Son (or is it the Crown Prince) proceeds to judge both the quick and the dead, to sort the citizens whose kingdom is said to have no end.

As an integral part of the architectural scheme, the space between the heavenly cupola and the earthly floor is studded with symbols, each and every one ordered according to the prevailing hierarchy. Shown next to the commanding Christ are therefore the archangels, usually four sometimes seven, a little further away the red seraphim with their three pairs of wings and the throne-bearing cherubim with their blue garments and rosy-faced expressions.[42] The level immediately below is taken up by the evangelists, each placed in his own corner of the Pentecost four-square. Further down still come pictures with scenes from Jesus' life on earth and the various feasts of the church year. Finally, closest to the floor, a collection of diverse figures, many of them patrons of the church in question.

The blueprint of this sacred building is clearly that of the Plotinian house, the host of the LORD pictured as a congregation of Vitruvian men, a trickery in everything as sophisticated as the Albertian codification of the Brunelleschian perspective, an elaborate technique for making what is unseen seem real. To the unaware a complete jumble, to the initiated yet another expression of the hierarchical thoughts-and-actions which from the outset has permeated every phase of Byzantine life, political as well as religious. Everything *in* a limit, everything *of* a limit.

*

***In* a limit and *of* a limit,** such is the nature of the iconostasis, the most conspicuous element of any Orthodox church, a hybrid accumulation of images which over the centuries has reached the extreme form it now exhibits in Russia. Already from its Byzantine beginnings it is this wall that has constituted the non-crossable boundary between the priests in the sanctuary and the laypersons in the nave, the performing actors on one side, the auditing audience on the other—Samuel Beckett's tympanum and the Saussurean Bar under another name, Plato's cave wall and the slanted line of the *Republic* maps as well. But the iconostasis is at the same time also the mappa of the iconodule's guide to the unknown, a projection screen *with* images rather than a wall *of* images.[43]

Not a solid wall, though, but a divider with three doors, two to the side, one at the center. While the opening to the left (as seen from the sanctuary) leads from the nave to a storage or "treasure" room, where the deacons keep the various implements for the liturgy, the door to the right takes the priests to the "prothesis" room, the place where the oblations for the eucharist are prepared. Exactly in between is the third door, the Great Entrance, a portal which leads directly from the nave to the sanctuary. Ontological transformations in the making, for it is in the hypostases of this architectural setting that bread turns oblation turns flesh turns Christ turns forgiveness turns eternal life—the gessoeing of the Christian mappa for each and everyone to witness.

Now, having readied themselves in the prothesis room, the clergy enter the nave in procession, chanting and throwing incense before the iconostasis. Passing through the gate, kissing the holy doors of the Great Entrance, they then proceed to the "inaccessible space" of the sanctuary itself. Once there, the double doors are closed and the priests perform the eucharist in isolation, no longer visible to the congregation. In the Russian church the sacrifice of the mass is consequently not for everyone, the laypersons not directly participating, merely a crowd taking in what is being said and sung. Throughout these performances the believers' eyes are fastened not

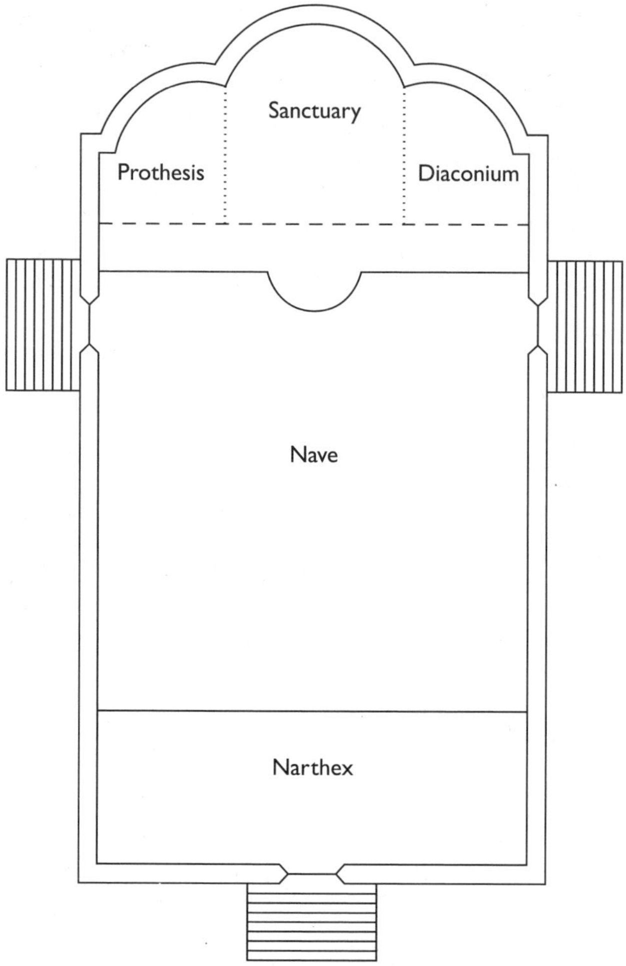


The layout of an Orthodox church. Adapted from Leonide Ouspensky, *Theology of the Icon* (Crestwood, N.Y.: St. Vladimir's Seminary Press, 1978), fig. 1.

on the altar, which they normally cannot see, but on the icons of the picture wall in front of them.

Given the geometry of the Orthodox liturgy, it is tempting to imagine the Great Entrance not only as the navel of a Vitruvian man but as a counterpart to the portal of the Battiseria di San Giovanni, *the* fix-point of Filippo Brunelleschi's famous experiment. That temptation must nevertheless be resisted, for an icon is in almost all respects the opposite of a Renaissance picture. Москва must consequently not be mistaken for Firenze, because even though the photographs of the Soviet Politburo on the balcony might have looked like family portraits of the Godfather and his lieutenants, they were in effect stylized icons of the Incarnated and his apostles; as I have repeatedly noted, there is always an intimate relation between politics and religion, between the state and the church, between the crowned Constantine on one throne and the anointed Yeshua on the other. Some icons from late Byzantium are in fact painted on both sides and although these bilaterals were mostly used in processions, some were actually mounted on the templon itself. From their Janus-position in between these images are obviously addressing two different worlds at the same time, one face turned to the laymen in the nave, the other to the clerics in the sanctuary.[44]

Now, if not before, it should be clear that an icon is not what it appears to be. Not a picture of the holy, but the holy in and of itself. With its antecedents in Egyptian mummy-portraiture, the icon is consequently not *about* something, it *is* that something: an image of the invisible; a pictorial oxymoron; a most peculiar self-portrait. It follows that for the true believer the message of an icon cannot be rephrased, translated and understood, for to the devout the real icon is not a material representation of a touchable reality but a shadowy amalgamation of Plato's Good and the Christians' God, in both cases a self-referential expression as close to a perfect sign as any sign may ever come. For that reason the icon-maker (in principle not an artist but a monk) prepares himself in accordance with a set of well-established canons and regulations, fasting and praying as he performs his tasks. Both the subject matter of the icons and the techniques for making them are in fact so rigorously preordained that the final product is never signed. The motivation is, of course, that it is not the painter, but some kind of holiness, that has held the brush. Like an echo from Homeric times, the brusher does not act on his own but as the extended arm of forces unknown.

A mystical and mysterious symbol from the borderlands of the Human Territory, Mindscape located on one side, Rockscape on the other. That is what an icon is. A symbolic imagination deeply rooted in a long history of ontological theory and aesthetic practice. And this is exactly why the critic of cartographical reason finds it so fascinating, for the fix-points, scales and mappae of these Russian images are entirely different from what the conventional mapmaker has been taught to practice. A life lived in the Light of Light, Very God of Very God, begotten not made. Such is the idea of the *acheiropoetos,* by definition an image not made by human hands.

*

**The Light of Light**, the Very God of the Very God, Christ himself not merely the logos of God, but "the real light which enlightens every man" (John 1:9). In the Russian church the mystique of this blessed medium is everywhere enhanced by the countless wax candles whose soot makes the air so thick that the visitor feels as if (s)he were physically embraced. In addition, the reflecting particles affect the illumination in ways that make it seem as if the pictures were painting themselves, another version of the ageing Henri Matisse experimenting with the color red in the Chapelle de

Francisco de Zurbarán, *The Veil of Veronica.* Ca. 1635. Oil on canvas, 70 × 51.5 cm. Nationalmuseum, Stockholm.

Rosaire, that crowning achievement in which the sun-rays are first filtered through the stained glass windows and then projected onto the white ceramic walls—an outstanding example of the taken-for-granted captured by its own means.[45] Furthermore, since the Orthodox church always smells of incense (a smell so penetrating that it invades the taste buds) and since the sounds of the chanting voices fill everyone's ears, it may be said that the liturgy plays over the entire register of bodily senses.[46] The use of the loose brush technique adds further to this experience of dematerialization and spiritualization. Beware, though, because the icon is always painted on a two-dimensional surface, never molded into a three-dimensional statue, a reminder that the reality which the monks are trying to represent is of another type than the symbolic imaginations which they are actually producing. Always veneration, never adoration.

All this granted, sight remains the privileged sense, the light of the icon seemingly emanating from everywhere and not, as in the paintings of the Renaissance, from one specific source. So striking are the eyes of the depicted figures that the aesthetic quality of the best icons is condensed into the angle between the nose and the eyes. Calling that conjunction the icon-maker's fix-point would nevertheless be wrong, for the truth is that the typical icon does not have a fix-point. And for that same reason it is difficult to determine the icon's scale, except to say that also the scale resides in an angle, best exemplified by the tilt of the virgin's head as she holds her child. For the critic of cartographical reason it is no coincidence that icons of the Virgin and Child are called *Hodegetria,* "She who points the way." Devotion, though, always comes with a price.

A sophisticated play of fix-points, scale and mappa—therein lies not only the hidden power of the Orthodox church but the emergence of a cartography which transcends the confines of the Brunelleschian perspective. And once it has been subjected to that twist of imagination the icon ceases to be a religious symbol and becomes a map of power structurally very similar to the *Ebstorfer Karte.* Much discussed in that context is the icon painter's reversal of the linear perspective, a trick by which the vanishing point is located not within or behind the picture, but within or behind the viewer—it is not me who is looking at the picture, it is the picture that is looking at me. So strong is in fact the pictorial manipulation that in a fully decorated Russian church it is virtually impossible to escape the gaze of some wide-open eyes. And so strange are those same eyes that they seem to be inward- and outward-looking at the same time—seeing is literally believing.

As a way of heightening this sense of omnipotence, the icons are almost never framed and frequently painted with tempera paint on a concave surface of shining gold, gradually also of beaten silver. One may even say that while western painting attempted to create the illusion of three-dimensionality, the eastern icons are approaching a kind of fourth dimension in which space is infinite and time eternal. Different skirmishes in the

*The Mother of God of Tenderness with Selected Saints on Borders*. Russian icon from last quarter of the 15th century. Egg tempe on linden wood, 57 × 44.5 cm. Nationalmuseum, Stockholm.

perpetual war, one force striving for Florentine realism, the other for Byzantine transcendentalism. The two contemporaries Filippo Brunelleschi and Andrei Rublov—Russia's greatest icon painter—consequently found themselves stationed at different frontiers of the Human Territory, the former trying to represent the stable things of Rockscape, the latter imitating a Mindscape beyond the reach of bodily senses. It cannot be said more clearly: while Renaissance art captures the world as the eye finds it, the icon serves as a bridgehead to the invisible.

No joking yoke. For now it should be clear that what our mission to Nicaea has demonstrated is that the major disputes have all centered on the nature of the mappa, the projection screen which like Malevich's *White Square on White* is one with the taken-for-granted. And so it is that the iconoclastic controversy (still kicking) was not primarily a quarrel about the second paragraph of the Constitutional Law, but an argument about the third paragraph and the associated techniques of legitimation. The result is a very special type of "self-portrait," not an image of one particular power-holder, but a portrait of Power in general. Evasive and precise, always on the move, still constantly the same. An image not made by human hands, yet signed with a miniature black square.

*

**Not made by human hands**—surely a major confrontation in the perpetual war of what it means to be human, the issue over which today's battles are fought most fiercely. Indeed it is not at the godly but at the rocky front that the heretic cartographer currently runs the greatest risk of being killed. It is there that the limits of the oikumene are presently pushed into the previously unknown, animal rights, quarks and stem-cells created in the process.

The examples are overwhelming, for just as the politically correct often claim that the otherness of gays and lesbians is really a likeness, so the trendsetters of contemporary ethics are arguing about the humaneness of dolphins and the beastliness of *Homo sapiens.* Many environmentalists follow suit, some of them claiming that the bacteria are more crucial to the humans than the humans are to themselves. To discuss abortion and organ transplants is in that perspective to engage in a quarrel about the distinction between meaningless flesh and meaningful life, by extension a problem of whether children issue from their parents' genitals or from God's benevolence. When does a fertilized egg cease to be an egg and when does it begin to be a human? How many new crutches, teeth, glasses, hearing aides, kidneys, hearts, noses, knees and hip-bones does it take before the ageing man is changed into a creature made by human hands?[47] When he touched the tentative of the impossible, René Magritte sensed what was coming: the remembrance of an arm not yet created.

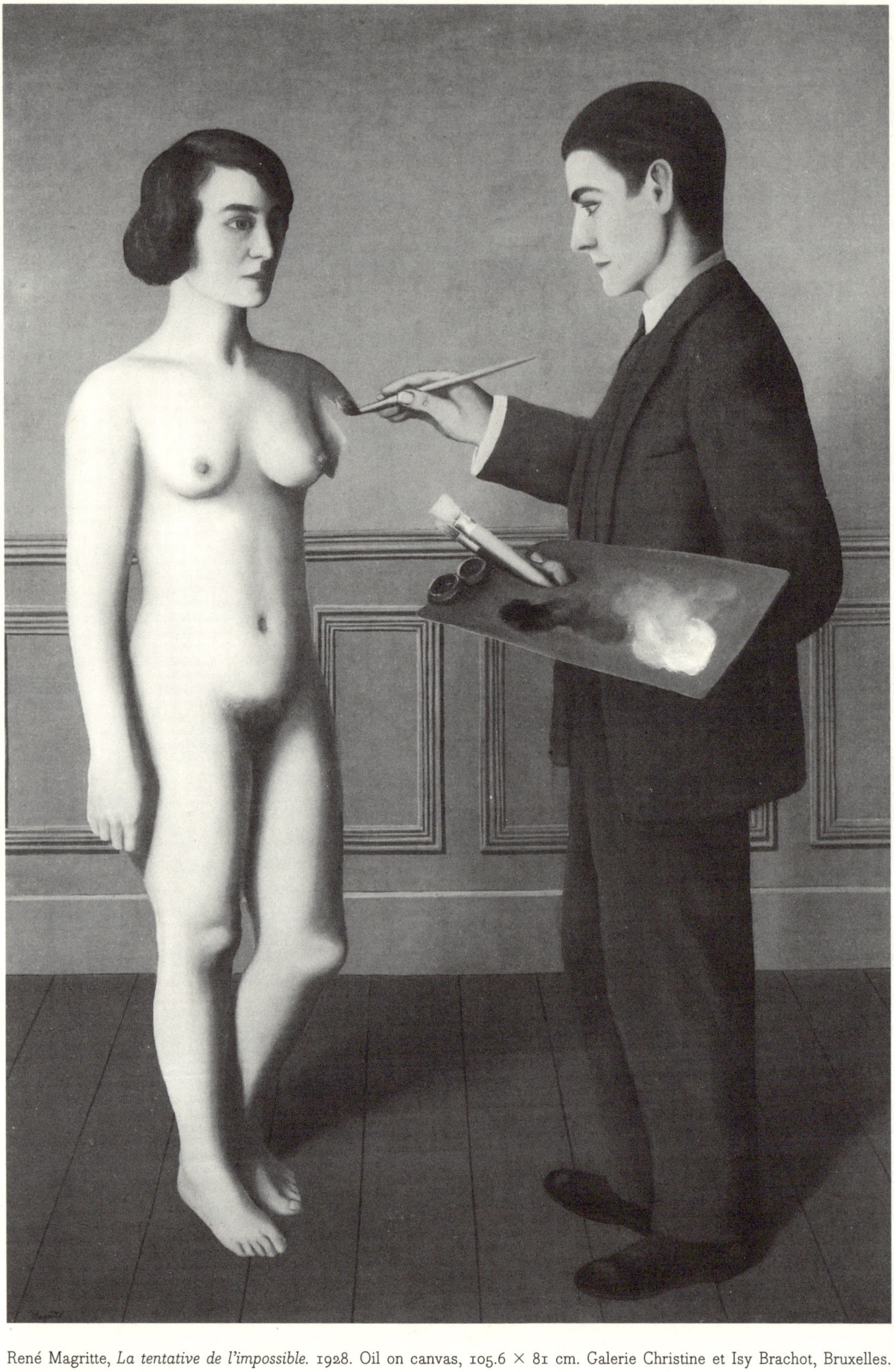

René Magritte, *La tentative de l'impossible.* 1928. Oil on canvas, 105.6 × 81 cm. Galerie Christine et Isy Brachot, Bruxelles.

The same question of creativity was doubtless also on Albrecht Dürer's mind when he signed his most famous painting, arguably the greatest self-portrait ever made: the artist rendering himself as Jesus Christ in half-figure, en face and painted in proper colors, in effect a version of Veronica's kerchief, the *vera icon non manufectum,* that most venerated of all images.[48] It has even been suggested that Dürer, by using Jesus Christ as his model, implied that the painting had been created out of nothing, in essence that he himself was another God. "Who held the brush?," such was his question.[49] It is said that by licking the picture's face, Dürer's dog provided the answer.[50]

Anecdotes aside, there is no doubt that the German artist thought of himself not only as a Christlike icon but as a Narcissus and an Actaeon too. Thus, already in 1503, at age thirty-one, Dürer drew himself again, on this occasion nude and visibly hit by an undiagnosed epidemic. Surely a dramatic change, for in this moving picture there is hardly a trace of the self-centered pride from three years earlier. If one accepts Joseph Leo Koerner's argument, as I think one should, then

> Dürer evinced here a kind of narcissistic crisis, in which the celebration of his own physical beauty in the 1500 *Self-Portrait* gives way to an awareness of the frailty of the flesh or, more radically, in which all that had to be expelled in order to sustain the narcissistic ideal—ugliness, particularity, sexuality, and death—comes back to haunt the self as it compares its own real body against its constructed ideal.[51]

*

**Coming back to haunt its self.** Such was the destiny of Dürer's art as he ventured into the border areas of the Human Territory, the forces of the Mind-Lord constituting the enemy at one frontier, those of the Rock-Lord at the other. Such too is the destiny of cartographical reason.[52] For even though the base map of the Human Territory poised between the non-crossable limits of the spiritual Mindscape and the material Rockscape is an extraordinary rhetorical success, it comes with a real problem. *It is not true!*

To be precise, the mapmaker's major mistake is that (s)he is treating the processes of deification and reification as if they were separable, while in fact they are two sides of the same coin of socialization. What on the base map is shown as two boundaries—two forms of silence—is consequently not two but one. Likewise with the mappa, the projection screen whose structure is not a reflection of the world as it is but a consequence of how the canvas has been gessoed. Surely the world is not a flat thing, but something far more abstract. And therein lies the real challenge, for it is in the nature of the semiotic animal to put more trust in the Q.E.F. than in the Q.E.I. A difference which makes a difference.

1500

(*facing*) Albrecht Dürer, *Self-Portrait in a Fur Trimmed Coat.* 1500. Oil on wood, 67 × 47 cm. Bayerische Staatsgemäldesammlungen, Alte Pinakothek, München.

Albrecht Dürer, *Nude Self-Portrait.* 1503. Pen and brush, heightened with white on green grounded paper, 29 × 15 cm. Klassik Stiftung Weimar.

Still! The trustworthy truth remains. And this truth is that the map is the most powerful of all power-filled rhetorical tools, *the* intermediary without which we would be madly lost in a world unknown. Its major strength is its ability to weave picture and story together into a suddarium so strong that it cannot be ripped apart, a loincloth which no bastard can do without. Its weaknesses will nevertheless continue to haunt us, for by all evidence the world of the twenty-first century is a creation in which the ontological fix-points are being unfixed and the epistemological scales made unstable. The mappa, though, is what it has always been and will always be: the metaphysical napkin of the shifting powers that weave it.

*

**Midnight mass in preparation**, another wake to be found again, a requiem waiting to be performed.

# REQUIEM

# PHILADELPHIA

**In the wake of modernity** we once again find ourselves in the abyss of the infra-thin, a concept that permeates Marcel Duchamp's entire oeuvre,[1] a shorthand label for his life-long desire to overcome the distance between Signifier and signified, thing and meaning, love and knowledge, inner and outer, up and down, appearance and apparition, mold and molded, *of* and *in*. To be precise, the word "infra-thin" was to him "not a noun but an adjective, although the *sign of the accordance* might be the perfect exemplification of the infra thin made into a noun. Even as an adjective, 'infra-thin' never qualifies a thing or an experience, but rather the difference between two things or experiences. This difference is at its thinnest when those two things are the same."[2]

About Duchamp's refusal to choose sides—a desire which focuses neither on the either-or nor on the both/and—so much has been written that I have no wish to add to the exegesis.[3] The professor should tread cautiously, especially as Monsieur Marcel himself was a person never to profess except to say that "there is no solution, because there is no problem."[4]

The reader must nevertheless be warned that my earlier remarks about Plato, Brunelleschi and Cézanne almost certainly have been influenced by my longstanding interest in Duchamp's oeuvre. Not, of course, as strange as it might initially appear, for since that Sunday afternoon in October 1963, when I first strolled into the Arensberg collection of the Philadelphia Museum of Art,[5] hardly a day has passed without some sort of interaction with this kindred spirit who never wasted his time but always cultivated

Man Ray, *Marcel Duchamp as Rrose Sélavy*. 1920–21. Gelatin silver print, 21.6 × 17.3 cm. Philadelphia Museum of Art: The Samuel S. White III and Vera White Collection, 1957. 

chance with exactitude, a man so obsessed by prime numbers that he became a prime himself. A paradox disguised as a human being. If anyone, this clever ironist, alchemist and nominalist devoted his life to the exploration of the unknown territory of the in-between, a reconciler of opposites, a latter-day Apsu constantly roaming the excluded middle of the categorial abyss. If anyone, this ultimate escapist was fascinated by the phenomenon of transition rites, especially the passage from virgin to bride, albeit with the proviso that in the privacy of his own world the bride never looses her virginity. Contrary to common belief the sexual passage from bride to wife never seems to have much concerned him.

The best known of Duchamp's obsessions is of course *La Mariée mise à nu par ses célibataires, même,* for short often referred to as the *Large Glass,* a work begun in earnest in 1915 and abandoned definitively unfinished eight years later. In his own words a "hilarious" picture which he insisted was not a picture at all, in my conception an elaborate map of what it means to be human. As the critic of cartography might have expected, its theme comes in two variations, one being the painting/sculpture itself, the other the notes that go with it;[6] while the former is essentially an installation to be seen and savored (an image constructed by an abdicated painter) the latter is a narrative to be read and understood (an imagination dreamt by a meticulous and obsessed engineer, a machinery of desires impossible to satisfy). The unity is so complete that "the assembled notes are a masterwork in themselves, as important as any of Duchamp's realized visual projects, perhaps more so since the latter take their cues from the notes."[7] Yet, while the Glass is for the eyes and the Notes are for the ears, their relation to painting is much like the relation of James Joyce's *Finnegans Wake* to literature, both works uniquely inimitable, both originals irresistibly tempting to copy.[8] Like the *Ebstorfer Karte* before it, the *Large Glass* (here understood as the Glass and the Notes taken together) is fruitfully approached as an illuminated manuscript.

While some have seen this work as nothing but an elaborate hoax, others—Octavio Paz most prominent among them—approach it as the last really significant work of the West, the end of a long tradition and perhaps the beginning of something radically new. As indicated by its nickname it consists of a large glass, rectangular in shape à la Alberti's instructions, 272.5 centimeters high and 175.8 centimeters wide. This total area is in turn divided into two panels, the upper somewhat smaller than the lower, three glass staffs inserted horizontally between them, effectively a Saussurean Bar drawn exactly in the boundary between the two territories. The structure is virtually identical with the maps of Plato's *Republic.*

The viewer who meets this work for the first time is likely to be overwhelmed by its stunning beauty and to be completely lost as to what (if anything) it might be meant to mean. As a guide, the notes of the *Green*

*Box* will eventually prove indispensable, even though the beginner may find Jean Suquet's *Mirroir de la mariée* even more helpful. Much like the construction drawing of an electrical engineer or the plan of a landscape architect, so Suquet's graph is an excellent introduction to a world which looks more complicated than it actually is.

The upper panel of Duchamp's construction is known as the Bride's domain. To the left there is something that seems like a cross between a living insect and an internal combustion engine (two-stroke model), extending towards the right a long horizontal cloud (a milky way or halo) with three windows or draft pistons—perhaps a banner fluttering in the wind, perhaps a foxtail on a motorcyclist's helmet, perhaps the image of a motorcar driving up a hill in first gear. Everything except the open windows is painted in a flesh-like color reminiscent of Brunelleschi's burnished silver, a technique which is here used for similar reasons and to similar effects. To the right of this blossoming are nine small holes cut with an ordinary drill at points given by ink-stained matches fired from a toy canon; "chance and an obvious thematic logic had come together: the shots come from below, but when they are fired from a perpendicular cannon, the geometric dimensions of the work are multiplied."[9]

Multiplied? Indeed! For while to the mathematician $9 = 3^2$, to Duchamp 3 was the first of all numbers, the beginning of the crowd, the precondition of power. And since to the perspectivist there can be no painting without a picture-plane, the toy-artillerist literally shoots it to pieces. To the bride, however, the cannonade does nothing, because the bride is not a body which can be killed or injured but the timid potency of a motor and a tank of love petrol: an imagination twice stripped bare, 1° by the bride's fantasy, 2° by the ejaculations of a bachelor gang. Out of these initial contacts there finally comes a blooming of which there is no determinable cause, hence no logical explanation. From beginning to end the bride floats in a heaven of imagination, thoroughly narcissistic and intent on nothing but her own pleasure, a wonderful version of Brunelleschi's clouds dripping their wet humor on whatever happens to lie below.

The lower level of the *Large Glass* is the bachelors' domain, in everything except its unquenchable desire different from the bride who hovers at the top. Shown to the far left is her harem of nine men dressed in mail-ordered uniforms or liveries, an assemblage best described either as a collection of beheaded, hence faceless, dummies, or as a set of hollowed forms, empty molds or husks filled with gas, their outlines delineated by thin lead wires of the type normally found in ordinary household fuses, their bodies painted in red lead.

The remainder of the lower realm is studded with various bits of machinery all marked on Suquet's graph and all designed to get the masturbating bachelors going. Most noteworthy are the following gadgets and the following processes:

1 Chariot
1a Water Mill
1b Pinion
1c Trap-Door opening to the Basement
1d Pulley
1e Revolution of the Bottle of Benedictine
1f Runners of the Sleigh
1g Sandows [Mechanical Chest-Expanders]
2 Cemetery of Uniforms and Liveries or Eros' Matrix
2a Priest
2b Department-Store Delivery Boy
2c Gendarme
2d Cuirassier
2e Policeman
2f Undertaker
2g Flunkey
2h Chasseur de Café [Messenger Boy]
2i Stationmaster
3 Capillary Tubes
4 Sieves
5 Chocolate Grinder
5a Nickeled Louis XV Chassis
5b Rollers
5c Necktie
5d Bayonet
6 Large Scissors
7 Bride
7a Ring for hanging the Pendu Femelle
7b Patella-Mortice
7c Pole carrying the Filament Substance
7d Wasp
7e Head or Eyes
7f Weathervane
8 Flesh coloured Milky Way
9 Draft Pistons
10 Churn-Ventilator
11 Slopes or Planes of Flow
12 Crashes-Splashes
13 Horizon – Bride's Clothes
13a Vanishing Point of the Bachelor's Perspective
13b Prism with the Wilson-Lincoln Effect and 9 Holes
14 Battering Rams
15 Oculist Charts
16 Kodak Lens
17 9 Shots
18 Tender of Gravity
18a Tripod
18b Rod
18c Black Ball

⟵ Journey of the Illuminating Gas
⟶ Language of the Bride

From Jean Suquet, *Le Miroir de la Mariée* (Ed. Flammarion).

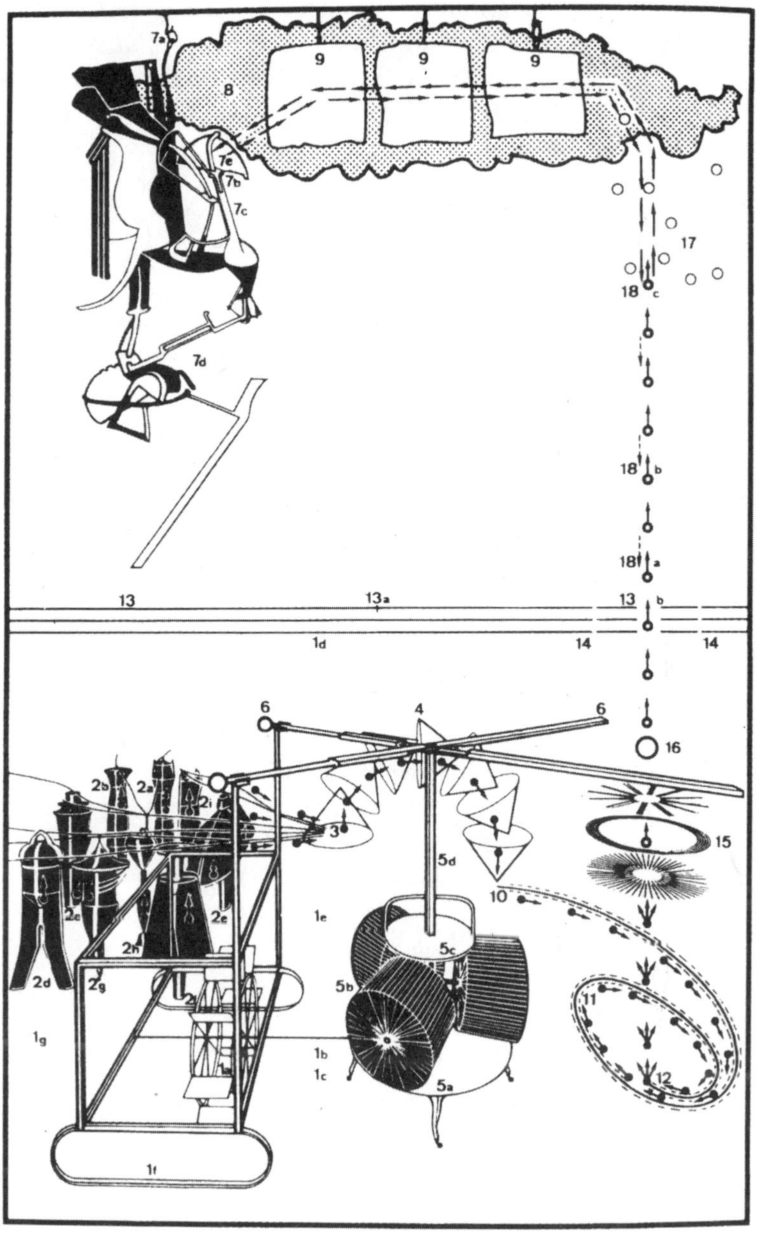


(*facing*) Marcel Duchamp, *Bruden avklädd av sina ungkarlar, t.o.m.* 1915–23. Oil varnish, lead foil on glass plate mounted between two glass panels, 283 × 189 cm. Copy made by Ulf Linde and the artists Henrik Samuelsson and John Stenborg, 1982. Based on Ulf Linde's replica from 1961, which was signed in conformance with the original by Marcel Duchamp in Stockholm, 1961. Moderna Museet, Stockholm. © Succession Marcel Duchamp/BUS 2005.

Jean Suquet, *Miroir de la Mariée* (Paris: Éditions Flammarion, 1974).

- a kind of trolley or sleigh, its convulsive movements to and fro driven by a water-wheel;
- a chocolate grinder—a symbol for the male genitals—which rests on a nickel-plated Louis XV tripod and is connected to the same water-wheel as the sleigh, a device which allows the bachelors to grind their own chocolate and drink their own chocolate milk;
- a pair of scissors mounted on the bayonet of the grinder,[10] the movements of its shanks directly linked to the movements of the sleigh;
- a set of capillary tubes which are connected to every bachelor mold, each tube in turn filled with a gas that under pressure solidifies and splits into a swarm of small needles or spangles, seemingly made of ice and lighter than air;
- seven sieves or parasols—outlined in lead wire and colored not with paint but with dust—instruments that steer the spangles into a labyrinth where they become so dazed that they loose all sense of left and right, up and down, their form changing from solid to fluid in the process;
- a ventilator—imagined in Duchamp's mind and mentioned in the notes but never pictured on the glass, a sort of signified without a Signifier—that catches the fluid from the sieves, churns it around and through a corkscrew motion carries it past the area of the three crashes to the region of the splash, turns it into drops of gas which in their turn are projected upwards in a vertical direction;
- three eye-witnesses, circular diagrams including a Kodak magnifying lens, stations which the ascending gas must pass before it can penetrate the three glass plates and enter the bride's domain, where it first reaches the milky way, then eventually finds its way into the pendulum of the bride herself.

Once inside this solipsistic and self-referential machinery the gas is processed into dripping desire and sent back the same route it came. Driven by the draft pistons and passing the holes from the canon shots, the bride and the bachelors engage in a sort of telegraphic exchange, the former actively issuing her commands, orders and authorizations, the latter passively obeying; according to the notes, this cinematic blossoming is the most important part of the entire work. Further down the bride has evidently had enough, takes her clothes off and hangs them to dry on the uppermost horizon-line, stripped of everything but her longing for the orgasm that never comes. Forever a bride never turned wife. An ontological transformation aborted, the glass of the Glass a reflecting shop window in which the onanistic play of fantasy and desire is more important than its consummation. In the Plato-sounding words of Jerrold Seigel:

> Duchamp's freedom requires that the inner play of fantasy meet the world of material things wholly on the former's terms: it is lost when one breaks

> the shop window and discovers that the objects which beckon there only yield to possession by imposing the actuality of their limitations on desire's infinite wish. Such freedom cannot be experienced through direct interaction with the world, but only at a remove, in the Large Glass's perpetual delay, the chessboard's abstraction from social relationships, or the protected enclosure of the *Boîtes en valise.* Such spaces are worlds in themselves, into which objects enter only as symbols, so that the ideas they stand for encounter no material, mundane resistances, but echo endlessly off each other in a kind of constant interior reverberation.[11]

*

**Such spaces are worlds in themselves,** things, humans and animals behaving as they never did before. For Duchamp, however, there seems never to have been any sharp line between such a condition and the normal one, perhaps because his "geometrical awareness immediately directs our thinking to the system of conical projection used in many maps of the world (*mappae mundi*) since the sixteenth century and taught in all French schools when Marcel was studying for his diploma. If we acknowledge this geographical source, it seems logical to suppose that the projective cones contain nine holes . . . which form a polygon, like an unusual 'continent,' which is similar to the malic moulds' genital areas."[12]

In this context of geographic way-finding, the glass of the Glass functions as a vertical window *through* which the artist sees the world and *onto* which his desire is projected. The connections to Dürer's woodcut are immediate, although in Duchamp's case the captured nude is not transposed to the horizontal paper on the designer's desk but left entangled in the framed, infra-thin and vertical screen of the accidentally shattered glass. Like Ptolemy who long ago had designed his imitations in drawing, so the abdicated painter is here struggling with the same problem of how best to map a multidimensional reality onto a flat surface. Duchamp's theatrical approach was to project whatever he thought was happening on the horizontal stage onto the *backside* of the pulled curtain,[13] the actors never shown to the audience in their full bodiness merely as images that cast their Platonist shadows on, or rather *through,* the intervening screen. To be precise, "the projection [of each part of the Glass] is a perfect example of classical perspective, I mean that I imagined the various elements of the bachelor machine first of all as arranged behind the Glass, on the ground, rather than distributed over a surface in two dimensions."[14] *Vive la Republic!*

The innovative idea of using glass instead of some other material struck Duchamp sometime in 1913. According to himself, it came

> [t]hrough color. When I had painted, I used a big thick glass as a palette, and seeing the colors from the other side, I understood there was some-

> thing interesting from the point of pictorial technique. After a short while, paintings always get dirty, yellow, or old because of oxidation. Now, my own colors were completely protected, the glass being a means for keeping them both sufficiently pure and unchanged for a long time. I immediately applied this glass idea to 'The Bride.'
>
> The glass has no other significance?
>
> No, no, none at all. The glass, being transparent, was able to give its maximum effectiveness to the rigidity of perspective. It also took away the idea of 'the hand' of materials. I wanted to change, to have a new approach.[15]

The parallels to Ptolemy, Brunelleschi, Alberti and Masaccio are almost too obvious to mention, the technical skill by which the machinery is constructed nothing but remarkable. "The problem was how to draw and yet avoid the old-fashioned form of drawing. Mechanical drawing was the saving clause. A straight line done with the ruler, not with the hand. . . . When you draw, no matter what you do, your taste comes in subconsciously. But in mechanical drawing you are directed by the impersonality of the ruler. . . . I actually had to forget with my hand."[16]

What this means is that the onanistic world of the bachelors was painted with the aid of distancing tools like the brush, the compass and the ruler, all images securely fastened to their proper places by threads and similar devices, especially by the ninety degree angle through which the vertical glass is attached to the horizontal floor. In contrast, the bride's domain grew out of the artist's fingers, nothing allowed to interfere with the tender touch; as everything in the bachelors' world bounces off the squares of the Florentine piazza, so the desirous bride appears as an amorphous cloud in the sky, her lusty motions faithfully reflected in Brunelleschi's mirror. The lower panel is consequently constructed with techniques that are far more rigorous (and far less sensitive) than the upper. Add to this Duchamp's obsession with the number three—*the* holy number without which there would be neither society nor power, neither triangulation nor mapping— and his relevance for the understanding of cartographical reason should be obvious.

And so it is that *La Mariée mise à nu par ses célibataires, même* turns into an outstanding map of what it means to be human, a machine that grinds and grinds but never comes to a stop, a Hegelian synthesis of things and relations, a modernist thesaurus sapientiae, a graveyard of ideas, a perpetual hanging on. The electricity (alternating current) that drives the nine cylinder engine is presumably generated by the waterfall so handsomely described in the notes but not included in the Glass itself, allegedly because its maker did not want to fall into the trap of becoming a landscape painter.[17] It has even been suggested that the waterfall flows out of a higher reality impossible to capture in the perspective of the bachelor machine, our bodies in reality three-dimensional "shadows" of our parents' four-dimensional desire.[18] In search of the next higher dimension, Duchamp was in fact working backwards from the shadows to the shadowed. A clear-cut case of the Geographical Inference problem, an accidental victim of the idea that to any particular reason another equally valid may be opposed. In his own words:

> Simply, I thought of the idea of a projection, of an invisible fourth dimension, something you couldn't see with your eyes. Since I found that one could make a cast shadow from a three-dimensional thing, any object whatsoever—just as the projecting of the sun on the earth makes two dimensions—I thought that, by simple intellectual analogy, the fourth dimension could project an object of three dimensions, or, to put it another way, any three-dimensional object, which we see dispassionately, is a projection of something four-dimensional, something we're not familiar with. It was a bit of sophism, but still it was possible. 'The Bride' in the *Large Glass* was based on this, as if it were the projection of a four-dimensional object.[19]

It seems, therefore, that the really given of the *Large Glass* is not in the waterfall as such but in the cascading spring of desire. And for exactly that reason the waterfall *must* be excluded, for if it were not, then the tautological world of the taken-for-granted would no longer be tautological but open to reformulation, the unconscious sucked into the conscious, the humble solipsist changed into a bragging politician. In the meantime the glider sings its melancholy litany of slow life and vicious circles, of motion for the sake of motion. And, as in Duchamp's apartment at 11 rue Larrey, the door is now open and now closed, the adverb *now* on this occasion functioning as the cornered hinge of the world.

*

**Have another apple, my dear**, and your eyes will be opened! Go bananas and the world will behave as it never did before! Paradise lost, self-reference regained. The dreaming lady up there on her balcony, the calculating troubadours down here on the ground. A clothes line and a horizon drawn in-between.

And yet. The very idea of the *Large Glass* is not to capture the world but to show that no matter how much she tries, and no matter how much gas the bachelors are transforming, the bride's desire can never be satisfied. The real reason for the blockage is itself dialectical, for the "freedom of choice in the upper half is offset by a grim determinism in the lower half; [as] the bride imagines and commands, [so] the bachelors react and obey."[20] The two worlds are nevertheless connected by the three glass staffs, two of them green (a color Marcel thoroughly disliked), the third colorless;[21] as indicated by the notes, the staffs represent the three brothers Duchamp, the green standing for Jacques and Raymond, the colorless for Marcel himself. And on the back, hiding in the void between the two uppermost isolation plates, is the vanishing point of the entire bachelor apparatus.

The numbers and the colors of the coolers suggest that the *Large Glass* may be approached also as a self-portrait, an interpretation which is further supported by the work's own title, *La **Mariée** mise à nu par ses **Célibataires**,* ***même.*** Pointing in the same direction is the circumstance that this salty sea *(mar-cel)* frequently presented himself as an androgynous figure dressed in women's clothes, in that guise better known as *Rrose Sélavy–Éros, c'est la vie.* In addition, the modifying adverb *même* sounds like *m'aime,* a connection which has tempted many critics to speculate about the possible relations between the master's art-work and his presumed love-life, including his well-attested closeness to his sister Yvonne, a Janus-head torn in tatters.

Nothing for the tabloids, even though there is much to indicate that the unfulfilled desires which permeate the *Large Glass* were deeply ingrained in Duchamp's equally charming and private personality.[22] Among the most intriguing and best documented of these affairs was his relation with Beatrice Wood, an attractive and wealthy young actress (eventually one of America's pioneering ceramic artists), who at their first meeting fell deeply in love with this most singular man alive. Whether their liaison ever went all the way has been much debated, less because the affair should be anybody's business, more because it may throw some light on the form and content of Duchamp's work. In a commentary which in many ways parallels the relations between the bride and her bachelors, Wood retells how

> like tides moving towards the moon, Marcel and I became closer. A physical relationship was inevitable. But he still continued to utter the same words, a phrase that he would repeat time and time again: *"Cela n'a pas d'importance."*
>
> There is a wonderful Indian saying that the eyes cannot see until they are incapable of tears. Marcel's saying "nothing had importance" somehow reflects the same idea. It was as if he had gone through all trials of the flesh and left that behind. With his grave perception, did he realize that in the long space of time, nothing really mattered? He had the objectivity of a guru. His mind touched stillness, beholding the unity of life. Yet with this understanding went a certain deadness. Many have observed it. The upper part of his face was alive, the lower part lifeless. It was as if he suffered an unspeakable trauma in his youth. . . . To him, it made no difference whether people responded to his work or not. Chess was his mistress, perhaps his escape from fame, from worldly stupidity. . . .
>
> I was in no way the great love of his life, but one who for a short time was close to him. That I am now sentimental about him is something that Marcel probably would not have liked. But that the man meant more to me than his art is something that he would have understood, always smiling and murmuring, *"Cela n'a pas d'importance."*[23]

Tides moving towards the moon! An image perfectly suited to Duchamp's oft-repeated remark that the single part of the *Large Glass* with which he was least satisfied was the three-layered penumbra that keeps the two panels together and apart, a line which by one of the more sensitive commentators has been characterized as a fold in his own personality.[24] After many years in the heady atmosphere of the Saussurean Bar, I have no difficulties recognizing both his sense of frustration and the attitude that nothing really matters. Perhaps this shared feeling also explains why I sometimes think that the really given in Duchamp's definitely unfinished is not (as so often suggested) in the unmarked waterfall, but in the receding horizon. Perhaps it is even in this lowest of the three coolers that we find the privileged state of rest which no one can ever reach, that complete merger of opposites which normally is relegated to the abyss of the excluded middle, the infra-thin line of imagination, the boundary between presence and absence.

The ghost of Euclid evidently refuses to die. For in this respect language, color and the vanishing point are alike: they all lie beyond the sense of touch. And yet it remains equally true that the limits of the body are the limits of the arts, imagination located not in the mind alone but in a crate of shattered glass as well.

The story is well known, its beginning dating back to January 1927, more exactly to the closing day of the International Exhibition of Modern Art at the Brooklyn Museum, the first and last time that the original version of the *Large Glass* was shown to the public; the first because *La Mariée* belonged to Katherine Dreier's private collection and was therefore normally available only to her friends, the last because immediately after the show was over the masterpiece was taken apart and packed into a wooden box, the bride placed on top of her bachelors. The whole thing was then trucked via a bumpy road to the Lincoln Warehouse on Manhattan, where it was stored and not reopened until the fall of 1931. Shocking moment, for it was then discovered that the glass had been shattered.

One of the century's most significant artworks severely damaged. Five years and no one had noticed! No wonder that Dreier was so upset that it took until March 18, 1936 (over lunch in Lille) before she brought the news, so moved that she could barely speak. "That's too bad. Too bad," was Duchamp's only response. Later, though, he admitted that "I was a little sorry, that big thing. I didn't know how much it was broken, whether it could be repaired or not. But on principle I was not going to cry. Because after all, it had no value in the artistic world at that time, nobody cared for it, nobody saw it or even knew about it."[25] *On principle* I was not going to cry!

Once the shock had subsided Duchamp moved decisively, perhaps because he came to image chance as another (non-material) "material," a happenstance which like the physical matter of wood, paint or glass might be turned into an artistic tool. Accordingly, and already the following summer (with the assistance of a skilled workman and the financial support of

Katherine Dreier) he did his utmost to put the broken Humpty together again. Not, however, by gluing the shattered pieces to each other, but by encasing them between two glass plates and a metal frame, effectively a kind of coffin destined for a mausoleum. The repair work took about two months, six to seven hours a day, almost certainly impossible had it not been for the fact that most of the lead wires were unbroken. At the end, he not only declared the previously abandoned work as definitively unfinished, but in a later interview he also remarked:

> I like these cracks because they don't look like a broken glass. They have a form, a symmetrical architecture. Also, I see them as a curious intention of something for which I am not responsible, an intention *already carried out* in a certain way, which I respect and like.[26]

The beauty is beyond words. And who knows, perhaps after the accident the bride and her bachelors are finally communicating, not, however, by the dint of logical necessity or by the artist's design, but by the fortunate play of objective chance. Dürer's orthogonal coordinates net transformed into a spider's web. A rare instance of Russian roulette with a happy end, the Mallarméan dice thrown by an unknown packer, Duchamp himself the self-appointed croupier with his pockets full of false Monte Carlo obligations. Like everyone else an offspring of his own time and place, on this occasion blessed by the goddess Fortuna.

The roots go deep, especially to January 25, 1881, when the French state officially, and in great detail, decreed how its children should be taught how to draw. The idea behind this *Republic*-sounding project was that every child should learn the difference between a reproduction of the way things appear to the eye and a representation of the way things really are. And for these reasons of socialization:

> Line lay, somewhat oddly, behind all the activity in Duchamp's *Large Glass*. His first drawings drafted the principals into ur-lines, orthogonals, ellipses, and loose spirals, perhaps platonic skeletons; whatever, the line itself was identifiable. It was geometrical; as such, steady and knowable, capable of being plotted in quadrants and measured to scale. At least that was how the line began. Then it sped off. It indulged itself in complex gyrations, flipped between dimensions, combined to make machine parts and then an apparatus, one for some bachelors, one for a bride. The parts spun, like motors, with desire; the bride was stripped bare, unfolded; she bloomed. The end. No recognizable body had fallen into view. All by itself the geometry had touched off a sexual plot.[27]

Viewed in that light the *Large Glass* stands out as an intricate family affair lived forwards and understood backwards, a readymade turned into a ready-maid, a mistress in search of some masters to strip her, incestuous parallels drawn from lines constructed. A historical consequence of this early training is that not even Marcel Duchamp—*the* artist of the mind—ever managed to free himself from the power of the eye which he so sincerely wanted to subdue. As a way out he consistently cultivated paradox, perhaps because he somehow understood that paradox is the only means powerful enough to reveal the contradictions of the taken-for-granted, the only approach to Euclid's fifth axiom, the only bridge between the Q.E.D. of the mind and the Q.E.F. of the body, the only road to Mandalay.

*

**Family resemblances, incestuous parallels,** the Euclidean steps from Plato's philosophy to Duchamp's semiotics shorter than anyone could have suspected. To be specific:

- The slanted line on Plato's base-map—the boundary between the two Realms of Cognition—now shows itself to be the conceptual twin of the Saussurean Bar, the visible objects to the left securely lodged in the materiality of the Signifier [in Plato's terminology an *object* of cognition], the untouchable relations to the right signifying the fleeting meanings of the signified [in Plato's terminology a *kind* of cognition]; the two prepositions *of* and *in* intertwined. In my own imagination it is in the seclusion of this ultra-thin line between the two Realms that "the bride reveals herself nude in 2 appearances: the first, that of the stripping by the bachelors. the second appearance that voluntary-imaginative one of the bride. On the coupling of these 2 appearances of pure virginity—On their collision, depends the whole blossoming, the upper part and crown of the picture. . . . Mixture, physical compound of 2 causes (bach. and imaginative desire) unanalyzable by logic."[28] Not unexpectedly (at least when you come to think of it), this anonymous marker has the same slope (23.5°) as the invisible line which in the *Large Glass* separates the erotic machinery to the left and the tracings of the dripping desires to the right. It is amusing (one of Duchamp's favorite expressions) that this hinge has the same tilt as the earth axis.
- The horizontal line on Plato's maps is by all evidence another version not only of the Saussurean Bar but also of the Duchamp coolers, the clothesline on which the bride hangs her white garment and the artist his uncolored self-portrait.
- Just as the bride of the Bride is a paradigmatic example of Plato's Form of Forms, so the malic molds are empty shells waiting to be filled with semiotic content, a sort of Saussurean Signifiers without any signifieds.
- The orthogonal frames are drawn by silent lines, and that is regardless of whether their antecedents are in Plato, Alberti, Kant, Duchamp

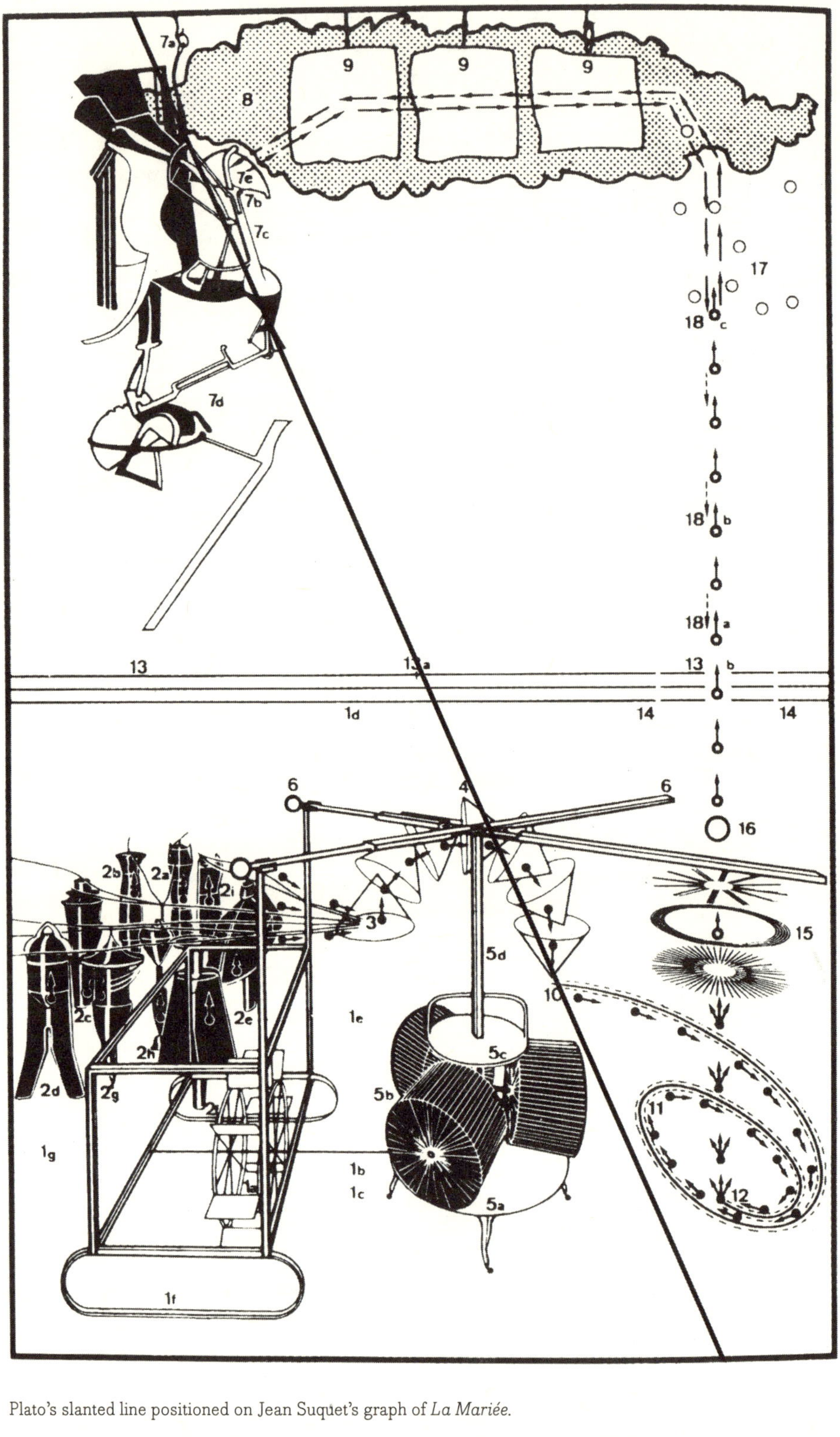


Plato's slanted line positioned on Jean Suquet's graph of *La Mariée*.

> or myself. On the inside of these non-crossable boundaries lies the taboo-ridden Realm of the taken-for-granted, on the outside the imaginary lands of the totally alien. While the landscapes of the former are studded with loudspeakers which cannot be silenced, the mindscapes of the latter are enwrapped in silences that cannot be expressed. As already suggested, inside the frame governs the Wittgensteinian motto *Wovon mann nicht sprechen kann, darüber muss mann schweigen,* outside it rules the chiastic retort *Wovon mann nicht schweigen kann, darüber muss man sprechen.*

Just another way of saying that the basic metaphor of Plato's Divided Line is that of an image to its object,[29] therein included the relations between ruler and ruled, the wisdom-loving philosophers (the bride) occupying the upper panel, the food- and sex-loving workers (the bachelors) the lower.

*

**Wisdom-loving philosophers and food-loving workers** are at opposite ends of the social hierarchy of which the *Republic* is equally famous and infamous. With the purpose of determining the boundaries of that universe, the mapping project will now be brought to its conclusion. Accordingly, a third map will be constructed, the only one in our series which explicitly focuses on the concept of power and on the socialization processes that make us so obedient and so predictable. Although the previous ties to Plato's Divided Line will remain strong, the more immediate connections are to the allegory of the Cave and, by extension, to the House of Plotinus.

Like its predecessors, this map is also structured around the slanted and horizontal lines, their intersection forming the pivot of the world; as recalled, the slanted line marks the boundary between the two Realms of Cognition, the horizontal between the *A* and *B* regions of the visible, on the one hand, and the *C* and *D* regions of the intelligible, on the other. In the socio-economic case, however, the associations run less to Plato directly and more to the Neoplatonic layout of Plotinus' House, the edifice in which the guardians occupy the upper floors and the workers the lower, the former focusing their desiring minds on the accumulation of knowledge about non-changeable forms, the latter on their bodily needs and fluctuating opinions. At issue is once again the fix-point of fix-points, more precisely the question of whether the critic should reason à la Jacques Lacan and let the Signifier search for its meaning or whether he should follow Ferdinand de Saussure and allow the signified to find its expression. Serious business, for it was exactly over this issue that Aristotle most sharply disagreed with his master, forcefully asserting that "the belief in forms came about in those who spoke about them, because, in regard to truth, they were persuaded by the Heraclitean argument that all perceptibles are

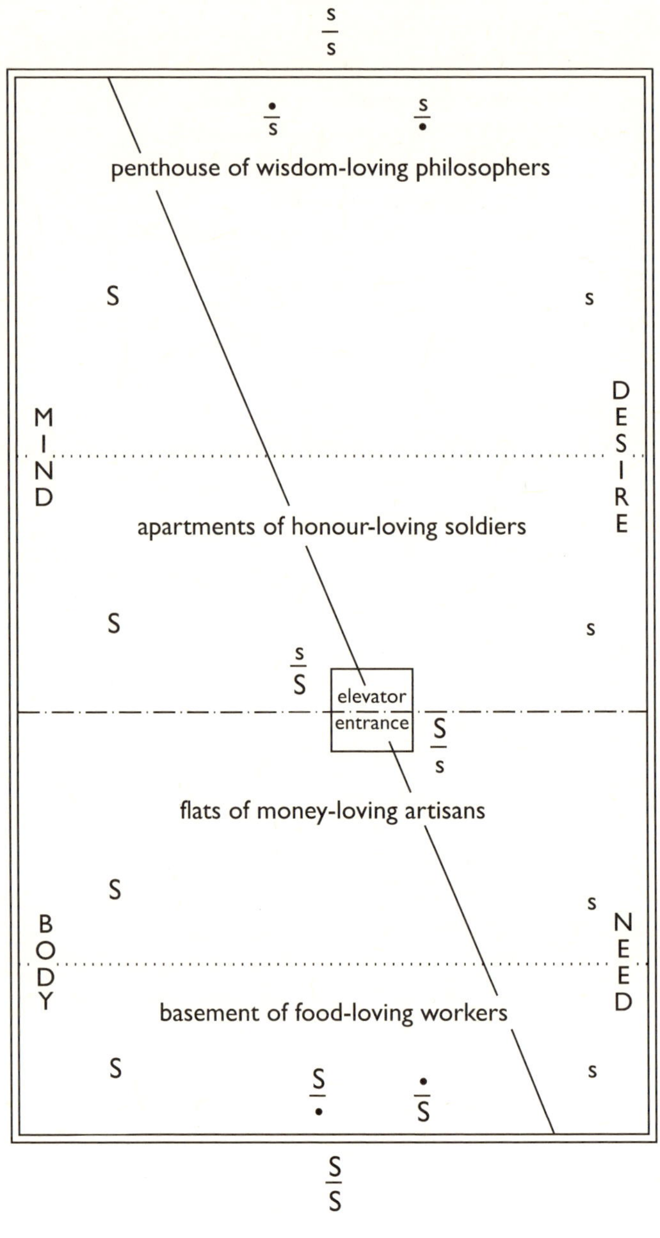


Socio-economic map of Plato's *Republic.*

always in flux, so that if scientific knowledge is to be about anything, there must, in their view, be some different natures, over and above perceptibles, which are permanent; for there is no scientific knowledge of things in flux. . . . [However,] of the ways in which it is proved that the forms exist, none is convincing."[30] A case of critical deconstruction in the making, Aristotle a budding Lacanian, Lacan a withering Aristotelian.

The socio-economic map of the *Republic* closely reflects the segregation of the Plotinian House, the food- and sex-loving workers relegated to the crowded quarters in the deep and windowless cellars, the money-loving artisans to the claustrophobic abodes on the entresol. Rising from the ground is an executive tower whose lower floors are assigned to the honor-loving soldiers, the penthouse at the top to the wisdom-loving philosophers; while the auxiliaries have been trained to look down at (or is it *on*) the street below, the real politicians are free to see everything and anything, including whatever hovers above the rooftops and whatever gleams beyond the horizon.

In this strange co-op the different population groups are communicating through some sort of language. Their respective idioms are nevertheless drastically different, especially in the ways they tie together the sensible (roughly the Signifiers) and the intelligible (roughly the signifieds), the workers and artisans emphasizing their bodily needs (the Saussurean *S*), the guardians their desire for knowledge (the Saussurean *s*). The fullest and best-balanced expressions are consequently exchanged in the entrance hall of the executive tower—effectively the agora of the Plotinian House, where the elevator boys are the only ones who can speak to and be understood by everyone else. In contrast, the basement floor and the penthouse roof constitute impenetrable limits of language such that the entire construction rests on a foundation of silent stones and reaches into a sky of non-expressible meaning, the former lodged in the pure and non-differentiable matter of

$$\frac{S}{S}$$

the latter in the equally pure and non-differentiable spirit of

$$\frac{s}{s}$$

Nothing new, of course. For already in the crucial passage that transfers the analogy of the Line into the allegory of the Cave, Socrates suggested that we should "take, as corresponding to the four sections, these four states of mind: *intelligence* [or understanding] for the highest, *thinking* [or reason] for the second, *belief* [or trust] for the third, and for the last *imagining* [or imagination]. These you may arrange as the terms in a proposition, assigning to each a degree of clearness and certainty corresponding to the measure in which the objects possess truth and reality."[31] That measurable measure of truth and reality is indeed to be found also in the layout of the Plotinian House, not the least in the relative proportions through which the *S* and the *s* enter the prevalent signs of their respective congregations. To be more precise, the two non-expressions

$$\frac{s}{\quad}$$

and

$$\frac{\quad}{S}$$

may be taken as impossible forms of silence, the former typical of the evasive philosophers, the latter of the thingifying workmen.

It requires little imagination (merely a turn from the vertical to the horizontal) to transform the Saussurean conception of the Plotinian House into an architectual drawing of Plato's Cave.[32] The latter is nowadays often pictured as an old-fashioned movie theater with the guardians seated in the back rows, ready to escape, and the workmen on benches up front, too close to the screen to grasp what is going on:

> Next, said I, here is a parable to illustrate the degrees in which our nature may be enlightened or unenlightened. Imagine the condition of men living in a sort of cavernous chamber underground, with an entrance open to the light and a long passage all down the cave. Here they have been from childhood, chained by the leg and also by the neck, so that they cannot move and can see only what is in front of them, because the chains will not let them turn their heads. At some distance higher up is the light of a fire burning behind them; and between the prisoners and the fire is a track with a parapet along it, like the screen at a puppet-show, which hides the performers while they show their puppets over the top.
>
> I see, said he.
>
> Now behind this parapet imagine persons carrying along various artificial objects, including figures of men and animals in wood and stone or other materials, which project above the parapet. Naturally, some of these people will be talking, others silent.
>
> It is a strange picture, he said, and a strange sort of prisoners.
>
> *Like ourselves,* I replied, for in the first place prisoners so confined would have seen nothing of themselves or of one another, except the shadows thrown by the fire-light on the wall of the cave facing them, would they? . . . Now, if they could talk to one another, would they not suppose that their words referred only to those passing shadows which they saw? . . .
>
> No doubt.
>
> In every way, then, such prisoners would recognize as reality nothing but the shadows of those artificial objects.
>
> Inevitably.
>
> Now consider what would happen if their release from the chains and the healing of their unwisdom should come about in this way. Suppose one of them set free and forced suddenly to stand up, turn his head, and

> walk with eyes lifted to the light; all these movements would be painful, and he would be so dazzled to make out the objects whose shadows he had been used to see. What do you think he would say, if someone told him that what he had formerly seen was meaningless illusion, but now, being somewhat nearer to reality and turned towards more real objects, he was getting a truer view? . . . . Would he not be perplexed and believe the objects shown him to be not so real as what he formerly saw? (*Republic* 514a–515d, trans. Cornford, emph. added)

Ludwig Wittgenstein—Surveyor General of Lands Unknown—would almost certainly have known what the released would have said: "I have gone mad!" Immediately after these words had left his mouth, however, he would have added that "this expression is merely another way of saying that I don't know my way about, that my social compass has lost its bearings, my map its scale. What I see looks so unfamiliar because my translation function has gotten all screwed up, nothing but aborted signs, nothing but mute Signifiers with no signifieds to unlock them." But as soon as the remaining cave-dwellers, chained by the leg and also by the neck, have heard this analysis from the strange aristrocat, then they, *like ourselves,* "would laugh at him and say that he ha[s] gone up [into the light] only to come back with his sight ruined; it was worth no one's while even to attempt the ascent. If they could lay hands on the man who was trying to set them free and lead them up, they would kill him" (*Republic* 517a).

Finally, Socrates' own summarizing conclusion, a rhetorical masterpiece starting with a well-defined beginning and leading up to a firm conclusion, like all analyses a travel story in disguise. To be exact:

> Every feature in this parable, my dear Glaucon, is meant to fit our earlier analysis. The prison dwelling corresponds to the region revealed to us through the sense of sight, and the free-light within it to the power of the Sun. The ascent to these things in the upper world you may take as standing for the upward journey of the soul into the region of the intelligible; then you will be in possession of what I surmise, since that is what you wish to be told. Heaven knows whether it is true; but this, at any rate, is how it appears to me. In the world of knowledge, the last thing to be perceived and only with great difficulty is the essential Form of Goodness. Once it is perceived, the conclusion must follow that, for all things, this is the cause of whatever is right and good; in the visible world it gives birth to light and to the lord of light; while it is itself sovereign in the intelligible world and the parent of intelligence and truth. Without having had a vision of this Form no one can act with wisdom, either in his own life or in matters of state.
>
> [And, Glaucon replied:] So far as I can understand, I share your belief. (*Republic* 517b–c)

*

**I too share Socrates' belief**, for I too have traveled from the depths of darkness to the heights of dialectical insight, a road built not on the quicksand of good intentions but on the solid foundation of a life-long education;[33] to Plato no one is a dialectician by nature, and that explains why the *Republic* may be better read as a practical manual of pedagogy than as a theoretical treatise of justice. The whole tenor of the allegory of the cave is in fact to equate the prisoners with the crowd, to downgrade the reader's ordinary beliefs and to show that the very purpose of the dialectic is to destroy the generally accepted hypotheses of mathematics and thereby the foundation of the taken-for-granted. Here, then,

> we come to the main theme, to be developed in philosophic discussion. It falls within the domain of the intelligible world; but its progress is like that of the power of vision in the released prisoner of our parable. When he had reached the stage of trying to look at the living creatures outside the Cave, then at the stars, and lastly at the Sun himself, he arrived at the highest object in the visible world. So here, the summit of the intelligible world is reached in philosophic discussion by one who aspires, through the discourse of reason *unaided by any of the senses,* to make his way in every case to the essential reality and perseveres until he has grasped by pure intelligence the very nature of Goodness itself. *This journey* is what we call Dialectic.
>
> Yes, certainly.
>
> There was also that earlier *stage* when the prisoner, set free from his chains, turned from the shadows to the images which cast them and to the fire-light, and climbed up out of the cavern into the sunshine. When there, he was still unable to look at the animals and plants and the sunlight; he could only see the shadows of things and their reflections in water, though these, it is true, are works of divine creation and come from real things, not mere shadows of images thrown by the light of the fire, which was itself only an image as compared with the Sun. Now the whole course of study in the arts we have reviewed has the corresponding effect of leading up the noblest faculty of the soul towards the contemplation of the highest of all realities, just as in our allegory the bodily organ which has the clearest perceptions was led up towards the brightest of visible things in the material world.
>
> I agree to what you are saying, Glaucon replied; I find it very hard to accept, but in another way no less hard to deny. [Now, however,] I want you to describe the function of philosophic discussion. Into what divisions it falls, and what are its methods; for here, it seems, we have come to the procedure which should lead to *the resting-place at our journey's end.*
>
> My dear Glaucon, said I, you will not be able to follow me farther, though not for want of willingness on my part. It would mean that, in-

> stead of illustrating the truth by an allegory, I should be showing you the truth itself, at least as it appears to me. (*Republic* 532a–533a, emph. added)

And so it is that Plato's dialectic—just as Duchamp's *Large Glass*—focuses less on *what* we see and more on *how* we see. Hence the dialectic is not a technique which, like geometry or mathematics, draws conclusions from well-defined premises, but an art in and of itself, the acknowledged ruler over all the other sciences.[34] Just as the angels long ago ascended and descended Jacob's ladder, so the modern dialectician now uses the elevator, its shaft constructed around the slanted line shown on the architectural drawings of the Plotinian House, its double doors giving access to every floor. Driven by their desires and constantly chatting with the operators, the philosophers are effectively carried from the workers' basement to the ruler's penthouse, the resting-place at their journey's end. Once there the doors slide open and the wisdom-lovers are free to enter the secret rooms of the taken-for-granted, some so enchanted by the alta vista that they never want to leave the camaraderie of the non-ending symposia, others so fundamentally changed by the experience that they return to the dungeons as self-appointed missionaries, Vitruvian or Wittgensteinian Men in disguise. At any rate,

> no one will maintain against us that there is any other method of inquiry which systematically attempts in every case to grasp the nature of each thing as it is in itself. The other arts are nearly all concerned with human opinions and desires, or with the production of natural and artificial things, or with the care of them when produced. There remain geometry and those other allied studies which, as we said, do in some measure apprehend reality; but we observe that they cannot yield anything clearer than a dream-like vision of the real so long as they leave the assumptions they employ unquestioned and can give no account of them. If your premise is something you do not really know and your conclusion and the intermediate steps are *a tissue of things you do not really know,* your reasoning may be consistent with itself, but how can it ever amount to knowledge?
>
> So, said I, the method of dialectic is the only one which takes this course, doing away with assumptions and traveling up to the first principle of all, so as to make use of confirmation there. When the eye of the soul is sunk in a veritable slough of barbarous ignorance, this method gently draws it forth and guides it upwards, assisted in this work of conversion by the arts we have enumerated. (*Republic* 533c–d, trans. Cornford, emph. added)

A tissue of things you do not really know, assumptions not left unquestioned! Surely a case of cartographical reason par excellence, for just as *mappa* means "tablecloth" or "napkin" and *charta* stands for "leaf of paper," so a *tissue* is, according to the *OED,* "a rich kind of cloth, often interwoven with gold or silver now applied to various rich or fine stuffs of delicate or gauzy structure, a 'fabric,' 'network,' 'web' (of things abstract, most usually of a bad kind, as absurdities, errors, falsehoods, etc.)." Surely the case also of Marcel Duchamp laying bare the given of taken-for-granted.

It cannot be said more clearly: if it were not for the resistance of the limestone wall, there would be no shadows in the Cave, hence nothing for the prisoners to see and nothing for the wisdom-lovers to deconstruct; if it were not for the fabric of the mappa, there would be nothing to capture the cartographer's travel story, hence no map to guide him; if there were no Saussurean Bar, there would be neither Signifier nor signified, hence no language to translate. Rephrased and condensed: if there were no imaginations, then there would be no humans. And without the glass of the *Large Glass* Marcel Duchamp would never have seen through what he saw, never understood that the semiotic animal is obsessed by a mimetic desire so strong that it can never be overcome.

The history of philosophy is one with the history of cartography, the history of cartography closely related to the theory of perspective and the practice of plasticity. For what ties the mapmaker and the painter together is that both are artists of deception, experts on the irresolvable problem of turning a three-dimensional original into a two-dimensional copy. Yet their respective approaches run in opposite directions, for whereas the traditional (Ptolemaic) aim of mapmaking is to flatten the globe, the purpose of painterly (Albertian) perspective is to fatten the nude. The letter *l* obviously makes a wor(l)d of difference, the present example itself an illustration of how "a pure imagination which conditions all *a priori* knowledge, is one of the fundamental faculties of the human soul."[35]

* * *

**If there were no imagination** Marcel Duchamp would never have completed his most dialectical and self-referential work, an entire wing of the Philadelphia Museum of Art, a kind of mausoleum erected in his own honor, a modernist version of the Egyptian Valley of the Kings.

As the most clever among ironists Duchamp used his Philadelphia connections to show how Plato's dialectic is a most profound affair, an engagement to be lived forwards and understood backwards, perhaps with the undefined goal of knowing thyself. Like Socrates and Kierkegaard before him, he too was obsessed by the idea of merging together the activities of love and knowledge, not, however, as a way of valorizing himself, but because "I had a certain love for what I was making, and this love was translated into that form [of a Museum wing]. I wanted the whole body of work to stay together. Moreover, I found that my works weren't numerous enough to make a profit painting after painting. And, above all, I wanted as much as possible not to make money."[36] Thanks to the generous support of the Arensbergs, Katherine Dreier and William Copley this dream of a

Paul Cézanne, *The Large Bathers*. 1906. Oil on canvas, 210.5 × 250.8 cm. Philadelphia Museum of Art: Purchased with the W. P. Wilsstach Fund, 1937.

private mausoleum eventually came true, the absent artist now more present in Philadelphia than most pharaohs in Luxor.

No visitor to the Quaker City can miss it, the imposing building which like an ancient temple lies at the end of the Benjamin Franklin Parkway. Once there, climb the wide stairs, enter the Main Hall, turn right and you find yourself surrounded by an outstanding collection of European paintings from 1850 to 1900, most prominent among them Paul Cézanne's *The Large Bathers* from 1906 (the year of the artist's death), a geometric triangulation which "in its unfinishedness *is* its definitiveness; and it is an unfinish that comes out of forty years spent meditating on what a conclusion in painting could be. . . . It states what the conditions of depicting the body in the world now amount to, and it does so with utter completeness."[37] A personal oikumene mapped through a set of triangulations à la Ptolemy and Hobbes, a revolutionary artist grabbed by his own obsession, a misunderstood man caught in the self-referential clutches of desires non-suppressed.

Spurred on by this sense of unfinished completeness, proceed and turn right once again, this time to enter a long corridor filled with some remarkable pieces of twentieth-century art, including several paintings by Picasso (who Duchamp never liked), Matisse (who he admired and whose daughter-in-law he eventually married), Dalí (whose company he enjoyed) and sixteen sculptures by Constantin Brancusi (one of his closest friends and allies), most striking among them *Princess X,* once (like Duchamp's *Fountain*) considered too obscene to be exhibited—precisely ambiguous, a stunning demonstration of the Brunelleschian revelation that what you see depends on where you stand. Fix-points on the move.

Finally, the galleries for which we have come: selected pieces from the largest Duchamp collection in the world.

Virtually everything is there, including the original, broken and repaired version of the *Large Glass,* its colors once so splendid now peeling and fading. When this work was put in place in October 1954 Duchamp insisted that the thick wall behind it should be broken and a rectangular window with a view to the courtyard be installed, a slit somehow reminiscent of a gunport. And out there in the unpleasant Philadelphia weather, clearly visible through the glass itself was at that time a sculpture of a nude female figure, her hands clutched in wonder, her eyes reaching out for our own. A wonderful instance of the saying that "between two beings in love, language is not the deepest. . . . The exchange occurs through the eyes."[38]

This particular eye-catcher is entitled *Yara* and was made by Maria Martins, wife of the Brazilian ambassador to the United States, a lady who according to her own daughter could seduce anybody. Evidently also the handsome Frenchman, who by several accounts fell deeply in love with her, seemingly the only woman he really desired, perhaps because he somehow sensed that she would never leave her husband, perhaps because he thought her an incarnation of the virgin who became a bride but never a wife. In one of the last letters he sent her he simply concluded that "I accept the situation as it is and no longer hope for a miracle. I feel happy when I think of you."[39] Whatever the nature of their relation, there is no doubt that she played a decisive role in his rebirth as an artist.

And when I stand there in admiration, I suddenly realize where I am. Not in a museum but in an Orthodox church, not in front of a piece of modern art but face to face with an iconostasis. With my eyes as feet I proceed first through the Great Entrance (in this case one with the vanishing point of the bachelor machine), then via the window into the sanctuary. In there—or rather out there—I find myself as close to Duchamp's altar that I may ever come, the now empty spot in which Maria Martins' statue used to stand, at the same time the locus of a sacrifice and a communion. The key to Duchamp's imagination inexplicably thrown away.

Too moved to focus, my perspective changes from naturalis to artificialis. And in the cupola of the church-turned-museum I imagine not a

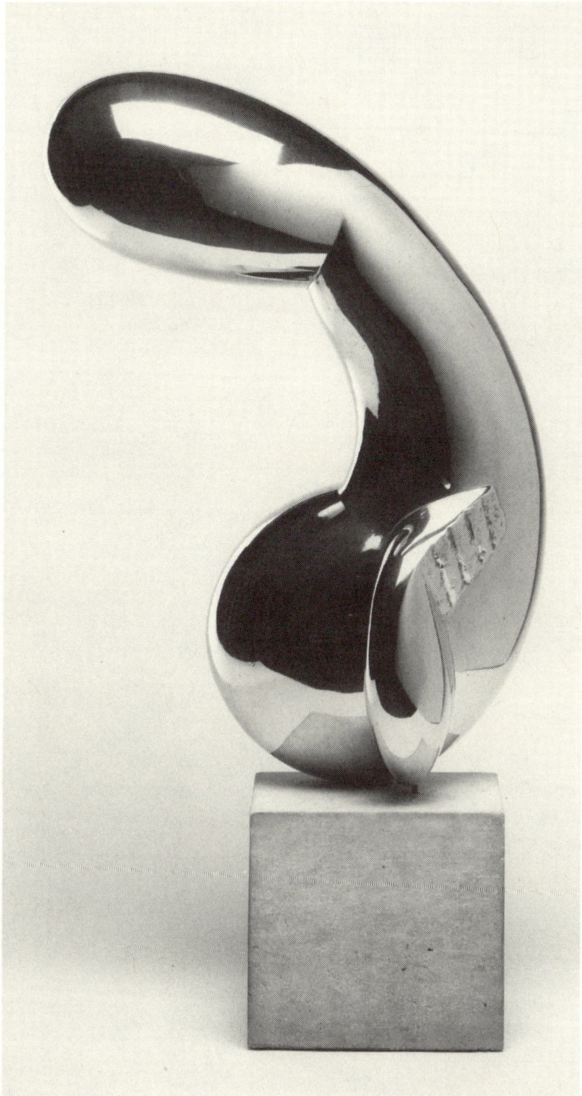

Constantin Brancusi, *Princess X.* 1915–16. Polished bronze, 57.4 × 22.7 × 41.3 cm; Limestone block, 18.5 cm. Philadelphia Museum of Art: The Louise and Walter Arensberg Collection, 1950. © Constantin Brancusi/BUS 2005.

circular picture of the Pantocrator but a milky bride throwing off her halo and descending to the real world of desires so desirous that they can never be fulfilled, language, thought and action all turned into mutual visions of each other. And close to the window, perpendicular to the wall, is a kind of screen which on one side shows the painting *The King and Queen Surrounded by Swift Nudes* from 1912 and on the other *Paradise (Adam and Eve)*, the latter mounted upside-down. A remark to be constantly kept in mind.

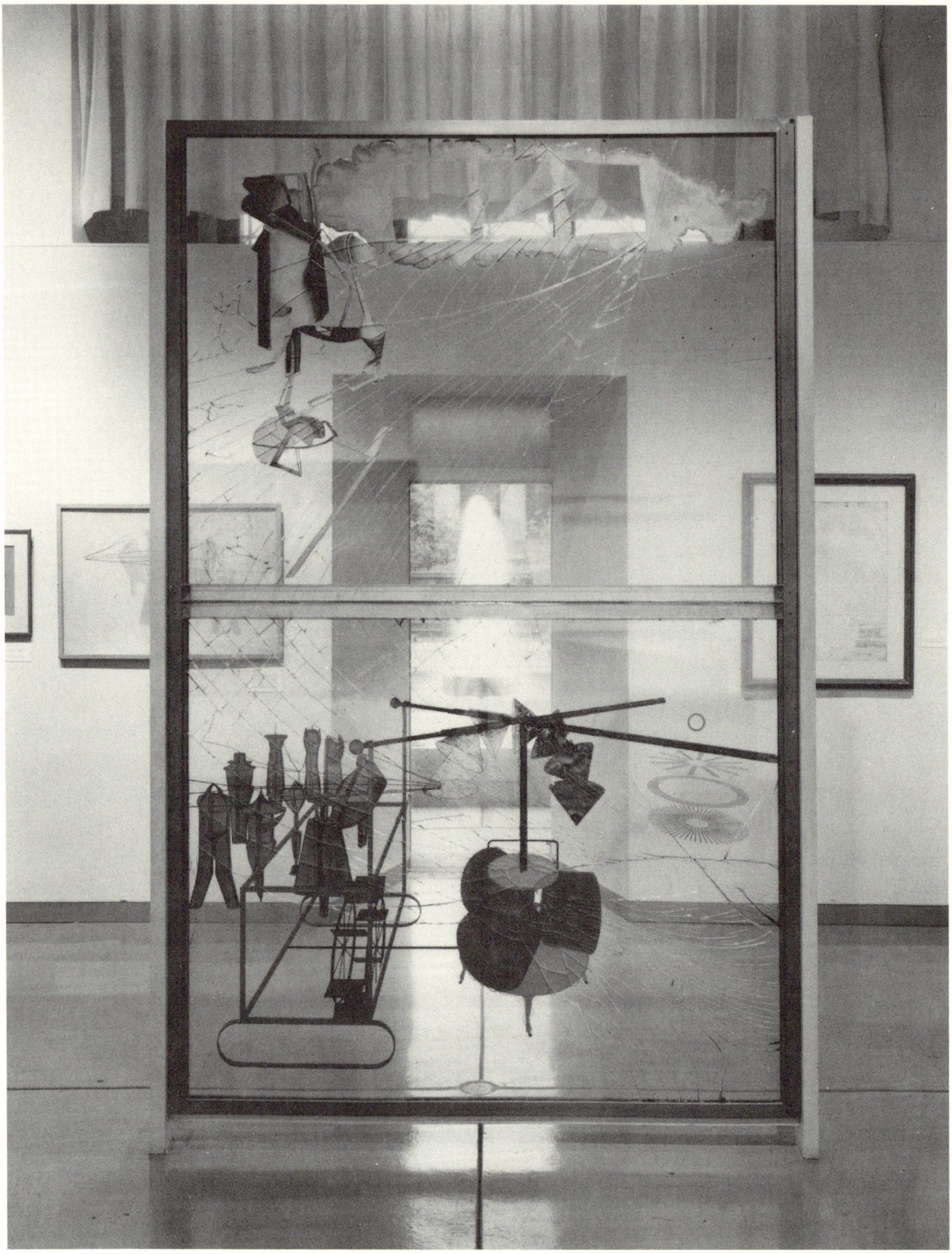

Marcel Duchamp, *The Bride Stripped Bare by Her Bachelors, Even. (The Large Glass.)* 1915–23. Oil, varnish, lead foil, lead wire and dust on two glass panels, 277.5 x 175.9 cm. Philadelphia Museum of Art: Bequest of Katherine S. Dreier, 1952. © Succession Marcel Duchamp/BUS 2005.

*

**With that remark constantly in mind** we now move on towards the hidden tomb for which we have come, Egyptian treasure hunters in disguise. Passing through a low doorway we finally enter a small room, dark and completely empty. No paintings on the walls, no windows for the sun, nothing but a brick portal with an old door, worm-eaten, patched, and closed by a rough crossbar. Dead end, for as your body bounces against the waiting-room walls you suddenly realize that the door cannot be opened. At that same moment, though, your eyes are caught by two small holes, apparently drilled for a purpose. No longer a closed door but an open window. Step up, peep in, and you will find yourself engrossed in a scene you are likely never to forget. In the words of Octavio Paz, perhaps the most enthusiastic—and therefore the most criticized—of all the self-appointed guides:

> First of all, a brick wall with a slit in it, and through the slit, a wide open space, luminous and seemingly bewitched. Very near the beholder—but also very far away, on the 'other side'—a naked girl, stretched on a kind of bed or pyre of branches and leaves, her face almost completely covered by the blond mass of her hair, her legs open and slightly bent, the pubes strangely smooth in contrast to the splendid abundance of her hair, her right arm out of the line of vision, her left slightly raised, the hand grasping a small gas lamp made of metal and glass. The little lamp glows in the brilliant light of this motionless, end-of-summer day. Fascinated by this challenge to our common sense—what is there less clear than light?—our glance wanders over the landscape: in the background, wooded hills, green and reddish; lower down, a small lake and a light mist on the lake. An inevitably blue sky. Two or three little clouds, also inevitably white. On the far right, among some rocks, a waterfall catches the light. Stillness: a portion of time held motionless. The immobility of the naked woman and of the landscape contrasts with the movement of the waterfall and the flickering of the lamp. The silence is absolute. All is real and verges on banality; all is unreal and verges—on what?[40]

Poetic description à la Paz, a literary translation of what another admirer has called "the strangest work of art in any museum,"[41] the most profound critique of cartographical reason yet to be formulated: Janus' temple with its doors open in times of war and closed in times of peace; the bathing Diana and the spying Actaeon caught in the middle of the act, vicious hounds ready to attack; Plato's cave turned into a stereoscope, a camera obscura or an elaborate apparatus for anamorphic (de)composition; a mixed media assemblage of "wooden door, bricks, velvet, wood, leather stretched over an armature of metal, twigs, aluminum, iron, glass, plexiglass, linoleum, cotton, electric lights, gas lamp (Bec Auer type), motor, etc.," in short a jumble of materials bound to self-destruct; a most personal

Marcel Duchamp, *Etant donnés: 1° la chute d'eau, 2° le gaz d'éclairage. (Given: 1. The Waterfall, 2. The Illuminating Gas.)* 1946–66. Mixed media assemblage: wooden door, bricks, velvet, wood, leather stretched over an armature of metal, twigs, aluminum, iron, glass, Plexiglas, linoleum, cotton, electric lights, gas lamp (Bec Auer type), motor, etc., 242.6 × 177.8 × 124.5 cm. Exterior view. Philadelphia Museum of Art: Gift of the Cassandra Foundation 1969. © Succession Marcel Duchamp/BUS 2005.

affair and as such a constellation of everything its maker knew and everything he had not yet discovered; a stripping bare not of the bride but of the taken-for-granted; a *coitus aeternalis* of love and knowledge; a large and complex tableau commonly known as

*Etant donnés:* *1° la chute d'eau*
*2° le gaz d'éclairage*

—the title here reproduced in the typographical form which Duchamp himself always used, the 2° placed underneath the 1°, not beside it, as it is usually shown in the English version of *Given: 1. The waterfall, 2. The illuminating gas.*[42]

Marcel Duchamp, *Etant donnés: 1° la chute d'eau, 2° le gaz d'eclairage. (Given: 1. The Waterfall, 2. The Illuminating Gas.)* 1946-66. Mixed media assemblage, 242.5 × 177.8 × 124.5 cm. Interior view. Philadelphia Museum of Art: Gift of the Cassandra Foundation, 1969. © Succession Marcel Duchamp/BUS 2005.

More immediately, the *Given* is the outward result of a sustained effort begun in 1946 and finished in 1966, two years before the artist passed away. Bristling with cross-references to his previous works, not the least to the *Large Glass,* it was conceived and executed in his New York studio at 210 West 14th Street, the last touches added in a small room in a commercial building at 80 East 11th Street. While still a work in progress it was known to only a handful of people, but when it approached completion it was purchased for an undisclosed sum (reputedly around $60,000) by the art dealer William Copley, all with the understanding that it was eventually to join the rest of Duchamp's oeuvre at the Philadelphia Museum of Art. And thus it was that the *Etant donnés* finally went on view in June 1969, albeit with the provision that for fifteen years its interior must not be photographed. Not as strange as it may seem, for the whole thing is in itself a sort of candid camera.

To facilitate the transfer from the studio to the museum, Duchamp left a set of detailed notes (long kept secret) on how the thing should be taken apart and put together again.[43] This means that just as the *Large Glass* comes in two parts—a text and an artifact—so does the *Given.*[44] There is nevertheless a decisive difference between the two alter egos, for while the *Green Box* is filled with a mishmash of ideas searching for their material expression, the *Manual* contains little on meaning and much on doing, nothing about dripping desire, everything about screws and light-bulbs, bricks, twigs and a discarded tin-can. Thus, if the *Large Glass* may be seen as a static diagram of a mobile construction impossible to decipher without the legend of the notes, then the *Given* is a physical model in which the relations between thing and text, picture and story, are exactly the reverse. The crucial difference is that while "the machine of the Large Glass is an enigma; the nude of the Given is the enigma in person, its incarnation."[45] Yet it should be noted that the only detail which the *Manual* explicitly leaves to the assemblers' discretion concerns the position of the clouds—in my ears a silent echo of Brunelleschi's burnished silver, a whisper that there are some things which cannot, indeed must not, be captured in the orthogonal net of Dürer's window.

As may be expected, the *Manual* has been subjected to much analysis, not the least in an influential text by Jean-François Lyotard. To my knowledge, however, the work as a whole has never been connected with the principles of cartography and this is despite the fact that Lyotard's creative recreations are deeply rooted in the three primitives of the map-maker's fix-point, scale and canvas. To be exact:

- the fix-point seemingly coincides with the vanishing point of the cunt, itself slippery;
- the scale is projected into the skewed perspective of the dismembered nude, turning and twisting;

- the canvas is one with the penetrated picture plane of the broken wall.

It is to these primitives that the discussion must now turn.

*

**The vanishing fix-point** of Duchamp's *Given* is easy to detect and difficult to grasp, its striking image as engaging to radical feminists as to pimpled masterbaitors, boys and girls alike.

Countless are those who have walked up to the famous wall with the strange door. Subsequently much altered, the latter originally came from the Spanish village of Cadaqués, where in the early 1960s the vacationing Duchamps had picked it up and got it shipped to New York. Although it shares many traits of the readymade, it is in fact a highly manipulated piece of art deliberately cut to fit the proportions of the golden section. Removed are the hinges, the handle, the key and the key-hole, added some newly cut wooden bars, many spikes and the two holes drilled at eye level, seven centimeters apart.

Eyes to the holes! Nose poking the slit-in-between! Bound to experience!

And when I stand there glued to the viewing-position someone else has chosen for me, then I realize that while the bride of the *Glass* can be stripped and shared by many bachelors at the same time, the nude of the *Given* is now-and-here and for me alone. When the exhibitionist issues her command(ment) "Thou shalt not touch!" the voyeur's only reply is "Here I stand"— an unexpected echo of Abraham's words "Here I am" (Gen. 22:1), the only answer which the father of nations could give to the LORD who had appointed him. Nothing left to do but to stare, especially as the French expression *étant donnés* is a distant relative of the simsalabim of the English "Let there be!"

But what is it that I see, when in shame I see? A dumped dummy, a raped corpse, a woman in postcoital repose, a pornographic whore? All and none, but most immediately an image impossible to resist, the sight of a bared cunt. (Pardon the term, but no other seems appropriate.)[46] And exactly in that most taboo-ridden of all places—the holey of the holy—lies the vanishing point, its prominence further enhanced by the combination of the slightly plunging perspective and the interplay of different light sources, strongest among them three fluorescents of 40 watts each, three century lights and a spotlight of 150 watts, the latter equipped with a reflector of aluminum; not surprisingly the generated heat is so intense that a transparent glass had to be inserted to keep the thing from catching fire. The rays fall vertically, accurately on the cunt, and, as a way of reinforcing the brightness, the antechamber between the door and the broken wall is upholstered with black velvet on four of its six sides. Seizing hold of the viewer's eyes, the force of the taken-for-granted is so strong that it leads them directly to the supposed focal point, leaving him no choice but to submit, the voyeur and the exhibitionist reaching out for one another. Plato's Sun in another context. Brunelleschi's experiment too, for also in the *Given* the viewpoint and the vanishing point are symmetrical. It follows that when your eyes "think they see the vulva, they are seeing themselves. A cunt is he who sees."[47] Beware of the embarrassing footsteps behind you, though, for what you are just about to discover is that the voyeur is not a desiring human being of four dimensions but a being without dimensions, nothing but a point!

Such is the immediate reaction. But as the excited teenager gets a little more experienced, (s)he gradually learns that what initially looked like a stable fix-point is neither fixed nor stable, not an end in itself but a means to another end. For instance, the professional critic is quick to inform the youngster that in the archives of art history the seeing cunt is treated less as a cunt and more as the artist's response to those who went before him; first to Albrecht Dürer's reclining nude, her clothes in a heap beside her, then to Gustave Courbet's *The Origin of the World,* its locus exactly delineated.

While the former work has already been dealt with, the latter is an excellent example of the retinal art which Duchamp so strenuously opposed, an unabashedly pornographic picture painted in 1866 and originally commissioned by the Turkish diplomat and bon-vivant Khalil Bey, later bought and cherished by none less than Jacques Lacan. In the palace of the former it was hung behind a curtain to be opened only on special occasions and for special guests; in the vacation house of the latter it was encased in a well-sealed wooden box, presumably the analyst's private answer to the nursery rhyme of pussycat, pussycat, where did you go.

In the case of the *Given* not so easy to tell, however, for at closer inspection what at first sight looks like a woman's genitals no gynecologist would ever recognize as such. Under the voyeur's eyes the shaven female is instead gradually changing into a dismembered (not castrated) male, the imagined clitoris taking on the traits of a symbolized scrotum, the real pouch which on ordinary men encloses the testicles. In this process of deconstruction, the cut-off penis literally moves from the nude's crotch to its left hand, where it can sometimes be seen flickering as a gas lamp (Bec Auer type), sometimes beaconing like the torch of the Statue of Liberty—in typical fashion neither either-or nor both/and. Who knows, perhaps the ready-maid is meant as just another portrait of the androgyne figure of Rrose Sélavy, a rendering of Lacan's *objet petit* defiled, deified and reified all at the same time.[48] In order to settle the issue the best approach is to look again, this time with one eye, close to, for almost an hour. And in the course of that mind-boggling exercise even the aroused teenager will eventually catch a glimpse of something strikingly similar to the artist's own mouth—somber, guarded, watchful, unsurprised. Smiling nevertheless, for even in this somewhat awkward position he sticks to humor as his preferred weapon.

Brunelleschi revitalized. You see what you see and you see what you see because you stand where you stand. Self-portrait in the making, for in everything he did Duchamp meticulously followed the textbook instructions, always more of a practicing land surveyor than a conceptual painter. As in the maps of the *Republic* he first placed the theodolite in the intersection of Magritte's *Ceci,* then aimed its sighting tube at the two landmarks of the sensible Cunt and the intelligible Desire, the former located in the lower left-hand corner of Plato's Realm of Objects, the latter in the upper left-hand corner of the Realm of Meaning. Like Plato in the *Symposium,* Duchamp here shows how a body can be purified and transformed into an ecstatic contemplation of the pure form of desire and beauty:[49]

> "But Diotima," said I, "who are those seekers of wisdom, if they are neither the wise nor the ignorant?"
>
> "Ah, even a child could tell you that! Those in-between, of course. And most prominent among the in-betweens is Love. For wisdom is one of the loveliest things and Love is love of the lovely. It follows that Love must be a philosopher, a seeker of truth who stands between the wise and the ignorant. . . . Such, my dear Socrates, is the spirit of love."

And such is also Duchamp's projection of perspectivism. Not a relativism but a strangely ordered translation function.

⁂

**The projection of perspective** is in this perspective not a relativism but a scale, in effect a verification of Lacan's proposal that "at the scopic level we are no longer at the level of demand, but of desire, of the desire of the Other."[50] And rarely has that desire been more revealingly unfolded than in the *Etant donnés,* art history's most exemplary example of how the geometrical perspective is a mapping of space, not of sight. Easy to see, because "the whole assemblage creates the effect of a diorama with its sharp distinctions between foreground (the hole in the brick wall), middleground (the nude on her layer of twigs), and background (the posterlike landscape). Stereoscopy is an effect that must have delighted Duchamp in permitting him to get away from painting: the visual elements of *Etant donnés* . . . appear to operate in deep space, as the eye continually moves from foreground to background in a vain effort to establish actual distances."[51] For the critic of cartographical reason it is even more remarkable that "the difference between the peepshow and an anamorphic image is that the viewer of the [perspective] box is required to approach the illusion from the point at which it works, whereas an anamorphosis deliberately presents a scrambled image from the anticipated viewpoint, and often leaves the viewer to find the secret of its optical trickery by searching out its wildly improbable viewing position."[52] It was Duchamp's genius that he never said, just showed, how these two techniques had been mixed.

Whatever trickery he might have used, Duchamp's move from painting to sculpture was directly related to the geographer's struggle with the two-dimensional map, on the one hand, and the three-dimensional globe, on the other. A more immediate inspiration was nevertheless the famous sculpture from 1890(?) in which Auguste Rodin depicted *Iris*—in Greek mythology the personification of the rainbow and a messenger of the gods—as a headless nude with open legs.[53] And sometimes in the late 1940s Duchamp made a small but elaborate study of the nude modeled in gesso and covered with vellum, a preliminary step toward the final version in which the maquette is dressed in a semi-translucent pigskin, a material chosen because of its likeness to human skin. In addition, the actual shape of the somewhat dingy "sculpture" is a mixture of two living beings, one the rather strong-bodied Maria Martins, the missing lady behind the waterfall in the Museum courtyard,[54] the other the blond Teeny Matisse, née Alexina Sattler, the woman Duchamp married at the age of sixty-six and whose characteristic hand holds the Bec Auer, the illuminating gas of his happy end.

But precisely because the anatomical eye of the perspectiva naturalis is at the same time the cultural eye of the perspectiva artificialis, the sighting organ must be further investigated, not the least in its relation to the concept of imagination, that uniquely human faculty by which the absent is made present and the present made absent. What in that context makes the *Given* so crucial is the fact that here the cultural imaginations have no parallels in the world of physical reality; even though the viewer automatically thinks that the nude figure has a face, a right arm, a neck and a back, in actuality there is none of it. Nothing hidden except the agenda! And so it is that the really given of the *Given* is certainly neither in the waterfall nor in the illuminating gas but more likely in the small peepholes. On angelic wings Filippo Brunelleschi flies from the Florentine portal to the Philadelphia mortuary, temporarily immortalized in the process.

Once sucked into the penetrated door we are reminded not only of Duchamp's early studies at the Bibliothèque de Sainte Geneviève but also, and in particular, of the officially sanctioned drawing lessons of his school years. Although these lessons always focused on the relations between true and apparent representation, and thereby potentially on the relations between retinal and nonretinal art, there were distinct differences between the instructions given to the boys and the girls, the former budding engineers, the latter prospective housewives. This split in the teaching program in fact "followed the split in the two kinds of representation, with the result that a hierarchy of drawing was established according to gender. The man had to master the drawing of things as they were, the nonretinal

Auguste Rodin, *Iris, Messenger of the Gods.* Modeled ca. 1895, cast 1965. Bronze on black marble base, h. 48.6 cm. Metropolitan Museum of Art, New York.

mechanical drawing; he had to learn to see these lines in objects, seeing through the object to its plan; he saw what passed in those days for truth. The woman had to work further on her perspective, on seeing things as they appeared; she had to learn how to apply what she saw to cloth. He mastered both perspective and projection, however, while she mastered only enough projection to know what she was missing. Projection was principally a male space. Perspective was common ground."[55]

Of great interest in this context of socialization is Duchamp's long-standing engagement with the curious and absurd perspectives of anamorphosis, a technique whose name stems from the Greek words *ana* and *morphe*, the former standing for "back towards," the latter for "form." The important difference between the "real" and the "absurd" is that while ordinary perspective is a technique for restoring the appearance of three-dimensionality, anamorphosis "plays havoc with elements and principles; instead of reducing forms to their visible limits, it projects them outside themselves and distorts them so that viewed from a certain point they return to normal. [It follows that perspective] is a science which fixes the exact dimensions and positions of objects in space, but it is also an art of illusion which recreates them. Its history is not only the history of artistic realism but the story of a dream."[56] As Jean-François Lyotard put it, "the 3-dim nude of *Given* is incomplete: the head is formed of two empty shells; the right upper arm, the ankle, the right foot, and the left foot are missing. However, the same object appears complete to the eye of the voyeur, the absent parts being hidden."[57]

Hiding the very object it seeks to represent. This is exactly the function of Duchamp's dummy as it projects itself out of the realistic perspectiva naturalis and into the imaginary perspectiva artificialis; what is absent to the eye of the natural becomes present in the mind of the artificial. No wonder, then, that Duchamp consistently criticized the cubists for their failure of not being abstract enough, indeed for being naive propagators of the retinal art which he himself so heartily despised. The very purpose of the *Given* is consequently not to bare a shaven cunt (although it does that as well), but to experiment with the faculty of imagination and, by extension, with the translation function of the mapmaker's scale. And on the stump of the right arm, completely outside the spectator's field of vision, the Franco-American inscribed, signed and dated his outstanding critique of cartographical reason:

*ETANT DONNÉS: 1° LA CHUTE D'EAU*
*2° LE GAZ D'ÉCLAIRAGE*

*Marcel Duchamp*

*1946–1966*

With that signing—a blessing gesture learned from Albrecht Dürer—the notary's son certified that throughout his career his main purpose had been to deconstruct the history of art. And in June 1969, nine months after the master's death, this final and most important of his works was put to rest in the Philadelphia tomb, its real subject being neither the vantage point of the pierced door nor the fix-point of the cunt, neither the anamorphic distortion of the nude nor the scale of the missing invisibles, but the mappa of the picture plane, the Museum itself. This screening screen is consequently one with the infra-thin boundary between Plato's *Objects* of Cognition and his *Kinds* of Cognition, a texture in which the *of* functions as the weft and the *in* as the warp. Without this tain of tains there is in fact nothing to see and nothing to understand, no Quai d'Orsay and no 11 West 53rd Street.

*

**The mappa of the picture plane**—the iconostasis—now emerges as the infra-thin given not only of the *Etant donnés,* where it appears as the airy nothingness of the hole in the brick wall, but also of *La Mariée,* in which the transparent projection screen is one with the rhizomatic napkin; as the upper cooler is the thin line on which the bride hangs her clothes, so the breach in the wall is the abysmal home of the wife beyond imagination. And to the extent that everything simultaneously appears *on* and *through* an infra-thin surface, the parallels between the glass of the *Glass* and the given of the *Given* are stunning.

The most mysterious part of the brick-wall hole is not its distance from the pierced door (a distance as carefully computed as Brunelleschi's position inside the portal of the Santa Maria dei Fiori) but its irregular shape (a clear violation of Alberti's original recommendation that the framed canvas should always be rectangular). Even though the contours suggest that the breach was struck by something like the iron ball of a wrecking machine (in my own case an American football),[58] there is no debris to indicate whether the blow came from the inside or the outside, whether the thrower was the exhibitionist, who wanted to escape, or the voyeur, who wanted to enter. Yet, since it is highly unlikely that the sixty-nine carefully numbered bricks were assembled with less thought and less care than the rest of the tableau, it is natural to hypothesize that the irregularity of the window (effectively the picture plane of the tableau or the mappa of the map) is somehow related to the anamorphic projection of the nude. Nothing but an unprovable hunch, of course. A hunch nevertheless.[59]

Much the same holds for the striking similarities between the nude and the amoeba-like picture which Duchamp sent to Maria Martins in 1946, an abstract landscape entitled *Paysage fautif,* a term best translated as "wayward," "faulty" or "offending" landscape. As it turns out, this is a most appropriate title, for later laboratory analyses have revealed that it was made with human semen presumably ejaculated by the artist himself, his penis a tube of readymade paint, the rectangular canvas by one account made of celluloid backed by black satin, by another a regular map cloth.[60] Of course "one might see the work as intended above all to poke fun at art, and especially at those who give it some kind of sacred status. But the picture is also, like many others, a kind of self-portrait, the readymade image of Duchamp as a lover of sorts, but a lover who makes his link to the woman on his mind, and creates an image of himself to send her, through masturbation."[61] If such is the shoot that created the nude (and by anamorphic extension the hole of the picture plane), then it is hard to imagine an onanist more creative. Peek-a-boo, Madame, and out they come, all the suckers of the shaven femme.

Mind-boggling fantasy? With ordinary people, yes. With Duchamp, impossible to tell. As a sobering antidote the critic should therefore shift her attention away from the wayward landscape of the hole in the wall, and focus instead on the more conventional landscape in the background, its glittering waterfall driven by an electric motor and a rotating tin-box. While some commentators have tied this kitschy picture to the Swiss Alps and others to a country restaurant outside Cadaqués, Duchamp himself took it as a pastiche on the advertisement for Coors beer so frequently drunk in the Rocky Mountains bars. Retinal art par excellence! Beware, though, for the vanishing point of the *Given* lies below the picture's edge, a fix-point not for the eye of the perspectiva naturalis but for the vision of the perspectiva artificialis.

Similarly hidden from the Peeping Tom is the floor, *the* foundation of the entire contraption. Yet the instructions of the *Manual* leave no doubt that the floor is not just any floor, but a surface covered by two pieces of checkered linoleum. Moreover—and just as the wooden door was picked up at Cadaqués and the sixty-nine bricks from various building sites on Manhattan—so also the linoes stem from a local habitation with a name: the kitchen at 210 West 14th Street, the studio where the *Etant donnés* was originally conceived and put together. Two pieces of black-and-white linoleum! Such was the non-visible homage which the devoted chess player paid to his artistic forefathers: Brunelleschi and Leonardo; Dürer, Velázquez and Vermeer; Cézanne, Seurat and Matisse.

And when I now stand there in the bare room at the dead end of the Philadelphia Museum, symbolically chained to the door which is not a door, then I suddenly realize that I am totally alone. In spite of everything given, indeed because of the *Given,* my mind is sucked away by Ludwig Wittgenstein's saying that "we cannot think what we cannot think; so that what we cannot think we cannot *say* either. [Indeed] this remark provides the key to the problem, how much truth there is in solipsism. For what the solipsist *means* is quite correct; only it cannot be *said,* but makes itself manifest. The world is *my* world: this is manifest in the fact that the limits of *language* (of that language which alone I understand) mean the limits of *my* world. The world and life are one."[62] To the punning Duchamp, for whom world and life by all evidence were one, the thought of finding

Richard Baquié, *Sans titre. Etant donnés: 1° la chute d'eau, 2° le gaz d'éclairage.* 1991. Mixed media installation, 251 × 204 × 206 cm. Musée d'Art Contemporain, Ville de Lyon. © Richard Baquié/BUS 2005.

the key to the problem was less important than the act of removing the keyhole; after all it was he who had said that there is no solution because there is no problem. And instead of handling contradictions by privately contemplating suicide, which was the Austrian's approach, the Frenchman chose to publicly laugh them away.

And yet. To any mapper of love and knowledge solipsism remains the only honest philosophy. For once inside the infra-thin he realizes that the subject who stands outside the Duchampian door does not belong to the world, but is a limit of the world. Since whatever the voyeur sees could be other than it is (or at least be other than it appears to be), "solipsism, when its implications are followed out strictly, coincides with pure realism. The self of solipsism shrinks to a point with extension, and there remains the reality co-ordinated with it. . . . What brings the self into philosophy is the fact that 'the world is my world.' The philosophical self is not the human being, not the human body, or the human soul, with which psychology deals, but rather the metaphysical subject, the limit of the world—not a part of it."[63]

Although I know of no incidence when Duchamp referred to Wittgenstein or Wittgenstein to Duchamp, it seems virtually certain that some of the most central propositions of the *Tractatus* could just as well have been included both in the *Green Box* of the *Glass* and in the *Manual* of the *Given*.[64] It should likewise be stressed that Wittgenstein's expression "the self shrunk to a point" has much in common with René Magritte's *Ceci* and thereby also with my own mappings of Plato's *Republic*. To be precise, the infra-thin of the infra-thin is exactly in the slanted line, *La Mariée* approaching it from one side, *Etant donnés* from the other. The transparent curtain makes no distinction between showing and telling, for projected onto one of its sides is the story of the bachelors stripped bare by their apparition, onto the other the picture of the nude covered up by her appearance.

*

**Appearance and apparition** were among the most frequent terms in Marcel Duchamp's vocabulary, a circumstance that ties his work directly to the theory and practice of cartography. The reason is that in spatial analysis there is nothing more crucial than the geographical inference problem of form and process, the mold and the molded. But which *are* the relations between the spatial patterns depicted on the map and the stories told of how they came about? And what does it mean to lose one's way?

Anyone who approaches these perennial questions should recall the standard definitions according to which an *appearance* is "the action of coming forward into view or becoming visible"; "a clear manifestation to the sight or understanding"; "that which appears—an object meeting the view, esp. a natural occurrence presenting itself to observation, a phenomenon." In contrast, *apparition* stands for "the action of appearing or becoming visible, the supernatural appearance of invisible beings"; "an immaterial appearance as of a real being; a spectre, phantom or ghost." Finally, *to save* or *to keep up an appearance* means "to maintain artificially the outward signs, so as to conceal the absence of the realities which they are assumed to represent." [65] Once again William Shakespeare demonstrates his genius, for when Brutus meets the Ghost of Caesar, he cries out:

> Ha, who comes here?
> I think it is the weakness of mine eyes
> That shapes this monstrous *apparition*.
> It comes upon me,—Art thou nothing?
> Art thou some god, some angel, or some devil,
> That mak'st my blood cold and my hair stare?
> Speak to me what thou art.
> (*Julius Caesar* 4.3.276–82)

In any language a set of questions easier to ask than to answer. And any English-speaking Frenchman, Duchamp certainly among them, knows that the expression *il n'a fait qu'une apparition* is best translated as "he stayed for just a moment."

Although Duchamp must have been aware that the relation between appearance and apparition was central to Plato, he probably did not know that it was crucial to Ludwig Wittgenstein as well. In different words the latter often insisted that "language disguises thought. So much so, that from the outward form of the clothing it is impossible to infer the form of thought beneath it, because the outward form of the clothing is not designed to reveal the form of the body, but for entirely different purposes. . . . Most of the propositions and questions of philosophers arise from our failure to understand the logic of our language."[66] Furthermore, and immediately after this bachelor-sounding passage, he first makes the observation that "it was Russell who performed the service of showing that the apparent logical form of a proposition need not be its real one," then adds the conclusion that "a proposition is a picture of reality [or, more correctly], a proposition is a model of reality as we *imagine* it."[67] Later the same idea reappears in the formulation that "ever and again comes the thought that what we see of a sign is only the outside of something within, in which the real operations of sense and meaning go on."[68]

It can hardly be stated any better: what we see with our eyes is the appearance of a phenomenon, what we grasp with our mind is the apparition of a noumenon. In my interpretation it is this Platonist intertwining of visible matter and intelligible meaning that permeates Duchamp's entire oeuvre, starting with the *Coffee Mill* and ending with the *Given*. The rhetorical relation between the metaphor of the mold (roughly the apparition) and the metonymy of the molded (the appearance) were constantly on his mind, the nine bachelors cast in husks filled with gas, a harem of

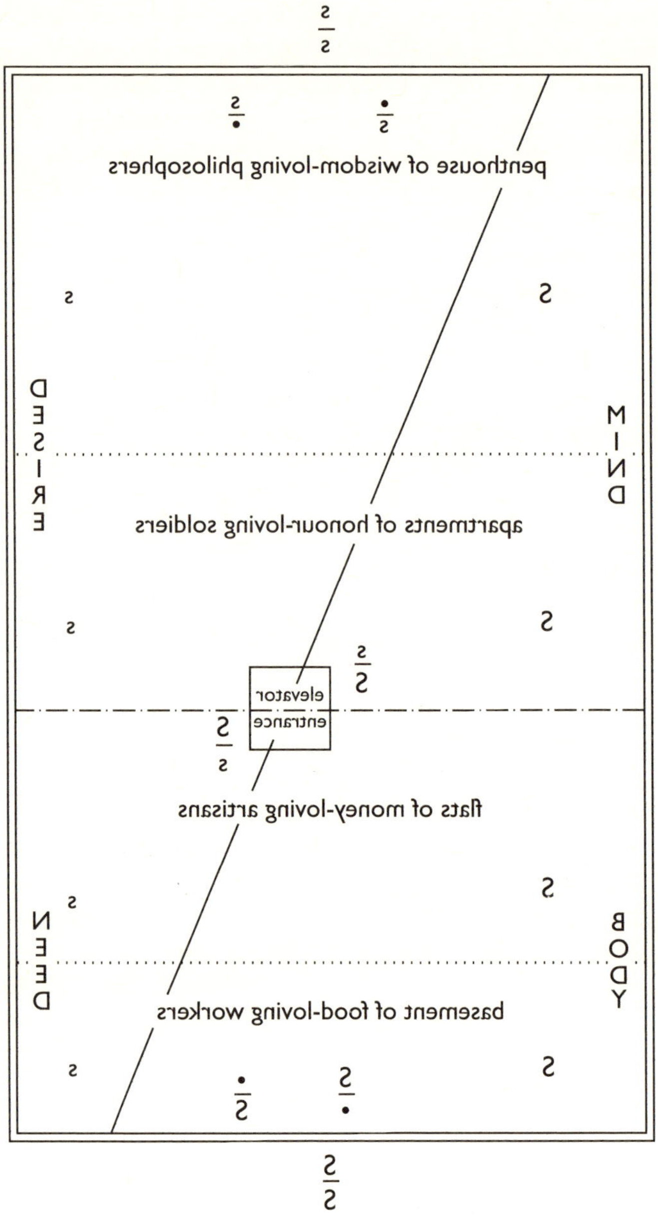


Socio-economic map of Plato's *Republic* from the reverse.

eroticized tin-soldiers dressed up in mail-ordered uniforms. Recall also *both* how the apparition of desire shows itself in the apparent striptease of the bride *and* how the entire tableau of the *Given* is an apparition of the *Glass* itself. In conclusion, the slanted line of the *Republic* is in the Duchampian context transformed into a transparent curtain, *La Mariée* projected onto one of its sides, the *Etant donnés* onto the other. Dialectics at work, Plato turned inside out, Saussure upside down, Ptolemy rotating in his grave.

Through this creative twist the maps of the *Republic* and the blueprint of *La Mariée* may be viewed not from the spectator's but from the actor's vantage point. As a result the slanted line of \ turns into the dash of /, the infra-thin sign of a dialectical relation, the conventional symbol of the inseparability of identity and difference.[69] The ensuing problems are thoroughly Socratic, for it cannot be repeated too often that what makes us see cannot itself be seen, just as what makes us understand must not itself be understood. What is reflected in the mirror is consequently less the object in front of the glass and more the tain behind it, less the visible cunt and more the intelligible desire. And so it is that the slash becomes the quintessential symbol of the excluded middle, the abysmal home of Apsu, the taboo-ridden void of categorial limits. Through the silence echoes once again the echo of Samuel Beckett's *Unnamable:*

> Perhaps that's what I feel, an outside and an inside and me in the middle, perhaps that's what I am. The thing that divides the world in two, on the one side the outside, on the other the inside, that can be thin as a foil, I'm neither one side nor the other, I'm the middle, I'm the partition, I've two surfaces and no thickness, perhaps that's what I feel, myself vibrating, I'm the tympanum, on the one hand the mind, the other the world, I don't belong to either.[70]

As so many times before, Immanuel Kant paved the way. The fog-banks are lifting as the winds of pure imagination bring "the manifold of intuition on the one side, into connection with the condition of the necessary unity of pure apperception on the other. The two extremes, namely sensibility and understanding, must stand in necessary connection with each other through the mediation of this transcendental function of imagination, because otherwise the former, through indeed yielding appearances, would supply no objects of empirical knowledge, and consequently no experience."[71]

More than anything else it is this ability to imagine a necessary connection between bodily appearance and mindful apparition that makes *Homo sapiens* both human and knowing. It is in the no-man's land between the *objects* of cognition and the *kinds* of cognition that we live and die. Mapping the exact location of the infra-thin boundary between these realms is an assignment impossible to carry out, a mission impossible to refuse.

*

**The point of the point,** a conclusion not to be missed: making a difference is by necessity an act of violence, the marking of a boundary. *The* expert on boundary-marking is the land-surveyor, by training cartographical reasoner par excellence, weaver of coordinates and shaper of the taken-for-granted. Another way of saying that even though Plato, Kant and Wittgenstein all were philosophers in theory they were geographers in practice, forgers of the uncreated consciousness of the utopian No-where of the present Now-here, sacralized baptizers of local habitations.

Such is the condition of being human that without names we are nothing, for without names the Kantian as-if has nothing but its own trickery to hook on to. It was Marcel Duchamp's genius to smash this readymade idol, slaying the ghost of cartographical reason in the process. In the emerging world of globalization the fix-points are in fact so invisible and unstable that they are not at all; the scaling lines so twisted and the angles so skewed that the threads which once were woven into a net of longitudes and latitudes now form a hopelessly tangled skein; the screening mappa not a flat surface but a set of warped and rhizomatic napkins, culture-stained and with a scent distinctly their own.

No wonder that people sometimes get lost. Not, however, because we are all mad (although that happens too), but because our navigational tools have become badly outdated, ordering directives designed for another time and another place, politics itself a clandestine case of anamorphic art.

# UPPSALA

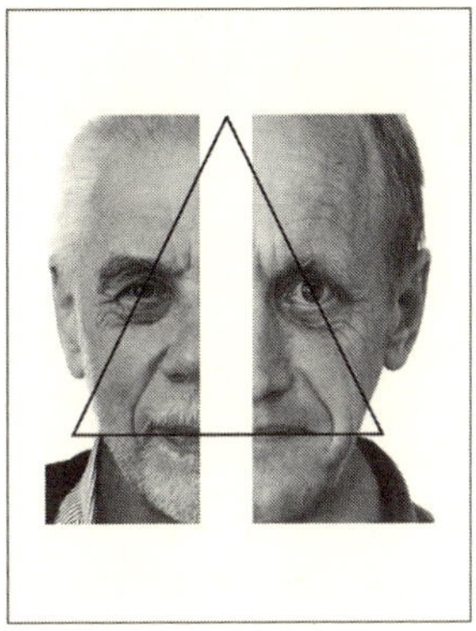

**The first week of the ninth month of the third millennium** brought some of the best moments of my life, the unveiling of a self-made map of the world. It had taken years to complete and in every respect it was the joint work of myself and my former student Ole Michael Jensen, a Dane whose first degrees were in the theory and practice of land surveying. So close was our cooperation that we did not report our findings under our individual names but under the amalgamated imprint of Gunnael Jensson.

The obsession reached deeply into our respective pasts, but most immediately to two papers which appeared next to each other in the 1993 issue of *Environment and Planning D: Society and Space.*[1] After that cooperation we went on to pursue our separate and more immediate interests, Jensen in his doctoral dissertation, I in the abstract mapping of power and knowledge.[2] Then, together with others from our Nordic group, we began in earnest to ask what we had actually learned from our descent into the forbidden abyss.[3]

In the early phases of that stock-taking there were long walks, countless words and heaps of drawings. Out of the stories and pictures then welled an avalanche of physical models which in the beginning were built from folded sheets of paper, broken matches, safety pins, screws, plywood, needles, steel wires, rubber bands and scratchy Plexiglas. And suddenly, out of the mishmash, grew first a pyramid with its mandatory tomb, then a transparent tetrahedron. Not until later did the two bricoleurs understand that what they had really embarked on was a self-referential mapping expedition into the lands of the taken-for-granted, an attempt to lay

*Gunnael Jensson.* Photo montage by Tommy Westberg.

bare the fix-points, sight-lines and projection-planes which together constitute the foundation of western culture, perhaps of all cultures.

When this work was well under way, the curator Luciano Escanilla graciously invited us to participate in the Uppsala International Contemporary Art Biennial Eventa 5. We were elated, for here was a rare opportunity to present our research in a setting where the artificial boundaries between the contemporary art world, the university and the church were all but erased. This circumstance was crucial, for it fit nicely into our own conception that even though knowledge by definition is an exercise in translation, no translation can ever be perfect. Indeed, we were convinced that in the myriad of existences there are essentially three modes of being and three modes of understanding—the artistic, the scientific and the religious. At the same time we were eager to acknowledge that every translation is in effect an act of violence, a ritual killing in which a primordial distinction is sacrificed and turned into a set of alternative identities; for instance, the original difference

$$\boldsymbol{a}$$

may be minimally transformed into a shadowy

$$a$$

more moderately translated into the tautological

$$a = a$$

or maximally falsified by the informative

$$a = b$$

In the cartographer's imagination these reformulations are readily turned into fix-points which may then be cast onto three different projection screens, each mappa specially prepared to receive its own translation—the first artistic $(a)$, the second religious $(a = a)$, the third scientific $(a = b)$.

Such is the theory. The practice, of course, is different. For as the projected images bounce off the projection screens, they normally do not return to their original source but are diverted to the other walls instead. In that transformation process the various translations are typically getting so mixed up that they can no longer be separated—even though Albert Einstein insisted that God does not play dice, he was still referring to God; and even though dice-throwing is the subject matter of scientific probability theory, it performs a significant role in the art of surrealism as well.

What eventually came out of our experiments was a beautiful crystal palace, an enchanting imagination at the same time a cultural melting-pot and a linguistic prison-house.

*

**A crystal palace,** that is exactly what Gunnael Jensson constructed. Not entirely by themselves, though, but with the skillful assistance of a well-equipped stone cutter, the best glazier in the land of Denmark, an instrument maker so meticulous that before he got to work he insisted that he must first design his own tools, a jeweler with hundreds of red rubies locked up in his safe. All of this and more to boot.

And finally. Like a Phoenix out of the ashes, the birth of the long-awaited *Mappa Mundi Universalis.* Seemingly not a map at all, but a tetrahedron of transparent glass grown out of a square slab of dark granite. Projected from the corners of that geometric volume are four threads—three made of pure gold, one of invisible imagination—which hit the opposite walls at right angles. At the center of the bottom triangle is a small cavity which is covered by a ruby, a symbol of the blood that welled up from the first sacrifice, the killing and resurrection of the original difference. The sides of the granite square are 64 centimeters long, the top of the tetrahedron 49 centimeters high.

One square piece of granite, three centimeters thick; three equilateral triangles of German Weissglass; three spoke-like threads of gold; one gem of a ruby. Nothing more and nothing less.

It was this minimalist sculpture that in September 2000 was shown and talked about in the Uppsala Cathedral, the major church in the Lutheran Kingdom of Sweden. For a full week it occupied this House of God, where it was prominently placed on top of an inlaid windrose located exactly where the transepts cross the nave. On their way from the entrance in the west to the altar in the east the visitors passed two smaller and slightly different versions of the object for which they had come. Since the light from the stained windows varied with the weather and the day, no spectator ever saw the same thing twice.

Although the sculpture was exhibited in the cathedral, the organizational setting was that of the art biennial, an event which in the course of seven days drew close to 10,000 visitors. In conjunction with the opening show, Gunnael Jensson—now in the double appearance of Ole Michael and myself—made a brief presentation of the work. For the following discussion we then moved across the street to the Anatomical Theater, the positivist temple which in the 1660s had been built on top of the Old University Building, a scandalous competitor to the church next door. Once inside that scientific laboratory, a duo of invited discussants—an art historian and a theologian—completed the ritual by cleverly dissecting our creation. And while the corpse ascended to heaven, the rest of us descended on a thanksgiving party that nothing could beat.

Throughout the festival week visitors were given copies of a text in which the Jenssons had done their best to explain what the commotion was all about. The following pages are given over not to a rephrasing of that highly condensed pamphlet, but to a facsimile reproduction thereof. In many ways a premature summary and conclusion of the present book, but at the same time a restatement of whatever preceded it. An exercise in minimalist writing, the Proustian madeleine cookie in another form.

A death notice nevertheless, for some time during the twenty-first century the mode of cartographical reason will most likely appear just as ghostlike as the marvelous individuals who first formulated and practiced it. In the meantime, the corpse keeps kicking.

*

# MAPPA MUNDI

# UNIVERSALIS

GUNNAEL JENSSON
alias
Ole Michael Jensen & Gunnar Olsson

# MAPPA MUNDI UNIVERSALIS

a commentary
on

THE POWER OF CARTOGRAPHICAL REASON

performed at the

Uppsala International Contemporary Art Biennial Eventa 5
Section II – OBOG
PER ASPERA AD ASTRA

Uppsala Cathedral

September 4–10, 2000

GUNNAEL JENSSON
alias
Ole Michael Jensen & Gunnar Olsson

# GENESIS

IN THE BEGINNING are the heavens and the earth. The earth is without form and void and there is darkness upon the face of the earth. Nothing to see, nothing to hear, nothing to touch, nothing to smell, nothing to taste. No distinction. No identity. No difference. In this spiritual land of silence there is a meaning so meaningful that it refuses to be expressed.

A wind moves across the void. Lightening. Thunder. Rainbow in the sky. Out of the mist rises a flat granite rock that gently slopes into the distant sea. In this physical land of silence is a matter so material that it emits no meaning whatsoever.

Strange creatures emerge out of Nowhere, quickly spreading across the rocky ground. A foot gets stuck in a crevice, others observe, and for the first time there is a difference important enough to make a difference. In the intercourse of the body and the stone, the origin of man is conceived.

The primordial distinction—**a**—splits into three: one a shadow of the shadow (a); one a tautological expression which keeps repeating that it is what it is (a = a); one an informative statement insisting that it is something else (a = b). Identity and difference separated and united. Atoms of understanding captured in a mushroom cloud of perpetual fission.

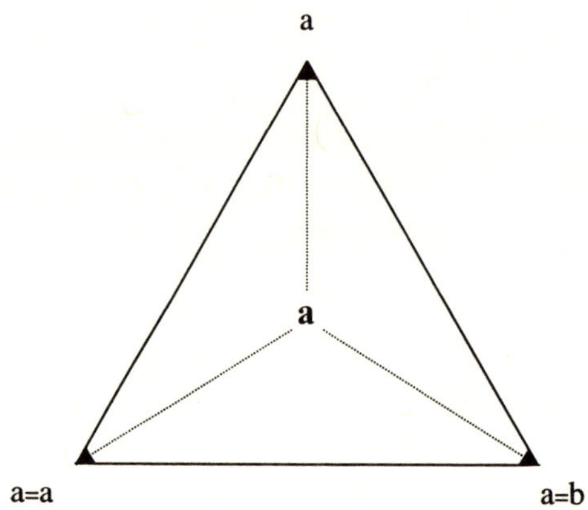


Points of distinction

WHEN THE TENSION reaches its limit, the cloud bursts into fire. Out of the ashes grows a crystal palace, sometimes known as the crucible of man, sometimes as the prison-house of language. Both the floor and the walls are built as equal-sized equilateral triangles, the walls invisible, the foundation sunk into the granite ground. At the center of the basement is a well of identities and differences, its opening covered by a red-colored lid, itself the memorial mark of the original distinction and the trace of the first sacrifice, the blood from the killing of an identical twin, the footprints of a deviance turned scapegoat, the navel of what it means to be human.

A twist of cultural survival and the four-cornered deities merge into one, the multitudes of polytheism concentrating in the singularity of monotheism. In the process of that unmooring, absolute power finds its place at the top of the pyramidal structure, a pivotal point which is the locus of a tautological and nameless entity that defines itself as that which it is: **a** = a. A contradictory condensation of identity and difference, one God one Being, the Almighty created in the image of man.

From its inception this Absolute speaks, its power one with its language, its language one with its power. "Let there be!" And there is. A universe flowing out of the actor's mouth.

In the coolness of the evening, the Absolute looks back at what he has uttered, claiming first that it is very good, then that he alone has the right to judge. Tolerating neither idols nor false prophets he declares that all usurpers will be killed. Impressed by his own achievements, he finally proclaims a day of rest, a sabbath without work, twenty-four hours devoted to the glorification of himself and his faithful. Such is the subjection of subjects, such is the structure of dictatorial power. Now as well as then, then as well as now.

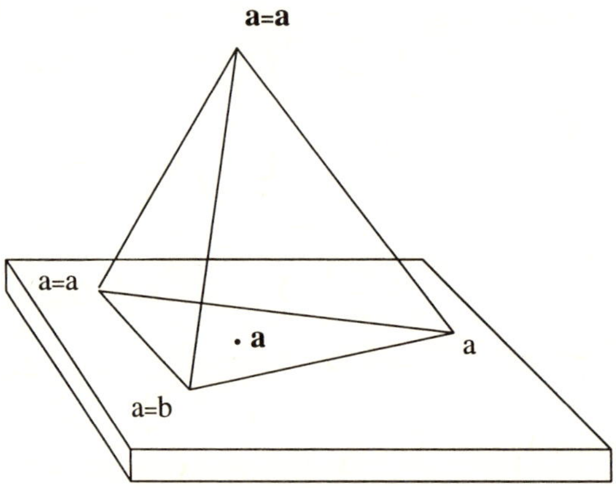


Crystal Palace

# LAWS

THE ALMIGHTY REALIZES that he leads a dangerous life. For that reason—and with the purpose of regulating the relations between ruler and ruled—he formulates a Constitutional Law, a power-filled document which begins with the reminder that it was he who liberated the suppressed, he who cut their chains, he who brought them out of the land of bondage. A rhetorical trick based on the false premise that old wolves automatically turn to peaceful sheep. And that is why the political atrocities of people like Lenin, Pol Pot and Fidel Castro are easy to explain, impossible to excuse.

*Paragraph 1:* You shall have no other ruler before me.

*What is that*? You shall fear and love your Leader above everything else, placing all your belief and trust in him. Martin Luther was rejected by the Pope in Rome before he was accepted by the king in Stockholm, the praying Mohammed faced Jerusalem before he turned to Mecca.

EVERY DICTATOR KNOWS that to stay in power he must squash all opposition. But he also knows that all power rests in language. This in turn explains why the operationalization of the Constitutional Law takes the form of three lines extending from the top of the crystal pyramid to the corners at the base. It is through these communication lines that the Lord and his people are tied together into a network of mutual dependence. Issued from the top are orders that follow no rules but their own, predictable only in their unpredictability. Echoing from the underground is nothing but legitimation. In the rooms of the Almighty anything goes; in the dungeon of the critics translation is strictly forbidden.

Coupled with each line is a segment of the Law, each one a different aspect of §1. While paragraphs 2a and 2b are alike in the sense that they block all roads towards representation and thereby to critical understanding, paragraph 3 rules that in conflicts between the forces of collective cohesion and individual rebellion, the former are always right. While the paragraphs 2a and 2b ensure that only the Absolute can eat from the tree of knowledge, paragraph 3 focuses on the relations between categorization and socialization.

*Paragraph 2a:* You shall not make for yourself a graven image or any likeness of the Almighty, for no power-wielder allows his secrets to be revealed. No statue, no picture.

*What is that?* You shall fear and love your Lord without questioning who he is or what he does, accepting every decree as an integral part of the taken-for-granted. Forbidden is every attempt to picture the invisible, every attempt to sculpture the untouchable. In the world of the crystal palace, the message of §2a is transmitted through the communication line that runs from the tautological $\boldsymbol{a} = \boldsymbol{a}$ to the informative $a = b$.

It is in the formulation of this paragraph that the law maker demonstrates how sophisticated his conception of power actually is. For the prohibition against images stems from the recognition that the likeness of metaphor plays the same role in the rhetorical art of ontological transformations as the igniting charge does in the engineering science of blasting. Since it is this device that turns untouchable ideology into touchable stone, visible structure into invisible meaning, the Absolute does his utmost to ensure that it will never fall into enemy hands; just as the ordinary dynamiter jealously guards his detonators, so does the Mosaic Absolute.

So important is this principle that any transgression is most severely punished—executed are not only the sinning fathers but also their children, grandchildren and great-grandchildren. Any ruler who resorts to penalties of that magnitude is scared out of his wits. Given the outrageous claims of §1, he ought to be.

*Paragraph 2b:* You shall not misuse the name of your Leader, never tie his proper name to a definite description. For every power-wielder is eager to punish the truth-sayers and to reward the ass-lickers.

*What is that?* You shall love and fear your Lord and never use his name for evil wishes, swearing, lying or deception. Forbidden is every mode of reformulation, every deconstruction of the sign, every visit to the Saussurean Bar. In the blueprint of the crystal pyramid, §2b is twined into the line that runs between the big tautology of $\boldsymbol{a} = \boldsymbol{a}$ and the small tautology of $a = a$, the former located at the top of the world, the latter stuck into a corner at the bottom.

The Almighty categorically refuses to be categorized, and that is why he chooses a tautology as his name: "I am who I am." Always true, never informative. Yet, for the ruler who rules by systematically contradicting himself, no name is more appropriate. Beatings at dusk, blessings at dawn. Fear institutionalized. For he who has hit you once is likely to hit you again.

Tautology is the name of the Almighty. It is also the fix-point of two-valued logic. And herein lies the paradox of the social sciences, for whereas the words and objects of power never sit still, the words and objects of science must not hop capriciously about. It follows that logical analyses will never lead to a proper understanding of power itself. In studies of Venus, the planet, Bertrand Russell's theory of proper names and definite descriptions is often helpful. To the understanding of Venus, the goddess, it contributes little but confusion. The concepts of transparency and obliqueness are like oil and vinegar.

*Paragraph 3:* You shall attend all party meetings, never enter into the no-man's land between clean and unclean, never bite the hand that feeds you.

*What is that?* We shall fear and love our Leader together, gratefully honoring and obeying all his commands. For united we stand, divided we fall. §3 is the line that runs from the elevated $\boldsymbol{a} = \boldsymbol{a}$ to the shadowy $a$.

WHILE THE REGULATIONS of §§2a and 2b address man in its appearance as a semiotic animal, §3 speaks to us as social and political beings. Of the three paragraphs, the third is in fact the most crucial, for without rules of conduct there is neither society nor individual, neither language nor power. It is by intercepting and decoding messages sent through this channel that we learn why and how we become so obedient and so predictable; the imperative of communication rests on a foundation of social control. The question is a question of socialization, the answer a bucket-full of insights lifted from the well of original distinctions.

# PROJECTIONS

LINES ARE NOT ONLY connectives between points, they are also dividers between planes. In the latter function they constitute the corners of the crystal palace, at the same time separating and uniting the adjacent walls.

Grasping the nature of the invisible walls is extremely difficult, for even though we seem to be standing on the outside looking in, we are in fact confined to a life on the inside. In this prison-house of language there are no windows and no escape routes, only a constant bouncing against unbreakable walls. Yet there are human lives and experiences of incredible richness, for the pyramid is in reality a gigantic movie theater, in each corner a projector, every wall a screen. The arrangement is familiar, its predecessors in the cave of Plato's *Republic,* in Fra Angelico's rendering of the *Annunciation,* in Marcel Duchamp's *La Mariée mise à nu par ses célibataires, même.* Without the limestone wall, the wood panel and the transparent glass, these artists would have nothing to show, no means for capturing the split-up versions of the original distinction ***a.***

When the golden rays hit the opposite plane at a right angle, they rebound to the point of origin, offering no news from the travel. This is the case of perfect translations and perfect signs, imaginable in theory impossible in reality. From the a = a projector come the mantras of tautology, which—when captured by the altarpiece of religion—appear as the perfect sign of $a = a$, in Charles Sanders Peirce's terminology an *icon.* From the corner of a = b beam a set of definite descriptions, informative statements cast onto the wall of science as the perfect sign of $a = b$, a so-called *index.* Finally, projected from the shadowy point of a are a series of artistic images thrown onto the canvas of aesthetics as *a,* that particular version of the perfect sign which in Peirce's philosophy is called a *symbol.*

As the tautology is the fix-point of Aristotelean logic and Old Testament power, so the ninety degree angle is the fix-point of Euclidean geometry and New Testament penance. But in the real world of imperfect communication, the projection lines never strike the knowledge planes straight on. Instead of bouncing back to their respective point of origin, they are reflected onto one of the other walls. This combined principle of uncertainty and complementarity explains why the three modes of knowledge—religion, science, and the arts—never appear in their pure form, always as a series of approximations, a chain reaction in which the grammar of religious belief turns to the rhetoric of scientific law, the rhetoric of scientific law to the aesthetics of artistic expression. And so on, and so on, one meaning colliding with another, the critical mass the trigger of itself.

And so it is that the corner lines of the crystal palace at the same time communicate the paragraphs of the Constitutional Law and mark the limits of the three planes of knowledge. Since limits by definition are taboo, the movements from one screen to another occur only through the intermediary of a transition rite. In Christianity these magic formulas are one with the sacraments, in Catholicism seven, with Luther condensed into two: baptism and communion. The latter ritual is intricately intertwined with §3 of the Law, the former with §§2a and 2b; while the naming ceremony of §2a is performed in the water of images, in §2b it is done in the medium of ordinary language.

No wonder that so many sinners pray that their trespasses be forgiven, no wonder that so many dictators kill those who trespass. Every wall is a wailing wall, a structure of exclusion which locks some people in and others out.

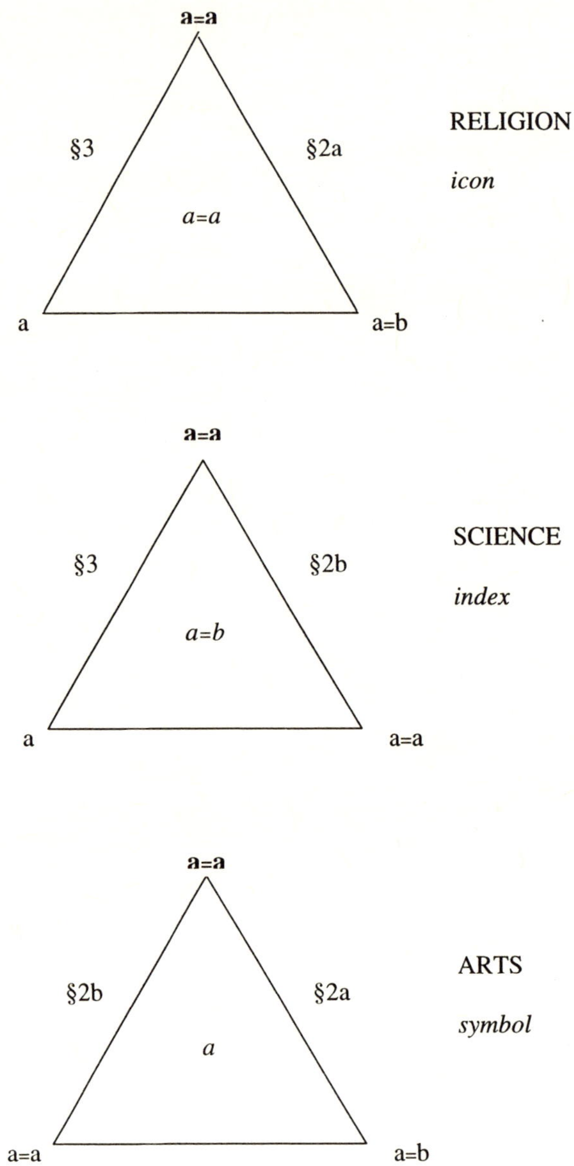


Planes of projection

# TRINITY

IN THE BLUEPRINT of the crystal palace one trajectory is left invisible, perhaps because it is too culturally specific to be noticed. This is the line that runs from the **a** = **a** of God the Father to the **a** of the Holy Spirit. As recalled, the **a** stands for the covered mark of the original distinction, the **a** = **a** for its tautological reformulation, the trace of an aborted attempt at reification, a desire to concretize the abstract. The King is dead, long live the King!

Impressed by his own omnipotence the Absolute wants to return to the crevice of the original distinction, no longer satisfied with being the tautological reformulation **a** = a, but obsessed with the desire of once again being one with the untouched purity of the **a;** as an early Narcissus also the Almighty strives to merge with his own image. To that end he tries to do away with himself by jumping out of his privileged position at the top. But in midair—moments before smashing through the lid and plunging into the covered well—he is caught in a semiotic rescue-net whose corners are fastened to the perfect signs of the icon, the index and the symbol. The prototype of this ontological surface was first invented in Nicaea in the year 325, whence it was formulated as a political/ theological commentary on the issues of representation in general and the prohibition against images in particular.

Floundering in the western net of perfect signs is nothing less than Jesus Christ, here known under the pseudonym **a** = **b,** by definition a complete merger of mind and matter, the logical union of **a** and **a** = **a;** as recalled, the only Son of God is defined as being of one substance *(homoousios)* with the Father, incarnated by the Holy Spirit and made man. Himself a sacrificed scapegoat, this epistemological savior eventually ascends to the heavens, gets appointed to the Highest Court, wherefrom he is now sentencing both the quick and the dead.

The most remarkable story of an ontological transformation ever to be told. The word made flesh, the unnameable named. Eucharist.

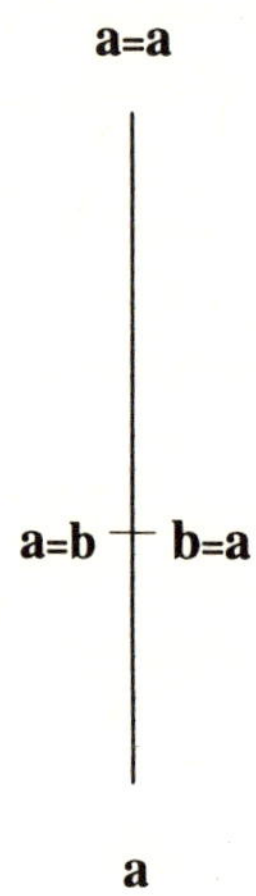


Line of trinity

# MAPPING

THE CRYSTAL PALACE is a well guarded castle, its ruling resident the tyrant of tyrants. Admittedly a rhetorical exaggeration, for no Absolute is absolutely absolute, no crook crooked enough to live on forever. And for that reason it should be recalled that just as the doctrine of *Imitatio Dei*—the imitation of God—plays a central role in Jewish piety, so the imitation of Christ fills the same function in Christianity. The difference is nevertheless crucial: to imitate God (**a** = **a**) is to recant a tautological contradiction (a = a), to pray the mantra of the icon (*a* = *a*); to imitate Christ (**a** = **b**) is to reformulate a definite description (a = b), to invent an index (*a* = *b*).

The architectural tour of the crystal palace should not be interpreted as a glorification of life in a monotheistic temple. Instead it is the drawing of a heretic (not blasphemous) map of power, an attempt to understand how and why we find our way in a universe of obedient submission and unpredictable cruelty. Not a *Mappa Europæa Mundi* but a *Mappa Mundi Universalis,* a universal means for catching the world in a net of points, lines and planes, the mandala constituting a typical nonwestern case. As usual, it is more difficult to understand the world than to change it, for the reality of being has a richness which no abstraction can harness, a living life that neither stories nor pictures manage to name.

THE NINETY DEGREE angle is the fix-point of fix-points, a human invention of the highest rank. Triangulation is the name of the game, for in order to make a map only three ingredients are needed: the scale, the pointer, the canvas. As triplets of the right angle these elements form the foundation on which the crystal palace is constructed, that power-filled edifice which at the same time is the crucible of man and the prison-house of language. The Library of Invisible Maps is the best-guarded part of the palace, for it is there that we learn how to learn, it is there that we find out both where we are and where we should go.

The **scale of scales** is by definition a translation function, the ruler over ontological transformations, the magic formula of "Let there be!—And there is." An early version is in the creation myth of the Genesis, the most crucial reformulations in the Divided Line of Plato's *Republic* and the speech acts of Euclid's *Elements.*

The **pointer of pointers** is the hook on which everything is hung, the invisible force which in ordinary maps corresponds to the magnetic North Pole. In the construction of the crystal palace its counterpart is in the collective unconscious and thereby in Plato's conception of the Sun, that medium which simultaneously lets us see and makes us blind. It should nevertheless be remembered that it was the exiled Jews who insisted that the Sun—which the idolatrous Babylonians took to be the god of gods—in fact is nothing but an illuminating lamp. A definite description of revolutionary importance, an act of cultural survival closely tied to §3 of the Constitutional Law. Attend the compulsory meetings and your eyes will be opened, the truth revealed!

But it must also be recalled that long before the events in Babylonia, the Sun's dethroner—the tautological Yahweh—had warned Moses that "you cannot see my face; for no one can see my face and live. . . . And I will cover you with my hand until I have passed by, then I will take away my hand, and you shall see my back; but my face shall not be seen." And yet, that is exactly the blasphemous act that Jacob (the crook of crooks) claimed to have committed: seen God's face and survived.

The prophet Job had much the same experience, for also he told the story of how his ears first had heard about the Absolute then how his eyes had seen him. But unlike the treasonous Jacob, the man of integrity from the land of Uz was ashamed of what he saw, despising himself, repenting in dust and ashes. And he lived on in this world for another one hundred and forty years.

The **canvas of canvases** is the background cloth, the receptor onto which everything is projected, the tain of the mirror, Plato's limestone wall, Fra Angelico's wood panel, Duchamp's large glass, the screen that captures all signs. As a matter of fact, the canvas is nothing less than the physical resistance without which nothing can be be noticed. The latter-day version of the granite rock that sprang a well of clear distinctions.

# REVELATION

IN THE PRACTICE of cartographical reason, man is once again put back at the center of the universe. No longer a noun, not even a verb. A pre-position, a place assumed in advance! A point at which lines are projected onto a plane of power. A coordinate cross of right angles.

Such is the nature of cartographical reason, a mode of understanding at the verge of realizing that the canvas of the world is not a smooth flatness but a wrinkled manifold, that the pointer is not straight but crooked, that the scale is not a suspended line between *alpha* and *omega* but a Moebius band of chiastic reversals. And it has no idea what will happen next.

# MAPPA MUNDI

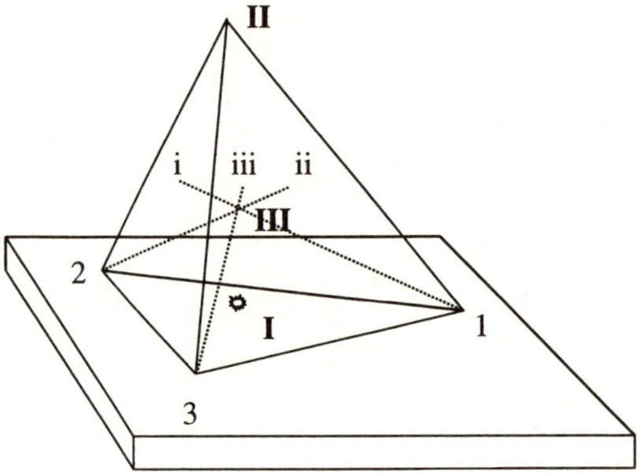


Compas

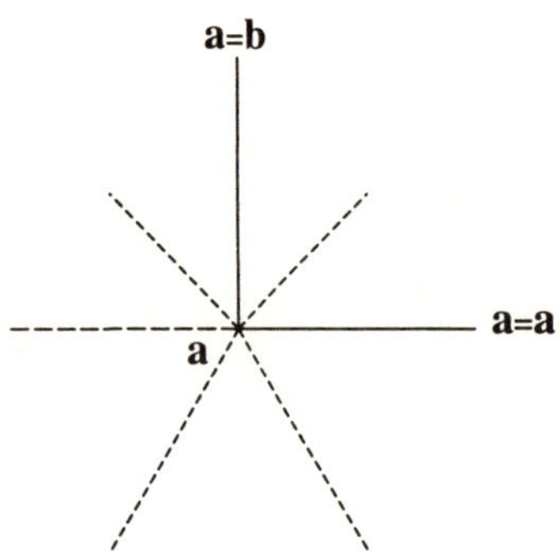


Scale

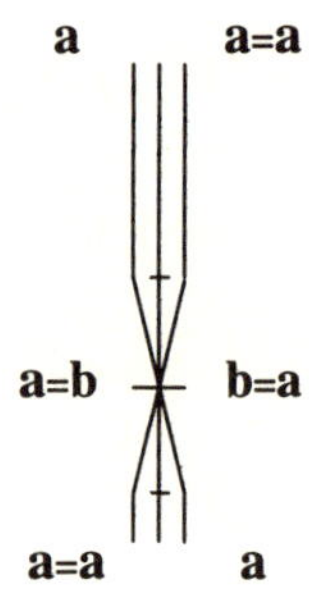


**1 : Divided Line**

**Points of Distinction:**

| | | |
|---|---|---|
| **I** | **a** | Foot in the crevice; Holy Spirit |
| **II** | **a** = a | Absolute power; God, the Father |
| **III** | **a** = b | Ontological transformation; Jesus Christ |

**Line of Trinity:**

| | |
|---|---|
| **II-III-I** | Plato's Divided Line; Nicaea 325 |

Points of distinction:

| | | |
|---|---|---|
| 1 | a | Shadow |
| 2 | a = a | Tautology |
| 3 | a = b | Definite description |

Lines of power:

| | |
|---|---|
| **II**-3 | §2a; Baptism by imaging |
| **II**-2 | §2b; Baptism by naming |
| **II**-1 | §3; Communion |

Plane of socialization:

| | |
|---|---|
| 1-2-3 | Stone tablet; Constitutional law |

Planes of knowledge:

| | |
|---|---|
| **II**-1-2 | Science |
| **II**-1-3 | Religion |
| **II**-2-3 | Art |

*Lines of projection:*

| | |
|---|---|
| 1-*i* | Aesthetics |
| 2-*ii* | Grammar |
| 3-*iii* | Rhetoric |

*Points of signs:*

| | | |
|---|---|---|
| *i* | *a* | *symbol* |
| *ii* | *a* = *a* | *icon* |
| *iii* | *a* = *b* | *index* |

MAPPING

MAPPING IS TRIANGULATION

TRIANGULATION IS THE GEOMETRY OF POWER

THE GEOMETRY OF POWER IS THE PRACTICE OF CARTOGRAPHICAL REASON

THE PRACTICE OF CARTOGRAPHICAL REASON IS THE CRITIQUE OF MAPPING

+

Gunnael Jensson, *Mappa Mundi Universalis.* 2000. Glass tetrahedron on granite base, 64 × 64 × 49 cm. Mixed media (Kalmar granite, Weissglass, gold, ruby). Museum Gustavianum, Uppsala. Photo by Teddy Thörnlund.

*

**Five years later** the *Mappa Mundi* is on permanent show at the Museum Gustavianum, where it stands in a window-bay overlooking the Uppsala Cathedral. The room immediately adjacent houses not only Gustav Adolphus's wonder-filled Curiosity Cabinet made in Augsburg 1625–1631,[4] but also a large collection of scientific instruments designed and used by Olof Rudbeck, Carl von Linné, Anders Celsius and Anders Berch, the models of the latter profoundly influenced by Christopher Polhem, the inventor who for a period was Emmanuel Swedenborg's supervisor. It is hard to imagine a better foster home for one's brainchild, especially as the smaller prototype is deposited in the Olsson dining room, where I am free to converse with it whenever I please.

And the most remarkable of everything remarkable is that this purely material thing never ceases to respond, never stops revealing secrets of which the Jenssons were naïvely unaware when they first met at the Bar de Saussure to plan their raids into the lands unknown. Indeed this entire volume may be read as a record of these silent conversations. Difficulties abound, however, most of them related to a confusion of standpoints. Thus, while it initially seems that I am standing outside the pyramid looking into it, the semiotic animal is in fact always already inside, a monkey in a cage, a fly in a fly-bottle. As a consequence, the critical surveyor realizes first that he is hitting his head against barriers he cannot see, then that his mind is occupying a Brunelleschian vantage-point which is radically separated from his body! After that maddening insight has taken hold, the truths which hitherto have governed my imaginations can no longer be trusted, for the body, which I have always assumed to be *the* fix-point par excellence, now proves not to be fixed at all. As with the famous Moebius band, it is impossible to distinguish the inside from the outside, the fix-points from the mappae, one scale from the other.

The conundrum is built into the sign itself, a fact which becomes most noticeable in the choreography of the Signifier and the signified. Thus, in the beginning of our exploration we were pondering a set of ideas (signifieds) that searched for their expression (Signifier), while in the end we were playing with a material construction that searched for its meaning. One of the most remarkable of those moments came when the dead plastic was replaced by the living Weissglass, for in that new material the projection rays were not absorbed by the mappa of the opposite wall but managed somehow to break through the iconostasis and continue into infinity. Likewise with the corners of the crystal palace, those ultrathin surfaces which at the same time separate and join together the adjacent walls. Even more surprising, though, was the discovery that my own image is not reflected in the glass—to the mystic perhaps a sign that the creator's face must not be seen, to the critical cartographer a reminder that the insider is just as securely locked in as the outsider is locked out.

Much more, of course. But I must not fall into he trap of becoming an exegete of Jenssoniana or an analyst of myself. Suffice it to say that the outstanding problems all seem related to the base on which everything is resting, the fourth plane of the tetrahedron; as Constantin Brancusi knew so well, the foundation is an integral part of any sculpture. And so it is with the entire oeuvre of Marcel Duchamp as well, the parallels between his givens and the Jenssons' endeavor too numerous to enumerate. And yet there is a crucial difference. For whereas the corpse of the Duchampian version of art history rests securely in the installations of the Philadelphia Mausoleum, whoever tries to map the Island of Truth is bound to hit his head against the walls of the Crystal Palace. Triangulation three times codified, first in the Laws of Thought, then in the commandments of the first stone tablet, eventually in the arts of mapping. Thrice thrice.

"Cool it," said the engineer from Blainville and remodeled the legendary Bar de Saussure into a self-portrait of three glass staffs, two disturbingly greenish, one as clear as the spring water in which Artemis took her bath and Narcissus caught his reflections, in both cases a deadly play of desire. And while the Duchampian bachelors are shooting their bullets into the screening charta, the Jenssons are pondering the consequences of the first sacrifice, the bloody ruby concealing a myriad of bubbling differences under a layer of soothing identities. The gesso of the taken-for-granted, a cover-up of what it means to be human.

*

**And thus it is** that we become so obedient and so predictable. For now, when this mission impossible is finally (in)completed, the ghost of cartographical reason has merged into one with its basic instruments—fixpoint, scale and mappa. Perhaps a case of Power disrobed, perhaps the Semiotic Animal parading as a dinosaur in disguise. At any rate an antiphony executed by the three voices of Q.E.D., Q.E.F. and Q.E.I. A requiem in commemoration of everything found again, all by way of a commodius vicus of recirculation. Less for the sake of understanding, more for the sake of living.

Sign off. Mission completed. Go home Professor,

GO

# MEMORIALS

# NOTES

## CONFESSION

1. Gunnar Olsson, "Chiasm of Thought-and-Action," *Environment and Planning D: Society and Space* 12 (1993): 221–25.

2. For details, see instead Gunnar Olsson, "Stadier på livets väg," in Gunnar Olsson, ed. *Att famna en ton* (Uppsala: Acta Universitatis Upsaliensis, 1998), and Gunnar Olsson, "Glimpses," in Peter Gould and Forrest R. Pitts, eds., *Geographical Voices: Fourteen Autobiographical Essays* (Syracuse, N.Y.: Syracuse University Press, 2002). Sensitive commentaries are in Chris Philo, "Reflections on Gunnar Olsson's Contribution to the Discourse in Contemporary Human Geography," *Environment and Planning D: Society and Space* 2 (1984): 217–40; Chris Philo, "Escaping Flatland: A Book Review Essay Inspired by Gunnar Olsson's *Lines of Power/Limits of Language*," *Environment and Planning D: Society and Space* 12 (1994): 229–52; Martin Gren, *Earth Writing: Exploring Representation and Social Geography In-Between Meaning/Matter* (Gothenburg: Department of Human and Economic Geography, University of Gothenburg, 1994); and Marcus A. Doel, "Gunnar Olsson's Transformers: The Art and Politics of Rendering the Co-Relation of Society and Space in Monochrome and Technicolor," *Antipode* 35 (2003): 140–67.

## BORDER-MAN

1. The exact dating is highly disputed. While the oldest manuscripts yet discovered stem from the ninth century BCE, it is generally agreed that the original must be considerably older, possibly even from the time of Hammurabi, who ruled from 1792 to 1750.

2. *Enuma elish,* in the translation by N. K. Sandars, *Poems of Heaven and Hell from Ancient Mesopotamia* (Baltimore, Md.: Penguin, 1971), 73. As will eventually become clear, the translations are many and sometimes quite diverging, not only in form but in content as well.

3. *Enuma elish,* in the translation by Alexander Heidel, *The Babylonian Genesis,* 2d ed. (Chicago: University of Chicago Press, 1951), Tablet I, 1–8.

4. Ibid., Tablet VII, 86, 88.

5. Edward S. Casey, *The Fate of Place: A Philosophical History* (Berkeley and Los Angeles: University of California Press, 1997), 16. On related themes see also Jonathan Z. Smith, *To Take Place: Toward Theory in Ritual* (Chicago: University of Chicago Press, 1987), as well as Edward S. Casey, *Getting Back Into Place: Toward a Renewed Understanding of the Place-World* (Bloomington: Indiana University Press, 1993).

6. The roots of the western unconscious reach deeper than we often care to admit, a point well made by Werner Jaeger, *Paideia: Die Formung des Griechischen Menschen.* 3 vols. (Berlin: Walter de Gruyter, 1934, 1944, 1947). More recent excavations are in Keld Zeruneith, *Træhesten: Fra Odysseus til Sokrates* (København: Gyldendal, 2002) and in Walter Burkert, *Babylon, Memphis, Persepolis: Eastern Contexts of Greek Culture* (Cambridge, Mass: Harvard University Press, 2004). At the same time I am fully aware that herein lies both insights and blindness. For instance, and for comparisons between Greek and Chinese modes of understanding, see François Jullien, *Un sage est sans idée: Ou l'autre de la philosophie* (Paris: Seuil, 1998), timely translated by Lars Fyhr as *En vis är utan idé: Eller det andra hos filosofin* (Gråbo: Anthropos, 1999).

7. Henri Bergson, *Matter and Memory,* trans. Nancy Margaret Paul and W. Scott Paul (New York: Zone Books, 1991), 13.

8. For an alternative interpretation witness the following quotation from Norman Malcolm, *Wittgenstein: A Religious Point of View* (London: Routledge, 1993), 65f.: "A conspicuous feature of philosophical thinking about memory has been the insistence that in remembering there must be an image, or copy, or picture, or mental representation, of what is remembered. Russell maintained that memory demands an image. In fact [though], images are sometimes present and sometimes not, when we remember or recall situations, events or objects that we have perceived, witnessed or experienced in the past. Why is there this demand that there must be an image, or 'something like an image,' in remembering?"

The question whether we can or cannot do without images is the quandary that permeated also the groundbreaking exhibition held in Karlsruhe in the summer of 2002 and lavishly documented in Bruno Latour and Peter Weibel, eds., *Iconoclash: Beyond the Image Wars in Science, Religion and Art* (Cambridge, Mass.: MIT Press, 2002). Although that enterprise has numerous points (and several illustrations) in common with my own, our respective vantage points are in reality widely apart. Our basic conclusion is nevertheless the same: every image-breaker is at the same time an image-maker, every maker a breaker. No end in sight and that explains why Gunnael Jensson's *Mappa Mundi Universalis* (originally shown in Uppsala in the year 2000 and reproduced in the final chapter of the present volume) would have fitted perfectly into the Karlsruhe exhibition two years later.

9. For a fascinating story of that development told by two cognitive scientists, see Peter Gärdenfors, *Hur Homo blev sapiens: Om tänkandets evolution* (Nora: Nya Doxa, 2000) as well as Michael Tomasello, *The Cultural Origins of Human Cognition* (Cambridge, Mass.: Harvard University Press, 1999). In addition the masterwork of popular science—published after the completion of my own manuscript—Felipe Fernández-Armesto, *So You Think You're Human? A Brief History of Humankind* (Oxford: Oxford University Press, 2005).

10. In deed I take the prepositions (literally *pre*-positions) to be the most power-

filled and most culture dependent of all parts of speech, impossible to master in a foreign language. For a useful overview see Annette Herskovits, *Language and Spatial Cognition: An Interdisciplinary Study of the Prepositions in English* (Cambridge: Cambridge University Press, 1986). Also note the groundbreaking study of Karl Gunnar Lindkvist, *Studies on the Local Sense of the Prepositions 'in', 'at', 'on', and 'to', in Modern English,* Lund Studies in English, no. 20 (Lund, 1968).

11. Aristotle, *Physics* 220a27. Quoted from the translation by R. P. Hardie and R. K. Gaye in *The Basic Works of Aristotle,* ed. Richard McKeon (New York: Random House, 1941).

12. Gunnar Olsson, *Lines of Power/Limits of Language* (Minneapolis: University of Minnesota Press, 1991), 16.

13. Immanuel Kant, *Critique of Pure Reason,* trans. F. Max Müller (Garden City, N.Y.: Anchor Books, 1966), 92, B 157.

14. David Papineau, *Thinking about Consciousness* (Oxford: Oxford University Press, 2002).

15. Kant, *Critique of Pure Reason,* B 276.

16. Ibid., B 694.

17. Michael Tomasello, *Cultural Origins of Human Cognition.* Also Christopher Boehm, *Hierarchy in the Forest: The Evolution of Egalitarian Behavior* (Cambridge, Mass.: Harvard University Press, 1999).

18. For elaborations see the sensitive analyses in Rodolphe Gasché, *The Tain of the Mirror: Derrida and the Philosophy of Reflection* (Cambridge, Mass.: Harvard University Press, 1986).

19. A wonderful variation on this theme is in the autobiographical tale of Carl Fredrik Reuterswärd, *Titta, jag är osynlig!* (Stockholm: Gedins Förlag, 1989).

20. On this incongruity see Arthur Schopenhauer, *The World as Will and Representation,* trans. E. F. J. Payne (New York: Dover, 1966), esp. vol. 2, chap. 37, ostensibly about poetry, in essence about tragedy. On related themes in comedy see Henri Bergson, *Le rire: Essay sur la signification du comique* (Paris: Félix Alcan, 1917). Also Gregor Malantschuk, *Kierkegaard's Thought* (Princeton, N.J.: Princeton University Press, 1971), and, especially, G. W. F. Hegel, *The Phenemenology of Mind,* trans. J. B. Baillie (New York: Harper and Row, 1967), 725–49.

21. Søren Kierkegaard, *Concluding Unscientific Postscript,* trans. D. F. Swenson and Walter Lowrie (Princeton, N.J.: Princeton University Press, 1944), 459.

22. For a discussion of the somatic model, see Aristotle, *Physics* 4.1–2. At this point I am reminded also of the Sinhalese god Kataragama, known by many names including that of Sanmugam, which means "The one with six faces"—a kind of multidimensional Janus. See Paul Wirz, *Kataragama: The Holiest Place in Ceylon* (Colombo: Lakehouse Investments, 1972), 3. Many thanks to Professor M. M. Karunanayake for the reference and for insightful excursions through the war- and tsunami-ridden landscapes of Sri Lanka, including not only a menagerie of chattering monkeys and fighting tigers but an exploration of the relation between the two terms "destiny" and "destination."

23. Anthony Grafton, *Leon Battista Alberti: Master Builder of the Italian Renaissance* (London: Penguin, 2002), 118f.

24. In this context it is interesting to note that the Egyptian sign for "city"—*nywt*—was a cross inscribed in a circle. While the circle signifies closure and unity, the cross stands for the meeting of two roads, the most basic of all city-forms.

## WHEN ABOVE

1. For an extreme position see the pamphlet by Rabbi Dr. I. Rapaport, OBE, entitled *The Babylonian Poem Enuma Elish and Genesis Chapter One: A New Theory on the Relationship Between the Ancient Cuneiform Composition and the Hebrew Scriptures* (Melbourne: Hawthorn Press, 1979).

2. For illustrative examples see the introduction in Alexander Heidel, *The Babylonian Genesis: The Story of Creation,* 2d ed. (Chicago: University of Chicago Press, 1951), 15–17. Since this type of poetry is literally non-translatable—and since my own interpretations build entirely on various renderings into English, Swedish, Danish and German—much will clearly be lost. Trying to find my way in the wilderness I have picked my quotations not from one but from several of the classical translations. About Babylonia I am of course saying nothing new. And here—as throughout this entire text—I hope the experts will not be too offended by the inevitable mistakes. A useful introduction to the historical setting of the various poems is in the introductory chapter of W. G. Lambert, *Babylonian Wisdom Literature* (Oxford: Clarendon Press, 1960). Also see Jean Bottéro, *Religion in Ancient Mesopotamia,* trans. Teresa Lavender Fagan (Chicago: University of Chicago Press, 2001), esp. chap. 5. In addition I have consulted the Swedish translations of Alfred Haldar, *Det babyloniska skapelseeposet "enuma elis"* (Uppsala: Almqvist and Wiksell, 1952), and also the more recent Swedish version, *Enuma elish: Det babyloniska skapelseeposet,* trans.Ola Wikander (Stockholm: Wahlström and Widstrand, 2005).

3. N. K. Sandars, *Poems of Heaven and Hell from Ancient Mesopotamia* (Harmondsworth: Penguin, 1971), 25f. In this context also compare the following quotation from Revelation 20:1–3: "And I saw an angel coming down out of heaven, having the key to the Abyss and holding in his hand a great chain. He seized the dragon, the ancient serpent, who is the devil, or Satan, and bound him for a thousand years. He threw him into the Abyss, and locked and sealed it over him, to keep him from deceiving the nations any more until the thousand years were ended. After that he must be free for a short time."

4. Already here—indeed only here—I must make a general reference to the highly creative works on the abyss by Kenneth Grant, especially in *Nightside of Eden* (London: Frederick Muller, 1977), and *Outside the Circles of Time* (London: Frederick Muller, 1980). Also see Clayton Eshleman, *Juniper Fuse: Upper Paleolithic Imagination and the Construction of the Underworld* (Middletown, Conn.: Wesleyan University Press, 2003), esp. 215–24.

5. Stephanie Dalley, *Myths from Mesopotamia: Creation, the Flood, Gilgamesh, and Others* (Oxford: Oxford University Press, 1989), 234.

6. Ibid.

7. In this context it is worth recalling the Gnostic world of Valentinus, in which the sacred place of radical transcendence is located in the Bythos, by definition a bottomless abyss containing nothing but the nothingness of radical silence; ignorance is here raised to an ontological position of the first order. According to this heretical theory formulated around 150 CE, the beginning of the beginning was in the primal Father named "Bythos" (another word for "abyss"), uncontainable and invisible. Sometimes this figure of Pre-Beginning acts alone, sometimes in concert with his own Thought variously called "Grace" and "Silence" but always given in the feminine gender. Together these forces eventually produce an extended family of offspring, among them

the monster Achamonth, herself the child of the Aeon Sophia and a power called "Limit," the latter sometimes confirming and strengthening, sometimes dividing and delimiting. It was this misfit who in her wickedness brought forth the Demiurge, purportedly made in the image of the primal Father, but in effect a false and ignorant copy of the original Bythos. In the mistaken belief that he himself was the abyss, the Demiurge (by the Valentinians understood not as the primal Father of Bythos but as a strange mixture of Plato's "creator of the world," as described in the *Timaeus,* and the Jewish God, as presented in the Old Testament) then crafted heaven and earth and everything else there is. Perhaps an echo from pre-Christian Greece, perhaps an attempt to tie the evil to a premundane fault, perhaps an early case of iconoclasm. In my own imagination a fantastic troupe of categorial jugglers performing their ontological tricks on the floor of the Saussurean Bar-in-Between.

On these relations more will be said later. For the moment see Robert M. Grant, ed., *Gnosticism: A Source Book of Heretical Writings from the Early Christian Period* (London: Collins, 1961), esp. chap. 5; Hans Jonas, *The Gnostic Religion: The Message of the Alien God and the Beginnings of Christianity* (Boston: Beacon Press, 1963); Anders Olsson, *Läsningar av INTET* (Stockholm: Albert Bonniers, 2000), esp. 44–54; Alain Besançon, *The Forbidden Image: An Intellectual History of Iconoclasm,* trans. Jane Marie Todd (Chicago: University of Chicago Press, 2000), esp. 86ff.

8. Dalley, *Myths from Mesopotamia,* 235f.

9. Sandars, *Poems of Heaven and Hell from Ancient Mesopotamia,* 81. Given the Old Testament warning that no one can see the LORD's face and survive, this wording is nothing but remarkable.

10. On this important point all translations seem to agree. As proof see Heidel, *The Babylonian Genesis,* 29, Tablet II: 102 and Tablet II: 120; James B. Pritchard, *Ancient Near Eastern Texts Relating to the Old Testament* (Princeton, N.J.: Princeton University Press, 1955), 64, Tablet II: 102 and Tablet II: 120; Sandars, *Poems of Heaven and Hell from Ancient Mesopotamia,* 81 and 82. In Danish the equivalent of the LORD is HERREN, which consequently is the term used in Ulla and Aage Westenholz, *Gilgamesh og Enuma Elish: Guder og mennesker i oldtidens Babylon* (København: Spektrum, 1997), 206.

11. Heidel, *The Babylonian Genesis,* 29–30, Tablet II: 122–29.

12. Ibid., 35–36, Tablet III: 129–38.

13. Dalley, *Myths from Mesopotamia,* 250.

14. The term here translated as "constellation" has often been rendered as "garment," sometimes even as "a kind of apparition," perhaps a warning against the temptations of nudity and a statement to the effect that neither gods nor emperors may be seen naked. I nevertheless prefer the word "constellation," because that term better reflects the importance of astronomy and astrology in Babylonian culture. The problem has its roots in the cuneiform signs *ba* and *ma,* which are exceedingly difficult to distinguish from each other.

15. Dalley, *Myths from Mesopotamia,* 250 and 275 n. 15.

16. Ibid., 253.

17. Heidel, *The Babylonian Genesis,* 41.

18. Sandars, *Poems of Heaven and Hell from Ancient Mesopotamia,* 91–92.

19. Ibid., 92.

20. Here it should be noted that Marduk's enormous ears were considered true signs of wisdom. In this respect there are obvious connections between the *Enuma elish* and the Hebrew Bible, for in both cases hearing was considered the outstanding mode

of understanding. With the Greeks and the advent of Jesus Christ this preference was changed and sight became the privileged sense, the knowledgeable seers literally leading the ignorant blind.

21. Sandars, *Poems of Heaven and Hell from Ancient Mesopotamia,* 95–96, emph. added.

22. Karen Armstrong, *Historien om Gud,* trans. Inger Johansson (Stockholm: Forum, 1995), 30; *A History of God* (New York: Alfred A. Knopf, 1993), 16.

23. For detailed and fascinating discussions of Marduk's fifty names—not the least as they prepared the way for the art of counting in fives and tens—see Franz M. Th. Böhl, "Die fünfzig Namen des Marduk," *Archiv für Orientforschung* 11 (1936): 191–218; and J. Bottéro, "Les noms de Marduk, l'écriture et 'la logique' en Mésopotamie ancienne," in Maria de Jong Ellis, ed., *Essays on the Ancient Near East in Memory of Jacob Joel Finkelstein* (Hamden, Conn.: Archon, 1977).

24. Gilles Deleuze and Félix Guattari: *A Thousand Plateaus: Capitalism and Schizophrenia,* trans. Brian Massumi (Minneapolis: University of Minnesota Press, 1987), 264, emph. added.

## AND BELOW

1. The mythological roots of this worldview are laid bare in Mary K. Wakeman, *God's Battle with the Monster: A Study in Biblical Imagery* (Leiden: E. J. Brill, 1973).

2. Perhaps Alexander was reacting also against Socrates' remark that the Greeks were sitting "like frogs around a frog-pond." Quotation from Robin Lane Fox, *The Search for Alexander* (Boston: Little, Brown and Company, 1980), 11.

3. Marco Picone, "Anaximander versus Anaximander," in Marco Picone, ed., *Bodies and Space: Gunnar's Travels* (Palermo: Università di Palermo, Laboratorio Geografico, 2002), 66.

4. The literature is large but for translations and general overviews see Francis MacDonald Cornford, *From Religion to Philosophy: A Study in the Origins of Western Speculation* (London: Edward Arnold, 1912), esp. chap. 2; Geoffrey Stephen Kirk et al., *The Presocratic Philosophers: A Critical History with a Selection of Texts,* 2d ed. (Cambridge: Cambridge University Press, 1983); Jonathan Barnes, *The Presocratic Philosophers* (London: Routledge, 1993). For our purposes the most pertinent treatments are Charles H. Kahn, *Anaximander and the Origins of Greek Cosmology* (New York: Columbia University Press, 1960), and Paul Seligman, *The Apeiron of Anaximander: A Study in the Origin and Function of Metaphysical Ideas* (London: Athlone Press, 1962).

5. Anaximander himself imagined the apeiron not only as a god but as a place. For further comments see Erik Stenius, *Tankens gryning: En studie över den västerländska filosofins ursprungsskede,* 2d ed., rev. (Helsingfors: Söderström, 1975), 34.

6. Michel Serres, *Les origines de la géométrie* (Paris: Flammarion, 1993), 71–110.

7. The Ocean River is a preconceptual symbol inherited from the Egyptians and Babylonians and one which Hepæstus forged into the outermost rim of Achilles' shield. On the latter see Homer, *Iliad* 18.607. For a popular yet useful overview of this fascinating topic see James S. Romm, *The Edges of the Earth in Ancient Thought: Geography, Exploration, and Fiction* (Princeton, N.J.: Princeton University Press, 1992).

8. Not surprisingly this fading of the Ocean was gradual. As evidence witness how Paul Seligman in *The Apeiron of Anaximander* noted that a "Babylonian world map on

a cuneiform tablet in the British Museum . . . , dated circa 500 BC, still shows the earth encircled by the 'earthly ocean' entitled 'Bitter River'" (142). A more general overview is in O. A. W. Dilke, *Greek and Roman Maps* (Ithaca, N.Y.: Cornell University Press, 1985).

9. For translations and critical comments on passages and fragments from all eighteen, see Christina Horst Roseman, "Pytheas of Massalia: A Critical Examination of the Texts" (Ph.D. diss., University of Washington, 1983). A slightly revised version is in Christina Horst Roseman, *Pytheas of Massalia: On the Ocean* (Chicago: Ares Publishers, 1994). The most convincing account of Pytheas' voyage is in Barry Suncliffe, *The Extraordinary Voyage of Pytheas the Greek, the Man Who Discovered Britain* (London: Allen Lane, 2001). Useful information is also in Richard Hennig, *Terræ Incognitæ* (Leiden: E. J. Brill, 1944), chap. 20; C. F. C. Hawkes, *Pytheas and the Greek Explorers* (Oxford: Blackwell, 1977); J. Oliver Thomson, *History of Ancient Geography* (Cambridge: Cambridge University Press, 1948), 143ff.; D. R. Dicks, *The Geographical Fragments of Hipparchus* (London: Athlone Press, 1960), 179ff.; and Germaine Aujac, "L'île de Thulé, de Pythéas à Ptolémée," in Monique Pelletier, ed., *Géographie du monde au Moyen Âge et à la Renaissance* (Paris: Édition du CTHS, 1989).

10. Roseman, *Pytheas of Massalia: On the Ocean,* 24. Regrettably, Strabo was not the last geographer to turn personal ignorance into social influence.

11. Ludwig Wittgenstein, *Zettel,* 2d ed., ed. G. E. M. Anscombe and G. H. von Wright, trans. G. E. M. Anscombe (Oxford: Basil Blackwell, 1980), 393.

12. This is a major argument in Suncliffe, *The Extraordinary Voyage of Pytheas the Greek.*

13. William Arthur Heidel, *The Frame of the Ancient Maps: With a Discussion of the Discovery of the Sphericity of the Earth* (New York: American Geographical Society, 1937), esp. chap. 5.

14. Roseman, *Pytheas of Massalia: On the Ocean,* 32.

15. Using the same technique Pytheas also worked out the latitude of Marseilles as 43°12′N, very near the correct value of 43°15′N.

16. To later commentators Pytheas' reference to honey has created problems. The reason is that while in Iceland there are no bees, in Norway they are plentiful, a fact that speaks against the opinion that his Thule was our Iceland. The counter-argument is that he might well have sighted the mighty peak of Mount Hekla but never gone ashore to check what was there.

17. Hawkes, *Pytheas: Europe and the Greek Explorers,* 36.

18. Thomas Pynchon, *Mason & Dixon* (New York: Henry Holt and Company, 1997), 151f., emph. added. In my estimation Pynchon's book is an outstanding exposition of cartographical reasoning, from beginning to end a stunning performance. For insightful comments see Joakim Sigvardson, *Immanence and Transcendence in Thomas Pynchon's "Mason and Dixon": A Phenomenological Study* (Stockholm: Almqvist and Wiksell, 2002). Also Brook Horvath and Irving Malin, eds., *Pynchon and "Mason and Dixon"* (Cranbury, N.J.: Associated University Press, 2000). For detailed information about the actual surveying of the line, see Edwin Danson, *Drawing the Line: How Mason and Dixon Surveyed the Most Famous Border in America* (New York: Wiley, 2001).

19. The most comprehensive treatment is in Klaus Geus, *Eratosthenes von Kyrene: Studien zur hellenistischen Kultur- und Wissenschaftsgeschichte* (München: Beck, 2002). Much supporting material can be found in William Arthur Heidel, *The Frame of the Ancient Greek Maps;* J. Oliver Thomson, *History of Ancient Geography;* Gerald R. Crone, *Maps and Their Makers: An Introduction to the History of Cartography* (Hamden, Conn.,

and London: Hutchinson, 1953); George Sarton, *A History of Science,* vol. 2: *Hellenistic Science and Culture in the Last Three Centuries BC* (Cambridge, Mass.: Harvard University Press, 1959); D. R. Dicks, *The Geographical Fragments of Hipparchus;* O. A. W. Dilke, *Greek and Roman Maps;* Christian Jacob, *Géographie et ethnographie en Grèce ancienne* (Paris: Armand Colin, 1991); John Noble Wilford, *The Mapmakers: The Story of the Great Pioneers in Cartography, From Antiquity to the Space Age* (London: Pimlico, 2002).

20. Sarton, *A History of Science,* vol. 2: *Hellenistic Science and Culture in the Last Three Centuries BC,* 102.

21. Crone, *Maps and Their Makers.* Also Evelyn Edson, *Mapping Time and Space: How Medieval Mapmakers Viewed Their World* (London: The British Library, 1999), 4. In addition see Harry Järv, "Biblioteket i Alexandria: Världens viktigaste kulturinstitution," in Stefan Hilding and Suzanne Unger Sörling, eds., *Alexandria: Musernas stad* (Stockholm: Rubicon, 1997), 88.

22. The face and the arse! Or should it rather be the ace and the farce? When the alphabet soup burns your tongue, stick it out and it cools down. A throw of dice in the casino of meaning, a letter rolling across the table. Reaching the edge Humpty Dumpty tumbles down. Egg yolk on the floor.

23. When the word "farce" appears as a noun, it boils down to "a dramatic work (usually short) which has for its sole object to excite laughter," by extension "a proceeding that is ludicrously futile or insincere, a hollow pretence, a mockery." When instead it enters the stage as a verb, it means not only "to stuff (an animal, a piece of meat) with force-meat, herbs, etc., to pad out, to interlard," but also "to cram the stomach with food, to fill out what is lean or shrunken." Quotations from *The Compact Edition of the Oxford English Dictionary* (Oxford: Oxford University Press, 1971).

24. Dilke, *Greek and Roman Maps,* 34.

25. This fascinating story is popularly retold in Dava Sobel, *Longitude: The Story of a Lone Genius Who Solved the Great Scientific Problem of His Time* (New York: Walker, 1995); also William J. H. Andrews, *The Quest for Longitude: The Proceedings of the Longitude Symposium, Harvard University, Cambridge, Massachusetts, November 4–6, 1993* (Cambridge, Mass.: Harvard University Press, 1996). For recollections of how Harrison's neglected clocks were restored by a man much like him, see Rupert T. Gould, *The Marine Chronometer: Its History and Development* (London: J. D. Potter, 1923); Rupert T. Gould, *John Harrison and His Timekeepers* (London: National Maritime Museum, 1978).

26. If they could talk, both parallels and meridians might say that to themselves they are straight lines returning to themselves, the former traveling around the globe without ever meeting another of its kind, the latter all converging at the poles. If they could listen, the same parallels and meridians might well hear the geometrician say that what is a straight line on a bent surface becomes a bent line on a flat surface. No wonder we sometimes get lost. For a general introduction see Lloyd A. Brown, *The Story of Maps* (Boston: Little, Brown and Company, 1949), chaps. 7 and 8.

27. A long-lived mistake that eventually influenced Columbus' attempt to reach India by going west rather than east.

28. Christian Jacob, "Mapping in the Mind: The Earth from Ancient Alexandria," in Denis Cosgrove, ed., *Mappings* (London: Reaktion Books, 1999), 41.

29. Dilke, *Greek and Roman Maps,* 35.

30. J. Lennart Berggren and Alexander Jones, *Ptolemy's "Geography": An Annotated Translation of the Theoretical Chapters* (Princeton, N.J.: Princeton University Press, 2000), 29 for the latitude and 110 for the longitude. While determining the latitude was quite

straightforward, to measure the longitude was much more complicated. Add to this that the zero meridian sometimes was taken through Alexandria, more often, as here, through the Islands of the Blest, at the time considered the westernmost limit of the oikumene.

31. The best introduction is still in Lloyd A. Brown, *The Story of Maps,* chap. 2. But there is also Gerald James Toomer, "Ptolemy," *Dictionary of Scientific Biography,* 11:186–206. For outstanding translations see Gerald James Toomer, *Ptolemy's "Almagest"* (London: Duckworth, 1984), and Berggren and Jones, *Ptolemy's "Geography."*

32. Berggren and Jones, *Ptolemy's "Geography,"* 59. In a footnote to this passage Berggren and Jones insert the following comment: "This rather obscure peroration entwines two ideas: that astronomy and geography are parts of a single rational science, and that whereas astronomy can make its demonstrations using the heavens themselves as a visible object of study, geography must make use of maps. We are inside a celestial sphere, and can behold half of it at once. By way of contrast, our position on the surface of the earth prevents us from taking in the earth's form at a glance, and it is too large for any single person to explore."

33. According to a slightly different translation, Ptolemy defined geography as "a representation in picture of the whole known world together with the phenomena which are contained therein," while the concern of chorography is "to paint a true likeness, and not merely to give exact positions and size." Quotations from Lloyd A. Brown, *The Story of Maps,* 61.

34. Berggren and Jones, *Ptolemy's "Geography,"* 57f., emph. added.

35. Pynchon, *Mason & Dixon,* 504f.

36. Wilford, *The Mapmakers,* 36. As already noted, the difficulty was not to measure the latitudes but to determine the longitudes.

37. While the Latin term *charta* denotes a "leaf of paper" or any kind of formal document (such as a deed), *mappa* means "napkin" or "table-cloth." To the ancient Greeks a map was a *pinax,* literally the piece of wood, metal, or stone onto which the mapmaker cast his message. For etymological details see Christian Jacob, *L'empire des cartes: Approche théorique de la cartographie à travers l'histoire* (Paris: Albin Michel, 1992), 37ff.

38. Berggren and Jones, *Ptolemy's "Geography,"* 83.

39. On the possible relations between Ptolemy's coordinate net and Brunelleschi's picture of the square-patterned piazza, see Samuel Y. Edgerton Jr, "From Mental Matrix to *Mappamundi* to Christian Empire: The Heritage of Ptolemaic Cartography in the Renaissance," in David Woodward, ed., *Art and Cartography: Six Historical Essays* (Chicago: University of Chicago Press, 1987). However, see also the more ambivalent remarks in Berggren and Jones, *Ptolemy's "Geography,"* 39, as well as the counterargument in Svetlana Alpers, "The Mapping Impulse in Dutch Art," in David Woodward, ed., *Art and Cartography.*

40. Berggren and Jones, *Ptolemy's "Geography,"* 83, emph. added.

41. Brown, *The Story of Maps,* 74 and 75.

42. The final transcendence of Ptolemy came in 1570 with the publication of the first atlas, the *Theatrum orbis terrarum.* For an accessible introduction to the latter see Paul Binding, *Imagined Corners: Exploring the World's First Atlas* (London: Headline, 2003).

43. Pynchon, *Mason & Dixon,* 350.

44. For an informative overview see Robert North, *A History of Biblical Map Making* (Wiesbaden: Ludwig Reichert, 1979).

45. C. Raymond Beazley, *The Dawn of Modern Geography: A History of Exploration and Geographical Science* (London: John Murray, 1897), 303.

46. Despite his nickname, Cosmas probably never set foot in India.

47. Most central here is Cosmas' mention of an inscription which he had seen on an elaborate marble throne, perhaps in the city of Adulis on the coast of the Red Sea. For a discussion of this inscription and its role in Ethiopian history, see L. P. Kirwan, "The Christian Topography and the Kingdom of Axum," *Geographical Journal* 138 (1972): 166–77.

48. The definite bilingual edition is Cosmas Indicopleustès, *Topographie Chrétienne: Introduction, texte critique, traduction et notes,* ed. and trans. Wanda Wolska-Conus, 3 vols. (Paris: Éditions du Cerf, 1968, 1970, 1973). This masterpiece represents the outcome of excellent research first reported in Wanda Wolska, *La "topographie chrétienne" de Cosmas Indicopleustès: Theologie et science au VIe siecle* (Paris: Presses Universitaires de France, 1962). Earlier and less satisfactory translations are in *The "Christian Topography" of Cosmas, an Egyptian Monk,* trans. J. W. McCrindle (London: Hakluyt Society, 1897) and *The "Christian Topography" of Cosmas Indicopleustes,* ed. and trans. E. O. Winstedt (Cambridge: Cambridge University Press, 1909). Extensive comments can be found also in Beazley, *The Dawn of Modern Geography.*

49. Beazley, *The Dawn of Modern Geography,* 278.

50. Virtually all educated people after Aristotle have known that the earth is a sphere. A least in this respect the so called "Dark Ages" were no exception. For more on this misunderstood issue see Jeffrey Burton Russell, *Inventing the Flat Earth: Columbus and Modern Historians* (New York: Praeger, 1991).

51. Berggren and Jones, *Ptolemy's "Geography,"* 57, emph. added.

52. For a fascinating account of this transition from paganism to Christianity, itself paving the way for the belief systems of the Middle Ages, see Ramsay Macmullen, *Christianity and Paganism in the Fourth to Eighth Centuries* (New Haven, Conn.: Yale University Press, 1997).

53. There is much to suggest that Cosmas' views about the earth's flatness are more complicated than they may at first appear, indeed that in this area his work is full of internal contradictions. For instance, while he rejected the idea about the globe, he retained the spherists' conception of the oikumene as rectangular. For details see Wolska, *La topographie chrétienne de Cosmas Indicopleustès,* 260.

54. Central to Cosmas' argument was the following quotation from Job 26:7: "He spreads out the northern skies over empty space; he suspends the earth over nothing."

55. Pynchon, *Mason & Dixon,* 482.

56. As the deranged Hölderin is reported to have said in an instance of "incoherent speech," "everything is rhythm, the entire destiny of man is one heavenly rhythm, just as every work of art is one rhythm, and everything swings from the poetizing lips of the god." Quoted from Giorgio Agamben, *The Man Without Content,* trans. Georgia Albert (Stanford, Calif.: Stanford University Press, 1999), 94.

57. Sometimes John Philoponus was also referred to as John the Grammarian, a designation he himself seems to have preferred.

## IN-BETWEEN

1. "Dancing upstairs—feeling / that sleepy is the house no more. / And in a flash it strikes me that my ceiling, / my ceiling, is another's floor." My translation. The Swedish original was first published in Nils Ferlin, *Barfotabarn* (Stockholm: Albert Bonnier, 1933).

2. The best introduction is in Richard Sorabji, ed., *Philoponus and the Rejection of Aristotelian Science* (London: Duckworth, 1987). Aspects of Philoponus' critique of Aristotle's physics are expertly discussed in Christian Wildberg, *John Philoponus' Criticism of Aristotle's Theory of Aether* (Berlin: Walter de Gruyer, 1988), while the theological debate is best captured in Uwe Michael Lang, *John Philoponus and the Controversies over Chalcedon in the Sixth Century: A Study and Translation of the "Arbiter"* (Sterling, Va.: Peeters, 2001). As could be expected, Philoponus' relations to Cosmas are detailed in Wanda Wolska, *La "Topographie chrétienne" de Cosmas Indicopleustès: Theologie et science au VIe siecle* (Paris: Presses Universitaires de France, 1962), 147–92.

3. Frans A. J. de Haas, *John Philoponus' New Definition of Prime Matter: Aspects of its Background in Neoplatonism and the Ancient Commentary Tradition* (Leiden: Brill, 1997), 295.

4. By the Nestorians Theodore was called "the Interpreter," for like Nestorius himself he too was an outstanding exegete, both of them arguing that the Holy Scripture should be read literally, not allegorically.

5. Nestorius' parents came from Persia, and this circumstance may help to explain why his influence was strongest among Christians in that country. For a long period he served as patriarch of Constantinople, a position from which he was eventually removed. Condemned for heresy by the Council of Ephesus in 431, he was later exiled to a number of far-away places, including the Elephantine Island opposite Syene, otherwise known for its shadowless well.

6. In my attempts at clarification I have been especially helped by Henry Chadwick, "Philoponus the Christian Theologian," in Richard Sorabji, ed., *Philoponus and the Rejection of Aristotelian Science,* chap. 2; and by Uwe Michael Lang, *John Philoponus and the Controversies over Calchedon in the Sixth Century.*

7. Franz Erdin, *Das Wort Hypostasis, seine bedeutungsgeschichtliche entwicklung in der altchristlichen Literatur bis zum Abschluss der trinitarischen auseinandersetzungen* (Freiburg im Breisgau: Herder and Co., 1939), 2. For more along similar lines see Ubaldo Ramón Pérez Paoli, *Der plotonische begriff von ʽΨΠΟΣΤΑΣΙΣ und die augustinische Bestimmung Gottes als Subiectum* (Würzburg: Augustinus Verlag, 1990).

8. In my understanding of Plotinus I have drawn on the general overviews in Povl Johs. Jensen, *Plotin* (København: Ejnar Munksgaard, 1948); J. M. Rist, *Plotinus: The Road to Reality* (Cambridge: Cambridge University Press, 1967); Kirsten Friis Johansen, *Den europæiske filosofis historie,* vol. 1: *Antiken* (København: Nyt Nordisk Forlag Arnold Busk, 1991); Philip Merlan, "Plotinus," in Paul Edwards, ed., *The Encyclopedia of Philosophy* (New York: Macmillan, 1967), 6:351–59; and Herbert Boeder, *Topologie der metaphysik* (München and Freiburg: Karl Alber, 1980), 213–78. For more pointed discussions see Frederic M. Schroeder, *Form and Transformation in the Philosophy of Plotinus* (Montreal: McGill-Queen's University Press, 1992); and Dmitri Nikulin, *Matter, Imagination and Geometry: Ontology, Natural Philosophy and Mathematics in Plotinus, Proclus and Descartes* (Aldershot: Ashgate, 2002).

9. The original passages are in Plato, *Republic* 509d–511e. My own renderings have been much aided by John Sallis, *Being and Logos: The Way of Platonic Dialogue* (Atlantic Higlands, N.J.: Humanities Press, 1986), 413ff.; Holger Thesleff, *Platon* (Lund: Pegas, 1990), 137ff.; Julia Annas, *An Introduction to Plato's "Republic"* (Oxford: Clarendon, 1981), chap. 10; and the interpretive essay by Allan Bloom in *The Republic of Plato* (New York: Basic Books, 1991).

10. Plotinus, *Enneads,* trans. A. H. Armstrong (Cambridge, Mass.: Harvard University Press, 1985), 5.2, emph. added.

11. Plotinus, *Enneads* 2.4.11.

12. Uwe Michael Lang, *John Philoponus and the Controversies over Chalcedon in the Sixth Century,* 138.

13. Arthur Hilary Armstrong, *The Architecture of the Intelligible Universe in the Philosophy of Plotinus* (Cambridge: Cambridge University Press, 1940).

14. For a wonderful rendering of the choreography between Plato's *chora* and Ptolemy's *chorology* see Alessandra Bonazzi, "Chora and Chorography," in Marco Picone, ed., *Bodies and Space: Gunnar's Travels* (Palermo: Università di Palermo, Laboratorio Geografico, 2002). Also Augustin Berque, *Écoumène: Introduction à l'étude des milieux humains* (Paris: Belin, 2000), chap. 1; and Jaroslav Pelikan, *What Has Athens to Do with Jerusalem? Timaeus and Genesis in Counterpoint* (Ann Arbor: University of Michigan Press, 1997).

15. In this context it is interesting to note that also to Immanuel Kant "human reason is by nature architectonic. That is to say, it regards all our knowledge as belonging to one possible system." Kant, *Critique of Pure Reason,* trans. Norman Kemp Smith (London: Palgrave Macmillan, 2003), A 474/B 502.

16. Chadwick, "Philoponus the Christian Theologian," in Sorabji, *Philoponus and the Rejection of Aristotelean Science,* 49.

17. Thomas Pynchon, *Mason & Dixon,* 404.

18. Quotation from Cosmas Indicopleustès, *The "Christian Topography" of Cosmas, an Egyptian Monk,* trans. J. W. McCrindle (London: Hakluyt Society, 1897), 1.

19. Unless otherwise noted, all biblical quotations in this chapter will be from *The Holy Bible,* New International Version (Grand Rapids, Mich.: Zondervan Publishing House, 1984).

20. Cosmas was evidently not the first to draw the analogy between the tabernacle of Moses and the structure of the material universe, a circumstance that may explain why he dedicated his book to a person called Pamphilus, a fellow Alexandrian, who, like Cosmas himself, was much influenced by Theodore of Mopsuestia. For details see C. Raymond Beazley, *The Dawn of Modern Geography: A History of Exploration and Geographical Science,* 287f., and Wolska, *La "Topographie chrétienne" de Cosmas Indicopleustès,* 1, 13, 68.

21. It is an interesting parallel to the Roman temple of Janus (closed in times of peace and open in times of war) that the Israelites set out on their travels only when the cloud lifted from the tabernacle; whenever the cloud did not lift, they stayed at home.

22. On the dialectics of these two roles see Jack Miles, *God: A Biography* (London: Simon and Schuster, 1995), 85–126.

23. Pynchon, *Mason & Dixon,* 542 and 615.

24. Ibid., 692.

## MAPPAE MUNDI MEDIEVALIS

1. David Woodward, "Reality, Symbolism, Time, and Space in Medieval World Maps," *Annals of the American Association of Geographers* 75 (1985): 515. Also note the considerable overlaps between that article and the more extensive treatments in David Woodward, "Medieval Mappaemundi," in J. B. Harley and David Woodward, eds., *The History of Cartography,* vol. 1: *Cartography in Prehistoric, Ancient, and Medieval Europe and the Mediterranean* (Chicago: University of Chicago Press, 1987). Woodward's way to this now standard interpretation was prepared especially by John Kirtland Wright, *The Geographical Lore of the Time of the Crusades: A Study in the History of Medieval Science and Tradition in Western Europe* (New York: American Geographical Society, 1925).

In this context it is important to stress the close cooperation between Woodward and J. B. Harley during the decade prior to the latter's untimely death in 1991. Several of Harley's seminal pieces—with a critical introduction by J. H. Andrews—have later been collected in J. B. Harley, *The New Nature of Maps: Essays in the History of Cartography,* ed. Paul Laxton (Baltimore, Md.: Johns Hopkins University Press, 2001). For further comments on Harley's conception of power and politics see Jeremy W. Crampton, "Maps as Social Constructions: Power, Communication and Visualization," *Progress in Human Geography* 25 (2001): 235–52.

2. In addition to the classical overviews by Woodward and Wright, see George H. T. Kimble, *Geography in the Middle Ages* (London: Methuen, 1938); Lloyd A. Brown, *The Story of Maps* (Boston: Little, Brown and Company, 1949), chap. 4; Joachim G. Leithäuser, *Mappae Mundi: Die geistige Eroberung der Welt* (Berlin: Safari, 1958); Patrick Gautier Dalché, *Géographie et culture: la représentation de l'espace du VIe au XIIe siècle* (Aldershot: Ashgate, 1997); Evelyn Edson, *Mapping Time and Space: How Medieval Mapmakers Viewed Their World* (London: British Library, 1999); and especially the detailed and insightful discussions in Jörg-Geerd Arentzen, *Imago Mundi Cartographia: Studien zur Bildlichkeit mittelalstericher Welt- und Ökumenekarten unter besonderer Berücksichtigung des Zusammanwirkens von Text und Bild* (München: Wilhelm Fink, 1984). More than a century after their original publication, much useful information may still be found in the six volumes of Konrad Miller, *Mappae mundi: Die ältesten Weltkarten* (Stuttgart: Jos. Roth'sche, 1895–98), updated and reworked in Marcel Destombes, *Mappemondes A.D. 1200–1500,* Catalogue préparé par la Commission des Cartes Anciennes de l'Union Géographique Internationale (Amsterdam: Israel, 1964). A most extensive bibliography is in Andrew Gow and Jolanta Pekacz, *A Bibliography of Scholarship on Mappaemundi and Early World Maps,* http://www.ualberta.ca/~agow/mapbib.html.

3. There is in fact much to suggest that in Plato's *Timaeus* the starry heaven was considered the tautological topos of time. For further elaborations see John Sallis, *Chorology: On Beginning in Plato's "Timaeus"* (Bloomington: Indiana University Press, 1999), chap. 2.

4. Augustine, *Confessions,* trans. R. S. Pine-Coffin (Harmondsworth: Penguin, 1961), 95, quoted in Woodward, "Medieval Mappaemundi," 326.

5. The long life of this image is nothing but remarkable, for already in the fifth century BCE Herodotus had written that "I laugh when I see the many men who draw maps of the world without using their heads; they make the earth a perfect circle, better even than one drawn with a compass, with Ocean running around it." And as an echo a hundred years later, Aristotle remarked that "they draw maps of the earth in a laughable manner; for they draw the oikoumene in a very round form, which is

impossible on the basis of both logic and observed facts." Quotations from James S. Romm, *The Edges of the Earth in Ancient Thought: Geography, Exploration, and Fiction* (Princeton, N.J.: Princeton University Press, 1992), 34 and 42. For a creative reinterpretation of Herodotus' laughter, see Franco Farinelli, "Squaring the Circle, or the Nature of Political Identity," in Franco Farinelli, Gunnar Olsson, and Dagmar Reichert, eds., *Limits of Representation* (München: Accedo, 1994).

6. Louis Marin, *Utopics: Spatial Play* (Atlantic Highlands, N.J.: Humanities Press, 1984), xv–xxvii.

7. For more on this exciting connection, see Giuliana Bruno, *Atlas of Emotion: Journeys in Art, Architecture, and Film* (London: Verso, 2002). Of special importance are Bruno's detailed analyses of Madeleine de Scudery's "Map of the Land of Tenderness"—*Carte du pays de Tendre*—from 1654. For us it is especially noteworthy that in the engraving which accompanies Scudery's original text, the main waterways are depicted in the shape of the letter T, its cross-bar forming "La Mer Dangereuse" beyond which lies the "Terres Inconnues"! More centrally located, and better suited for bathing, are the "Sea of Intimacy" and the "Lake of Indifference." For details see Bruno, plate 4 and 217–29.

8. Hugh of St. Victor, *The "Didascalicon" of Hugh of St. Victor: A Medieval Guide to the Arts,* trans. and ed. Jerome Taylor (New York: Columbia University Press, 1991), 121. On similar themes also see Ivan Illich, *In the Vineyard of the Text: A Commentary to Hugh's* Didascalicon (Chicago: University of Chicago Press, 1993). Hugh's role as a geographer is handsomely treated in Patrick Gautier Dalché, *La 'Descriptio Mappe Mundi' de Hugues de Saint Victor* (Paris: Études Augustiennes, 1988).

9. Quoted in Evelyn Edson, *Mapping Space and Time,* 190, n. 47.

10. This conception of memory has been dominant at least since Plato. Although sometimes criticized as too simplistic, it has recently received much support from experimental work in cognitive psychology. For details see John T. E. Richardson, *Mental Imagery and Human Memory* (London: Macmillan, 1980).

11. Quoted from the translation in Mary J. Carruthers, *The Book of Memory: A Study of Memory in Medieval Culture* (Cambridge: Cambridge University Press, 1990), 264 as well as 9; for a historical discussion of this passage, see *op. cit.,* 80ff. In addition Grover A. Zinn Jr, "Hugh of St.. Victor and the Art of Memory," *Viator* 5 (1974): 211–34. The classic work on mnemonics is Frances Yates, *The Art of Memory* (Chicago: University of Chicago Press, 1966).

12. The most pervasive memory device in medieval rhetoric was in fact the monastic buildings themselves. For further elaborations on this most Vitruvian theme, see Mary Carruthers, *The Craft of Thought: Meditation, Rhetoric and the Making of Images, 400–1200* (Cambridge: Cambridge University Press, 1998), esp. chap. 5.

13. Grover A. Zinn Jr, "Hugh of St Victor, Isaiah's Vision, and *De arca Noe,*" in Diana Wood, ed., *The Church and the Arts* (Oxford: Blackwell, 1992).

14. Danielle Lecoq, "La 'mappemonde' du *De arca noe mystica* de Hugues de Saint-Victor (1128–1129)," in Monique Pelletier, ed., *Géographie du monde au moyen âge et à la renaissance* (Paris: Éditions du CTHS, 1989).

15. Friedrich Ohly, "Deus Geometra: Skizzen zur Geschichte einer vorstellung von Gott," in Norbert Kamp and Joachim Wollasch, eds., *Tradition als historische Kraft: Interdisziplinäre Forschungen zur Geschichte des früheren Mittelalters* (Berlin: Walter de Gruyter, 1982).

16. David Woodward, "Reality, Symbolism, Time, and Space in Medieval World Maps," *Annals of the American Association of Geographers* 75 (1985): 513.

17. The work of Deleuze and Guattari represents one of the most sustained critiques of cartographical reason that have yet appeared. See in particular Gilles Deleuze, *The Fold: Leibniz and the Baroque,* trans. Tom Conley (Minneapolis: University of Minnesota Press, 1993); Gilles Deleuze and Claire Parnet, *Dialogues,* trans. Hugh Tomlinson and Barbara Habberjam (London: Athlone Press, 1987); Gilles Deleuze and Félix Guattari, *A Thousand Plateaus: Capitalism and Schizophrenia,* trans. Brian Massumi (Minneapolis: University of Minnesota Press, 1987); Gilles Deleuze and Félix Guattari, *What is Philosophy?,* trans. Hugh Tomlinson and Graham Burchell (New York: Columbia University Press, 1994). For rare discussions of these connections, see Marcus A. Doel, *Poststructuralist Geographies: The Diabolical Art of Spatial Science* (Edinburgh: Edinburgh University Press, 1999); and Marcus A. Doel, "Gunnar Olsson's Transformers: The Art and Politics of Rendering the Co-Relation of Society and Space in Monochrome or Technicolor," *Antipode* 35 (2003): 140–67. See also Mark Bonta and John Protevi, *Deleuze and Geophilosophy: A Guide and Glossary* (Edinburgh: Edinburgh University Press, 2004).

18. Armin Wolf, "News on the Ebstorf World Map: Date, Origin, Authorship," in Monique Pelletier, *Géographie du monde au moyen âge at à la renaissance.*

19. Ernst Sommerbrodt, *Die Ebstorfer Weltkarte,* Text und Tafelband (Hannover: Historischen Vereins für Niedersachsen, 1891); Konrad Miller, *Mappae mundi: Die ältesten Weltkarten,* vol. 5: *Die Ebstorfkarte* (Stuttgart: Jos. Rot'sche, 1896). Attached to Miller's treatise is a wonderfully colored and easily readable reproduction of the map.

20. Most standard works mention the map, some in more detail than others. In depth treatments are in Uwe Ruberg, "Mappae mundi des Mittelaters in Zusammenwirkung von Text und Bild," in Christel Meier and Uwe Ruberg, eds., *Text und Bild: Aspekte des Zusammenwirkens zweier Künste in Mittelalter und früher Neuzeit* (Wiesbaden: Dr. Ludwig Reichert Verlag, 1980); Jörg-Geerd Arentzen, *Imago Mundi Cartographia;* Hartmut Kugler, "Die Ebstorfer Weltkarte: Ein europäisches Weltbild in deutschen Mittelater," *Zeitschrift für deutsches Altertum und deutsche Literatur* 116 (1987): 1–29; Birgit Hahn-Woernle, *Die Ebstorfer Weltkarte* (Ebstorf: Das Kloster, 1989); Hartmut Kugler, ed., *Ein Weltbild vor Columbus: Die Ebstorfer Weltkarte Interdisciplinäres Colloquium 1988* (Weinheim: VCH, Acta Humaniora, 1991); Jürgen Wilke, *Die Ebstorfer Weltkarte* (Bielefeld: Verlag für Regionalgeschichte, 2001).

21. Michael Camille, *Image on the Edge: The Margins of Medieval Art* (London: Reaktion Books, 1992).

22. Here it is instructive to recall that Aristotle thought of history as the study of what *has* happened and of poetry as the exploration of what *might* happen.

23. Anna-Dorothee von den Brinken, "Mappa mundi und Cronographia: Studien zur imago mundi des abendländischen Mittelalters," *Deutsches Archiv für Erforschung des Mittelalters* 24 (1968): 118–86.

24. Among all mappae mundi, the Hereford map is the best known. The literature is consequently large, but here it suffices to mention P. D. A. Harvey, *Medieval Maps* (London: British Library, 1991); P. D. A. Harvey, *Mappa Mundi: The Hereford World Map* (London: British Library, 1996); and Scott D. Westrem, *The Hereford Map: A Transcription and Translation of the Legends with Commentary* (Turhout: Brepols, 2001).

25. Samuel Beckett et al., *Our Exagmination Round His Factification for Incamination of Work in Progress* (Paris: Shakespeare and Co., 1929), 14.

26. For a historical overview of this idea, see Richard Padovan, *Proportion: Science, Philosophy, Architecture* (London: E. and F. N. Spon, 1999).

27. Umberto Eco, *Art and Beauty in the Middle Ages* (New Haven, Conn.: Yale University Press, 1986), 35f.

28. Vitruvius, *The Ten Books on Architecture,* trans. M. H. Morgan (New York: Dover, 1960), book 3, chap. 1, p. 72f.

29. What was held to be right and what left was evidently not always so easy to remember. For example, Revelation 10:12 tells about an angel who "planted his right foot on the sea and his left foot on the land."

30. The most thorough and impressive treatment of this problem is in Anna-Dorothee von den Brinken, *Fines Terrae: Die Enden der Erde und der vierte Kontinent auf Mittelalterlichen Weltkarten* (Hannover: Hahnsche Buchhandlung, 1992).

31. David John Athole Ross, *Alexander Historiatus: A Guide to Medieval Illustrated Alexander Literature* (London: Warburg Institute, 1963); W. J. Aertz et al., eds., *Alexander the Great in the Middle Ages: Ten Studies on the Last Days of Alexander in Literary and Historical Writing* (Nijmegen: Alfa Nijmegen, 1978); and A. B. Bosworth and E. J. Baynham, eds., *Alexander the Great in Fact and Fiction* (Oxford: Oxford University Press, 2000).

32. Also, Babylonia appears in two places, on one occasion as a bridge across the Euphrates, on another as an island in the Nile.

33. Alesssandro Scafi, "Mapping Eden: Cartographies of the Earthly Paradise," in Denis Cosgrove, ed., *Mappings* (London: Reaktion Books, 1999), 60.

34. Sir John Mandeville's highly influential *Travels and Voyages* did not appear until 1356. Although phrased as a first-person narrative, it is uncertain whether the author traveled at all or whether he conceived the world entirely in his study. At any rate, Mandeville claimed to have visited not only the realm of Prester John and the mythical Land of Darkness, but also the ten lost tribes of Israel imprisoned by Alexander the Great in the Caspian mountains. On the history of Mandeville see Josephine Waters Bennett, *The Rediscovery of Sir John Mandeville* (New York: Modern Language Association of America, 1954). On the history of the lost tribes, see Charles F. Beckingham and Bernard Hamilton, eds., *Prester John, the Mongols and the Ten Lost Tribes* (Aldershot: Variorum, 1996) as well as Tudor Parfitt, *The Lost Tribes of Israel: The History of a Myth* (London: Weidenfeld and Nicolson, 2002).

35. Hartmut Kugler, "Die Ebstorfer Weltkarte," 24f.

36. Richard Hennig, *Terrae incognitae,* vol. 2 (Leiden: Brill, 1950), chap. 89, which is entitled "Sallam der Dolmetscher und der 'Gog und Mogog-Mauer.'"

37. Andrew Runni Anderson, *Alexander's Gate, Gog and Magog, and the Inclosed Nations* (Cambridge, Mass.: Medieval Academy of America, 1932). Also see Andrew Gow, "Gog and Magog on Mappaemundi and Early Printed World Maps: Orientalizing Ethnography in the Apocalyptic Tradition," *Journal of Early Modern History* 2 (1998): 61–88.

38. These monstrous races constitute merely a sample of what there was thought to be. For details see John Block Friedman, *The Monstrous Races in Medieval Art and Thought* (Cambridge, Mass.: Harvard University Press, 1981), chap. 1; Claude Lecouteux, *Les monstres dans la pensée médiévale européenne: Essai de présentation* (Paris: Presses de l'Université de Paris-Sorbonne, 1993); and Michael Camille, *Mirror in the Parchment: The Luttrell Psalter and the Making of Medieval England* (London: Reaktion Books, 1998), esp. chaps. 5 and 6. It is significant that the Luttrell Psalter, the real subject of Camille's book, was made in order to consolidate the social position of the Luttrell family after its members had been accused of incest.

39. Friedman, *The Monstrous Races in Medieval Art and Thought,* 61.

40. So unclean were the dogs considered that David, in a psalm sung to the tune of "Doe in the Morning," once lamented that "Dogs have surrounded me; a band of evil men has encircled me, they have pierced my hands and my feet" (Ps. 22:16). In this connection also see Mary K. Wakeman, *God's Battle with the Monster: A Study in Biblical Imagery* (Leiden: E. J. Brill, 1973).

41. Although drawing on material from other parts of the world, the outstanding treatment of ceremonial centers and cardinal directions is still in Paul Wheatley, *The Pivot of the Four Quarters: A Preliminary Enquiry into the Origins and Character of the Ancient Chinese City* (Chicago: Aldine, 1971). Also, and perhaps even more of a masterpiece, Paul Wheatley, *The Places Where Men Pray Together: Cities in Islamic Lands, Seventh through the Tenth Centuries* (Chicago: University of Chicago Press, 2001).

42. The key passage is in Ezekiel 5:5, according to which "the Sovereign LORD says: This is Jerusalem, which I have set in the center of the nations, with countries all around her."

43. The importance of the main meridian—one end in the Far East, the other at the Straits of Gibraltar—can hardly be overemphasized. In that connection it is fascinating to recall also Joseph Conrad's novel *The Secret Agent* in which the terrorist plot is to shatter the world by bombing the Greenwich Observatory and thereby the main meridian.

44. Hoping that one day my granddaughter Idun Wahlgren might come across this passage, I fall to the temptation of mentioning Olof Rudbeck (1630–1702). It was this scientist / entrepreneur (among other achievements discoverer of the lymphatic vessels and builder of Uppsala's Anatomical Theater), who in his *Atlantica* developed a theory that Sweden was the lost Atlantis, hence the cradle of Western civilization. As an ingredient in that fantastic brew of mythology and homemade etymology, he "proved" not only that the letters of the Greek alphabet were younger than the Nordic runes but that the former were outgrowths of the latter. As part of this heroic argument he then claimed that the golden apples guarded by the three nymphs of the Hesperidian Garden in actuality were the same apples—somehow turned into runes, primary means of remembrance—that kept the Nordic gods forever young. The guardian of these marvels was none less than Idun, namesake of my own grandchild, her basket serving as a kind of thesaurus sapientiae. Had Rudbeck known of the Ebstorf map, he most likely would have stressed that the Mercurian serpent which frames the Hesperidian Garden in fact was the tail-biting snake that structures virtually all runestones. At the same time it would have upset him that even though both Norway and Sameland are clearly shown on the map, Sweden is harder to find. For a magisterial treatment of Rudbeck and his epoch see Gunnar Eriksson, *Rudbeck 1630–1702: Liv, lärdom, dröm i barockens Sverige* (Stockholm: Atlantis, 2002), esp. 332ff.

45. Anna-Dorothee von den Brinken, "Ebstorf Map," in John Block Friedman and Kristen Mossler Figg, eds., *Trade, Travel, and Exploration in the Middle Ages: An Encyclopedia* (New York: Garland, 2000), 162.

46. For more on the widespread use of the Vitruvian Man, see Herbert von Einem, *Der Mainzer Kopf mit der Binde: Arbeitsgemeinschaft für Forschung des Landes Nordrhein-Westfalen, Geisteswissenschaften,* Heft 3 (Köln: Westdeutscher Verlag, 1955).

47. Thomas Pynchon, *Mason & Dixon,* 345f.

## SAUSSUREAN BAR

1. For another attempt to stop the sign in its own tracks see Umberto Eco, *Semiotics and the Philosophy of Language* (London: Macmillan, 1984), chap. 1.

2. Ferdinand de Saussure, *Course in General Linguistics,* trans. and annotated by Roy Harris (London: Duckworth, 1983), [98].

3. Immanuel Kant, *Critique of Pure Reason,* trans. F. Max Müller (Garden City, N.Y.: Doubleday, 1966), B203.

4. Kris Kristoferson, *Border Lord* (New York: Resaca, 1972).

5. Roland Barthes, *A Lover's Discourse: Fragments,* trans. Richard Howard (New York: Hill and Wang, 1978), 73.

6. Ulf Linde, *Efter hand: Texter 1950–1985* (Stockholm: Bonniers, 1985), 308.

7. Kant, *Critique of Pure Reason,* trans. Müller, A51/B75.

8. Gunnar Olsson, *Birds in Egg/Eggs in Bird* (London: Pion, 1980), chap. 12.

9. Hans Bohr, "My Father," in S. Rozenthal, ed., *Niels Bohr* (New York: Wiley, 1967), 328.

10. Harald Høffding, *A History of Philosophy,* trans. B. E. Meyer (New York: Dover, 1955), 288.

11. For a penetrating rendering of these relations see the drama of Michael Frayn, *Copenhagen* (New York: Methuen, 1998).

12. In this context it is telling that the traditional DNA-alphabet of genetics consists of only four letters, A, C, G and T. See, e.g., Jacques Monod, *Chance and Necessity: An Essay on the Natural Philosophy of Modern Biology,* trans. Austryn Wainhouse (New York: Knopf, 1971). Also the classical tale in James D. Watson, *The Double Helix* (New York: New American Library, 1969).

13. Anders Piltz, "Ansikte, mask, person: Om den europeiska människobilden," in Olle Matsson et al., eds., *Libens merito: Festskrift till Stig Strömholm* (Uppsala: Kungliga Vetenskapsamhällets i Uppsala Handlingar, 2001), vol. 21, 357. On this point Piltz refers explicitly to Albertus Magnus.

14. For a creative and highly significant treatment of these issues see F. G. Asenjo, *In-Between: An Essay on Categories* (Washington, D.C.: University Press of America, 1988).

15. This image ties neatly into Saussure's comment that "just as it is impossible to take a pair of scissors and cut one side of a paper without at the same time cutting the other, so it is impossible in language to isolate sound from thought, thought from sound." Saussure, *Course in General Linguistics,* [157].

16. Mircea Eliade, *The Sacred and the Profane,* trans. Willard R. Trask (New York: Harcourt, Brace and Company, 1959), 34.

## A = B

1. For the background of my own views, see Gunnar Olsson, *Birds in Egg/Eggs in Bird* (London: Pion, 1980), chaps. 3 and 16. Some sentences of the present section are more or less lifted from that discussion.

2. My own views of dialectics were formed long ago through intensive readings of G. W. F. Hegel, *The Phenomenology of Mind,* trans. J. B. Baillie (New York: Harper Torchbooks, 1967). In these studies I was much helped by the bat-like analyses of Bertell Ollman, *Alienation: Marx's Conception of Man in Capitalist Society* (Cambridge: Cambridge

University Press, 1971). For elaborations see also the highly influential interpretation in Alexandre Kojève, *Introduction to the Reading of Hegel: Lectures on "The Phenomenology of Spirit,"* trans. James H. Nichols Jr (Ithaca, N.Y.: Cornell University Press, 1980). The most clearheaded account of the Hegelian (including the Kierkegaardian and Heideggerian) legacy in French philosophy is Mark C. Taylor, *Altarity* (Chicago: University of Chicago Press, 1987).

3. Gottlob Frege, "Über Sinn und Bedeutung," *Zeitschrift für Philosophie und philosophische Kritik* 100 (1892): 25–50. Translated by Herbert Feigl as "On Sense and Nominatum," in Herbert Feigl and W. S. Sellars, eds., *Readings in Philosophical Analysis* (New York: Appleton-Century-Crofts, 1949). Also translated by Max Black originally as "On Sense and Reference," in a later edition changed into "On Sense and Meaning." The latter version is in Gottlob Frege, *Translations from the Philosophical Writings of Gottlob Frege,* ed. Peter Geach and Max Black (Oxford: Blackwell, 1980).

4. In Geach and Black, eds., *Translations from the Philosophical Writings of Gottlob Frege,* 80.

5. See especially the highly influential works by Michael Dummett, in particular *Frege: Philosophy of Language* (New York: Harper and Row, 1973), *Frege: Philosophy of Mathematics* (Cambridge, Mass.: Harvard University Press, 1991), and *The Interpretation of Frege's Philosophy* (London: Duckworth, 1981). However, also note the alternative interpretation in Saul A. Kripke, *Naming and Necessity* (Oxford: Blackwell, 1981), perhaps the most influential work in philosophy since the death of Wittgenstein; a comprehensive discussion is in Christopher Hughes, *Kripke: Names, Necessity and Identity* (Oxford: Oxford University Press, 2004).

6. And this explains not only Umberto Eco's scathing critique of Frege but also his warm embrace of Peirce. See Umberto Eco, *Den frånvarande strukturen: Introduktion till den semiotiska forskningen,* trans. Estrid Tenggren (Lund: Bo Cavefors, 1971). Also Umberto Eco, *A Theory of Semiotics* (Bloomington: Indiana University Press, 1976).

7. Alfred North Whitehead and Bertrand Russell, *Principia Mathematica,* 2d ed. (Cambridge: Cambridge University Press, 1925); and Bertrand Russell, *Introduction to Mathematical Philosophy* (London: George Allen and Unwin, 1920).

8. Ludwig Wittgenstein, *Tractatus Logico-Philosophicus,* trans. D. F. Pears and B. F. McGuiness (London: Routledge and Kegan Paul, 1961), 3.203.

9. Bertrand Russell, "Meinong's Theory of Complexes and Assumptions," *Mind* 13 (1904): 204.

10. William Shakespeare, *Romeo and Juliet,* 2.2.33–34, 37–48. Quoted from *The Complete Works of William Shakespeare,* rev. ed., ed. Hardin Craig and David Bevington (Glenview, Ill: Scott, Foresman and Co., 1973).

11. Reviel Netz, "The Aesthetics of Mathematics: A Study," in Paolo Mancosu et al., eds., *Visualization, Explanation and Reasoning Styles in Mathematics* (Dordrecht: Springer, 2005), 281.

12. This is in fact another point of contact between Kierkegaard and myself. See especially Søren Kierkegaard, *Repetition,* trans. Walter Lowrie (Princeton, N.J.: Princeton University Press, 1962). It should be explicitly noted that Kierkegaard's pseudonym in that work was "Constantin Constantinus."

13. For the early history of the techniques of remembering see, for instance, Frances A. Yates, *The Art of Memory* (Chicago: University of Chicago Press, 1966) as well as the more focused discussion in Frances A. Yates, *Giordano Bruno and the Hermetic Tradition* (Chicago: University of Chicago Press, 1978).

14. For an up-to-date discussion see Lawrence Sklar, *Theory and Truth: Philosophical Critique within Foundational Science* (Oxford: Oxford University Press, 2000).

15. In my own view actually a triple, for the sign consists not only of Signifier and signified but equally much of the bar in-between. While mapping by definition is an exercise in the art of triangulation, so meaning is a triadic relation à la Peirce. On that last point see, e.g., Charles Sanders Peirce, "The Principles of Phenomenology," *Philosophical Writings of Peirce,* ed. Justus Buchler (New York: Dover, 1955), 91.

16. Saussure, *Course in General Linguistics,* [166].

17. Ibid., [169].

18. Ibid.

19. The most comprehensive theory of translation I know is in Dinda L. Gorlé, *Semiotics and the Problem of Translation: With Special Reference to the Semiotics of Charles S. Peirce* (Amsterdam: Rodopi, 1994). Gorlé's point of departure is in Peirce's statement that "the meaning of a sign is its translation into another equivalent or perhaps more developed sign" (quoted at 27).

20. Homer, *The Odyssey,* trans. Robert Fitzgerald (Garden City, N.J.: Doubleday, 1963), book 9, p. 157.

21. Jacques Derrida, *Speech and Phenomena: And Other Essays on Husserl's Theory of Signs,* trans. David B. Allison (Evanston, Ill.: Northwestern University Press, 1973), 142f. In this context of *spacing* it is extremely important to note that Derrida repeatedly acknowledged how the lessons from his first book—an introduction to and translation of Husserls' *The Origin of Geometry*—had a decisive influence on whatever came thereafter, partly because for Husserl there could be no truth without judgment. For elaborations see especially Jacques Derrida, "The Time of a Thesis: Punctuations," in Alan Montefiore, ed., *Philosophy in France Today* (Cambridge: Cambridge University Press, 1983). For another testimonial on the formative influence of an early engagement with the axiomatics of geometry, see Michel Serres, *Les origines de la géométrie* (Paris: Flammarion, 1993). My own roots in the same tradition are explored in Gunnar Olsson, "Glimpses," in Peter Gould and Forrest R. Pitts, eds., *Geographical Voices* (Syracuse, N.Y.: Syracuse University Press, 2002).

22. Robert Kaplan, *The Nothing That Is: A Natural History of Zero* (Oxford: Oxford University Press, 2001); and Charles Seife, *Zero: The Biography of a Dangerous Idea* (New York: Penguin, 2000). The intricate relations between cartographical reason, the zero, the map and the labyrinth are creatively explored in Franco Farinelli, "The Geography of Zero," in Marco Picone, ed., *Bodies and Space: Gunnar's Travels* (Palermo: Università di Palermo, Laboratorio Geografico, 2002).

23. Especially Jacques Derrida, *Dissemination,* trans. Barbara Johnson (London: Athlone Press, 1981).

24. For a detailed discussion see Derrida, *Dissemination,* 130ff. As an amplification I will simply add that once placed on the altar of metaphysics, sacrificial offerings turn to cleansing smoke, rites of purification to hegemonic acts of social survival. To illustrate, the Swedish for "poison" is *gift,* a word which also means "married," itself clearly connected with the English "gift" meaning "present," originally in the form of "payment for a wife."

25. For an interesting attempt to rewrite the early history of rhetoric see Edward Schiappa, *The Beginnings of Rhetorical Theory in Classical Greece* (New Haven, Conn.: Yale University Press, 1999).

26. Friedrich Nietzsche, "On Truth and Lie in an Extra-Moral Sense," *The Portable Nietzsche,* ed. Walter Kaufmann (Harmondsworth: Penguin, 1976), 46f.

27. Charles Sanders Peirce, "How to Make our Ideas Clear," *Philosophical Writings of Peirce,* 38.

28. Charles Sanders Peirce, "Logic as Semiotic: The Theory of Signs," *Philosophical Writings of Peirce,* 99, 103.

29. A similar point was persuasively argued in Stephen Toulmin, *The Uses of Argument* (Cambridge: Cambridge University Press, 1964).

## QUOD ERAT

1. Not so easy to translate, but to mind comes a mélange of sentences like "Thinking freely is great, but thinking correctly is greater"; "Thinking free is big, but thinking right is bigger"; etc.

2. Aristotle, *De memoria et reminiscentia,* trans. J. I. Beare, in *The Basic Works of Aristotle,* ed. Richard McKeon (New York: Random House,1941), 450a.

3. Immanuel Kant, *Critique of Pure Reason,* trans. Norman Kemp Smith (London: Palgrave Macmillan, 2003), B xi–B xii.

4. For detailed and most instructive discussions, see David Rapport Lachterman, *The Ethics of Geometry: A Genealogy of Modernity* (New York: Routledge, 1989), esp. 50–67.

5. Ludwig Wittgenstein, *Philosophical Investigations,* trans. G. E. M. Anscombe (Oxford: Basil Blackwell, 1953), 219.

6. Ludwig Wittgenstein, *Tractatus Logico-Philosophicus,* trans. D. F. Pears and B. F. McGuiness (London: Routledge and Kegan Paul, 1961), 6.124 and 6.1251.

7. "Everything that makes philosophical beginning possible we must first acquire by ourselves." Edmund Husserl, *Cartesian Meditations: An Introduction to Phenomenology,* trans. Dorion Cairns (The Hague: Martinus Nijhoff, 1973), 13.

8. In the words of the young Socrates responding to Zeno: "What is there surprising in someone pointing out that I am one thing and also many? When he wants to show that I am many things, he can say that my right side is a different thing from my left, my front from my back, my upper parts from my lower, since no doubt I partake of plurality. When he wants to prove that I am one thing, he will say that I am one person among the seven of us, since I partake also of unity. So both statements are true." Plato, *Parmenides* 129c, trans. F. M. Cornford. Quotation from *Collected Dialogues of Plato,* ed. Edith Hamilton and Huntington Cairns, Bollingen Series 71 (Princeton, N.J.: Princeton University Press, 1961).

9. José Ortega y Gasset, *Man and People,* trans. Willard R. Trask (New York: Norton, 1957), chap. 6. For an instructive discussion of the differences between Husserl and Ortega y Gasset, see Sharon P. Rider, *Avoiding the Subject: A Critical Inquiry into Contemporary Theories of Subjectivity* (Stockholm: Thales, 1998), 47ff.

10. For an early treatment of how our use of language embodies the power relations between the sexes, see Robin Lakoff, *Language and Woman's Place* (New York: Harper and Row, 1975).

11. In the former case the elevation to the rank of Commander of the Order of Geometry (COG) is appropriate, in the latter the more common distinction of Knight

of the Order of Geometry (KOG) is sufficient. The main difference is that while the COG is conferred in recognition of unusual bravery, the KOG is in recognition of long and loyal service.

12. David Rapport Lachterman, *The Ethics of Geometry,* 66.

13. Ludwig Wittgenstein, *Zettel,* trans. G. E. M. Anscombe and G. H. von Wright (Berkeley and Los Angeles: University of California Press, 1979), 393.

14. In this rare case, elevation to the rank of Commander of the Order of Invention (COI) is most appropriate. Chain around the neck, band across the chest, star on the uniform. The COI is the highest honor a thought can ever receive. *Honi soit qui mal y pense,* the fitting motto of THE NOBLE ORDER OF THE GARTER.

15. The rhetorical power of visual display is brilliantly shown (self-exemplifying) in three wonderful volumes by Edward R. Tufte, the first devoted to the depiction of numbers (statistical data), the second to the picturing of nouns (including the things on the ground that go into conventional maps and aerial photographs), the third to the picturing of verbs (in my terminology the mapmaker's scale or the scientist's causal explanation). See Edward R. Tufte, *The Visual Display of Quantitative Information.* 2d ed. (Cheshire, Conn.: Graphics Press, 2001); *Envisioning Information* (Cheshire, Conn.: Graphics Press, 1990); and *Visual Explanations: Images and Quantities, Evidence and Narrative* (Cheshire, Conn.: Graphics Press, 1997).

16. These comments are essentially in line with the views of the so-called new rhetoric, a field defined as a theory of argumentation which has for its objects of study the various discursive techniques through which we are tied to the collective unconscious of the socially taken-for-granted. The most influential presentation is in Chaïm Perelman and L. Olbrechts-Tyteca, *The New Rhetoric: A Treatise on Argumentation,* trans. John Wilkinson and Purcell Weaver (Notre Dame, Ind., and London: University of Notre Dame Press, 1971). See also Mats Rosengren, *Psychagōgia—konsten att leda själar: Om konflikten mellan retorik och filosofi hos Platon och Chaïm Perelman* (Göteborg: Filosofiska institutionen, Göteborgs universitet, 1997).

17. In the development of this attitude I have worked closely with my former student and colleague, a man who in Sweden is usually named José Ramirez, sometimes José L. Ramírez, by his Spanish friends José Luis Ramírez, by his close family José Luis Ramírez y Gonzàlez. *Kärt barn har många namn.* For his characteristic take on these issues—especially as they relate to the heritage from Aristotle—see José L. Ramírez, *Skapande mening: En begreppsgenealogisk studie om rationalitet, vetenskap och planering* (Stockholm: Nordplan, 1995).

18. Christopher Norris, *Derrida* (London: Fontana, 1987), 19.

19. In Derrida's terminology: "We can no more imagine effective speech without there being self-representation than we can imagine a representation of speech without there being effective speech. . . . Speech represents itself; it *is* its representation. Even better, speech is *the* representation of itself." Jacques Derrida, *Speech and Phenomena,* trans. David B. Allison (Evanston, Ill.: Northwestern University Press, 1973), 57.

20. The three quotations all come from Ignatius Loyola, *The Spiritual Exercises* (Wheathampstead: Anthony Clark, 1973), 191, 192, 194.

21. The multiple texts of *The Spiritual Exercises* are beautifully delineated in Roland Barthes, *Sade, Fourier, Loyola,* trans. Richard Miller (New York: Hill and Wang, 1976). For instance, Barthes notes that the whole point of the well-designed exercise is to solicit God for a determining sign. "The invention of a language, this then is the object of the *Exercises*" (48).

22. For a detailed discussion see Umberto Eco, *A Theory of Semiotics* (Bloomington: Indiana University Press, 1976), 279ff.

23. Lewis Carroll, *Alice in Wonderland,* in *The Annotated Alice,* intro. and notes by Martin Gardner (New York: Clarkson N. Potter, 1960), chap. 6, 90f.

24. Wittgenstein, *Tractatus Logico-Philosophicus,* 4.0031. For more on Mauthner himself—and not only on Wittgenstein's interpretation—see Fritz Mauthner, *Beiträge zu einer Kritik der Sprache* (Stuttgart: J. G. Gotta, 1901–1902); and Helmut Henne und Kristine Kaiser, eds., *Fritz Mauthner—Sprache, Literatur, Kritik: Festakt und Symposion zu seinem 150. Geburtstag* (Tübingen: Niemeyer, 2000).

25. Jacques Lacan, *Écrits: A Selection,* trans. Alan Sheridan (London: Tavistock, 1977), 157.

26. Michel Foucault, *The History of Sexuality,* vol. 1: *An Introduction,* trans. Robert Hurley (Harmondsworth: Penguin, 1981), 93.

27. Michel Foucault, *Discipline and Punish: The Birth of the Prison,* trans. Alan Sheridan (Harmondsworth: Penguin, 1979), 30.

28. Julia Kristeva, "The Speaking Subject," in Marshall Blonsky, ed., *On Signs* (Oxford: Basil Blackwell, 1985), 217f.

29. Pierre Bourdieu, *Outline of a Theory of Practice,* trans. Richard Nice (Cambridge: Cambridge University Press, 1977). In this context see Edward S. Casey, "Between Geography and Philosophy: What Does It Mean to Be in the Place-World," *Annals of the Association of American Geographers* 91 (2001): 683–93. Also Casey's response to his critics (Nicholas Entrikin, Theodore Schatzki, and Barbara Hooper) all printed in the same issue of the same journal.

30. Perelman and Olbrechts-Tyteca, *The New Rhetoric,* pars. 22–28.

31. For insightful comparisons between these crucial figures in linguistic thought, see Roy Harris, *Language, Saussure and Wittgenstein: How to Play Games with Words* (London: Routledge, 1988).

32. Ludwig Wittgenstein, *Philosophical Grammar,* trans. Anthony Kenny (Oxford: Oxford University Press, 1974), 40.

33. Maurice Merleau-Ponty, *The Visible and the Invisible,* trans. Alphonso Lingis (Evanston, Ill.: Northwestern University Press, 1968), 155. Also the in-depth treatment of the body in Maurice Merleau-Ponty, *Phenomenology of Perception,* trans. Colin Smith (London: Routledge and Kegan Paul, 1962).

34. Roman Jakobson, "Two Aspects of Language and Two Types of Aphasic Disturbances," in Roman Jakobson and Morris Halle, *Fundamentals of Language* (The Hague: Mouton, 1956).

35. See for instance James Drake, "The Naming Disease: How Jakobson's Essay of Aphasia Initiated Postmodernist Deceits," *Times Literary Supplement,* September 4, 1998.

36. Jakobson and Halle, *Fundamentals of Language,* 71.

37. Wittgenstein, *Philosophical Investigations,* 123 and 203.

38. Gaston Bachelard, *The Poetics of Space,* trans. Maria Jolas (Boston: Beacon, 1969), 6 and 9. For a history of how to stabilize memory, see especially Frances A. Yates, *The Art of Memory* (Chicago: University of Chicago Press, 1966). While Quintilian is the classical reference point, a more recent illustration is in Pauli Tapani Karjalainen, "Place and Intimate Sensing," *Nordisk Geografisk Tidskrift* 27 (1998): 3–16. The latter text is essentially an analysis of Bo Carpelan, *Urwind* (Esbo: Schildt, 1993).

39. Umberto Eco, *Semiotics and the Philosophy of Language* (London: Macmillan, 1984), 45, emph. added.

40. Paul Ricoeur, *Memory, History, Forgetting,* trans. Kathleen Blamey and Paul Pellauer (Chicago: University of Chicago Press, 2004), 43.

41. Here quoted in English from Mikkel Borch-Jacobsen, *Lacan: The Absolute Master* (Stanford, Calif.: Stanford University Press, 1991), 194. French original in Maurice Blanchot, *De Kafka à Kafka* (Paris: Gallimard, 1982), 37.

42. Gunnar Olsson, "On Yearning for Home: An Epistemological View of Ontological Transfomations," in D. C. D. Pocock, ed., *Humanistic Geography and Literature* (London: Croom Helm, 1981). Also Karin Johannisson, *Nostalgia: En känslans historia* (Stockholm: Bonniers, 2001).

43. Here quoted from Peter Barber and Christopher Board, *Tales from the Map Room: Fact and Fiction About Maps and Their Makers* (London: BBC Books, 1993), 20.

## PLATO

1. Quoted in John Sallis, *Force of Imagination: The Sense of the Elemental* (Bloomington: Indiana University Press, 2000), 90.

2. Mary Carruthers, *The Craft of Thought: Meditation, Rhetoric, and the Making of Images, 400–1200* (Cambridge: Cambridge University Press, 1998), 8.

3. William Shakespeare, *A Midsummer Night's Dream,* 5.1.7–8 and 14–17. From *The Complete Works of William Shakespeare,* ed. Hardin Craig and David Bevington (Glenview, Ill.: Scott, Foresman and Company, 1973). Text references are to act, scene, and line of this edition.

4. The problems are formidable, but general overviews are in Jean-Paul Sartre, *Imagination,* trans. W. Williams (Ann Arbor: University of Michigan Press, 1962); Edward S. Casey, *Imagining: A Phenomenological Study* (Bloomington: Indiana University Press, 1976); Edward S. Casey, *Remembering: A Phenomenological Study* (Bloomington: Indiana University Press, 1987); Eva T. H. Brann, *The World of the Imagination: Sum and Substance* (Lanham, Md.: Rowman and Littlefield, 1991); J. M. Cokking, *Imagination: A Study in the History of Idea* (London: Routledge, 1991); Richard Kearney, *The Wake of Imagination* (London: Routledge, 1994); Richard Kearney, *Poetics of Imagining: Modern to Post-modern* (Edinburgh: Edinburgh University Press, 1998); John Sallis, *Force of Imagination.* Also several of the contributions to Kenneth Maly, ed., *The Path of Archaic Thinking: Unfolding the Work of John Sallis* (Albany: State University of New York Press, 1995).

5. Harold Bloom, *Shakespeare: The Invention of the Human* (New York: Riverhead Books, 1998), 517.

6. Immanuel Kant, *Critique of Pure Reason,* trans. F. Max Müller (Garden City, N.Y.: Doubleday, 1966), B 151.

7. Ibid., B 157.

8. Brian Rotman, *Signifying Nothing: The Semiotics of Zero* (London: Macmillan, 1987).

9. G. Wilson Knight, *The Wheel of Fire: Essays in the Interpretation of Shakespeare's Sombre Tragedies* (London: Oxford University Press, 1930), 174, emph. added.

10. Leif Zern, *Älskaren och mördaren* (Stockholm: Alba, 1984), 184.

11. Marilyn French, *Shakespeare's Division of Experience* (New York: Ballantine, 1981), 247 and 251.

12. Richard Lovelace, "To Althea from Prison," *The Poems of Richard Lovelace,* ed. C. H. Wilkinson (Oxford: Clarendon Press, 1930), 79.

13. Murray Wright Bundy, *The Theory of Imagination in Classical and Medieval Thought* (Urbana: Illinois Studies in Language and Literature, vol. 12, nos. 2–3, 1927), 18.

14. Throughout the previous pages I have searched for a place to refer to the exciting work of Eugenio Trías, in some ways very similar to my own, in some other ways very different; while for both of us the concept of limits lies at the heart of our respective investigations, and while for both of us the synthesis of the Jewish and the Greek is pivotal, his approach is essentially philosophical, mine cartographical. Among his most relevant studies are Eugenio Trías, *La filosofía y su sombra* (Barcelona: Editorial Seix Barral, 1969), English trans. *Philosophy and its Shadow,* trans. Kenneth Krabbenhoft (New York: Columbia University Press, 1983); *Lógica del límite* (Barcelona: Destino, 1991); *La razón fronteriza* (Barcelona: Destino, 1999); *Los límites del mundo* (Barcelona: Destino, 2000); *Pensar la religión* (Barcelona: Destino, 1997), Swedish trans., *Att tänka det religiösa,* trans. Anna-Lena Renqvist (Stockholm: Thales, 2003).

15. For insightful comments on the differences between Plato's and Aristotle's conceptions of memory and imagination, see Paul Ricoeur, *Memory, History, Forgetting,* trans. Katleen Blamey and David Pellauer (Chicago: University of Chicago Press, 2004), chap. 1.

16. Aristotle, *De anima* 433b25. From the translation of J. A. Smith in *The Basic Works of Aristotle,* ed. Richard McKeon (New York: Random House, 1941).

17. Aristotle, *Metaphysics* 1078b14–19. From the translation of W. D. Ross in *The Basic Works of Aristotle.*

18. For further discussions of the various desires that rule the wisdom-lovers (the philosopher-kings), the honor-lovers (the guardians), the money-lovers and the food/drink/sex-lovers (the producers) see C. D. C. Reeve, *Philosopher-Kings: The Argument of Plato's "Republic."* (Princeton, N.J.: Princeton University Press, 1988), chap. 2.2.

19. Ludwig Wittgenstein, *Zettel,* 2d ed., ed. G. E. M. Anscombe and G. H. von Wright (Oxford: Basil Blackwell, 1981), 621, 628 and 632.

20. Plato, *Republic* 518a. Quotation From *The Republic of Plato,* trans. Francis MacDonald Cornford (Oxford: Oxford University Press, 1945), 232.

21. Jeremy W. Crampton, "Thinking Philsophically in Cartography: Toward a Critical Politics of Mapping," *Cartographic Perspectives* 42 (2002): 12–31.

22. Plato, *Republic* 506d–e. In this and subsequent quotations I draw on a range of translations, each of them published in several editions. Especially valuable are *The Republic of Plato,* trans. Allan Bloom (New York: Basic Books, 1991); *The Republic of Plato,* trans. Francis MacDonald Cornford (Oxford: Oxford University Press, 1945); *Plato: The Republic,* trans. Paul Shorey (Cambridge, Mass.: Harvard University Press, 1956); *Plato: The Republic,* trans. Desmond Lee (London: Penguin, 2003); *Plato's Republic,* trans. G. M. A. Grube (Indianapolis, Ind.: Hackett, 1974); *The Republic,* trans. Tom Griffith and Giovanni R. F. Ferrari (Cambridge: Cambridge University Press, 2000). In addition, the excellent Swedish version in *Staten,* trans. Jan Stolpe (Stockholm: Atlantis, 2003).

23. Adi Ophir, *Plato's Invisible Cities: Discourse and Power in the "Republic"* (London: Routledge, 1991), 8.

24. Mark S. Monmonier, *How to Lie with Maps,* 2d ed. (Chicago: University of Chicago Press, 1996).

25. Franco Farinelli, *Geografia: Un introduzione ai modelli del mondo* (Torino: Einaudi, 2003), 26–33. The key reference is C. K. Ogden and I. A. Richards, *The Meaning of Meaning: A Study of The Influence of Language upon Thought and of The Science of Symbolism,* 4th ed. rev. (London: Kegan Paul, Trench, Trubner and Co., 1936), chap. 1.

26. Not only was there the dystopia of George Orwell's *Animal Farm,* but also the ideal city of Plato's *Republic.* In the latter, Socrates told Glaucon, "we'll need more servants too. Or doesn't it seem there will be need of teachers, wet nurses, governesses, beauticians, barbers, and, further, relish-makers and cooks? And what's more, we're in addition going to need swineheards. This animal wasn't in our earlier city—there was no need—but in this one there will be need of it in addition. And there'll also be need of very many other fatted beasts if someone will eat them, won't there?" (373c). Then, in a slightly different context, he tells Adeimantus, his own and Glaucon's brother, that what is to be blamed first and foremost is "when a man in speech makes a bad presentation of what gods and heroes are like, just as a painter who paints something that doesn't resemble the things whose likeness he wished to paint. . . . [Not even small lies] should so easily be told to thoughtless young things; best would be to keep quiet, but if there were some necessity to tell, as few as possible ought to hear them as unspeakable secrets, after making a sacrifice, not of a pig but of some great offering that's hard to come by" (377d–78a, trans. Bloom).

27. Plato, *Republic* 507b.

28. Aristotle, *De Anima* 412b–413 a. From the translation of J. A. Smith in *The Basic Works of Aristotle.*

29. Plato, *Republic* 509b, trans. Cornford.

30. Bloom, *The Republic of Plato,* xvii.

31. A total of eight pages in *The Oxford English Dictionary.*

32. Ogden and Richards: *The Meaning of Meaning,* esp. chap. 9.

33. Christine Buci-Glucksmann, *L'œil cartographique de l'art* (Paris: Galilée, 1996), 60.

34. Philip Steadman, *Vermeer's Camera: Uncovering the Truth Behind the Masterpieces* (Oxford: Oxford University Press, 2001). Also Arthur K. Wheelock Jr, *Vermeer and the Art of Painting* (New Haven, Conn.: Yale University Press, 1995), chap. 13; and Bryan Jay Wolf, *Vermeer and the Invention of Seeing* (Chicago: University of Chicago Press, 2001).

35. For an outstanding treatment of the relations between perspective and composition, see Thomas Puttfarken, *The Discovery of Pictorial Composition: Theories of Visual Order in Painting, 1400–1800* (New Haven, Conn.: Yale University Press, 2000), esp. part 2.

36. Samuel Y. Edgerton Jr, *The Renaissance Rediscovery of Linear Perspective* (New York: Basic Books, 1975).

37. For comments on the original text, see David C. Lindberg, *Roger Bacon and the Origins of Perspectiva in the Middle Ages: A Critical Edition and English Translation of Bacon's "Perspectiva"* (Oxford: Clarendon Press, 1996). In addition, Lindberg's introduction serves as a useful presentation of Bacon's time and life as a whole.

38. Roger Bacon, *The Opus Majus of Roger Bacon,* trans. John Henry Bridges (Oxford: Clarendon Press, 1897), 1:211.

39. Leon Battista Alberti, *On Painting,* trans. C. Grayson, intro. M. Kemp (London: Penguin, 1991), 94–95, emph. added.

40. Anthony Grafton, *Leon Battista Alberti: Master Builder of the Italian Renaissance* (London: Penguin, 2002), 119.

41. Samuel Y. Edgerton Jr, *The Heritage of Giotto's Geometry: Art and Science on the Eve of the Scientific Revolution* (Ithaca, N.Y.: Cornell University Press, 1991).

42. The very word luxury is a reflection of God's *lux.*

43. Indications are that the actual invention might have occurred already in 1413.

44. A prototype which a decade later was codified in Alberti's *Della pittura*. For an English version see Alberti, *On Painting*. Many relevant comments are in James S. Ackerman, *Distance Points: Essays in Theory and Renaissance Art and Architecture* (Cambridge, Mass.: MIT Press, 1991), chaps. 4 and 6.

45. The threshold reference is Erwin Panofsky, *Perspective as Symbolic Form*, trans. Christopher S. Wood (New York: Zone Books, 1991), a book which itself is best characterized as a long footnote to Ernst Cassirer, *The Philosophy of Symbolic Forms*, trans. Ralph Manheim (New Haven, Conn.: Yale University Press, 1953–57). The most up-to-date treatments are Hubert Damisch, *The Origin of Perspective* (Cambridge, Mass.: MIT Press, 1994); Martin Kemp, *The Science of Art: Optical Themes in Western Art from Brunelleschi to Seurat* (New Haven, Conn.: Yale University Press, 1990), esp. chap. 1 and the two appendicies; and Thomas Puttfarken, *The Discovery of Pictorial Composition*, esp. chaps. 3 and 10. However also see John White, *The Birth and Rebirth of Pictorial Space*, 2d ed. (London: Faber, 1967), chap. 8; Samuel Y. Edgerton Jr, *The Renaissance Rediscovery of Linear Perspective*, esp. 124–52; Martin Kemp, "Science, Non-Science and Nonsense: The Interpretation of Brunelleschi's Perspective," *Art History* 1 (1978), 134–61; Martin Kemp, *Behind the Picture: Art and Evidence in the Italian Renaissance* (New Haven, Conn.: Yale University Press, 1997), esp. chap. 3.

46. One *braccio* is equivalent to about 58 centimeters.

47. Antonio di Tuccio Manetti, *The Life of Brunelleschi by Antonio di Tuccio*, intro. and ed. Howard Saalman, trans. Catherine Engass (University Park: Pennsylvania State University Press, 1970), 42–44, emph. added.

48. This issue of the mirror has created much controversy centered on the question of whether the panel painting was made via a mirror, as is normally the case with self-portraits. Unfortunately Manetti's description cannot be used to settle the controversy, because even "if Brunelleschi had painted his perspective directly on a mirror, or with the aid of a mirror, or even . . . by using the *velum* procedure as described by Alberti, we would still be at a loss as to how . . . he would have been able to determine the location of the vanishing point to pierce the hole necessary for the experiment. Thus Brunelleschi's first experiment did indeed correspond to a kind of mirror stage of painting, but one that implied no magic" (Damisch, *The Origin of Perspective*, 138). For discussions about these matters I am much indebted to Jette Hansen-Møller and Ole Michael Jensen.

49. Antonio di Tuccio Manetti, *The Life of Brunelleschi by Antonio di Tuccio*, 44.

50. Damisch, *The Origin of Perspective*, 91.

51. Puttfarken, *The Discovery of Pictorial Composition*, 76.

52. Damisch, *The Origin of Perspective*, 119.

53. Michel Serres, *Le Système de Leibniz et ses modèles mathémathiques* (Paris: Presses Universitaires de France, 1968), 693, quoted in Damisch, *The Origin of Perspective*, 49.

54. David Rapport Lachterman, *The Ethics of Geometry: A Genealogy of Modernity* (New York: Routledge, 1989).

55. On this observation and its relation first to Brunelleschi's experiments, then to the mysticism of Nicolas Cusanus, see the sensitive discussion in Alessandra Bonazzi, "Per una storia delle origini della prosettiva: il mondo e l'immagine," in Giulia de Spuches, ed., *Atlante virtuali* (Palermo: Università di Palermo, Laboratorio Geografico, 2002).

56. Ulf Linde, "Perifera notiser om centralperspektivet," *Artes* 70, no. 3 (1991): 45–64.

57. "Therefore, those who ask for an irresistible argument, and at the same time

demand to be called to account for their views, must guard themselves by saying that the truth is not what appears to exist, but that what appears exists *for him to whom* it appears, and *when,* and *to the sense to which,* and *under the conditions under which* it appears. And if they give an account of their view, but do not give it in this way, they will soon find themselves contradicting themselves. For it is possible that the same thing may appear to be honey to the sight, but not to the taste, and that, since we have two eyes, things may not appear the same to each, if their sight is unlike. . . . [For] things do not appear either the same to all men or always the same to the same man, but often have contradictory appearances at the same time (for touch says there are two objects when we cross our fingers, while sight says there is one" (Aristotle, *Metaphysics* 1011a, trans. of W. D. Ross in *The Basic Works of Aristotle*).

58. For fascinating explorations of this theme see Joseph Leo Koerner, *The Moment of Self-Portraiture in German Renaissance Art* (Chicago: University of Chicago Press, 1993).

59. More exactly by the sight-line that connects the vanishing point of the seen and the standpoint of the seer.

60. Keimpe A. Algra, *Concepts of Space in Greek Thought* (Leiden: E. J. Brill, 1994); also Max Jammer, *Concept of Space: The History of Theories of Space in Physics,* 2d ed. (Cambridge, Mass.: Harvard University Press, 1969).

61. See *The Compact OED.*

62. Ross King, *Brunelleschi's Dome: The Story of the Great Cathedral in Florence* (London: Chatto and Windus, 2000).

63. General introductions are in Göran Schildt, *Cézanne* (Stockholm: Wahlström and Widstrand, 1946); Kurt Badt, *The Art of Cézanne,* trans. Sheila Ann Ogilvie (New York: Hacker Art Books, 1985); Meyer Schapiro, *Paul Cézanne* (New York: Harry N. Abrams, 1990); John Rewald, *Cézanne: A Biography* (New York: Harry N. Abrams, 1990); Ulrike Becks-Malorny, *Paul Cézanne, 1893–1906: Pioneer of Modernism,* trans. Phil Goddard (Köln: Taschen, 1995). More detailed analyses are in John Rewald, *The Paintings of Paul Cézanne: A Catalogue Raisonné,* vol. 1: *The Texts;* vol. 2: *The Plates* (New York: Harry N. Abrams, 1996).

64. The following paragraphs are slight modifications of what has previously appeared in Gunnar Olsson, "Constellations," *Sistema Terra* 8, issue 1–3 (1999).

65. The classical and still very useful treatment is in Fritz Novotny, *Cézanne und das Ende der Wissenschaftlichen Perspektive* (Wien: Verlag von Anton Schroll, 1938).

66. Émile Bernard, *Souvenirs sur Paul Cézanne et lettres* (Paris: A la rénovation esthéthique, 1924), 39, my trans. Also Émile Bernard, *Paul Cézanne: Minnen och brev,* trans. Gustaf Engwall (Stockholm: Föreningen för nutida konst, 1947), 42–43.

67. For a detailed study of these relations—a large number of paintings juxtaposed to photographs taken from the same spot—see Pavel Machotka, *Cézanne: Landscape into Art* (New Haven, Conn.: Yale University Press, 1996).

68. Bernard, *Souvenirs de Cezanne et lettres,* 29–30 and 79.

69. Maurice Merleau-Ponty, *Sense and Non-Sense,* trans. Hubert L. Dreyfus and Patricia A. Dreyfus (Evanston, Ill.: Northwestern University Press, 1964), 15 and 19.

70. The original—nonparaphrased—is in Stéphane Mallamré's "Un Coup de dés" as printed in the bilingual edition of *Mallarmé,* ed. with an intro. and prose translation by Anthony Hartley (Harmondsworth: Penguin, 1965), 233.

71. It is paradoxical that a man with such a tremendous sense of touch himself was completely paranoid about being "grabbed"; indeed his housekeeper was instructed never to let even her skirt touch him. As he put it himself, "isolation is what I am wor-

thy of. Thus, at least, no one gets me in his clutches" (quoted in John Rewald, *Cézanne: A Biography,* 185).

72. Although there are clear affinities between the two masters, it is uncertain whether they actually ever met. If they did, it would most likely have been in the house of Claude Monet.

73. Quoted in Evmarie Schmitt, *Cézanne in Provence* (New York: Prestel, 1995), 62–63, emph. added.

74. For technical details see Gunnar Olsson, "Complementary Models: A Study of Colonization Maps," *Geografiska Annaler,* ser. B, vol. 50 (1968): 115–32; Gunnar Olsson, "Inference Problems in Locational Analysis," in K. R. Cox and R. G. Golledge, eds., *Behavioral Problems in Geography: A Symposium* (Evanston, Ill.: Northwestern University, Studies in Geography, no. 17, 1969); Gunnar Olsson, *Birds in Egg/Eggs in Bird* (London: Pion, 1980). The wider implications for my own work are treated in Gunnar Olsson, "Glimpses," in Peter Gould and Forrest R. Pitts, eds., *Geographical Voices: Fourteen Autobiographical Essays* (Syracuse, N.Y.: Syracuse University Press, 2002).

75. Quoted in Becks-Malorny, *Paul Cézanne 1839–1906: Pioneer of Modernism,* 48, emph. added.

76. Quoted in Hajo Düchting, *Cézanne, 1839–1906: Nature into Art,* trans. Michael Hulse (Köln: Taschen, 1991), 198.

77. For special treatment of the latter theme see Düchting, *Paul Cézanne, 1839–1906: Nature into Art,* 193–217; and esp. Machotka, *Cézanne: Landscape into Art.*

78. Wallace Stevens, *Collected Poems* (New York: Vintage, 1990), 512.

79. John Rajchman, *The Deleuze Connections* (Cambridge, Mass.: MIT Press, 2000), 5–6.

80. Meyer Schapiro, *Words, Script, and Pictures: Semiotics of Visual Language* (New York: George Braziller, 1996). Also Gilles Deleuze and Félix Guattari, *Nomadology: The War Machine,* trans. Brian Massumi (Edinburgh: AK Press, 1986).

81. The literature is large and of varying quality. Useful and readily available overviews include Suzi Gablik, *Magritte* (London: Thames and Hudson, 1970); Michel Foucault, *This is not a Pipe,* trans. James Harkness (Berkeley and Los Angeles: University of California Press, 1983); Jacques Meuris, *René Magritte, 1898–1967,* trans. Michael Scuffil (Köln: Taschen, 1997). An elementary but explicitly semiotic analysis is in Annette Landen, *Läsvägar* (Lund: Institutionen för konstvetenskap, 1979).

82. Quoted from a speech given by Magritte in 1938 and reprinted in David Sylvester et al., *René Magritte: Catalogue raisonné,* 5, Supplement (Antwerp: Fonds Mercator, 1997), 20.

83. Sylvester et al.: *René Magritte: Catalogue raisonné,* 21.

84. Damisch, *The Origin of Perspective,* 157f. and 161.

85. Quoted and discussed in Kemp, *The Science of Art: Optical Themes in Western Art from Brunelleschi to Seurat,* 136ff.

86. Alberti, *On Painting,* para. 19, emph. added.

87. For further elaborations on this theme, see the various contributions to Paul Duro, ed., *The Rhetoric of the Frame* (Cambridge: Cambridge University Press, 1996); and, especially, the most penetrating discussions in Ulrike Oudée Dünkelsbühler, *Reframing the Frame of Reason: 'Trans-lation' In and Beyond Kant and Derrida,* trans. Max Statkiewicz (Amherst, Mass.: Humanity Books, 2002).

88. These elements of Kandinsky's thoughts are most explicitly formulated in Wassily Kandinsky, *Point and Line to Plane,* trans. H. Dearstyne and H. Rebay (New York:

Dover, 1979). However, there are many connections also to Wassily Kandinsky, *Concerning the Spiritual in Art,* trans. with an intro. by M. T. H. Stadler (New York: Dover, 1977). On the close relations between Kandinsky's conception of painting and my own approach to geography, see Gunnar Olsson, *Lines of Power/Limits of Language* (Minneapolis: University of Minnesota Press, 1991), esp. 5–8. Also the somewhat parallel discussion in John Sallis, *Shades—Of Painting at the Limit* (Bloomington: Indiana University Press, 1998).

89. Ludwig Wittgenstein, *Tractatus Logico-Philosophicus,* trans. D. F. Pears and B. F. McGuiness (London: Routledge and Kegan Paul, 1961), 5.143.

90. T. E. Hulme, "Romanticisim and Classicism," in *The Collected Writings of T. E. Hulme,* ed. Karen Csengeri (Oxford: Clarendon Press, 1994), 68–69. For insightful comments see A. Alvarez, "Making It New," *The New York Review of Books* 50, no. 8 (May 15, 2003), 28–30; and Robert Ferguson, *The Short Sharp Life of T. E. Hulme* (London: Allen Lane, 2002).

91. Still a third mapping will be discussed in the chapter entitled "Philadelphia."

92. Wittgenstein, *Tractatus Logico-Philosophicus,* 4.114, emph. added.

93. Arthur Bannister, Stanley Raymond and Raymond Baker, *Surveying,* 7th ed. (Harlow: Longman, 1998); Raymond E. Davis, *Surveying: Theory and Practice,* 6th ed. (New York: McGraw-Hill, 1981); M. J. T. Lewis, *Surveying Instruments of Greece and Rome* (Cambridge: Cambridge University Press, 2001).

94. The extensive quotations in this and the following paragraphs are all taken from *Republic* 506–11, the exact wordings usually, but not always, borrowed from *The Republic of Plato,* trans. Cornford, 217–26. The parenthetical translations of the Greek terms *eikasia, pistis, dianoia* and *noesis* are from C. D. C. Reeve, *Philosopher-Kings: The Argument of Plato's "Republic,"* 58.

95. In a footnote to this phrase Cornford adds the following commentary: "Conversely, the fact that the mathematician can use visible objects as illustrations indicates that the realities and truths of mathematics are embodied, though imperfectly, in the world of visible and tangible things; whereas the counterparts of the moral Forms can only be beheld by thought" (225).

96. Eric A. Havelock, *The Greek Concept of Justice: From Its Shadow in Homer to Its Substance in Plato* (Cambridge, Mass.: Harvard University Press, 1978), 329.

97. The complex argument to this effect comes towards the end of Book 5, especially in 475a–480a. For detailed comments on these passages see C. D. C. Reeve, *Philosopher-Kings: The Argument of Plato's "Republic,"* 58–79.

98. Wittgenstein, *Tractatus Logico-Philosophicus,* 6.54.

99. In the following comments on Duchamp I borrow freely from my earlier treatments in Gunnar Olsson, "Die Projektion des Begehrens/Das Begehren der Projektion," in Dagmar Reichert, ed., *Räumliches Denken* (Zürich: vdf Hochschulverlag AG an der ETH Zürich, 1996); and Gunnar Olsson, "Washed in a Washing Machine™," in Claudio Minca, ed., *Postmodern Geography: Theory and Praxis* (Oxford: Basil Blackwell, 2001).

100. Drawing on his visits to Duchamp's library in Villiers-sur-Grèz, Ulf Linde noticed that it included several of Plato's dialogues, some of them in more than one edition. For details see Ulf Linde, *Marcel Duchamp* (Stockholm: Rabén and Sjögren, 1986), 156ff., where the emphasis is more on the *eros* of the *Symposium* than on the ontological transformations of the *Republic.* Not so surprising, of course, for it may be reasonably argued that Duchamp—just as Diotima—conceived of love as unsatisfied desire.

101. Allan Megill, *Prophets of Extremity: Nietzsche, Heidegger, Foucault, Derrida* (Berkeley and Los Angeles: University of California Press, 1985).

102. The relevant literature is extensive and steadily growing. Among the general overviews I have found the following to be especially useful: Marcel Duchamp, *The Complete Works of Marcel Duchamp,* ed. Arturo Schwarz. 2 vols. (London: Thames and Hudson, 1997); Marcel Duchamp, *Duchamp du signe: Écrits,* réunis et présentés par Michel Sanouillet (Paris: Flammarion, 1975), herein quoted mostly from Marcel Duchamp, *The Writings of Marcel Duchamp,* ed. Michel Sanouillet and Elmer Peterson (New York: Da Capo Press, 1973); Anne d'Harnancourt and Kynaston McShine, eds., *Marcel Duchamp* (New York: Museum of Modern Art, 1973); Jennifer Cough-Cooper and Jacques Caumont, eds., *Marcel Duchamp: Ephemerides On and About Marcel Duchamp and Rrose Sélavy, 1887–1968,* exhibition catalogue, Palazzo Grassi, Venezia (Milan: Bompiani, 1993); John Golding, *Duchamp: The Bride Stripped Bare by Her Bachelors, Even* (London: Penguin, 1973); Jean Clair, *Marcel Duchamp ou le grand fictif: Essai de mythanalyse du grand verre* (Paris: Éditions Galilée, 1975); Jean-François Lyotard, *Les TRANSformateurs DUchamp* (Paris: Éditions Galilée, 1977); Lyotard, *Duchamp's TRANS/formers,* trans. Ian McLeod (Venice, Calif: Lapis Press, 1990); Lawrence D. Steefel Jr, *The Position of Duchamp's Glass in the Development of His Art* (New York: Garland, 1977); Octavio Paz, *Marcel Duchamp: Appearance Stripped Bare,* trans. Rachel Philips and Donald Gardiner (New York: Viking Press, 1978); Ulf Linde, *Marcel Duchamp;* Thierry de Duve, *Pictorial Nominalism: On Marcel Duchamp's Passage from Painting to the Readymade* (Minneapolis: University of Minnesota Press, 1991); de Duve, *The Definitively Unfinished Marcel Duchamp* (Cambridge, Mass.: MIT Press, 1991); de Duve, *Kant after Duchamp* (Cambridge, Mass.: MIT Press, 1996); Dieter Daniels, *Duchamp und die anderen: Der Modelfall einer künstlerischen Wirkungsgeschichte in der Moderne* (Köln: DuMont, 1992); Amelia Jones, *Postmodernism and the En-Gendering of Marcel Duchamp* (Cambridge: Cambridge University Press, 1994); Jerrold Seigel, *The Private Worlds of Marcel Duchamp: Desire, Liberation, and the Self in Modern Culture* (Berkeley and Los Angeles: University of California Press, 1995); Calvin Tomkins, *Duchamp: A Biography* (London: Pimlico, 1996); Juan Antonio Ramírez, *Duchamp: Love and Death, Even* (London: Reaktion Books, 1998); David Joselit, *Infinite Regress: Marcel Duchamp 1910–1941* (Cambridge, Mass.: MIT Press, 1998).

103. It is often argued that Duchamp's entire oeuvre should be understood as a life-long study of the Renaissance perspective. For a comprehensive overview see especially Jean Claire, "Duchamp and the Classical Perspective," *Artforum* 16, no. 7 (1978): 40–49.

104. Quoted in d'Harnancourt and McShine, eds., *Marcel Duchamp,* 256.

105. Pierre Cabanne, *Dialogues with Marcel Duchamp,* trans. Ron Padgett (London: Thames and Hudson, 1971), 31.

106. Cough-Cooper and Caumont, eds., *Marcel Duchamp: Ephemerides On and About Marcel Duchamp and Rrose Sélavy,* 18 March.

107. Cabanne, *Dialogues with Marcel Duchamp,* 31.

108. Sanouillet and Peterson, eds., *The Writings of Marcel Duchamp,* 126.

109. Ibid., 125.

110. Quoted in "Marcel Duchamp" (an interview by James Johnson Sweeney) in James Nelson, ed., *Wisdom: Conversations with the Elder Wise Men of Our Day* (New York: W. W. Norton, 1958), 93.

111. Marcel Duchamp, "I Propose to Strain the Laws of Physics." Interview by Francis Roberts, *Art News* 67, no. 8 (December 1968): 47.

112. Thierry de Duve, "The Readymade and the Tube of Paint," *Artforum* 24, no. 9 (May 1986): 110–21.

113. Katharine Kuh, *The Artist's Voice* (New York: Harper and Row, 1960), 92.

114. For elaborations see Carol P. James, "Reading Art Through Duchamp's *Glass* and Derrida's *Glas,*" *Sub-Stance* 10, no. 2 (1981): 105–28.

115. The most sustained analyses are in William A. Camfield, *Marcel Ducamp / Fountain* (Houston, Tex.: Houston Fine Art Press, 1989); and de Duve, *The Definitively Unfinished Marcel Duchamp.*

116. Arthur C. Danto, *The Philosophical Disenfranchisement of Art* (New York: Columbia University Press, 1986), 14–15.

117. For further elaborations see Thierry de Duve, *Kant after Duchamp,* 398.

118. D'Harnancourt and McShine, *Marcel Duchamp,* 283.

119. Duchamp himself reputedly liked breathing better than working.

## ABR(AH)AM

1. Erich Auerbach, *Mimesis: The Representation of Reality in Western Literature,* Trans. Willard R. Trask (Princeton, N.J.: Princeton University Press, 1953). Reissued in a Fiftieth-Anniversary Edition with the same page-numbering as in the 1953 edition (Princeton University Press, 2003). Included in the latter edition is also an informative foreword by Edward W. Said and an appendix ("Epilegomena to Mimesis") with Auerbach's own response to his critics. For additional evaluations see Geoffrey Green, *Literary Criticism and the Structures of History: Erich Auerbach and Leo Spitzer* (Lincoln: University of Nebraska Press, 1982); and Seth Lerer, ed., *Literary History and the Challenge of Philology* (Stanford, Calif.: Stanford University Press, 1996).

2. Edward W. Said, *The World, the Text, and the Critic* (Cambridge, Mass.: Harvard University Press, 1983), 8.

3. Erich Auerbach, *Literary Language and Its Public in Late Latin Antiquity and in the Middle Ages,* trans. Ralph Manheim (Princeton, N.J.: Princeton University Press, 1965), 18.

4. Auerbach, *Mimesis,* 43. Throughout Auerbach's analysis the figure of Dante is continuously present, most prominently in the remarkable Chapter Eight which is devoted to the tenth canto of the *Inferno,* a stunning reading of how "the image of man eclipses the image of God . . . ; of how man's self-fulfillment, which comprises the individual's entire past—objectively as well as in memory—involves ontogenetic history, the history of an individual's growth; the resultant of that growth, it is true, lies before us as a finished product; but in many cases we are given a detailed portrayal of several phases; it is never entirely withheld from us." Thus, when Dante and Virgil come across Farinata and Cavalcante in their flaming tombs whose lids open they discover that the damned sinners have the same characters and passions in the underworld as they once had in their lifetimes—they behave in Hell *as if they were* in Florence. Rephrased, the here-and-now is transported to the beyond exactly as the principles of cartographical reasoning require it to be; "[for it] is precisely the absolute realization of a particular earthly personality in the place definitively assigned to it, which constitutes the Divine Judgment. . . . [This in turn] enables us to understand that the beyond is eternal and yet phenomenal; that it is changeless and of all time and yet full of history. It also enables us to show in what way this realism in the beyond is distinguished from every

type of purely earthly realism." Quotations from *Mimesis,* 202, 193, and 197 respectively. For even deeper analyses of these themes of prefiguration, see Erich Auerbach, *Dante, Poet of the Secular World,* trans. Ralph Manheim (Chicago: University of Chicago Press, 1961).

5. Although our respective readings focus on quite different issues, see also Johan Asplund, *Rivaler och syndabockar* (Göteborg: Korpen, 1989).

6. All subsequent quotations are from Homer, *The Odyssey,* trans. Robert Fitzgerald (Garden City, N.Y.: Doubleday, Anchor Books, 1963). Since I do not myself read Greek, I have done my best to check all quotations against a number of other translations, esp. *The Odyssey: The Story of Odysseus,* trans. W. H. D. Rouse (New York: The New American Library, 1937); *The Odyssey,* trans. E. V. Rieu (Harmondsworth: Penguin, 1946); *Odyssée,* trans. Erland Lagerlöf (Lund: Gleerups, 1908); *Sången om Odysseus,* trans. Tord Bæckström (Stockholm: Forum, 1975); *Odissea,* trans. Rosa Calzecchi Onesti (Torino: Einaudi, 1963).

On Homer himself the literature is immense. My own readings include Cedric Whitman, *Homer and the Heroic Tradition* (Cambridge, Mass.: Harvard University Press, 1958); Geoffrey Stephen Kirk, *The Songs of Homer* (Cambridge: Cambridge University Press, 1962); W. B. Stanford, *The Ulysses Theme: A Study in the Adaptability of a Traditional Hero,* 2d ed. (Ann Arbor: University of Michigan Press, 1978); Alfred Heubeck, *Die homerische Frage: Ein Bericht über die Forschung der letzten Jahrzehnte* (Darmstadt: Wissenschaftliche Buchgesellschaft, 1974); Sture Linnér, *Homeros* (Stockholm: Norstedt, 1985); Michael Clarke, *Flesh and Spirit in the Songs of Homer: A Study of Words and Myths* (Oxford: Clarendon Press, 1999); Keld Zeruneith, *Træhesten: Fra Odysseus til Sokrates. En bevidsthedshistorie* (København: Gyldendal, 2002). Much helpful information is in Joseph Russo et al., eds., *A Commentary on Homer's Odyssey.* 3 vols. (Oxford: Clarendon Press, 1992).

7. *The Odyssey* brims with epithets, many of them so rich that they defy translation. Nevertheless the epithets, more often used for metrical than for semantic purposes, are so important that after Parry they have been called "The Homeric Formula." Crucial here is the term *polytropos* which is used for denoting the cunning trickster of Odysseus himself, the well-traveled adventurer so skilled in adapting his figures of speech (*tropes*) that he always gets his way. Thus, the *polytropos* man "is a man of the moment a man of change: now one thing, now another; he slides and shifts from one extreme to the other. . . . [He is] distinguished by the control he possesses: supple and shifting as he is, he is always master of himself and is only unstable in appearance. His *volte-faces* are a trap—the net in which his adversary becomes entangled. He is not the plaything of movement but its master. . . . It is this ability of the octopus and the polútropos one, the man of thousand tricks, to assume every form without becoming imprisoned within any." Quotation from Marcel Detienne and Jean-Pierre Vernant, *Cunning Intelligence in Greek Culture and Society,* trans. Janet Lloyd (Chicago: University of Chicago Press, 1991), 40. Also see Milman Parry, *The Making of Homeric Verse: The Collected Papers of Milman Parry,* ed. Adam Parry (Oxford: Clarendon Press, 1971); Norman Austin, *Archery at the Dark of the Moon: Poetic Problems in Homer's "Odyssey"* (Berkeley and Los Angeles: University of California Press, 1975), chap. 1; Pietro Pucci, *Odysseus Polutropos: Intertextual Readings in the "Odyssey" and the" Iliad"* (Ithaca, N.Y.: Cornell University Press, 1987); Stanford, *The Ulysses Theme,* 99; and Zeruneith, *Træhesten: Fra Odysseus til Sokrate,* 171 and 619.

8. "But first my name," says the wanderer in another context, "let that be known

to you. . . . I am Laertes' son, Odysseus. Men hold me *formidable for guile in peace and war:* this fame has gone abroad to the sky's rim" (9.17–20, emph. added.) The hero's personality has deep roots, however, for the name of the grandfather—*Autolykos*—means "the wolf himself" or "the very wolf," a man known for stealing horses, cattle and sheep, a most fitting description also of the grandson; furthermore, not only was Hermes the patron of Autolykos, he was his father. Joseph Russo et al., eds., *A Commentary on Homer's "Odyssey,"* 3:96. In later traditions, though, Odysseus was the son of Sisyphus. At any rate, through his wanderings and "in accepting the implications of his name, 'Trouble,' Odysseus establishes his identity in harmony with the nature of things. In the ultimate sense he is 'Zeus-sprung,' one whose existence is rooted in life itself." Quotation from George E. Dimock Jr, "The Name of Odysseus," in Charles E. Taylor, ed., *Essays on the "Odyssey": Selected Modern Criticism* (Bloomington: Indiana University Press, 1969).

9. The term "flashback" seems more appropriate than "digression," for the purpose of this scene is clearly to doubly ground Odysseus' name and destiny, first in the transition rite of the hunt, then in the boy's specific relation to Autolykos, a family whose future was always precarious, because the wives were rumored never to get more than one son. The significance of the mark is consequently not confined to the surface of the trickster's body but reaches deeply into his mind. Thus, through the name "Odysseus," the physicality of the Signifier (the scar) and the meaning of the signified (the grandfather and the inherited character) are unbreakably brought together. As so often, the signs of disguise and recognition are the same.

10. More literally: "I myself [am] that very man, returned home (that is, here before your eyes)." Russo et al., eds., *A Commentary of Homer's "Odyssey,"* 3:170.

11. Auerbach, *Mimesis,* 6, emph. added.

12. What we now take to be the Hebrew Bible (which is not the same as the Christian Old Testament) was a long time in the making. The final result is a set of texts with significant differences in style, vocabulary and content as well as a history which remains disputed. However, it is generally agreed that the oldest portions (perhaps first formulated around 925 BCE) were authored by a person identified as J or the Yahwist, so called because its God was usually named *Yahweh* or *YHWH,* in English translation customarily rendered as *The Lord*. The next layer (perhaps two generations younger) is credited to E or the Elohist, where God tends to be called *Elohim,* in English simply *God.* While the two writers J and E composed major parts of what is now known as Genesis, Exodus and Numbers, Deuteronomy (sometimes labeled The Fifth Book of Moses) was added some time in the fifth century BCE by a compiler identified as D. Then followed an edition with many additions of ritual material, now known as the P version, the Priestly code, begun after the fall of Jerusalem in 587 and continued for many years of exile. Finally, there was the formidable R or Redactor, most likely a committee of learned men, who around 400 BCE gave to the Torah the shape and content it still has.

Of these authors J is generally considered the master of narrative, a creative writer of Shakespearean proportions, an artist who presented God as someone who acts and speaks like a human person. In contrast, E is more reflective and theological, an interpreter who conceives the potentate as a more remote and less directly accessible being. The P code, although essentially a combination of the J and E versions, is nevertheless very different from the other two, mainly because the narrative is frequently interrupted by detailed ritual instructions. Finally, the Deuteronomist, or D version, is characterized by a hortatory style and by a vocabulary which is designed to hammer home the message of Israel's special role in the history of nations. In that latter context, it is telling that the ten commandments, now found in Exodus 20, were originally formulated not by J or E but by P, but are now repeated also in Deuteronomy 5.

Each of these versions possesses different qualities and each is favored by different commentators: for *The Book of J* there is Harold Bloom, who argues that J might well have been a woman, in that case probably a princess at King Solomon's court; for R there is especially Northrop Frye and Robert Alter, excellent advocates in themselves.

In my own approach to Genesis I have found much useful information in Mark S. Smith, *The Early History of God: Yahweh and the Other Deities in Ancient Israel* (New York: Harper and Row, 1990); Gary A. Rendsburg, *The Redaction of Genesis* (Winona Lake, Ind.: Eisenbrauns, 1986); *The Book of J*, trans. David Rosenberg and interpreted by Harold Bloom (New York: Faber and Faber, 1990); Bruce Feiler, *Abraham: In Search of the Father of Civilisation* (London: Judy Piatkus, 2002); Umberto Cassuto, *A Commentary on the Book of Genesis*, vol. 2: *Noah to Abraham*, trans. Israel Abrahams (Jerusalem: Magnes Press, 1964); Northrop Frye, *The Great Code: The Bible and Literature* (New York: Harcourt Brace Jovanovich, 1982); Northrop Frye, *Words with Power: Being a Second Study of The Bible and Literature* (New York: Harcourt Brace Jovanovich, 1990); Robert Alter, *The Art of Biblical Narrative* (New York: Basic Books, 1981); Robert Alter, *Canon and Creativity: Modern Writing and the Authority of Scripture* (New Haven, Conn.: Yale University Press, 2000); Robert Alter and Frank Kermode, eds., *The Literary Guide to the Bible* (Cambridge, Mass.: Harvard University Press, 1987); Jack Miles, *God: A Biography* (New York: Simon and Schuster, 1995); Avivah Gottlieb Zornberg, *The Beginning of Desire: Reflections on Genesis* (New York: Doubleday, 1995); Karen Armstrong, *In the Beginning: A New Interpretation of Genesis* (New York: Alfred A. Knopf, 1996); Carol Lowery Delaney, *Abraham on Trial: The Social Legacy of Biblical Myth* (Princeton, N.J.: Princeton University Press, 1998); A. Wénin, ed., *Studies in the Book of Genesis: Literature, Redaction and History* (Leuven: Leuven University Press, 2001); Karl-Josef Kuschel, *Abraham: A Symbol of Hope for Jews, Christians and Muslims* (London: SCM Press, 1995). In addition, I have often consulted the *Encyclopaedia Judaica* (Jerusalem: Keter Publishing House, 1971).

13. Most of the following quotations are from *The Holy Bible*, New International Version (Grand Rapids, Mich: Zondervan Publishing House, 1973). Since the translations are legion, I have done my best to check the exact wordings against a number of alternatives. In particular I have consulted the up-to-date Swedish edition from 1999, the work of an expert committee whose efforts most tellingly were blessed not by the Archbishop of the Lutheran Church, but by the Cultural Minister of the Swedish State: *Bibeln*, Statens Offentliga Utredningar, 1999:100.

14. Emphasis added as a reminder of the Baylonian, Greek and Christian cartographers already mentioned.

15. To modern readers the expression "wild donkey" or "wild ass" sounds pejorative, but most linguists agree that its real meaning is that Ishmael was predicted to be a Bedouin, not an outcast.

16. Carol Newsom and Sharon Ringe, *Women's Bible Commentary* (Louisville, Ky.: Westminster John Knox Press, 1998), 19.

17. Feiler, *Abraham: In Search of the Father of Civilization*, 66f.

18. The translation is disputed, and one alternative is "I have now seen *the back of* the One who sees me." In Northrop Frye's interpretation, Hagar's naming may well be a veiled question, "Have I really seen God and lived?" See Frye, *The Great Code*, 116.

19. Phyllis Trible, *Texts of Terror: Literary-Feminist Readings of Biblical Narratives* (Philadelpia: Fortress Press, 1984), 28. Also see Sharon Pace Jeansonne, *The Women of Genesis: From Sarah to Potiphar's Wife* (Minneapolis, Minn.: Fortress Press, 1990), chap. 4.

20. It is significant that in the earliest versions, i.e. in the Books of J and E, neither Abram nor Sarai are assigned new names.

21. The name *Isaac* means "he laughs."

22. Yet one should always remember that according to the story of Eden the first thing to be produced was a loincloth, presumably Adam's and Eve's way of telling the Creator that they were well aware of how their children were made.

23. It says much about the rhetoric of organized religion that the story about the *akedah* is not the work of J but a text that was added, in bits and pieces, by the later versions.

24. The literature is immense. For my own purposes I have drawn especially on Søren Kierkegaard, *Fear and Trembling: A Dialectical Lyric,* trans. Walter Lowrie (Princeton, N.J.: Princeton University Press, 1941); Robert L. Perkins, ed., *Kierkegaard's Fear and Trembling: Critical Appraisals* (Tuscaloosa: University of Alabama Press, 1981); Shalom Spiegel, *The Last Trial: On the Legends and Lore of the Command to Abraham to Offer Isaac as a Sacrifice: The Akedah* (Philadelphia: Jewish Publication Society of America, 1967); Bruce Vawter, *On Genesis: A New Reading* (London: Chapman, 1977); Jack Miles, *God: A Biography* (New York: Simon and Schuster, 1995); Avivah Gottlieb Zornberg, *The Beginning of Desire: Reflections on Genesis;* Heinz-Dieter Neef, *Die Prüfung Abrahams: Eine exegetisch-theologische Studie zu Gen. 22, 1–19* (Stuttgart: Calwer, 1998); Carol Delaney, *Abraham on Trial: The Social Legacy of Biblical Myth;* Edward Noort and Eibert Tigchelaar, *The Sacrifice of Isaac: The Aqeda (Genesis 22) and its Interpretations* (Leiden: Brill, 2002); Bruce Feiler, *Abraham: In Search of the Father of Civilisation;* Donald Capps, "Abraham and Isaac: The Sacrificial Impulse," in J. Harold Ellens, ed., *The Destructive Power of Religion: Violence in Judaism, Christianity and Islam,* vol. 1: *Sacred Scriptures, Ideology and Violence* (Westport, Conn.: Praeger, 2004).

25. It was not until the mid-nineteenth century that the term "philistine" came to be used for persons who were considered deficient in liberal culture and whose main interests were material and prosaic. For details see *The Oxford English Dictionary.*

26. Auerbach, *Mimesis,* 14–15, emph. added.

27. For technical and insightful discussions of this matter see Johan Brinkman, *The Conception of Space in the Old Testament: An Exploration of the Methodological Problems of Investigation, Exemplified by a Study of Exodus 25 to 31* (Kampen: Kok Pharos Publishing House, 1992).

28. *Bereshit Rabbah* 68:10. Quotation from Zornberg, *The Beginning of Desire,* 187.

29. Here it is significant to repeat that the name Isaac means "laughter," unclear whether for mockery or for joy. The ambiguity obviously reflects the uncertainty of Isaac's own ancestry; according to one Midrash he was not the real son of Abraham, but the offspring of Abimelech, in whose palace Sarah had been raped and made pregnant, much to her surprise and much against her will. For more on this interpretation see Zornberg, *The Beginning of Desire: Reflections on Genesis,* 112–14.

30. Feiler, *Abraham: In Search of the Father of Civilisation,* 80f.

31. Ibid., 48f.

32. Words later put in Jesus' mouth by both Matthew (27:46) and Mark (15:34).

33. Kierkegaard, *Fear and Trembling,* 89.

34. Mark Dooley, *The Politics of Exodus: Søren Kierkegaard's Ethics of Responsibility* (New York: Fordham University Press, 2001), xxi.

35. See for instance the detailed analyses in Ronald M. Green, *Kierkegaard and Kant: The Hidden Debt* (Albany: State University of New York Press, 1992); and the more passing remarks in Rodolphe Gasché, *The Tain of the Mirror: Derrida and the Philosophy of Reflection* (Cambridge, Mass.: Harvard University Press, 1986).

36. For insightful analyses see M. Jamie Ferreira, *Transforming Vision: Imagination and Will in Kierkegaard's Faith* (Oxford: Clarendon Press, 1991).

## MOSES

1. Erich Auerbach, *Mimesis: The Representation of Reality in Western Literature,* trans. Willard R. Trask (Princeton, N.J.: Princeton University Press, 1953), 11–12. For a well-taken critique of Auerbach's point that the *Odyssey* is all foreground and no background, see Keld Zeruneith, *Træhesten: Fra Odysseus til Sokrates. En bevidsthedshistorie* (København: Gyldendal, 2002), 620. Zeruneith's main argument here is that "Auerbach overlooks, or does not understand, that the Greek aesthetics of beauty springs directly from a deep form of religious feelings" (my translation). Also Alan Griffiths, "Euenius the Negligent Nightwatchman (Herodotus 9.92–6)," in Richard Buxton, ed., *From Myth to Reason? Studies in the Development of Greek Thought* (Oxford: Oxford University Press, 1999), 178 n. 34.

2. For brief and nontechnical introductions see Stephan Körner, "Laws of Thought," in Paul Edwards, ed., *The Encyclopedia of Philosophy* (New York: Macmillan, 1967), 4:414–17; and Stephan Körner, *Conceptual Thinking: A Logical Enquiry* (New York: Dover, 1959), 74–79.

3. Aristotle, *Metaphysics* 1011b, trans. Ross in *The Basic Works of Aristotle,* ed. Richard McKeon (New York: Random House, 1941).

4. For instructive discussions of this theme in relation to the Neoplatonism of the sixth century CE, see Gerhard Seel, ed., *Ammonius and the Seabattle: Texts, Commentary and Essays* (Berlin: Walter de Gruyter, 2001).

5. Aristotle, *De Interpretatione* 19a, trans. Edghill in *The Basic Works of Aristotle.*

6. For a wonderful analysis see F. G. Asenjo, "In Search of the Present," *Studies in Humanities and Social Sciences: Journal of Inter-University Centre,* I.I.A.S. Shimla, 10 (2003): 50–69.

7. Quoted in Nicholas Rescher, *Many-valued Logic* (New York: McGraw-Hill, 1969), 5. Peirce's letter was written in 1909.

8. An up-to-date introduction is in J. C. Beall and Bas C. van Fraassen, *Possibilities and Paradox: An Introduction to Modal and Many-valued Logic* (Oxford: Oxford University Press, 2003).

9. The inserted quotation from Matthew is Jesus' commentary on the eighth commandment, according to which (Exod. 20:16) "You shall not give false testimony against your neighbor." Clearly an issue of trust and verification, for according to the old Israeli law all serious cases involving your neighbor should be grounded in the testimony of at least two witnesses. In dealings with the Other, the issue of conflicting truths was evidently of lesser concern.

10. The latter appear in slightly different versions, first in Exod. 20:1–17, then in

Deut. 5:6–21. In general, Deuteronomy (Greek for the Second Law) offers an interpretation of the four books that precede it.

11. Georg Henrik von Wright, *An Essay in Deontic Logic and the General Theory of Action,* Acta Philosophica Fennica, Fasc. XXI (Amsterdam: North-Holland, 1968). Also Gunnar Olsson, *Birds in Egg/Eggs in Bird* (London: Pion, 1980), chap. 9.

12. Erhard Gerstenberger, *Wesen und Herkunft des 'Apodiktischen Rechts'* (Neukirchen-Vlyun: Neukirchener, 1965); Erhard Gerstenberger, "Covenant and Commandment," *Journal of Biblical Literature* 84 (1965): 38–51. For more on the same theme see Calum M. Carmichael, *The Origins of Biblical Law: The Decalogues and the Book of the Covenant* (Ithaca, N.Y.: Cornell University Press, 1992); and the detailed overview in Johan Jacob Stamm with Maurice Edward Andrew, *The Ten Commandments in Recent Research* (London: SCM Press, 1967), esp. 44–75. Although I will often twist his well-balanced account to fit my own purposes, I am much indebted to Tryggve Kronholm, *De tio orden* (Stockholm: Verbum, 1992).

13. The Bible explicitly treats man's everyday relations to the world as an integral part of his relation to the LORD; the master/slave dialectics in another light.

14. In Jewish thinking the Decalogue is split into two equal parts with five words on each tablet, the instruction to honor your father and your mother counting as the fifth. This is the only paragraph (in Lutheran tradition §4) which might just as well be included in the "Constitutional Law" as in the "Civil," where I place it.

15. It should be noted first that the division into two tablets does not occur in Exodus 20 but in Exodus 24 and Deuteronomy 5, and then that the assignment of the "Constitutional Law" to the first tablet and the "Civil/Criminal Law" to the second is a matter of interpretation.

16. Characterization borrowed from the title page.

17. Exod. 20:1 and Deut. 5:6, both with exactly the same wording. For excellent overviews of the book of Exodus in general, see Brevard S. Childs, *The Book of Exodus: A Critical, Theological Commentary* (Louisville, Ky.: Westminster Press, 1974), chap. 17; and Umberto Cassuto, *A Commentary on the Book of Exodus,* trans. Israel Abrahams (Jerusalem: Magnes Press, 1983), 235–53. Corresponding treatments of the book of Deuteronomy are in Peter C. Craigie, *The Book of Deuteronomy* (Grand Rapids, Mich.: Eerdmans, 1976), 149–64; and Ian Cairns, *Word and Presence: A Commentary on the Book of Deuteronomy* (Grand Rapids, Mich.: Eerdmans, 1992).

18. Here quoted from Exod. 20:3, the wording slightly, but not significantly, different from Deut. 5:7. It should be noted that in Judaism the Prologue and §1 are not split apart but kept together in the same First Commandment.

19. Henning Graf Reventlow, *Gebot und Predigt im Dekalog* (Gütersloh: G. Mohn, 1962), 25ff.

20. Exod. 20:4–6, the wording virtually identical to that of of Deut. 5:8–10.

21. Stamm with Andrew, *The Ten Commandments in Recent Research,* 82.

22. In this context of legitimation it says much about Luther's relations to the Protestant kings, who depended on him for their legitimation, that his *Small Catechism* never mentions the prohibition against graven images. The consequence is that for twenty generations of Northern Europeans the crucial message of §2a was effectively silenced. No wonder that the Swedes "have been quick to agree and late to question," a phrase which Gustavus Vasa used in his farewell speech to Parliament in 1560, a rare admission which perhaps was prompted by his insight that "I can feel in my own body the signs that I will soon be taken away."

In this context it is important to repeat that "the impact of Martin Luther's interpretation of the Decalogue for the subsequent history of Protestantism can hardly be exaggerated. By giving it a central role in both his larger and smaller catechism he assured that it entered into the blood stream of the church. Luther's exegesis gave to the Decalogue a profoundly comprehensive role, covering every area of life, which has seldom been matched. At the same time it clothed it with such a concrete form of flesh and blood that it carried a tremendous inpact in shaping the Christian life." Brevard S. Childs, *The Book of Exodus,* 433.

23. Childs, *The Book of Exodus,* 408.

24. Kronholm, *De tio orden,* 144. Not surprisingly, this is the paragraph that during the last century has received the most attention.

25. Homer, *The Odyssey,* trans. Robert Fitzgerald (Garden City, N.Y.: Doubleday, Anchor Books, 1963).

26. Emphasis added, the relations of Isaac and Abraham immediately recalled.

27. Søren Kierkegaard, *Fear and Trembling: A Dialectical Lyric,* trans. Walter Lowrie (Princeton, N.J.: Princeton University Press, 1941), 102.

28. It has in fact been suggested that the words of the first stone tablet may be best understood as one single elaborate response to the incident of the golden calf. For details see Calum M. Carmichael, *Law and Narrative in the Bible: The Evidence of Deuteronomic Laws and the Decalogue* (Ithaca, N.Y.: Cornell University Press, 1985), 313–26.

29. For more on Moses as an explorer of the No-Man's Land between God and man, see Nanette Stahl, *Law and Liminality in the Bible* (Sheffield: Sheffield Academic Press, 1995).

30. The Swedish Archbishop is not alone in having suggested that if Jesus had lived in the twentieth century, then he would have been a Social Democrat! In the bishop's defense it must be said both that his theology is best described as a form of applied semiotics and that his official office is adorned with two beautiful icons. For his pastoral letter see K. G. Hammar, *Samtal om Gud: Herdabrev* (Stockholm: Verbum, 1997).

31. Norman O. Brown, *Closing Time* (New York: Vintage Books, 1974), 18. In connection with this explicit connection between Vico and James Joyce see especially Donald Phillip Veren, *Knowledge of Things Human and Divine: Vico's New Science and Finnegans Wake* (New Haven, Conn.: Yale University Press, 2003).

32. Already during his lifetime Malevich was a legend, the charismatic founder of the Suprematist movement in art, a highly productive painter and a most prolific writer. Most of his essays are now available in translation first into Danish in 1963, then into English as *K. S. Malevich: Essays on Art 1915–1933,* 4 vols., ed. Troels Andersen (København: Borgen, 1968–78). Among the several catalogues and overviews I have drawn especially on Larissa A. Zhadova, *Malevich: Suprematism and Revolution in Russian Art, 1910–1930* (London: Thames and Hudson, 1982); Evgenija Petrova et al., *Malevich: Artist and Theoretician,* trans. Sharon McKee (Paris: Flammarion, 1990); Jeanne d'Andrea, ed., *Kazimir Malevich, 1878–1935* (Los Angeles: The Armand Hammer Museum of Art, 1990); Rainer Crone and David Moos, *Kazimir Malevich: The Climax of Disclosure* (London: Reaktion Books, 1991); Charlotte Douglas, *Kazimir Malevich* (London: Thames and Hudson, 1994); John Milner, *Kazimir Malevich and the Art of Geometry* (New Haven, Conn.: Yale University Press, 1996); John Golding, *Paths to the Absolute: Mondrian, Malevich, Kandinsky, Pollock, Newman, Rothko and Still* (London: Thames and Hudson, 2000); and Matthew Drutt, *Kazimir Malevich: Suprematism* (New York: Solomon R. Guggenheim Foundation, 2003).

33. Kazimir S. Malevich, "On New Systems in Art" (1919), *Essays on Art 1915–1933*, ed. Troels Andersen, trans. Xenia Glowacki-Prus and Arnold McMillin, 1:90–91.

34. When this work was first exhibited it was not called *Black Square* but *Quadrilateral.* For details see John Milner, *Kazimir Malevich and the Art of Geometry,* 122.

35. "From Cubism and Futurism to Suprematism," written in 1915. Quotation from Malevich, *Essays on Art 1915–1933*, 1:38. For detailed analyses of the geometric forms and the series that spring from this "royal infant," see Milner, *Kazimir Malevich and the Art of Geometry,* chap. 5.

36. See especially Michel Serres, *Les origines de la géométrie* (Paris: Flammarion, 1993).

37. The connection between Whitman and Malevich is cleverly pursued in W. Sherwin Simmons, *Kasimir Malevich's Black Square and the Genesis of Suprematism, 1907–1915* (New York: Garland, 1981), 246–53. The poem itself is included in Walt Whitman, *Leaves of Grass,* "Death-Bed Edition" from 1881 (New York: New American Library, Signet Classic, 1955).

38. In a much debated passage in the *Nicomachean Ethics* (1.10.1100b21) Aristotle argued that the most happy men are "foursquare beyond reproach." This term *foursquare*—a translation of the Greek *tetragonos*—does itself contain a hidden reference to Plato's extensive discussion in *Protagoras* (339a-347a) of a famous poem by Simonides: "Hard is it on the one hand to become/ A good man truly, hands and feet and mind/ Foursquare, wrought without blame."

More remarkably still, a Google search of "foursquare" results in no less than 146,000 hits in 0.30 seconds! The vast majority of these are references to the so-called "Church of the Foursquare Gospel," a Pentecostal congregation founded in 1923. Figures from March 2004 claim that there are a total of 4,113,991 worshippers distributed across 38,127 churches or meeting places divided among 138 countries. The four corners of this belief system are named à la Whitman as (a) the Savior; (b) the Baptizer with Holy Spirit; (c) the Healer; and (d) the Soon-Coming King. In an example taken from 1845–46, the *Oxford English Dictionary* refers to "a Gospel which stands *four-square,* with a side facing each side of the spititual world." Cosmas Indicopleustes might have felt at home.

Other Internet sites mention the Foursquare football game, the Foursquare American house and the Foursquare spiced rum, the latter a concoction of "aged rum with a little bit of cinnamon, vanilla and nutmeg gently blended in at the distillery, the actual recipe a family secret pased down through five generations." One can but wonder what Whitman's and Malevich's reactions might have been.

39. Malevich, "I am the Beginning . . ." (ca. 1915), *Essays on Art, 1913–33*, 4:12–13.

40. The parallels to Greimassian semiotics, especially to the semiotic square, are striking. In that hieroglyphic figure—nothing less than a map of meaning production—the movement between the corners is not haphazardous but determined by the narrative of logic itself; by no coincidence Whitman began his circling of the square deific in the corner of the Father and ended up in the corner of the Rebel. Even the equal sign has direction built into it, for just as the journey from here to there is not the same as the move from there to here—so that $(a=b) \neq (b=a)$—here is an order in every order, the production of meaning always a transformation of meaning.

In hindsight, the square which I once used for mapping a particular stage in my own epistemological journey (Gunnar Olsson, *Lines of Power/Limits of Language* [Minneapolis: University of Minnesota Press, 1991], 183–96) fits well into Greimas's framework. Of this connection I was not aware at the time, but was later helped to by Fredric Jameson's introduction to Algirdas Julien Greimas, *On Meaning: Selected Writings in Semiotic Theory,* trans. Paul J. Perron and Frank H. Collins (Minneapolis: University of Minnesota Press, 1987). On the spatial properties of the semiotic square see especially Jean Petitot-Cocorda, *Morphogenèse du sens: Pour un schématisme de la structure* (Paris: Presses Universitaires de France, 1985), chap. 3; and Jean Petitot-Cocorda, *Physique du sens: De la théorie des singularités aux sémio-narrative* (Paris: CNRS, 1992).

41. Malevich, "On New Systems in Art," *Essays on Art 1915–1933,* 1:113–14, emph. added.

42. Wassily Kandinsky, *Point and Line to Plane,* trans. H. Dearstyne and H. Rebay (New York: Dover, 1979), 62–65.

43. As so typical of Borges the argument is not only built around references to references to references but repeated in a number of places. Thus, in the essay "Partial Magic in Quixote," Borges quotes Josiah Royce's *The World and the Individual,* in which we are asked to "imagine that a portion of the soil of England has been levelled off perfectly and that on it a cartographer traces a map of England. The job is perfect; there is no detail of the soil of England, no matter how minute, that is not registered on the map; everything has there its correspondence. This map, in such a case, should contain a map of the map, which should contain a map of the map of the map, and so on into infinity." Borges then continues: "Why does it disturb us that the map be included in the map and the thousand and one nights in the book of the *Thousand and One Nights?* Why does it disturb us that Don Quixote be a reader of *Quixote* and Hamlet a spectator of *Hamlet?* I believe I have found the reason: these inversions suggest that if the characters of the fictional work can be readers or spectators, we, its readers and spectators, can be fictitious. In 1833, Carlyle observed that the history of the universe is an infinite sacred book that all men write and read and try to understand, and in which they are also written." Jorge Luis Borges, *Labyrinths: Selected Stories and Other Writings,* ed. Donald A. Yates and James E. Irby, trans. by James E. Irby (New York: New Directions, 1964), 195–96.

Borges' own(?) version of the same story is in his text "On Exactitude in Science," where he tells how "in that Empire, the craft of Cartography attained such Perfection that the Map of a Single province covered the space of an entire City, and the Map of the Empire itself an entire Province. In the course of Time, these Extensive maps were found somewhat wanting, and so the College of Cartographers evolved a Map of the Empire that was of the same Scale as the Empire and that coincided with it point for point. Less attentive to the Study of Cartography, succeeding Generations came to judge a map of such Magnitude cumbersome, and, not without Irreverence, they abandoned it to the Rigors of sun and Rain. In the western Deserts, tattered Fragments of the Map are still to be found, Sheltering an occasional Beast or beggar, in the whole Nation, no other relic is left of the Discipline of Geography." Jorge Luis Borges, *A Universal History of Infamy,* trans. Norman Thomas di Giovanni (New York: E. P. Dutton, 1972), 141; also in Jorges Luis Borges, *Dreamtigers,* trans. Mildred Boyer and Harold Morland (Austin: University of Texas Press, 1964), 90.

44. Malevich, "An Attempt to Determine the Relation Between Colour and Form in Painting" (1930), *Essays on Art 1915–1933,* 2:126–27. On the same theme see also Wassily Kandinsky, "The Language of Form and Colour," *Concerning the Spritual in Art,* trans. with an intro. by M. T. H. Stadler (New York: Dover, 1977), 27–45. By Malevich's own admission Kandinsky's book, originally published in 1911, played a crucial role in his thinking.

45. A number of Malevich's late paintings are mysteriously antedated, perhaps the artist's way of misleading his censors, perhaps a sign that he wanted to rewrite his own history so that he would appear even more ahead of his contemporaries than he actually was.

46. In this context it should be noted that Malevich was deeply interested in the new physics of his time, especially in the conception of a fourth dimension. See in particular Linda Dalrymple Henderson, *The Fourth Dimension and Non-Euclidean Geometry in Modern Art* (Princeton, N.J.: Princeton University Press, 1983), chap. 5.

47. Malevich, "Suprematism: 34 Drawings" (1920), *Essays on Art 1915–1933,* 1:125.

48. Rainer Crone and David Moos, *Kazimir Malevich: The Climax of Disclosure,* 128.

49. Now replaced by a concrete structure of the same type but with the original black square colored red.

50. Quoted in Charlotte Douglas, *Kazimir Malevich,* 33.

51. In this connection I cannot refrain from drawing attention to Jette Hansen-Møller, *Den skjulte diagonal: En landskabsfortelling i ord og billeder* (København: Christian Ejlers' Forlag, 1995). In my own mind, Hansen-Møller's study is an attempt to capture, in the same Mondrian-like coordinate net, both the words of Sophocles' *Antigone* and the pictures of Pablo Picasso's *Las Meninas;* a different form of cartographical reason in practice, a dance of life and death performed on a stage of color. In addition one of the most daring dissertations that ever came my way.

52. René Girard, *Job: The Victim of His People,* trans. Yvonne Freccero (Stanford, Calif.: Stanford University Press, 1985), 38.

53. Karl Marx, *Capital,* trans. S. Moore and E. Aveling (New York: International Publishers, 1964), 1:71.

54. When Erich Auerbach argued that the *Odyssey* is only foreground, then he was clearly overstating his case.

55. This is a point which the anonymous compilers J, E, D, P and R all share in common with Homer, Sophocles, Plato, Aristotle and Shakespeare.

56. Ludwig Wittgenstein, *Culture and Value,* ed. G. H. von Wright in collabration with Heikki Nyman, trans. Peter Winch (Oxford: Basil Blackwell, 1980), 19e.

## KANT

1. For brief comments about Kant as a practicing rhetorician, see Howard Caygill's introduction to Immanuel Kant, *Critique of Pure Reason,* trans. Norman Kemp Smith (New York: Palgrave Macmillan, 2003), xiii–xiv.

2. Ibid., A 474/B 502. In this context remember also the Plotinian House, the Vitruvian palaces, and Ludwig Wittgenstein's deep involvement in the design and building of his sister Gretl's house on Kundmannsgasse in Vienna. Also Kojin Karatani, *Architecture as Metaphor: Language, Number, Money,* trans. Sabu Kohso (Cambridge, Mass.: MIT Press, 1995).

3. Franco Farinelli, *I segni del mondo: Imagine cartografica e discorso geografico in età moderna* (Firenze: La Nuova Italia, 1992). In this connection also note the very special session held in conjunction with the meeting of the Association of American Geographers, Boston, 1998: Franco Farinelli, "Did Anaximander ever Say (or Write) any Words? – The Nature of Cartographical Reason"; Gunnar Olsson, "Towards a Critique of Cartographical Reason"; and Dagmar Reichert, "Obituary," all three contributions

subsequently published in *Ethics, Place and Environment* 1 (1998). Many thanks to the editor Tim Unwin for attending our session and for inviting us to his new journal. Parts of the present chapter have their roots in my own Boston paper, while other parts have previously been treated in Gunnar Olsson, "Horror vacui," in Martin Gren and P. O. Hallin, eds., *Svensk kulturgeografi: En exkursion inför 2000-talet* (Lund: Studentlitteratur, 1998); Gunnar Olsson, "Detta är detta," *Nordisk Estetisk Tidskrift* 20–21 (2000): 39–54; and Gunnar Olsson, "Skattkammarön," in Martin Gren et al., eds., *Kulturens plats/Maktens rum* (Stehag: Brutus Östling, 2000).

Then, on the same day that this footnote was first composed, I received a copy of the most wide ranging (post)modern introduction to cartographical reason that has yet come my way—John Pickles, *A History of Spaces: Cartographic Reason, Mapping and the Geo-Coded World* (London: Routledge, 2004). Pickles' generous response to my own question of what it means to draw a line moved me deeply. Peter Gould—John's teacher and my own friend—would have been doubly pleased, certainly by his student's achievement, hopefully by his comrade's reaction.

In this connection mention must also be made of the quite different approach of Karl Schlögel, *Im Raume lesen wir die Zeit: Über Ziviliationsgeschihte und Geopolitk* (München: Carl Hanser Verlag, 2003). In this engaging study history—especially the history of twentieth-century Europe—is seen as a mappable play of concrete places such that the Holocaust is treated not as an abstract instance of the banality of evil à la Hannah Arendt, but as a set of concrete actions which took place in the particular place of Birkenau. In many ways Schlögel's study does to history what Edward Casey's works have done to philosophy, both authors providing welcome additions to the literature of human geography, albeit not to a deeper understanding of cartographical reason per se.

4. Norman Kemp Smith, *A Commentary on Kant's "Critique of Pure Reason,"* 2d ed. (London: Macmillan, 1923), xxiii.

5. Ludwig Wittgenstein, *Philosophical Investigations,* trans. G. E. M. Anscombe (Oxford: Basil Blackwell, 1953), p. 223e.

6. In my estimation the best and for my own purposes most helpful critiques of the *Critique* are Stephan Körner, *Kant* (Harmondsworth: Penguin, 1955); Karl Jaspers, *Kant,* trans. Ralph Manheim (New York: Harcourt, Brace and World, 1962); Gilles Deleuze, *Kant's Critical Philosophy,* trans. Hugh Tomlinson and Barbara Habberjam (Minneapolis: University of Minnesota Press, 1984); Jonathan Bennett, *Kant's Analytic* (Cambridge: Cambridge University Press, 1966); Jonathan Bennett, *Kant's Dialectic* (Cambridge: Cambridge University Press, 1974); B. F. Strawson, *The Bounds of Sense: An Essay on Kant's "Critique of Pure Reason"* (London: Methuen, 1966); Henry E. Allison, *Kant's Transcendental Idealism: An Interpretation and Defense* (New Haven, Conn.: Yale University Press, 1983); Henry E. Allison, *Idealism and Freedom: Essays on Kant's Theoretical and Practical Philosophy* (Cambridge: Cambridge University Press, 1996); Karl Aschenbrenner, *A Companion to Kant's "Critique of Pure Reason": Transcendental Aesthetic and Analytic* (New York: University Press of America, 1983); Paul Guyer, *Kant and the Claims of Knowledge* (Cambridge: Cambridge University Press, 1987); Paul Guyer, *The Cambridge Companion to Kant* (Cambridge: Cambridge University Press, 1992); Michael Friedman, *Kant and the Exact Sciences* (Cambridge, Mass.: Harvard University Press, 1992); James van Cleve, *Problems from Kant* (Oxford: Oxford University Press, 1999); and Predrag Cicovacki, *Between Truth and Illusion: Kant at the Crossroads of Modernity* (Lanham, Md.: Rowman and Littlefield, 2002).

7. Immanuel Kant, *Critique of Pure Reason,* trans. F. Max Müller (Garden City, N.Y.: Doubleday, Anchor Books, 1966), A 236/B 295.

8. Immanuel Kant, *Critique of Pure Reason,* trans. Norman Kemp Smith, A 735/B 763.

9. Kant, *Critique of Pure Reason,* trans. F. Max Müller, A 795/B 823.

10. Kant, *Critique of Pure Reason,* trans. Norman Kemp Smith, A 255/B 311.

11. Quoted in Paul Guyer, *Kant and the Claims of Knowledge,* 15.

12. Quotation from the translation by Paul Carus reprinted in Beryl Logan, ed., *Immanuel Kant's "Prolegomena to Any Future Metaphysics" in Focus* (London: Routledge, 1966), §32, 78–79, emph. added.

13. Quotation from Ernst Cassirer, *Kant's Life and Thought,* trans. James Haden (New Haven, Conn.: Yale University Press, 1981), 40. Together with Cassirer's book the most useful introductions to Kant's life and work are Arsenij Gulyga, *Immanuel Kant: His Life and Thought,* trans. M. Despalatovic (Boston: Birkhäuser, 1987) and Manfred Kuehn, *Kant: A Biography* (Cambridge: Cambridge University Press, 2001). For a handy introduction to the often cumbersome terminology see Howard Caygill, *A Kant Dictionary* (Oxford: Oxford University Press, 1995).

14. Kant himself never published the notes for these courses, but they were later included in *Immanuel Kant's sämmtliche Werke, herausgegeben von Friedrich Wilhelm Schubert,* Teil 6: *Immanuel Kant's Schriften zur Physischen Geographie* (Leipzig: L. Voss, 1839). Now available in French translation as *Géographie (Physische Geographie)* (Paris: Aubier, 1999). For an ideology critique prompted by the latter edition see David Harvey, "Cosmopolitanism and the Banality of Geographical Evils," *Public Culture* 4 (2000): 529–64.

15. The only sustained study of Kant's concept of geography is in J. A. May, *Kant's Concept of Geography and Its Relation to Recent Geographical Thought* (Toronto: University of Toronto Press, 1970). In an appendix (pp. 255–64) May also provides an English translation of the introduction to Kant's *Physische Geographie.*

16. For insightful and highly readable accounts of these relations see Bryan Magee, *Confessions of a Philosopher: A Journey Through Western Philosophy* (New York: Random House, 1997). In a crucial footnote Magee in fact writes that one of his own favorites, Karl Popper, "once remarked to me: 'It was through Schopenhauer that I understood Kant.' Since it was also through Schopenhauer that Wittgenstein understood Kant, this means that Schopenhauer has been one of the most important influences on twentieth-century philosophy, while himself languishing unread by most professional philosophers" (159).

17. In his early work—and later in his senility—Kant devoted much energy to the study of religious mysticism. A notorious example is his scathing criticism of Emanuel Swedenborg, originally published in Immanuel Kant, *Träume eines Geistersehers–erläutert durch Träume der Metaphysik* in 1766, available in English translation in John Manoleso, *"Dreams of a Spirit Seeker" by Immanuel Kant and Other Related Writings* (New York: Vintage Press, 1962). For a recent critique of Kant's critique see Gottlieb Florschütz, *Swedenborg and Kant: Emanuel Swedenborg's Mystical View of Humankind, and the Dual Nature of Humankind in Immanuel Kant* (West Chester, Penn.: Swedenborg Foundation, 1993); and Hans Radermacher, *Kant, Swedenborg, Borges: Mit Paraphrasen von Jorge Luis Borges* (Bern: Peter Lang, 1986).

Swedenborg was a fascinating thinker, born in Sweden in 1688 as the son of a prominent theologian eventually made bishop, dead in London in 1772. Originally educated as a geologist, he grew into a highly accomplished cosmologist, geometrician, kabbalist and physiologist. In 1744 he went through a deep crisis that changed his life and led him to compose his famous dream book, a report of his conversations with the angels, by definition ambassadors appointed by God. Throughout his career, the world was for him a heavenly constellation of signs waiting to be interpreted. For that reason it is not surprising that his theosophic studies have influenced a range of authors including Carl Jonas Love Almqvist, August Strindberg, Gunnar Ekelöf and Lars Gyllensten, Baudelaire and Balzac, Blake, Emerson and Whitman, Dostoyevsky and Joyce, Milosz and Borges. Carl Jung and Arnold Schönberg too.

Excellent treatments of Swedenborg's life and work are in Inge Jonsson and Olle Hjern, *Swedenborg: Sökaren i naturens och andens världar. Hans verk och efterföljd* (Stockholm: Proporius, 1976); Olof Lagercrantz, *Dikten om livet på den andra sidan: En bok om Emanuel Swedenborg* (Stockholm: Wahlström och Widstrand, 1996); Lars Bergquist, *Biblioteket i lusthuset: Tio uppsatser om Swedenborg* (Stockholm: Natür och Kultur, 1996); Lars Bergquist, *Swedenborgs hemlighet: Om ordens betydelse, änglarnas liv och tjänsten hos Gud. En biografi* (Stockholm: Natur och Kultur, 1999). Also see James F. Lawrence, ed., *Testimony to the Invisible: Essays on Swedenborg* (West Chester, Penn.: Swedenborg Foundation, 1995); and Lynn R. Wilkinson, *The Dream of an Absolute Language: Emanuel Swedenborg and French Literary Culture* (Albany, N.Y.: State University of New York Press, 1996).

As testimony to Swedenborg's importance, Jorge Luis Borges once wrote that "Voltaire said that Charles XII was the most remarkable man that history knows. I would rather say—if such superlatives are not forbidden—that the most remarkable was the most mystical of Charles XII's subjects, Emanuel Swedenborg." Not a man apart, but a lucid and laborious investigator of Heaven and Hell.

18. Ludwig Wittgenstein, *Tractatus Logico-Philosophicus,* trans. D. F. Pears and B. F. McGuiness (London: Routledge and Kegan Paul, 1961), 2.1, 2.11, 2.12, 2.151, 2.1511, 2.1512, 2.15121, 2.1514, 2.1515, 2.171, 2.182, 2.221, 3.001, 3.031, 3.032, 3.11, 3.144, 3.203, 6.342, 6.35.

19. G. E. M. Anscombe, *An Introduction to Wittgenstein's "Tractatus"* (Philadelphia: University of Pennsylvania Press, 1971), 68.

20. Ibid., 72.

21. Wittgenstein, *Philosophical Investigations,* p. 187e.

22. There are even those who argue that Wittgenstein's mature work is less philosophy and more creative literature. In my view a well-taken position, but only as long as one admits that the *Tractatus* may be read as a piece of minimalistic poetry as well. Of course one can have neither literature, poetry nor philosophy without language. For a related argument see Marcus B. Hester, *The Meaning of Metaphor: An Analysis in the Light of Wittgenstein's Claim that Meaning is Use* (The Hague: Mouton, 1967); and Marjorie Perloff, *Wittgenstein's Ladder: Poetic Language and the Strangeness of the Ordinary* (Chicago: University of Chicago Press, 1996).

23. Wittgenstein, *Philosophical Investigations,* 373 and 329.

24. Wittgenstein, *Tractatus Logico-Philosphicus,* p. 1. (". . . And everything one knows, not only what can be heard humming and droning, may be said in three words.") My translation.

25. Ibid., p. 3. Wittgenstein's conception of nonsense has later become a matter of considerable controversy. For some of the most recent contributions (especially from James Conant and Cora Diamond, who strongly argue for the close relations between the *Tractatus* and Wittgenstein's later writings) see Alice Crary and Rupert Read, eds., *The New Wittgenstein* (London: Routledge, 2000); and Erich H. Reck, ed., *From Frege to*

*Wittgenstein: Perspectives on Early Analytic Philosophy* (Oxford: Oxford University Press, 2002).

26. Quotation in Logan, ed., *Immanuel Kant's "Prolegomena to Any Future Metaphysics" in Focus*, §58, p. 111.

27. Kant, *Critique of Practical Reason*, trans. Mary Gregor (Cambridge: Cambridge University Press, 1997), 5:162.

28. My own interest in Wittgenstein's life and work goes a long way back. Over the years I have found the following to be closest to my own concerns: Norman Malcolm, *Ludwig Wittgenstein: A Memoir* (Oxford: Oxford University Press, 1958); Erik Stenius, *Wittgenstein's "Tractatus": A Critical Exposition of its Main Lines of Thought* (Ithaca, N.Y.: Cornell University Press, 1964); Marcus B. Hester, *The Meaning of Poetic Metaphor: An Analysis in the Light of Wittgenstein's Claim that Meaning is Use;* P. M. S. Hacker, *Insight and Illusion: Wittgenstein on Philosophy and the Metaphysics of Experience* (Oxford: Oxford University Press, 1972); Hanna Fenichel Pitkin, *Wittgenstein and Justice: On the Significance of Ludwig Wittgenstein for Social and Political Thought* (Berkeley and Los Angeles: University of California Press, 1972); Allan Janik and Stephen Toulmin, *Wittgenstein's Vienna* (New York: Simon and Schuster, 1973); Anthony Kenny, *Wittgenstein* (Cambridge, Mass.: Harvard University Press, 1973); Jacques Bouveresse, *Le mythe de l'intériorité: Expérience, signification et langage privé chez Wittgenstein* (Paris: Minuit, 1976); Rush Rhees, ed., *Ludwig Wittgenstein: Personal Recollections* (Oxford: Basil Blackwell, 1981); David Bloor, *Wittgenstein: A Social Theory of Knowledge* (London: Macmillan, 1983); David Bloor, *Wittgenstein, Rules and Regulations* (London: Routledge, 1997); Henry Statten, *Wittgenstein and Derrida* (Oxford: Basil Blackwell, 1985); Norman Malcolm, *Wittgenstein: Nothing is Hidden* (Oxford: Basil Blackwell, 1986); Stephen S. Hilmy, *The Later Wittgenstein: The Emergence of a New Philosophical Method* (Oxford: Basil Blackwell, 1987); David Pears, *The False Prison: A Study of the Development of Wittgenstein's Philosophy*, 2 vols. (Oxford: Clarendon Press, 1987 and 1988); Brian McGuinness, *Wittgenstein: A Life. Young Ludwig (1889–1921)* (Berkeley and Los Angeles: University of California Press, 1988); Ray Monk, *Ludwig Wittgenstein: The Duty of Genius* (London: Jonathan Cape, 1990); Stephan Mulhall, *On Being in the World: Wittgenstein and Heidegger on Seeing Aspects* (London: Routledge, 1990); Cora Diamond, *The Realistic Spirit: Wittgenstein, Philosophy and the Mind* (Cambridge, Mass.: MIT Press, 1991); Marjorie Perloff, *Wittgenstein's Ladder: Poetic Language and the Strangeness of the Ordinary;* Stephen Mulhall, *Inheritance and Originality: Wittgenstein, Heidegger and Kierkegaard* (Oxford: Clarendon Press, 2001); Oswald Hanfling, *Wittgenstein and the Human Form of Life* (London: Routledge, 2002).

29. Kant, *Critique of Pure Reason*, trans. Norman Kemp Smith, A 51/B 75.

30. Wittgenstein, *Philosophical Investigations*, 219.

31. Kant, *Critique of Pure Reason*, trans. Norman Kemp Smith, B xvii and A 26/B 42.

32. The established image of Kant as an idealist philosopher who ignored embodied human existence is forcefully and correctly rejected by Helge Svare, *In situ corpori: Kant's Theory of Experience Interpreted in the Context of His Theory of the Embodied Mind* (Trondheim: Norwegian University of Science and Technology, 2003). Along the same lines also Arthur Melnick, *Space, Time and Thought in Kant* (Dordrecht: Kluwer, 1989) and Susan Meld Shell, *The Embodiment of Reason: Kant on Spirit, Generation, and Community* (Chicago: University of Chicago Press, 1996). The truth is that Kant suffered from a serious case of hypochondria.

33. For a clear and detailed analysis of Kant's ambiguous views on this complicated matter see Paul Guyer, *Kant and the Claims of Knowledge*, pt. 2; also Pierre Keller, *Kant*

*and the Demand of Self-Consciousness* (Cambridge: Cambridge University Press, 1998), chaps. 9 and 10.

34. Kant, *Critique of Pure Reason,* trans. Norman Kemp Smith, B 14.

35. Quoted in Guyer, *Kant and the Claims of Knowledge,* 364.

36. Rainer Enskat, *Kant's Theorie des geometrischen Gegenstandes: Untersuchungen über die Voraussetzungen der Entdeckbarkeit geometrischer Gegenstände bei Kant* (Berlin: Walter de Gruyter, 1978). Also P. F. Strawson, *The Bounds of Sense: An Essay on Kant's "Critique of Pure Reason,"* 277–92; and especially Michael Friedman, *Kant and the Exact Sciences,* 55–95.

37. On the corporeal turn in linguistics see especially Horst Ruthrof, *The Body in Language* (London: Cassell, 2000).

38. Ludwig Wittgenstein, *Tractatus Logico-Philosophicus,* 5.631. In my mind this passage is a back-reference to the title page of the book Wittgenstein *did* write, especially to the dedication to his friend David Pinsent and the motto that everything can be said in three words, those presumably being "I love you." For a fictional development of the same theme see Bruce Duffy, *The World As I Found It* (New York: Ticknor and Fields, 1987).

39. Wittgenstein, *Tractatus Logico-Philosophicus,* 6.1, 6.11, 6.1251, 6.522.

40. Kant, *Critique of Pure Reason,* trans. Norman Kemp Smith, B 276.

41. Kant, *Grounding for the Metaphysics of Morals,* trans. James W. Ellington (Indianapolis, Ind.: Hackett, 1981), 30.

42. Kant, *Critique of Judgment,* trans. James Creed Meredith (Oxford: Clarendon Press, 1952), 147.

43. For a similar argument see Thierry de Duve, *Kant after Duchamp* (Cambridge, Mass.: MIT Press, 1996), chap. 5.

44. Stephan Körner, *Kant,* 62.

45. Quoted from Erik van der Heeg and Sven Olov Wallenstein, eds., *Tankens arkipelag: Moderna kantläsningar* (Stehag: Brutus Östling, 1992). My translation.

46. Clayton Eshleman, *Juniper Fuse: Upper Paleolithic Imagination and the Construction of the Underworld* (Middletown, Conn.: Wesleyan University Press, 2003).

47. The classic treatment is in Hans Vaihinger, *The Philosophy of 'As-if': A System of the Theoretical, Practical and Religious Fictions of Mankind,* trans. C. K. Ogden (London: Routledge and Kegan Paul, 1924).

48. The crucial role of imagination in Kant's philosophy was first treated in Martin Heidegger, *Kant and the Problem of Metaphysics,* trans. Richard Taft (Bloomington: Indiana University Press, 1984). For later developments see Rudolf A. Makkreel, *Imagination and Interpretation in Kant: The Hermeneutical Import of "The Critique of Judgment"* (Chicago: University of Chicago Press, 1990); and Sarah L. Gibbons, *Kant's Theory of Imagination: Bridging Gaps in Judgment and Experience* (Oxford: Clarendon Press, 1994).

49. Wittgenstein, *Tractatus Logico-Philosophicus,* 6.13.

50. Barnett Newman, *Selected Writings and Interviews,* ed. John P. O'Neill (New York: Alfred A. Knopf, 1990), 247 and 304.

51. Kant, *Critique of Pure Reason,* trans. F. Max Müller, B 157 and B 158.

52. Quoted in John Sallis, *Spacings—Of Reason and Imagination in Texts of Kant, Fichte, Hegel* (Chicago: University of Chicago Press, 1987), 53.

53. For an up-to-date summary see James van Cleve, *Problems from Kant,* chap. 10.

54. In a passage that fits beautifully the mood also of the present chapter—indeed of this entire book—Howard Caygill makes the following point: "A confessional note

which Kant jotted down while writing CPR reveals the motivation behind the work in progress. It is the same project of defending metaphysics as had already been announced in [*Dreams of a Spirit-Seer elucidated by Dreams of Metaphysics*]: human beings are impelled to ask metaphysical questions such as 'From whence am I? What is the origin of all this?' In order to make any judgment in response to these questions, it is necessary to be oriented in the realm of metaphysics. But this realm does not lie without, but within us, and so 'the critique of pure reason thrusts a torch into this gloom but it illuminates the dark spaces of our understanding, not the things unknown to us beyond the world of sense.' The image captures nicely the equivocal character of CPR: the torch illuminates our own dark spaces but it also puts whatever is in them to the flame. God, the world and the soul are revealed as our own shadowy projections, products of our own lack of enlightenment or, in Kant's terms, our 'self-imposed tutelage.' The impact of CPR has been commensurate with its ambition." Howard Caygill, *A Kant Dictionary,* 149.

55. For an alternative interpretation of the Island-of-Truth-quotation see Michèle Le Doeuff, *The Philosophical Imaginary,* trans. Colin Gordon (London: Athlone Press, 1989), 1–20.

56. For the most influential texts of this group see George Lakoff and Mark Johnson, *Metaphors We Live By* (Chicago: University of Chicago Press, 1980); George Lakoff, *Women, Fire, and Dangerous Things: What Categories Reveal About the Mind* (Chicago: University of Chicago Press, 1987); George Lakoff and Mark Turner, *More than Cool Reason: A Field Guide to Poetic Metaphor* (Chicago: University of Chicago Press, 1989); George Lakoff and Mark Johnson, *Philosophy in the Flesh: The Embodied Mind and Its Challenge to Western Thought* (New York: Basic Books, 1999); George Lakoff and Rafael Núñez, *Where Mathematics Comes From: How the Embodied Mind Brings Mathematics into Being* (New York: Basic Books, 2000).

57. Quoted in Hajo Düchting, *Paul Cézanne, 1839–1906: Nature into Art* (Köln: Taschen, 1999), 214, emph. added.

58. These courageous explorers of limits—well worth entire chapters of their own—have taught me much about what it means to be human. For some of the most insightful studies of Bacon see David Sylvester, *The Brutality of Facts: Interviews with Francis Bacon* (London: Thames and Hudson, 1987); David Sylvester, *Looking Back at Francis Bacon* (London: Thames and Hudson, 2000); Ernst van Alphen, *Francis Bacon and the Loss of Self* (London: Reaktion Books, 1992); Achille Bonito Oliva, ed., *Figurabile: Francis Bacon* (Milano: Electa, 1993); Daniel Farson, *The Gilded Gutter Life of Francis Bacon* (London: Vintage, 1994); Marente Bloemheuvel, ed., *Det unika med att vara människa/The Particularity of Being Human: Marlene Dumas, Francis Bacon* (Malmö: Malmö Konsthall, 1995); Michael Peppiatt, *Francis Bacon: Anatomy of an Enigma* (London: Weidenfeld and Nicholson, 1996); Philippe Sollers, *Les passions de Francis Bacon* (Paris: Gallimard, 1996); Hugh M. Davies, *Francis Bacon: The Papal Portraits* (San Diego, Calif.: Museum of Contemporary Art, 1999); Gilles Deleuze, *Francis Bacon: The Logic of Sensation,* trans. Daniel W. Smith (London: Continuum, 2003). Corresponding treatments of Rothko include Lee Seldes, *The Legacy of Mark Rothko* (New York: Holt, Rinehart and Winston, 1978); Dore Ashton, *About Rothko* (Oxford: Oxford University Press, 1983); James E. B. Breslin, *Mark Rothko: A Biography* (Chicago: University of Chicago Press, 1993); Jeffrey Weiss, ed., *Mark Rothko* (New Haven, Conn.: Yale University Press, 1998); David Anfam, *Mark Rothko: The Works on Canvas. Catalogue Raisonné* (New Haven, Conn.: Yale University Press, 1998); Diane Waldman, *Mark Rothko, 1903–1970* (London: Thames and

Hudson, 2001); Marjorie Cohn, ed., *Mark Rothko.* Exhibition, 18 February–29 April 2001, Fondation Beyeler (Ostfildern-Ruit: Hatje Cantz, 2001). Also the newly found manifesto, Mark Rothko, *The Artist's Reality: Philosophies of Art,* ed. with an intro. by Christopher Rothko (New Haven, Conn.: Yale University Press, 2004).

59. Rothko, *The Artist's Reality: Philosophies of Art,* 22.

60. Sylvester, *The Brutality of Fact: Interviews with Francis Bacon,* 166.

61. Thomas Hobbes, *Leviathan,* ed. with an intro. by C. B. Macpherson (London: Penguin, 1968), 82–83.

62. Job 41:24 and 41:33. At the upper edge of the engraving these words are inscribed: *Non est potestas Super Terram quæ Comparetur ei Job 41.24.* It should be noted that the verse-numbering varies between different translations.

## MISSION IMPOSSIBLE

1. "At road's end I see power / and it looks like an onion / with overlapping faces / that peel off one after the other." My translation. The Swedish original was first published in Tomas Tranströmer, *Den stora gåtan* (Stockholm: Bonnier, 2004).

## URUK

1. Thorkild Jacobsen, "The Gilgamesh Epic: Romantic and Tragic Vision," in Tzvi Abusch et al. eds., *Lingering over Words: Studies in Ancient Near Eastern Literature in Honor of William L. Moran* (Atlanta, Ga.: Scholars Press, 1990), 249.

2. Unless otherwise noted all quotations in this chapter will be taken from Andrew R. George, *The Babylonia Gilgamesh Epic: Introduction, Critical Edition and Cuneiform Texts* (Oxford: Oxford University Press, 2003). The reader should be cautioned, however, that in the following quotations I follow the accepted practice of the Assyriologists, such that [square brackets] are enclosing words that are restored in passages where the tablet is broken, while *italics* are indicating insecure decipherments. (Round brackets) mark words which are inserted to give a better rendering in English.

3. E. A. Wallis Budge, *The Rise and Progress of Assyriology* (London: Martin Hopkinson, 1925), 153.

4. Since I cannot read any of the original texts, my fascination is due entirely to this well-connected group of world Assyriologists. Still, the material is so exceptionally relevant for my own concerns that it cannot be left to the experts alone. My only hope is once again that they—like all other experts whose territories this book is transgressing—will forgive my clumsy transgressions and realize that my interest is not in the epic per se, but in the many ways it ties in with the theme of cartographical reason. As much as exegesis is tempting, exegesis is out of the question.

The general introductions, translations and transformations are many and often quite diverging, partly because new material is constantly being discovered, partly because the original itself is full of word plays. In English alone there are at least twenty different versions. Personally I have worked within the orbits of the following: G. Contenau, *L'épopée de Gilgamesh: Poème babylonien* (Paris: L'artisan de livre, 1939); Knut Tallqvist, *Gilgamesh-eposet* (Stockholm: Geber, 1945); Alexander Heidel, *The Gilgamesh Epic and Old Testament Parallels* (Chicago: University of Chicago Press, 1946);

James B. Pritchard, ed., *Ancient Near Eastern Texts Relating to the Old Testament* (Princeton, N.J.: Princeton University Press, 1969); Nancy K. Sandars, *The Epic of Gilgamesh* (Harmondsworth: Penguin, 1960); Geoffrey S. Kirk, *Myth: Its Meaning and Functions in Ancient and Other Cultures* (Berkeley and Los Angeles: University of California Press, 1970), chap. 4; Jeffrey H. Tigay, *The Evolution of the Gilgamesh Epic* (Philadelphia: University of Pennsylvania Press, 1982); John Gardner and John Maier, *Gilgamesh: Translated from the Sin-leqi-sunnini Version* (New York: Knopf, 1984); Hope Nash Wolf, *A Study in the Narrative Structure of Three Epic Poems: Gilgamesh, The Odyssey, Beowulf* (New York: Garland, 1987); Stephanie Dalley, *Myths from Mesopotamia: Creation, the Flood, Gilgamesh, and Others* (Oxford: Oxford University Press, 1989); Jean Bottéro, *Mesopotamia: Writing, Reasoning, and the Gods,* trans. Zainab Bahrani and Marc van de Mieroop (Chicago: University of Chicago Press, 1992); Benjamin R. Foster, *Before the Muses: An Anthology of Akkadian Literature.* 2 vols. (Bethesda, Md.: CDL Press, 1996); Simo Parpola, *Assyrian Prophesies* (Helsinki: Helsinki University Press, 1997); Ulla and Aage Westenholz, *Gilgamesh—Enuma Elish: Guder og mennesker i oldtidens Babylon* (København: Spektrum, 1997); John Maier, ed., *Gilgamesh: A Reader* (Wauconda, Ill.: Bolchay-Carducci Publishers, 1997); Wayne Horowitz, *Mesopotamian Cosmic Geography* (Winona Lake, Ind.: Eisenbrauns, 1998); Andrew George, *The Epic of Gilgamesh: The Babylonian Epic Poem and Other Texts in Akkadian and Sumerian* (London: Penguin, 1999); Lennart Warring and Taina Kantola, *Gilgamesheposet: 'Han som såg Djupet'* (Stockholm: Natur och Kultur, 2001); Benjamin R. Foster, *The Epic of Gilgamesh: A New Translation, Analogues, Criticism* (New York: Norton, 2001); Jean-Daniel Forest, *L'épopée de Gilgamesh et sa postérité: Introduction au langage symbolique* (Paris: Paris Méditerranée, 2002); Andrew R. George, *The Babylonian Gilgamesh Epic: Introduction, Critical Edition and Cuneiform Texts.* 2 vols. (Oxford: Oxford University Press, 2003). And then—in time to be admired but too late to influence my own reading—there was the wonderful translation of Stephen Mitchell, *Gilgamesh: A New English Version* (New York: Free Press, 2004). In addition there is the brief but magisterial overview in Walter Burkert, *Babylon, Memphis, Persepolis: Eastern Contexts of Greek Culture* (Cambridge, Mass.: Harvard University Press, 2004).

The most ambitious and up-to-date of these works, including an extensive bibliography, is the George edition from 2003. Although most of the following interpretations are mine rather than his, this is the edition from which I have learned the most.

5. Alexander Heidel, *The Gilgamesh Epic and Old Testament Parallels,* tablet I, column i, lines 1–2.

6. The wording is essentially the same in the Swedish translation from 2001. In contrast, the Danish version from 1997 opts for the older version.

7. Notice here that Nuddimund is in fact a pseudonym of the god Ea, the wisest of the gods and the deity that ruled over the fresh-water Ocean below, otherwise known as Apsu.

8. Skakkan was the god of gazelles, wild asses and other animals.

9. For details see the Introduction of John Maier, ed., *Gilgamesh: A Reader,* 33ff.

10. Stephanie Dalley, *Myths for Mesopotamia: Creation, the Flood, Gilgamesh, and Others,* quotation from *Gilgamesh,* tablet I, iv.

11. Anu was not only the father of the gods and the special god of the sky but also a resident of the city of Uruk. Ishtar, on her part, was Anu's daughter, the deity of Uruk and the goddess of sexual love. Both Anu and Ishtar will later play important roles in Gilgamesh's search for himself.

12. For a general overview of the role of dreams in the Babylonian society, see

A. Leo Oppenheim, *The Interpretations of Dreams in the Ancient Near East: With a Translation of an Assyrian Dreambook* (Philadelphia: American Philosophical Society, 1956); and S. A. L. Butler, *Mesopotamian Conceptions of Dreams and Dream Rituals* (Münster: Ugarit-Verlag, 1998).

13. At this point it should be noted that there are several parallels between the treatment of love in *Gilgamesh* and Plato's discussion of the same them in the *Symposium.* See for instance George F. Held, "Parallels between the *Gilgamesh Epic* and Plato's *Symposium*," *Journal of Near Eastern Studies* 42 (1983): 133–141.

14. George, *The Babylonian Gilgamesh Epic,* 460.

15. In typical fashion "the thigh" may well be a euphemism.

16. For a sense of how the Babylonians viewed the problem of death, see Bendt Alster, ed., *Death in Mesopotamia: Papers read at the XXVIe Rencontre assyriologique internationale* (København: Akademisk Forlag, 1980).

17. As will soon be clear, Uta-napishti means "I Found Life," a most appropriate name for the only man who survived the Deluge and was made immortal in the process. Sin was the Moon God.

18. George, *The Babylonian Gilgamesh Epic,* 494–97.

19. On many of these issues see Wayne Horowitz, *Mesopotamian Cosmic Geography.*

20. George, *The Babylonian Gilgamesh Epic,* 527–28.

21. For the what, when and why of Tablet XII, see George, *The Babylonian Gilgamesh Epic,* 47–54.

## PENIEL

1. The first part of this chapter is a much reworked version of Gunnar Olsson, "Landscape—Border Station Between Stonescape and Mindscape," in Girolamo Cusimano, ed., *La construzione del Paesaggio Siciliano: Geografi e scrittori a confronto* (Palermo: Università degli Studi di Palermo, 1999).

2. Homeros, *Odysséen,* trans. Erland Lagerlöf (Lund. Gleerups, 1957), 1.1–6.

3. Karl Marx, *Capital,* trans. Samuel Moore and Edward Aveling (New York: International Publishers, 1967), 1:71, 72.

4. This conception of the relation between the Lord and his people—where at the beginning it was Abram who belonged to the Lord, while at the end it is the Lord who belongs to Abraham—is central also to the influential analysis in Jack Miles, *God: A Biography* (New York: Simon and Schuster, 1995), 64ff.

5. Once again the literature is immense. My own interpretations draw especially on Bruce Vawter, *On Genesis: A New Reading* (London: Chapman, 1977); Robert Alter, *The Art of Biblical Narrative* (New York: Basic Books, 1981); Northrop Frye, *Words with Power: Being a Second Study of the Bible and Literature* (New York: Harcourt Brace Jovanovich, 1990); Jack Miles, *God: A Biography;* Avivah Gottlieb Zornberg, *The Beginning of Desire: Reflections on Genesis* (New York: Doubleday, 1995), 144–252; Karen Armstrong, *In the Beginning: A New Interpretation of Genesis* (New York: Alfred A. Knopf, 1996).

6. The crucial role of Rebekah is sensitively discussed in Sharon Pace Jeansonne, *The Women of Genesis: From Sarah to Potiphar's Wife* (Minneapolis, Minn.: Fortress Press, 1990), chap. 5. It is especially noteworthy that when a wife was to be chosen for Isaac, Abraham sent a servant to find one, not the bridegroom to be; the relation between father and son was obviously never one of trust. And how could it have been after

the events on Mount Moriah! In addition, it should be noted that Rebekah is the first woman that the biblical narrator allows to speak directly to God except to answer a question.

7. Emphasis added, my way of stressing that Jacob did not ask to be blessed by *the* God, but by *Isaac's* God, by implication not necessarily his own. It may even be argued that in this passage it is not Isaac who is fooled, but the LORD himself. Here, as in the rest of the Jacob discussion, most quotations are from *The Holy Bible,* The Revised Standard Edition.

8. "Bethel" is actually the same word as "Babel," a term which derives from *bala,* "confusion" or "the gate of God."

9. *Bereshit Rabbah* 70:17. Quoted from Avivah Gottlieb Zornberg, *The Beginning of Desire: Reflections on Genesis,* 144.

10. That is, "God of Bethel."

11. As usual a host of people have guided me through a terrain of wonderful complexity, a remarkable illustration of the fact that even though power never sits still, its strategies and rhetorical techniques have remained essentially the same. For showing the way in the unknown I have drawn especially on Edouard Dhorme, *A Commentary on the Book of Job,* trans. Harold Knight (London: Nelson, 1967); Norman H. Snaith, *The Book of Job: Its Origin and Purpose* (London: SCM Press, 1968); Carl Gustav Jung, *Answer to Job,* trans. R. F. C. Hull (Princeton, N.J.: Princeton University Press, 1973); Robert Gordis, *The Book of Job: Commentary, New Translation, and Special Studies* (New York: Jewish Theological Seminary, 5738/1978); Robert Alter, *The Art of Biblical Poetry* (New York: Basic Books, 1985), chap. 4; René Girard, *Job: The Victim of His People,* trans. Yvonne Freccero (Stanford, Calif.: Stanford University Press, 1987); Moshe Greenberg, "Job," in Robert Alter and Frank Kermode, eds., *The Literary Guide to the Bible* (London: Collins, 1987); David J. A. Clines, *Job 1–20* (Dallas, Tex.: Word Books, 1989); Edwin M. Good, *In Turns of Tempest: A Reading of Job with a Translation* (Stanford, Calif.: Stanford University Press, 1990); Leo G. Perdue and W. Clark Gilpin, eds., *The Voice from the Whirlwind: Interpreting the Book of Job* (Nashville, Tenn.: Abingdon Press, 1992); Jonathan Lamb, *The Rhetoric of Suffering: Reading the Book of Job in the Eighteenth Century* (Oxford: Clarendon Press, 1995); Pieter van der Lugt, *Rhetorical Criticism and the Book of Job* (Leiden: E. J. Brill, 1995); Jack Miles, *God: A Biography;* David Wolfers, *Deep Things out of Darkness: The Book of Job. Essays and a New English Translation* (Grand Rapids, Mich.: Eerdmans, 1995); Yair Hoffman, *The Blemished Perfection: The Book of Job in Context.* Sheffield: Sheffield Academic Press, 1996; Scott B. Noegel, *Janus Parallelism in the Book of Job* (Sheffield: Sheffield Academic Press, 1996); Bill Thomason, *God on Trial: The Book of Job and Human Suffering* (Collegeville, Minn.: Liturgical Press, 1997); Norman Whybray, *Job* (Sheffield: Sheffield Academic Press, 1998); Raymond P. Scheindlin, *The Book of Job* (New York: W. W. Norton, 1998); David J. A. Clines, *On the Way to the Postmodern: Old Testament Essays, 1967–1998* (Sheffield: Sheffield Academic Press, 1998), 2:717–819; Philippe Nemo, *Job and the Excess of Evil,* trans. Michael Kigel (Pittsburgh, Penn.: Duquesne University Press, 1998); Joseph F. Kelly, *The Problem of Evil in the Western Tradition: From the Book of Job to Modern Genetics* (Collegeville; Minn.: Liturgical Press, 2002); Carol A. Newsom, *The Book of Job: A Contest of Moral Imaginations* (Oxford: Oxford University Press, 2003).

12. Unless otherwise stated all quotations will be from *The Holy Bible,* New International Version (Grand Rapids, Mich.: Zondervan, 1973).

13. For a most innovative study of the unspeakable acts of torture—especially of

the ways in which it destroys language—see Elaine Scarry, *The Body in Pain: The Making and Unmaking of the World* (Oxford: Oxford University Press, 1985). More straightforward historical accounts are in Page DuBois, *Torture and Truth* (New York: Routledge, 1991); Peter Edwards, *Torture,* new ed. (Philadelphia: University of Pennsylvania Press, 1996); George Ryley Scott, *The History of Torture Throughout the Ages* (1940; rpt., London: Kegan Paul, 2004). Also see William T. Cavanaugh, *Torture and Eucharist: Theology, Politics and the Body of Christ* (Oxford: Basil Blackwell, 1998).

14. René Girard, *Job: The Victim of His People,* 11–13.

15. Philippe Nemo, *Job and the Excess of Evil,* 90.

16. Here it should be noted that in the translation by Raymond P. Scheidlin (*The Book of Job,* 63), Eliphaz says to Job that he should entrust his affairs not to *the* God but to *a* god, perhaps a polytheistic bait that he hopes Job will swallow. In the version by David Wolfers (*Deep Things out of Darkness: The Book of Job. Essays and a New English Translation,* 322) the same passage is rendered as "for my part I would apply to *the deity;* / Yes, to God would I commit my cause."

17. In the biblical religion ascetism is virtually unthinkable, wealth considered a good thing.

18. It is worth noting that in René Girard's reading of Job as a scapegoat this passage is singled out as one of the high points of the entire book. See Girard, *Job: Victim of his People,* 141.

19. For a range of insights into this very special chapter, see Ellen van Wolde, ed., *Job 28: Cognition in Context* (Leiden: Brill, 2003).

20. Franz Kafka, *The Trial,* trans. Villa and Edwin Muir (New York: Schocken, 1968), 1 and 228–29.

21. For detailed analyses of the sublime in the Joban conception of power, see especially Carol A. Newsom, *The Book of Job: A Contest of Moral Imaginations,* 241–56. Also the classical discussion in Edmund Burke, *A Philosophical Enquiry into the Origin of our Ideas of the Sublime and Beautiful,* ed. J. T. Boulton (London: Routledge and Kegan Paul, 1958), 63–70.

22. Perhaps this is the real, hence unconscious, reason why Immanuel Kant devoted so much of the third *Critique* to the concept of sublimity.

23. Here quoted from *The Jerusalem Bible* (London: Darton, Longman and Todd, 1966).

24. The linguistic controversies are highly technical and rarely intended for the layman. For a brief taste see, for instance, Jack Miles, *God: A Biography,* 425–30.

25. Ibid., 318.

26. Ibid., 326–28.

## THEBES

1. My conception of tragedy has formed over a long period, but most decisively during the first half of the 1970s, especially in Epilogue I of Gunnar Olsson, *Birds in Egg/Eggs in Bird* (London: Pion, 1980). Of all my writings none has influenced me more and that explains why the original chapter appears also in Gunnar Olsson, *Lines of Power/Limits of Language* (Minneapolis: University of Minnesota Press, 1991). As with so much of the material for the present book the literature is enormous, but in my selective search I have drawn especially on Friedrich Nietzsche, *The Birth of Tragedy,*

trans. Walter Kaufmann (New York: Vintage Books, 1967); C. M. Bowra, *Sophoclean Tragedy* (Oxford: Clarendon, 1944); Ingemar Hedenius, *Fyra dygder* (Stockholm: Bonnier, 1955), 161–93; Bernard M. W. Knox, *The Heroic Temper: Studies in Sophoclean Tragedy* (Berkeley and Los Angeles: University of California Press, 1964); Bernard M. W. Knox, *Oedipus at Thebes: Sophocles' Tragic Hero and His Time,* 2d ed. (New Haven, Conn.: Yale University Press, 1966); Gerald F. Else, *The Origin and Early Form of Greek Tragedy* (Cambridge, Mass.: Harvard University Press, 1965); Gerald F. Else, *Aristotle's Poetics: The Argument* (Cambridge, Mass.: Harvard University Press, 1967); H. D. F. Kitto, *Poiesis: Structure and Thought* (Berkeley and Los Angeles: University of California Press, 1966); Walter Kaufmann, *Tragedy and Philosophy* (Garden City, N.Y.: Doubleday, 1968); Alister Cameron, *The Identity of Oedipus the King: Five Essays on the "Oedipus Tyrannus"* (New York: New York University Press, 1968); Karl Reinhardt, *Sophocles,* trans. H. and D. Harvey (Oxford: Basil Blackwell, 1979); Jean-Pierre Vernant and Pierre Vidal-Naquet, *Myth and Tragedy in Ancient Greece,* trans. Janet Lloyd (New York: Zone Books, 1988); Frederick Ahl, *Sophocles' Oedipus: Evidence and Self-Conviction* (Ithaca, N.Y.: Cornell University Press, 1991); Charles Segal, *Oedipus Tyrannus: Tragic Heroism and the Limits of Knowledge* (New York: Twayne Publishers, 1993); Jasper Griffin, ed., *Sophocles Revisited: Essays Presented to Sir Hugh Lloyd-Jones* (Oxford: Clarendon, 1999); Felix Budelmann, *The Language of Sophocles: Communality, Communication, and Involvement* (Cambridge: Cambridge University Press, 2000), esp. 206–31.

2. Ludwig Wittgenstein, *Tractatus Logico-Philosophicus,* trans. D. F. Pears and B. F. McGuiness (London: Routledge and Kegan Paul, 1961), 4.112.

3. Homer, *The Iliad,* trans. Richmond Lattimore (Chicago: University of Chicago Press, 1962), 394.

4. Sophocles, *Oedipus the King,* trans. Robert Fagles (Harmondsworth: Penguin, 1984), p. 241, emph. added.

5. Charles Segal, *Oedipus Tyrannus: Tragic Heroism and the Limits of Knowledge,* 35. Segal's reference is to E. R. Dodds, "On Misunderstanding the *Oedipus Rex,*" *Greece and Rome* 13 (1966): 37–49.

6. G. W. F. Hegel, *Phenomenology of Spirit,* trans. J. B. Baillie (New York: Harper Torchbooks, 1967), 237.

7. Unless otherwise noted all quotations (cited by page-number rather than line-number and with all emphases added) are taken from Sophocles, *The Oedipus Cycle,* trans. Dudley Fitts and Robert Fitzgerald (New York: Harcourt, Brace and World, Harvest Edition, 1949). I have compared this translation with several others, including *The Plays of Sophocles,* trans. J. C. Kamerbeck, part 4 (Leiden: E. J. Brill, 1967); Philip Vellacott, *Sophocles and Oedipus* (Ann Arbor: University of Michigan Press, 1971); Sofokles, *Kung Oidipus,* trans. Björn Collinder (Stockholm: Forum, 1972); *Oedipus the King,* trans. Stephen Berg and Diskin Clay (Oxford: Oxford University Press, 1978); and *The Three Theban Plays,* trans. Robert Fagles (Harmondsworth: Penguin, 1984). Although I know of no interpretation that focuses on the same issues that I shall be discussing, the play is so many-faceted that I would be surprised if a parallel reading did not exist. I am well aware that my own fascination with the play is related to my long-standing critique of the Swedish welfare state, a paradigm of good intentions turned to their opposites.

8. The word here translated as "children" is ambiguous as it refers both to Oedipus' subjects in the *polis* and to his children in marriage.

9. Bernard M. W. Knox, *Oedipus at Thebes,* 184.

10. Louis Gernet, *The Anthropology of Ancient Greece* (Baltimore, Md.: Johns Hopkins University Press, 1981), 225–26.

11. For insightful comments on the parallels between Oedipus and Gilgamesh see Jean-Daniel Forest, *L'Épopée de Gilgamesh et sa postérité: Introduction au langage symbolique* (Paris: Méditerranée, 2002), 468–93.

12. For explorations of this theme see Elefteria A. Bernidaki-Aldous, *Blindness in a Culture of Light: Especially the Case of "Oedipus at Colonus" of Sophocles* (New York: Peter Lang, 1990).

13. Jack Miles, *Christ: A Crisis in the Life of God* (London: William Heinemann, 2001), 251.

## NICAEA

1. The history of early Christendom is a real quagmire, deadly dangerous even for the exegeting experts. Yet the issues are too important to be left to them alone, which is not to deny that without their help the present chapter could never have been written. In forming my own understanding I have drawn especially on André Grabar, *Christian Iconography: A Study of Its Origins* (Princeton, N.J.: Princeton University Press, 1968); John Meyendorff, *Christ in Eastern Christian Thought* (Washington, D.C.: Corpus Books, 1969); Jaroslav Pelikan, *The Christian Tradition: A History of the Development of Doctrine,* vol. 1: *The Emergence of the Catholic Tradition (100–600)* (Chicago: University of Chicago Press, 1971); Jaroslav Pelikan, *The Christian Tradition: A History of the Development of Doctrine,* vol. 2: *The Spirit of Eastern Christendom (600–1700)* (Chicago: University of Chicago Press, 1974); Jaroslav Pelikan, *Imago Dei: The Byzantine Apologia for Icons* (Princeton, N.J.: Princeton University Press, 1990); J. N. D. Kelly, *Early Christian Creeds,* 3d ed. (New York: David McKay Company, 1972); Christoph von Schönborn, *L'icône de Christ* (Fribourg: Éditions Universitaires Fribourg Suisse, 1976); R. P. C. Hanson, *The Search for the Christian Doctrine of God: The Arian Controversy 318–381* (Edinburgh: T. and T. Clark, 1988); Judith Herrin, *The Formation of Christendom* (Princeton, N.J.: Princeton University Press, 1989); David Freedberg, *The Power of Images: Studies in the History and Theory of Response* (Chicago: University of Chicago Press, 1989), chap. 14; Kenneth Parry, *Depicting the World: Byzantine Iconophile Thought of the Eight and Ninth Centuries* (Leiden: E. J. Brill, 1996); Henry Maguire, *The Icons of Their Bodies: Saints and Their Images in Byzantium* (Princeton, N.J.: Princeton University Press, 1996); Richard E. Rubenstein, *When Jesus Became God: The Epic Fight over Christ's Divinity in the Last Days of Rome* (New York: Harcourt Brace and Company, 1999); Alain Besançon, *The Forbidden Image: An Intellectual History of Iconoclasm,* trans. Jane Marie Todd (Chicago: University of Chicago Press, 2000); H. A. Drake, *Constantine and the Bishops: The Politics of Intolerance* (Baltimore, Md.: Johns Hopkins University Press, 2000); Peter Brown, *The Rise of Western Christendom: Triumph and Diversity, A.D. 200–1000,* 2d ed. (Oxford: Blackwell Publishing, 2003); Thomas F. Mathews, *The Clash of Gods: A Reinterpretation of Early Christian Art,* rev. ed. (Princeton, N.J.: Princeton University Press, 2003).

2. As will eventually be clear, the current formulation is actually *not* from the council of Nicaea in 325, but from the council of Constantinople in 381. However, the document is still referred to as the Nicene (sometimes the Nicene-Constantinople) Creed.

3. Not merely the four that are included in the present redaction of the New Testament, but several others as well. For insights into the intricate development of the final document see Bert T. Ehrman, *Lost Scriptures: Books that Did Not Make It into the New Testament* (Oxford: Oxford University Press, 2003); and Marvin Meyer, with Esther A. De Boer, *The Gospels of Mary: The Secret Tradition of Mary Magdalene, the Companion of Jesus* (New York: Harper San Francisco, 2004).

4. The following quotations from the Creed (emphases added) all reach back to the first English translation from 1549. With minor variations it has been used ever since.

5. Quoted from *The New Oxford Annotated Bible,* 3d ed., ed. Michael D. Coogan, New Revised Standard Version (Oxford: Oxford University Press, 1989).

6. Held in place by the square brackets is the so-called Filioque clause, not included in the original Creed but introduced in the western church at the beginning of the sixth century, although not formally accepted by the papacy until some time in the eleventh century. It is difficult to find a theological formula of greater dogmatic and historical importance, for eventually it was this word that led to the splitting of the church into an eastern orthodox and a western catholic part. Serious business, of course, but a story brimming of humor as well. For when the Swedes of the eleventh to the fifteenth century listened to their priest speaking Latin, they obviously could not translate his words into their own idiom. Indeed they mistook *filioque* for *filiokus,* a term which subsequently became the Swedish counterpart of "abracadabra," more precisely *hokus, pokus, filiokus.* Perhaps the peasants really *did* understand. Perhaps Martin Luther too, because it was he who often made the point that the iconoclasts themselves took the natural object so seriously that they themselves became idolaters.

7. Indeed there is much to suggest that being human is to be a trinity. For an extended treatment of this conception of the semiotic animal, see Dany-Robert Dufour, *Les mystères de la trinité* (Paris: Gallimard, 1990).

8. While the Latin *catholikus* means "universal" or "general," the Greek *katholikos* combines the two terms *kata* and *holos,* "complete," and "whole."

9. John R. Searle, *The Construction of Social Reality* (New York: Free Press, 1995), 210.

10. See especially Christopher Stead, *Divine Substance* (Oxford: Clarendon Press, 1977); also R. P. C. Hanson, *The Search for the Christian Doctrine of God: The Arian Controversy 318–381,* chap. 7.

11. J. N. D. Kelly, *Early Christian Creeds,* 243. In addition to the vagueness of the term *ousia* the unaware should be cautioned not to confuse the Greek *homo* and the Latin *homo.* For whereas the former means "same," as for instance in "homosexual," the latter means "man," as in "Homo sapiens."

12. Unfortunately, the authentic account of the actual proceedings (if such an account was ever produced) is not extant.

13. For an up-to-date introduction to the conflicting schools see Christopher Stead, *Doctrine and Philosophy in Early Christianity: Arius, Athanasius, Augustine* (Aldershot: Variorum, 2000).

14. The wording comes from Arius' influential text *Thalia* or *The Banquet,* here quoted from Richard E. Rubenstein, *When Jesus Became God: The Epic Fight over Christ's Divinity in the Last Days of Rome,* 55.

15. John Henry Newman, *Arians of the Fourth Century,* new ed. (Westminster, Md.: Christian Classics, 1968), 109.

16. "St. Athanasius," *New Catholic Encyclopedia,* 2d ed. (Washington, D.C.: Catholic University of America, 2003).

17. Rubenstein, *When Jesus Became God,* 206, 209 and 210.

18. Peter Brown, *The Rise of Western Christendom: Triumph and Diversity, A.D. 200–1000,* 116.

19. Recall here my earlier discussion of John Philoponus.

20. Alain Besançon, *The Forbidden Image: An Intellectual History of Iconoclasm,* 119.

21. St. John of Damascus, *On the Divine Images: Three Apologies Against Those Who Attack the Divine Images,* trans. David Anderson (Crestwood, N.Y.: St. Vladimir's Seminary Press, 1980), 2.7, p. 54.

22. Peter Brown, *The Rise of Western Christendom: Triumph and Diversity, A.D. 200–1000,* 387 and 391.

23. According to bishop Eusebius of Caesarea, who claimed that he had heard the story from the monarch himself: "When the day was already beginning to decline, he [the emperor] saw with his own eyes the trophy of the cross of light in the heavens, above the sun, and bearing the inscription, *Conquer By This.* At this sight he himself was struck with amazement, and his whole army also, which followed him on this expedition, and witnessed the miracle. . . . [And on the following day the triumphant Constantine turned this vision into a luminous and talismatic weapon, such that] a long spear, overlaid with gold, formed the figure of the cross by means of a traverse bar laid over it. On the top of the whole was fixed a wreath of gold and precious stones; and within this, the symbol of the Saviour's name, two letters indicating the name of Christ by means of its initial characters, the letter P being intersected by an X in its centre; and these letters the emperor was in the habit of wearing on his helmet at a later period." Eusebius: *Life of Constantine,* quoted in Ewa Kuryluk, *Veronica and Her Cloth: History, Symbolism, and Structure of a 'True' Image* (Oxford: Basil Blackwell, 1991), 48. Later Constantine even had a cloth made that showed himself, not Christ, nailed to the cross.

24. Peter Brown, "A Dark-Age Crisis: Aspects of the Iconoclastic Controversy," *The English Historical Review* 88 (1973): 1–34, at 28. For a detailed critique of this article see Patrick Henry, "What Was the Iconoclastic Controversy About?" *Church History* 45 (1976): 16–31.

25. In a clear parallel to the incident of Rachel sitting on the stolen household gods, refusing to get up from the camel's saddle because the way of women was upon her, Theodora was hiding holy pictures under her clothes. When interrogated by her iconoclast husband she reportedly replied: "These are my dolls; they are pretty and I love them very much." Quoted in Kuryluk, *Veronica and Her Cloth,* 80.

26. Judith Herrin, *The Formation of Christendom,* 309.

27. A most useful overview is in Moshe Barasch, *Icon: Studies in the History of an Idea* (New York: New York University Press, 1992), chap. 10.

28. Brown, *The Rise of Western Christendom: Triumph and Diversity, A.D. 200–1000,* 398–99.

29. St. John of Damascus, *On the Divine Images: Three Apologies Against Those Who Attack the Divine Images,* 1.8, p. 18 and 1.16, p. 23.

30. Ibid., 3.26, p. 81.

31. This in turn helps to explain why the way Christ has been pictured has varied significantly over time. Thus, in the early renderings he was usually shown as a young man with a triangular head and a short beard, then as an old man with a white beard,

eventually as the so-called Pantocrator-type with broad forehead, dark long hair and a full but rather short beard. For details see for instance Kurt Weitzman, *The Monastery of Saint Catherine at Mount Sinai: The Icons,* Vol. 1: *From the Sixth to the Tenth Century* (Princeton, N.J.: Princeton University Press, 1976), 26–27.

32. St. John of Damascus, *On the Divine Images,* 1.17, p. 25. The words within brackets are my additions.

33. Ibid., 2.19, p. 64.

34. Theodore of Studion, quoted in Mark Patrick Hederman, "Cinema and the Icon," *Crane Bag* 8 (1984): 90–99, at 93.

35. Here I simply have to skip any treatment of the last—perhaps the greatest—flourishing of artistic creativity in Byzantium, the period which began about sixty years after Constantinople had been ransacked by the Latin West in 1204 and ended when the city finally fell to the Ottoman Turks in 1453. Nevertheless I cannot refrain from mentioning the outstanding exhibition held at New York's Metropolitan Museum of Art in 2004. For recollections of this great event see Helen C. Evans, ed., *Byzantium: Faith and Power (1261–1557)* (New York: Metropolitan Museum of Art, 2004).

36. For a brief but informative introduction see Geoffrey Hosking, *Russia and the Russians: A History* (Cambridge, Mass.: Harvard University Press, 2001), esp. 21–26 and 36–41. Also Per-Arne Bodin, *Världen som ikon: Åtta föredrag om den ryskortodoxa andliga traditionen* (Skellefteå: Artos Bokförlag, 1988); and Oleg Tarasov, *Icon and Devotion: Sacred Spaces in Imperial Russia,* trans. Robin Milner-Gulland (London: Reaktion Books, 2002).

37. For a semiotic analysis of the binary nature of Russian culture see chap. 1 of Ju. M. Lotman and B. A. Uspenskij, *The Semiotics of Russian Culture,* ed. Ann Shukman, trans. N. F. C. Owen (Ann Arbor: Department of Slavic Languages and Literature, University of Michigan, 1984).

38. Geoffrey Hosking, *Russia and the Russians,* 39–40.

39. Leonide Ouspensky, *Theology of the Icon* (Crestwood, N.Y.: St. Vladimir's Seminary Press, 1978), 31.

40. For detailed descriptions of the different picture programs, see Otto Demus, *Byzantine Mosaic Decoration: Aspects of Monumental Art in Byzantium,* 2d ed. (Boston: Boston Book and Art Shop, 1955); Leonide Ouspensky, *Theology of the Icon,* chap. 1; Konrad Onasch, *Die Ikonenmalerei: Grundzüge einer systematischen Darstellung* (Leipzig: Koehler und Armelang, 1968), chap. 2; Ulf Abel, *Ikonen—bilden av det heliga* (Hedemora: Gidlunds Bokförlag, 1989), 77–101.

41. Leonide Ouspensky, *Theology of the Icon,* 34.

42. For a fascinating illustration of the angelic world—by definition the world of messengers, ambassadors, and other intermediaries—see Michel Serres, *La légende des anges* (Paris: Flammarion, 1993).

43. Hans Belting, *Likeness and Presence: A History of the Image Before the Era of Art,* trans. Edmund Jephcott (Chicago: University of Chicago Press, 1994), 225.

44. For a discussion of this and related issues see the excellent discussion in Annmarie Weyl Carr, "Images: Expressions of Faith and Power," in Helen C. Evans, ed., *Byzantium: Faith and Power (1261–1557).*

45. For insightful comments on this masterpiece see Xavier Girard, *La chapelle de Rosaire* (Paris: Éditions de la Réunion des Musée Nationaux, 1992).

46. Here it should be noted that in the early church smell was considered anathema.

47. On these fundamentally moral issues no one has written more and with greater passion than Peter Singer. For a comprehensive selection of his best and most provocative writings see Peter Singer, *Writings on an Ethical Life* (London: Fourth Estate, 2001).

48. For one of many treatments see Ewa Kuryluk, *Veronica and Her Cloth: History, Symbolism, and Structure, of a 'True' Image.*

49. It was in the Mandylions of the late Byzantine period that Christ's portrait was distilled into its purest and most irreducible form. The background is in the legend of King Abgar of Edessa who, having heard of Jesus' miraculous powers, sent his ambassadors to Jerusalem to obtain a cure for his own illness. Jesus was too busy to go, but when he discovered the image of his face on a towel with which he had wiped himself, he sent that cloth instead. The king was healed and the towel with the non-painted image became a standard motif within the art of icon making. In western tradition the legend is instead tied to the figure of Veronica, who reputedly gave Jesus her kerchief to dry himself with when he was on his way to Golgotha. And Veronica, of course, then discovered that his image was on her cloth. In both cases an instance of *acheriopoetos,* an image not made by human hand.

50. The outstanding analysis of Dürer's development—not the least of how he was transcended by his pupil Hans Baldung Grien—is Joseph Leo Koerner, *The Moment of Self-Portraiture in German Renaissance Art* (Chicago: University of Chicago Press, 1993).

51. Koerner, *The Moment of Self-Portraiture in German Renaissance Art,* 268.

52. In this context it should be noted that the two terms "destiny" and "destination" both stem from the Latin *destinare,* "to fix in position," "to fasten," "to secure," "to determine," "to make firm," "to destine," "to fix for a particular purpose"—in essence a merger of the artist's standpoint and the logician's travel story, the beginning and the end of the cartographer's attempt to screen the world. The magic of ontological transformation at work. As usual the etymology comes from the *Compact OED.*

## PHILADELPHIA

1. The French term is *inframince,* here translated as "infra-thin," by some other commentators sometimes rendered as "infrafine," "infradelicate" or "superthin."

2. Thierry de Duve, *Kant after Duchamp* (Cambridge, Mass.: MIT Press, 1998), 173, n. 31. For further comments see Jerrold Seigel, *The Private Worlds of Marcel Duchamp: Desire, Liberation, and the Self in Modern Culture* (Berkeley and Los Angeles: University of California Press, 1995), 101–102; and Juan Antonio Ramírez, *Duchamp: Love and Death, Even,* trans. Alexander R. Tulloch (London: Reaktion Books, 1998), 192–94.

3. For its place in the overall development of twentieth century art, see for instance John Golding, *Visions of the Modern* (London: Thames and Hudson, 1994).

4. Quoted in Arturo Schwarz, *The Complete Works of Marcel Duchamp* (New York: Harry N. Abrams, 1969), 201. As we will eventually see, this is not the only time that Duchamp sounds like Ludwig Wittgenstein, who in one of his propositions declared that "it is not surprising that the deepest problems are in fact *not* problems at all." Ludwig Wittgenstein, *Tractatus Logico-Philosophicus,* trans. D. F. Pears and B. F. McGuiness (London: Routledge and Kegan Paul, 1963), 4.003.

5. It is a point of some significance that one of Walter Arensberg's Harvard classmates was Wallace Stevens, a person to whom Duchamp was introduced shortly after his arrival in New York. "Marcel speaks very little English, and when the three of them

speak French, Stevens finds that 'it sound[s] like sparrows around a pool of water.'" Quotation from Jennifer Couch-Cooper and Jacques Caumont, eds., *Marcel Duchamp–Ephemerides On and About Marcel Duchamp and Rrose Sélavy, 1887–1968*. Exhibition catalogue, Palazzo Grassi, Venezia (Milan: Bompiani, 1993), August 2, 1915.

6. Marcel Duchamp, *La Mariée mise à nu par ses célibataires, même* (Paris: Éditions Rrose Sélavy, 1934) (*The Green Box*). Translated as *The Bride Stripped Bare by Her Bachelors, Even: A Typographic Version by Richard Hamilton of Marcel Duchamp's "Green Box"* (New York: Wittenborn, 1960). Also see Marcel Duchamp, *In the Infinitive,* typotranslation by Richard Hamilton and Ecke Bonk of Marcel Duchamp's *White Box* (European distribution, Köln: Buchhandlung Walter König, 1999).

7. Anne d'Harnacourt and Walter Hopps, *"Etant Donnés: 1° la chute d'eau, 2° le gaz d'eclairage:* Reflections On a New Work by Marcel Duchamp," *Bulletin of the Philadelphia Museum of Art* 64, nos. 299–300 (1969): 15.

8. Thus far there are four replicas, most famous among them one made by Ulf Linde for the groundbreaking exhibition *Rörelse i konsten* held at the Moderna Museet in Stockholm in 1961 (later reworked with the assistance of many friends and reinstalled into the Museum's standing collection in 1982), another by Richard Hamilton, now at the Tate Modern in London. The former is in many respects a remarkable feat, not the least because the copier has never seen the original. In return Duchamp himself not only signed it but added the remark "certifié pour copie conforme." When Linde first saw pictures of Hamilton's rendering, however, he was so impressed that he immediately withdrew his own version from public inspection. Perhaps not as surprising as it may sound, for it was after having studied Linde's version at the famous retrospective at the Pasadena Art Museum in 1963 that Hamilton concluded that Linde had not been precise enough and that he could do better himself. After actually having seen Hamilton's version, though, Linde revised his opinion and thought it was "*too* good, the bachelors somewhat slack as if they had been copied merely on the surface and not rendered as what they must look like in three dimensions." For details see Ulf Linde, *Marcel Duchamp* (Stockholm: Rabén och Sjögren, 1986), 60–61; also Calvin Tomkins, *Duchamp: A Biography* (London: Pimlico, 1996), 418ff. It is a sobering (perhaps shame-filled) thought that even though I must have seen Linde's version when it was first exhibited in 1961, I can really not remember it!

9. Juan Antonio Ramírez, *Duchamp: Love and Death, Even,* 131.

10. Along with its ordinary meanings, the word *bayonet* is also French slang for "penis."

11. Jerrold Seigel, *The Private Worlds of Marcel Duchamp: Desire, Liberation, and the Self in Modern Culture,* 239.

12. Juan Antonio Ramírez, *Duchamp: Love and Death, Even,* 105.

13. Ulf Linde, *Marcel Duchamp,* 152.

14. Quoted in Jean Clair, "Duchamp and the Classical Perspectivists," *Artforum* 16, no. 7 (1978): 40.

15. Pierre Cabanne, *Dialogues with Marcel Duchamp,* trans. Ron Padgett (London: Thames and Hudson, 1971), 41–42; also Calvin Tomkins, *Duchamp: A Biography,* 124. In this connection it is interesting to note that when Ulf Linde had finished his own copy of the *Large Glass* he was particularly pleased with the color of the chocolate grinder. Over time, however, copy and original have aged according to different processes and what once was considered close to perfect no longer is. The vagaries of translation in a new light.

16. Quoted in Calvin Tomkins, *Duchamp: A Biography,* 127.

17. Marcel Duchamp, *Duchamp du signe: Ecrits,* réunis et présentés par Michel Sanouillet (Paris: Flammarion, 1974), 225. However, it should be noted that in the notes of the *Green Box* Duchamp himself described the Glider or Chariot as "hiding in its bosom the landscape of the water mill."

18. Ulf Linde, *Marcel Duchamp,* passim. On this much-debated issue also see Linda Dalrymple Henderson, *The Fourth Dimension and Non-Euclidean Geometry in Modern Art* (Princeton, N.J.: Princeton University Press, 1983), esp. chap. 3.

19. Cabanne, *Dialogues with Marcel Duchamp,* 40.

20. Tomkins, *Duchamp: A Biography,* 7.

21. And in that sense closely related to the draft pistons of the bride's halo.

22. The most sensitive treatment is in Jerrold Seigel, *The Private Worlds of Marcel Duchamp,* esp. chap. 7.

23. Beatrice Wood, "Marcel," in Rudolf Kuenzli and Francis M. Naumann, eds., *Marcel Duchamp: Artist of the Century* (Cambridge, Mass.: MIT Press, 1989), 16–17.

24. Rosalind E. Krauss, *Bachelors* (Cambridge, Mass.: MIT Press, 1999), 47.

25. Quoted in Tomkins, *Duchamp: A Biography,* 288.

26. Quoted in Craig E. Adcock, *Marcel Duchamp's Notes from the Large Glass: An N-dimensional Analysis* (Ann Arbor, MI: UMI Research Press, 1983), 7 and 83.

27. Molly Nesbit, "The Language of Industry," in Thierry de Duve, ed., *The Definitively Unfinished Marcel Duchamp* (Cambridge, Mass.: MIT Press, 1991), 351.

28. This syntactically awkward, and barely understandable, expression comes directly from the *Green Box,* here quoted in the translation of Michel Sanouillet and Elmer Peterson, eds., *The Writings of Marcel Duchamp* (New York: Da Capo Press, 1973), 42.

29. Robert J. Fogelin, "Three Platonic Analogies," *Philosophical Review* 80 (1971): 371–82.

30. Aristotle, *Metaphysics* 1078b14–19 and 1079a4-5. Here quoted in the translation of C. D. C. Reeve, *Substantial Knowledge: Aristotle's Metaphysics* (Indianapolis, Ind.: Hackett, 2000), 10.

31. Plato, *Republic* 511d, my own preferred vocabulary in brackets. Quotation from the translation of Francis MacDonald Cornford, *The Republic of Plato* (Oxford: Oxford University Press, 1945), 226.

32. This, however, is not to say that there is a perfect parallel between the segments of the Divided Line and the hierarchy of prisoners in the Cave. For detailed analyses of these discrepancies see Richard Robinson, *Plato's Earlier Dialectic,* 2d ed. (Oxford: Clarendon, 1953), chap. 11.

33. For details see Gunnar Olsson, "Glimpses," in Peter Gould and Forrest R. Pitts, eds., *Geographical Voices: Fourteen Autobiographical Essays* (Syracuse, N.Y.: Syracuse University Press, 2002). In that autobiographical context it should be stressed that for me too—as for Plato in the *Republic*—the form of the Good was revealed through studies of mathematical models, not through the practice of self-centered psychoanalysis. Yet it should be recalled that Plato was the first to argue not only that justice in the city is like health in the body, but also that the philosopher is the only person who can be truly just.

34. For detailed discussions see Richard Robinson, *Plato's Earlier Dialectic,* esp. chaps. 6, 7, and 10; as well as J. C. B. Gosling, *Plato* (London: Routledge and Kegan Paul, 1973), 108–14. It is a well-fitting irony that Socrates himself was sentenced to death for the alleged crime of corrupting the young.

35. Immanuel Kant, *Critique of Pure Reason,* trans. Norman Kemp Smith (London: Macmillan, 1973), A 124.

36. Pierre Cabanne, *Dialogues with Marcel Duchamp,* 74.

37. T. J. Clark, *Farewell to an Idea: Episodes from a History of Modernism* (New Haven, Conn.: Yale University Press, 1999), 153.

38. Duchamp, quoted in Dieter Daniels, *Duchamp und die anderen: Der Modellfall einer künstlerischen Wirkungsgeschichte in der Moderne* (Köln: DuMont, 1992), 287, my trans.

39. Quoted in Calvin Tomkins, *Duchamp: A Biography,* 366.

40. Octavio Paz, *Marcel Duchamp: Appearance Stripped Bare,* trans. Rachel Phillips and Donald Gardiner (New York: Viking Press, 1978), 95f.

41. Tomkins, *Duchamp: A Biography,* 451. The admirer was Jasper Johns, the painter.

42. Most books on Duchamp contain chapters on the *Etant donnés,* some more focused than others. I have found the following to be especially helpful: Anne d'Harnancourt and Walter Hopps, "*Etant donnés: 1° la chute d'eau, 2° le gaz d'éclairage:* Reflections On a New Work by Marcel Duchamp"; Octavio Paz, *Marcel Duchamp: Appearance Stripped Bare;* Jean-François Lyotard, *Les TRANSformateurs Duchamp* (Paris: Éditions Galilée, 1977), trans. by Ian McLeod as *Duchamp's TRANS/formers* (Venice, Calif.: Lapis Press, 1990); Pia Høy, *Det dekonstruerede maleri: Marcel Duchamps Etant donnés* (København: Rævens Sorte Bibliotek, 2001); Dalia Judovitz, "Rendezvous with Marcel Duchamp: *Given,*" in Rudolf E. Kuenzli and Francis M. Naumann, eds., *Marcel Duchamp: Artist of the Century;* Amelia Jones, *Postmodernism and the En-Gendering of Marcel Duchamp* (Cambridge: Cambridge University Press, 1994); Jerrold Seigel, *The Private Worlds of Marcel Duchamp: Desire, Liberation and the Self in Modern Culture;* Calvin Tomkins, *Duchamp: A Biography;* Juan Antonio Ramírez, *Duchamp: Love and Death, Even.*

43. Marcel Duchamp, *Manual of Instructions for Étant donnés: 1° la chute d'eau, 2° le gas d'éclairage* (Philadelphia: Philadelphia Museum of Art, 1978).

44. Like all other instances of cartographical reason both *La Mariée* and *Etant donnés* are picture and story turned into one.

45. Octavio Paz, *Marcel Duchamp: Appearance Stripped Bare,* 106.

46. Quite understandably Duchamp's oeuvre, not the least the *Etant donnés,* has stirred strong reactions among women in general and by feminists in particular. For personal testimonies, see e.g., Molly Nesbit, "The Language of Industry," in Thierry de Duve, ed., *The Definitively Unfinished Marcel Duchamp,* 351–94; Karen M. Olvera, *Marcel Duchamp's "The Large Glass" as "Negation of Women"* (Ann Arbor, MI: UMI, 1991); Dalia Judovitz, "Rendezvous with Marcel Duchamp: *Given,*" in Kuenzli and Nauman, eds., *Marcel Duchamp: Artist of the Century;* Amelia Jones, *Postmodernism and the En-Gendering of Marcel Duchamp;* Pia Høy, *Det dekonstruerede maleri: Marcel Duchamps Etant donnés.*

47. Lyotard, *Duchamp's TRANS/formers,* 175.

48. Here it should be stressed that the French word *rose* has exactly the same vulgar connotations as the English "pink." As if to prove the point "pornography shops have books and magazines organized according to the classification 'pink,' meaning that they contain images that show women with their legs spread, revealing the color pink." Quotation from a discussion in de Duve, *The Definitively Unfinished Marcel Duchamp,* 297.

49. Plato, *Symposium* 204a–b. Here quoted as a blend of two translations, the first by Michael Joyce and published in *The Collected Dialogues of Plato,* ed. Edith Hamilton and Huntington Cairns (Princeton, N.J.: Princeton University Press, 1961); the other by

myself based on the Swedish version by Jan Stolpe and published in *Platon: Gästabudet* (Stockholm: Atlantis, 2000).

50. Jacques Lacan, *The Four Fundamental Concepts of Psycho-Analysis,* ed. Jacques-Alain Miller, trans. Alan Sheridan (New York: W. W. Norton, 1981), 104.

51. d'Harnoncourt and Hopps, *"Etant Donnés: 1° la chute d'eau, 2° le gaz d'éclairage:* Reflections On a New Work by Marcel Duchamp," 33.

52. Martin Kemp, *The Science of Art: Optical Themes in Western Art from Brunelleschi to Seurat* (New Haven, Conn.: Yale University Press, 1990), 208.

53. Ramírez, *Duchamp: Love and Death, Even,* 238f.

54. The very first study of what was eventually to become the *Given* (a small pencil drawing of the nude with her legs open, dated Dec. 47) bears the title *Etant donnés: Maria, la chute d'eau et le gaz d'éclairage.* For further comments see Tomkins, *Duchamp: A Biography,* 357.

55. Molly Nesbit, "The Language of Industry," in de Duve, *The Definitively Unifinished Marcel Duchamp,* 366.

56. Jurgis Baltrušatis, *Anamorphic Art,* trans. W. J. Strachan (Cambridge: Chadwick-Healey, 1977), 1 and 4. Baltrušatis's work is the standard reference, but for a good introduction see also Martin Kemp, *The Science of Art,* 203–17.

57. Lyotard, *Duchamp's TRANS/formers,* 181.

58. Gunnar Olsson, "Chiasm of Thought-and-Action," *Environment and Planning D: Society and Space* 11 (1993) 279–94.

59. As far as I know, nothing in the *Manual,* nor in the secondary literature, addresses this hypothesis. Since the tableau itself is too fragile for the necessary measurements and experiments, the hunch is most likely to remain just that.

60. The first specification is given by Tomkins, *Duchamp: A Biography,* 353, the second by Ramírez, *Duchamp: Love and Death, Even,* 233.

61. Seigel, *The Private Worlds of Marcel Duchamp,* 197.

62. Ludwig Wittgenstein, *Tractatus Logico-Philosophicus,* 5.61, 5.62 and 5.621.

63. Ibid., 5.64 and 5.641.

64. One reason for the parallels is of course that the two geniuses were roughly the same age, Duchamp born in 1887, Wittgenstein in 1897. In addition, both of them were decisively, although completely differently, marked by their experiences of World War I, Wittgenstein fighting in the trenches, Duchamp partying in New York, the former with the *Tractatus* in his rucksack, the latter with *La Mariée* on his mind. Perhaps even more importantly, both of them lived in private worlds of their own making, their respective lovers more or less betwixed, their doings more or less clandestine. It is strangely revealing that whereas Duchamp's tombstone carries the epitaph "D'AILLEURS C'EST TOUJOURS LES AUTRE QUI MEURENT," Wittgenstein's last words were "Tell them I've had a wonderful life." As may be expected, the bibliographical literature is extensive. For Duchamp see the quoted works of Tomkins, Seigel and Ramírez; for Wittgenstein in particular Norman Malcolm, *Ludwig Wittgenstein: A Memoir,* with a Biographical Sketch by Georg Henrik von Wright (Oxford: Oxford University Press, 1958); Rush Rhees, ed., *Ludwig Wittgenstein: Personal Recollections* (Oxford: Basil Blackwell, 1981); Bruce Duffy, *The World as I Found It* (New York: Ticknor and Fields, 1987); Brian McGuiness, *Wittgenstein: A Life. Young Ludwig 1889–1921* (Berkeley and Los Angeles: University of California Press, 1988); and Ray Monk, *Ludwig Wittgenstein: The Duty of Genius* (London: Jonathan Cape, 1990).

65. As so often the *Oxford English Dictionary* is indispensable, especially to someone for whom English is not the native language.

66. Wittgenstein, *Tractatus Logico-Philosophicus,* 4.002 and 4.003.

67. Ibid., 4.0031 and 4.01, emph. added.

68. Ludwig Wittgenstein, *Zettel,* 2d ed., ed. G. E. M. Anscombe and G. H. von Wright, trans. G. E. M. Anscombe (Oxford: Basil Blackwell, 1981), 140.

69. For more on the three symbols of mimetic desire, "=", "/" and "—" (the equal sign of logics, the slash of dialectics and the Saussurean Bar of semiotics), see Gunnar Olsson, *Lines of Power/Limits of Language* (Minneapolis: University of Minnesota Press, 1991), 167–81.

70. Samuel Beckett, *The Beckett Trilogy: Molloy, Malone Dies, The Unnamable* (London: Picador, 1979), 352.

71. Kant, *Critique of Pure Reason,* A 124.

## UPPSALA

1. Gunnar Olsson, "Chiasm of Thought-and-Action," *Environment and Planning D: Society and Space* 11 (1993): 279–94; and Ole Michael Jensen, "Red River Valley: Geo-Graphical Studies in the Landscape of Language," *Environment and Planning D: Society and Space* 11 (1993): 295–301. For a most violent reaction, especially to my own article, see Matthew Sparke, "Escaping the Herbarium: A Critique of Gunnar Olsson's 'Chiasm of Thought-and-Action,'" *Environment and Planning D: Society and Space* 12 (1994): 207–20. The unavoidable response is in Gunnar Olsson, "Job and the Case of the Herbarium," *Environment and Planning D: Society and Space* 12 (1994): 221–25.

2. Ole Michael Jensen, *Vægge: Kundskabens projektioner/Walls: Projections of Knowledge* (Copenhagen: Scandinavian University Press, 1995).

3. Gunnar Olsson, ed., *Chimärerna: Porträtt från en forskarutbildning* (Stockholm: Nordplan, 1996).

4. Hans-Olof Boström, *Det underbara skåpet: Philipp Hainhofer och Gustav II Adolfs konstskåp,* Acta Universitatis Upsaliensis, ser. C, vol. 70 (2001).

# BIBLIOGRAPHY


Abel, Ulf. *Ikonen—bilden av det heliga*. Hedemora: Gidlunds Bokförlag, 1989.

Ackerman, James S. *Distance Points: Essays in Theory and Renaissance Art and Architecture.* Cambridge, Mass.: MIT Press, 1991.

Aertz, W. J., et al., eds. *Alexander the Great in the Middle Ages: Ten Studies on the Last Days of Alexander in Literary and Historical Writing.* Nijmegen: Alfa Nijmegen, 1978.

Agamben, Giorgio. *The Man Without Content.* Translated by Georgia Albert. Stanford, Calif.: Stanford University Press, 1999.

Ahl, Frederick. *Sophocles' Oedipus: Evidence and Self-Conviction.* Ithaca, N.Y.: Cornell University Press, 1991.

Alberti, Leon Battista. *On Painting.* Translated by C. Grayson and with an introduction by M. Kemp. London: Penguin, 1991.

Algra, Keimpe A. *Concepts of Space in Greek Thought.* Leiden: E. J. Brill, 1994.

Allison, Henry E. *Kant's Transcendental Idealism: An Interpretation and Defense.* New Haven, Conn.: Yale University Press, 1983.

———. *Idealism and Freedom: Essays on Kant's Theoretical and Practical Philosophy.* Cambridge: Cambridge University Press, 1996.

Alpers, Svetlana. "The Mapping Impulse in Dutch Art." In *Art and Cartography: Six Historical Essays,* edited by David Woodward. Chicago: University of Chicago Press, 1987.

Alphen, Ernst. *Francis Bacon and the Loss of Self.* London: Reaktion Books, 1992.

Alster, Bendt, ed. *Death in Mesopotamia: Papers read at the XXVIe Rencontre assyriologique internationale.* København: Akademisk Forlag, 1980.

Alter, Robert. *The Art of Biblical Narrative.* New York: Basic Books, 1981.

———. *The Art of Biblical Poetry.* New York: Basic Books, 1985.

———. *Canon and Creativity: Modern Writing and the Authority of Scripture.* New Haven, Conn.: Yale University Press, 2000.

Alter, Robert, and Frank Kermode, eds. *The Literary Guide to the Bible.* Cambridge, Mass.: Harvard University Press, 1987.

Alvarez, A. "Making It New." *The New York Review of Books* 50, no. 8 (May 15, 2003): 28–30.

Anderson, Andrew Runni. *Alexander's Gate, Gog and Magog, and the Inclosed Nations.* Cambridge, Mass.: Medieval Academy of America, 1932.

Andrews, William J. H. *The Quest for Longitude: The Proceedings of the Longitude Symposium, Harvard University, Cambridge, Massachusetts, November 4–6, 1993.* Cambridge, Mass.: Harvard University Press, 1996.

Anfam, David. *Mark Rothko: The Works on Canvas. Catalogue Raisonné.* New Haven, Conn.: Yale University Press, 1998.

Annas, Julia. *An Introduction to Plato's "Republic."* Oxford: Clarendon Press, 1981.

Anscombe, G. E. M. *An Introduction to Wittgenstein's "Tractatus."* Philadelphia: University of Pennsylvania Press, 1971.

Arentzen, Jörg-Geerd. *Imago Mundi Cartographia: Studien zur Bildlichkeit mittelalterlicher Welt- und Ökumenekarten unter besonderer Berücksichtigung des Zusammenwirkens von Text und Bild.* München: Wilhelm Fink, 1984.

Aristotle. *The Basic Works of Aristotle.* Edited by Richard McKeon. New York: Random House, 1941.

Armstrong, Arthur Hilary. *The Architecture of the Intelligible Universe in the Philosophy of Plotinus.* Cambridge: Cambridge University Press, 1940.

Armstrong, Karen. *A History of God.* New York: Alfred A. Knopf, 1993. Translated by Inger Johansson as *Historien om Gud.* Stockholm: Forum, 1995.

———. *In the Beginning: A New Interpretation of Genesis.* New York: Alfred A. Knopf, 1996.

Aschenbrenner, Karl. *A Companion to Kant's "Critique of Pure Reason": Transcendental Aesthetic and Analytic.* New York: University Press of America, 1983.

Asenjo, F. G. *In-Between: An Essay on Categories.* Washington, D.C.: University Press of America, 1988.

———. "In Search of the Present." *Studies in Humanities and Social Sciences: Journal of Inter-University Centre, I.I.A.S. Shimla* 10 (2003): 50–69.

Ashton, Dore. *About Rothko.* Oxford: Oxford University Press, 1983.

Asplund, Johan. *Rivaler och syndabockar.* Göteborg: Korpen, 1989.

Auerbach, Erich. *Mimesis: The Representation of Reality in Western Literature.* Translated by Willard R. Trask. Princeton, N.J.: Princeton University Press, 1953. Reprinted with a new introduction by Edward W. Said. Princeton, N.J.: Princeton University Press, 2003.

———. *Dante, Poet of the Secular World.* Translated by Ralph Manheim. Chicago: University of Chicago Press, 1961.

———. *Literary Language and Its Public in Late Latin Antiquity and in the Middle Ages.* Translated by Ralph Manheim. Princeton, N.J.: Princeton University Press, 1965.

Augustine. *Confessions.* Translated by R. S. Pine-Coffin. Harmondsworth: Penguin, 1961.

Aujac, Germaine. "L'île de Thulé, de Pythéas à Ptolémée." In *Géographie du monde au Moyen Âge et à la Renaissance,* edited by Monique Pelletier. Paris: Éditions du CTHS, 1989.

Austin, Norman. *Archery at the Dark of the Moon: Poetic Problems in Homer's "Odyssey."* Berkeley and Los Angeles: University of California Press, 1975.

Bachelard, Gaston. *The Poetics of Space.* Translated by Maria Jolas. Boston: Beacon Press, 1969.

Bacon, Roger. *The Opus Majus of Roger Bacon.* Translated by John Henry Bridges. Oxford: Clarendon Press, 1897.

Badt, Kurt. *The Art of Cézanne.* Translated by Sheila Ann Ogilvie. New York: Hacker Art Books, 1985.

Baltrušaitis, Jurgis. *Anamorphic Art.* Translated by W. J. Strachan. Cambridge: Chadwick-Healey, 1977.

Bannister, Arthur, Stanley Raymond and Raymond Baker. *Surveying.* 7th ed. Harlow: Longman, 1998.

Barasch, Moshe. *Icon: Studies in the History of an Idea.* New York: New York University Press, 1992.

Barber, Peter, and Christopher Board. *Tales from the Map Room: Fact and Fiction About Maps and Their Makers.* London: BBC Books, 1993.

Barnes, Jonathan. *The Presocratic Philosophers.* London: Routledge, 1993.

Barthes, Roland. *Sade, Fourier, Loyola.* Translated by Richard Miller. New York: Hill and Wang, 1976.

———. *A Lover's Discourse: Fragments.* Translated by Richard Howard. New York: Hill and Wang, 1978.

Bataille, Georges. "Molloy's Silence." In *On Beckett: Essays and Criticism,* edited by Stanley E. Gontarski. New York: Grove Press, 1986.

Beall, J. C., and Bas C. van Fraassen. *Possibilities and Paradox: An Introduction to Modal and Many-valued Logic.* Oxford: Oxford University Press, 2003.

Beazley, C. Raymond. *The Dawn of Modern Geography: A History of Exploration and Geographical Science.* London: John Murray, 1897.

Beckett, Samuel. *The Beckett Trilogy: Molloy, Malone Dies, The Unnamable.* London: Picador, 1979.

Beckett, Samuel, et al. *Our Exagmination Round His Factification for Incamination of Work in Progress.* Paris: Shakespeare and Co., 1929.

Beckingham, Charles F., and Bernard Hamilton, eds. *Prester John, the Mongols and the Ten Lost Tribes.* Aldershot: Variorum, 1996.

Becks-Malorny, Ulrike. *Paul Cézanne, 1893–1906: Pioneer of Modernism.* Translated by Phil Goddard. Köln: Taschen, 1995.

Belting, Hans. *Likeness and Presence: A History of the Image Before the Era of Art.* Translated by Edmund Jephcott. Chicago: University of Chicago Press, 1994.

Bennett, Jonathan. *Kant's Analytic.* Cambridge: Cambridge University Press, 1966.

———. *Kant's Dialectic.* Cambridge: Cambridge University Press, 1974.

Bennett, Josephine Waters. *The Rediscovery of Sir John Mandeville.* New York: Modern Language Association of America, 1954.

Berggren, J. Lennart, and Alexander Jones. *Ptolemy's "Geography": An Annotated Translation of the Theoretical Chapters.* Princeton, N.J.: Princeton University Press, 2000.

Bergquist, Lars. *Biblioteket i lusthuset: Tio uppsatser om Swedenborg.* Stockholm: Natur och Kultur, 1996.

———. *Swedenborgs hemlighet: Om ordens betydelse, änglarnas liv och tjänsten hos Gud. En biografi.* Stockholm: Natur och Kultur, 1999.

Bergson, Henri. *Le rire: Essay sur la signification du comique.* Paris: Félix Alcan, 1917.

———. *Matter and Memory.* Translated by Nancy Margaret Paul and W. Scott Paul. New York: Zone Books, 1991.

Bernard, Émile. *Souvenirs sur Paul Cézanne et lettres.* Paris: A la rénovation esthéthique, 1924.

———. *Paul Cézanne: Minnen och brev.* Translated by Gustaf Engwall. Stockholm: Föreningen för nutida konst, 1947.

Bernidaki-Aldous, Elefteria A. *Blindness in a Culture of Light: Especially the Case of "Oedipus at Colonus" of Sophocles.* New York: Peter Lang, 1990.

Berque, Augustin. *Écoumène: Introduction à l'étude des milieux humains.* Paris: Belin, 2000.

Besançon, Alain. *The Forbidden Image: An Intellectual History of Iconoclasm.* Translated by Jane Marie Todd. Chicago: University of Chicago Press, 2000.

*Bible:*

*Bibeln.* Statens Offentliga Utredningar, 1999:100.

*The Holy Bible.* King James Version. Grand Rapids, Mich.: Zondervan Publishing House, 1962.

———. Revised Standard Version. Dallas, Tex.: Melton Book Company, 1971.

———. New International Version. Grand Rapids, Mich.: Zondervan Publishing House, 1973.

*The Jerusalem Bible.* London: Darton, Longman and Todd, 1966.

*The New Oxford Annotated Bible.* New Revised Standard Version. 3d. ed. Edited by Michael D. Coogan. Oxford: Oxford University Press, 1989.

Binding, Paul. *Imagined Corners: Exploring the World's First Atlas.* London: Headline, 2003.

Blanchot, Maurice. *De Kafka à Kafka.* Paris: Gallimard, 1982.

Bloemheuvel, Marente, ed. *Det unika med att vara människa/The Particularity of Being Human: Marlene Dumas, Francis Bacon.* Malmö: Malmö Konsthall, 1995.

Bloom, Harold. *Shakespeare: The Invention of the Human.* New York: Riverhead Books, 1998.

Bloor, David. *Wittgenstein: A Social Theory of Knowledge.* London: Macmillan, 1983.

———. *Wittgenstein, Rules and Regulations.* London: Routledge, 1997.

Bodin, Per-Arne. *Världen som ikon: Åtta föredrag om den ryskortodoxa andliga traditionen.* Skellefteå: Artos Bokförlag, 1988.

Boeder, Heribert. *Topologie der metaphysik.* München and Freiburg: Karl Alber, 1980.

Boehm, Christopher. *Hierarchy in the Forest: The Evolution of Egalitarian Behavior.* Cambridge, Mass.: Harvard University Press, 1999.

Bohr, Hans. "My Father." In *Niels Bohr,* edited by S. Rozenthal. New York: Wiley, 1967.

Bonazzi, Alessandra. "Chora and Chorography." In *Bodies and Space: Gunnar's Travels,* edited by Marco Picone. Palermo: Università di Palermo, Laboratorio Geografico, 2002.

Bonta, Mark, and John Protevi. *Deleuze and Geophilosophy: A Guide and Glossary.* Edinburgh: Edinburgh University Press, 2004.

*The Book of J.* Translated by David Rosenberg and interpreted by Harold Bloom. New York: Faber and Faber, 1990.

Borch-Jacobsen, Mikkel. *Lacan: The Absolute Master.* Stanford, Calif.: Stanford University Press, 1991.

Borges, Jorge Luis. *Labyrinths: Selected Stories and Other Writings.* Edited by Donald A. Yates and James E. Irby. Translated by James E. Irby. New York: New Directions, 1964.

———. *Dreamtigers.* Translated by Mildred Boyer and Harold Morland. Austin: University of Texas Press, 1964.

———. *A Universal History of Infamy.* Translated by Norman Thomas di Giovanni. New York: E. P. Dutton, 1972.

Boström, Hans-Olof. *Det underbara skåpet: Philipp Hainhofer och Gustav II Adolfs konstskåp.* Uppsala: Acta Universitatis Upsaliensis, Ser. C, 70 (2001).

Bosworth, A. B., and E. J. Baynham, eds. *Alexander the Great in Fact and Fiction.* Oxford: Oxford University Press, 2000.

Bottéro, Jean. "Les noms de Marduk, l'écriture et 'la logique' en Mésopotamie ancienne." In *Essays on the Ancient Near East in Memory of Jacob Joel Finkelstein,* edited by Maria de Jong Ellis. Hamden, Conn.: Archon, 1977.

———. *Mesopotamia: Writing, Reasoning, and the Gods.* Translated by Zainab Bahrani and Marc van de Mieroop. Chicago: University of Chicago Press, 1992.

———. *Religion in Ancient Mesopotamia.* Translated by Teresa Lavender Fagan. Chicago: University of Chicago Press, 2001.

Bourdieu, Pierre. *Outline of a Theory of Practice.* Translated by Richard Nice. Cambridge: Cambridge University Press, 1977.

Bouveresse, Jacques. *Le mythe de l'intériorité: Expérience, signification et langage privé chez Wittgenstein.* Paris: Minuit, 1976.

Bowra, C. M. *Sophoclean Tragedy.* Oxford: Clarendon Press, 1944.

Brann, Eva T. H. *The World of the Imagination: Sum and Substance.* Lanham, Md.: Rowman and Littlefield, 1991.

Breslin, James E. B. *Mark Rothko: A Biography.* Chicago: University of Chicago Press, 1993.

Brinken, Anna-Dorothee von den. "Mappa mundi und Cronographia: Studien zur imago mundi des abendländischen Mittelalters." *Deutsches Archiv für Erforschung des Mittelalters* 24 (1968): 118–86.

———. *Fines Terrae: Die Enden der Erde und der vierte Kontinent auf Mittelalterlichen Weltkarten.* Hannover: Hahnsche Buchhandlung, 1992.

———. "Ebstorf Map." In *Trade, Travel, and Exploration in the Middle Ages: An Encyclopedia,* edited by John Block Friedman and Kristen Mossler Figg. New York: Garland, 2000.

Brinkman, Johan. *The Conception of Space in the Old Testament: An Exploration of the Methodological Problems of Investigation, Exemplified by a Study of Exodus 25 to 31.* Kampen: Kok Pharos Publishing House, 1992.

Brown, Lloyd A. *The Story of Maps.* Boston: Little, Brown and Company, 1949.

Brown, Norman O. *Closing Time.* New York: Vintage Books, 1974.

Brown, Peter. "A Dark-Age Crisis: Aspects of the Iconoclastic Controversy." *The English Historical Review* 88 (1973): 1–34.

———. *The Rise of Western Christendom: Triumph and Diversity, A.D. 200–1000.* 2d ed. Oxford: Blackwell Publishing, 2003.

Bruno, Giuliana. *Atlas of Emotion: Journeys in Art, Architecture, and Film.* London: Verso, 2002.

Buci-Glucksmann, Christine. *L'œil cartographique de l'art.* Paris: Galilée, 1996.

Budelmann, Felix. *The Language of Sophocles: Communality, Communication, and Involvement.* Cambridge: Cambridge University Press, 2000.

Budge, E. A. Wallis. *The Rise and Progress of Assyriology.* London: Martin Hopkinson, 1925.

Bundy, Murray Wright. *The Theory of Imagination in Classical and Medieval Thought.* Urbana: Illinois Studies in Language and Literature, vol. 12, 1927.

Burkert, Walter. *Babylon, Memphis, Persepolis: Eastern Contexts of Greek Culture.* Cambridge, Mass.: Harvard University Press, 2004.

Burke, Edmund. *A Philosophical Enquiry into the Origin of our Ideas of the Sublime and Beautiful.* Edited by J. T. Boulton. London: Routledge and Kegan Paul, 1958.

Butler, S. A. L. *Mesopotamian Conceptions of Dreams and Dream Rituals.* Münster: Ugarit-Verlag, 1998.

Böhl, Franz M. Th. "Die fünfzig Namen des Marduk." *Archiv für Orientforschung* 11 (1936): 191–218.

Cabanne, Pierre. *Dialogues with Marcel Duchamp.* Translated by Ron Padgett. London: Thames and Hudson, 1971.

Cairns, Ian. *Word and Presence: A Commentary on the Book of Deuteronomy.* Grand Rapids, Mich.: Eerdmans, 1992.

Cameron, Alister. *The Identity of Oedipus the King: Five Essays on the "Oedipus Tyrannus."* New York: New York University Press, 1968.

Camfield, William A. *Marcel Ducamp/Fountain.* Houston, Tex.: Houston Fine Art Press, 1989.

Camille, Michael. *Image on the Edge: The Margins of Medieval Art.* London: Reaktion Books, 1992.

———. *Mirror in the Parchment: The Luttrell Psalter and the Making of Medieval England.* London: Reaktion Books, 1998.

Capps, Donald. "Abraham and Isaac: The Sacrificial Impulse." In *The Destructive Power of Religion: Violence in Judaism, Christianity, and Islam,* vol. 1, *Sacred Scriptures, Ideology, and Violence,* edited by J. Harold Ellens. Westport, Conn.: Praeger, 2004.

Carmichael, Calum M. *Law and Narrative in the Bible: The Evidence of Deuteronomic Laws and the Decalogue.* Ithaca, N.Y.: Cornell University Press, 1985.

———. *The Origins of Biblical Law: The Decalogues and the Book of the Covenant.* Ithaca, N.Y.: Cornell University Press, 1992.

Carpelan, Bo. *Urwind.* Esbo: Schildt, 1993.

Carr, Annmarie Weyl. "Images: Expressions of Faith and Power." In *Byzantium: Faith and Power (1261–1557),* edited by Helen C. Evans. New York: Metropolitan Museum of Art, 2004.

Carroll, Lewis. *The Annotated Alice.* With an introduction and notes by Martin Gardner. New York: Clarkson N. Potter, 1960.

Carruthers, Mary J. *The Book of Memory: A Study of Memory in Medieval Culture.* Cambridge: Cambridge University Press, 1990.

———. *The Craft of Thought: Meditation, Rhetoric and the Making of Images, 400–1200.* Cambridge: Cambridge University Press, 1998.

Casey, Edward S. *Imagining: A Phenomenological Study.* Bloomington: Indiana University Press, 1976.

———. *Remembering: A Phenomenological Study.* Bloomington: Indiana University Press, 1987.

———. *Getting Back into Place: Toward a Renewed Understanding of the Place-World.* Bloomington: Indiana University Press, 1993.

———. *The Fate of Place: A Philosophical History.* Berkeley and Los Angeles: University of California Press, 1997.

———. "Between Geography and Philosophy: What Does It Mean to Be in the Place-World." *Annals of the Association of American Geographers* 91 (2001): 683–93.

Cassirer, Ernst. *The Philosophy of Symbolic Forms.* Translated by Ralph Manheim. New Haven, Conn.: Yale University Press, 1953–57.

———. *Kant's Life and Thought*. Translated by James Haden. New Haven, Conn.: Yale University Press, 1981.

Cassuto, Umberto. *A Commentary on the Book of Genesis*. Vol. 2, *Noah to Abraham*. Translated by Israel Abrahams. Jerusalem: Magnes Press, 1964.

———. *A Commentary on the Book of Exodus*. Translated by Israel Abrahams. Jerusalem: Magnes Press, 1983.

Cavanaugh, William T. *Torture and Eucharist: Theology, Politics and the Body of Christ*. Oxford: Basil Blackwell, 1998.

Caygill, Howard. *A Kant Dictionary*. Oxford: Oxford University Press, 1995.

Chadwick, Henry. "Philoponus the Christian Theologian." In *Philoponus and the Rejection of Aristotelean Science*, edited by Richard Sorabji. London: Duckworth, 1987.

Childs, Brevard S. *The Book of Exodus: A Critical, Theological Commentary*. Louisville, Ky.: Westminster Press, 1974.

Cicovacki, Predrag. *Between Truth and Illusion: Kant at the Crossroads of Modernity*. Lanham, Md.: Rowman and Littlefield, 2002.

Clair, Jean. *Marcel Duchamp ou le grand fictif: Essai de mythanalyse du grand verre*. Paris: Éditions Galilée, 1975.

———. "Duchamp and the Classical Perspectivists." *Artforum* 16 (1978): 40–49.

Clark, T. J. *Farewell to an Idea: Episodes from a History of Modernism*. New Haven, Conn.: Yale University Press, 1999.

Clarke, Michael. *Flesh and Spirit in the Songs of Homer: A Study of Words and Myths*. Oxford: Clarendon Press, 1999.

Cleve, James van. *Problems from Kant*. Oxford: Oxford University Press, 1999.

Clines, David J. A. *Job 1–20*. Dallas: Word Books, 1989.

———. *On the Way to the Postmodern: Old Testament Essays, 1967–1998*. Vol. 2. Sheffield: Sheffield Academic Press, 1998.

Cohn, Marjorie, ed. *Mark Rothko*. Exhibition, 18 February–29 April 2001, Fondation Beyeler. Ostfildern-Ruit: Hatje Cantz, 2001.

Cokking, J. M. *Imagination: A Study in the History of Ideas*. London: Routledge, 1991.

Conrad, Joseph. *The Secret Agent: A Simple Tale*. London: David Campbell, 1992.

———. *Heart of Darkness*. Harmondsworth: Penguin, 1994.

Cornford, Francis MacDonald. *From Religion to Philosophy: A Study in the Origins of Western Speculation*. London: Edward Arnold, 1912.

Cortázar, Julio. *Hopscotch*. Translated by Gregory Rabassa. New York: Random House, 1966.

Cosmas Indicopleustes. *The "Christian Topography" of Cosmas, an Egyptian Monk*. Translated by J. W. McCrindle. London: Hakluyt Society, 1897.

———. *The "Christian Topography" of Cosmas Indicopleustes, edited with Geographical Notes*. Edited and translated by E. O. Winstedt. Cambridge: Cambridge University Press, 1909.

———. *Topographie Chrétienne: Introduction, texte critique, traduction et notes*. Edited and translated by Wanda Wolska-Conus. 3 vols. Sources chrétiennes, nos. 141, 159, 197. Paris: Éditions du Cerf, 1968, 1970, 1973.

Cough-Cooper, Jennifer, and Jacques Caumont, eds. *Marcel Duchamp—Ephemerides On and About Marcel Duchamp and Rrose Sélavy, 1887–1968*. Exhibition catalogue, Palazzo Grassi, Venezia. Milano: Bompiani, 1993.

Craig E. *Marcel Duchamp's Notes from the Large Glass: An N-dimensional Analysis*. Ann Arbor, Mich.: UMI, 1983.

Craigie, Peter C. *The Book of Deuteronomy.* Grand Rapids, Mich.: Eerdmans, 1976.

Crampton, Jeremy W. "Maps as Social Constructions: Power, Communication and Visualization." *Progress in Human Geography* 25 (2001): 235–52.

———. "Thinking Philosophically in Cartography: Toward a Critical Politics of Mapping." *Cartographic Perspectives* 42 (2002): 12–31.

Crary, Alice, and Rupert Read, eds. *The New Wittgenstein.* London: Routledge, 2000.

Crone, G. R. *Maps and Their Makers: An Introduction to the History of Cartography.* Hamden, Conn. and London: Hutchinson, 1953.

Crone, Rainer, and David Moos. *Kazimir Malevich: The Climax of Disclosure.* London: Reaktion Books, 1991.

Dalley, Stephanie. *Myths from Mesopotamia: Creation, the Flood, Gilgamesh, and Others.* Oxford: Oxford University Press, 1989.

Damisch, Hubert. *The Origin of Perspective.* Cambridge, Mass.: MIT Press, 1994.

D'Andrea, Jeanne, ed. *Kazimir Malevich, 1878–1935.* Los Angeles: The Armand Hammer Museum of Art, 1990.

Daniels, Dieter. *Duchamp und die anderen: Der Modellfall einer künstlerischen Wirkungsgeschichte in der Moderne.* Köln: DuMont, 1992.

Danson, Edwin. *Drawing the Line: How Mason and Dixon Surveyed the Most Famous Border in America.* New York: Wiley, 2001.

Danto, Arthur C. *The Philosophical Disenfranchisement of Art.* New York: Columbia University Press, 1986.

Davies, Hugh M. *Francis Bacon: The Papal Portraits.* San Diego, Calif.: Museum of Contemporary Art, 1999.

Davis, Raymond E. *Surveying: Theory and Practice.* 6th ed. New York: McGraw-Hill, 1981.

Delaney, Carol Lowery. *Abraham on Trial: The Social Legacy of Biblical Myth.* Princeton, N.J.: Princeton University Press, 1998.

Deleuze, Gilles. *Kant's Critical Philosophy.* Translated by Hugh Tomlinson and Barbara Habberjam. Minneapolis: University of Minnesota Press, 1984.

———. *The Fold: Leibniz and the Baroque.* Translated by Tom Conley. Minneapolis: University of Minnesota Press, 1993.

———. *Francis Bacon: The Logic of Sensation.* Translated by Daniel W. Smith. London: Continuum, 2003.

Deleuze, Gilles, and Félix Guattari: *A Thousand Plateaus: Capitalism and Schizophrenia.* Translated by Brian Massumi. Minneapolis: University of Minnesota Press, 1987.

———. *Nomadology: The War Machine.* Translated by Brian Massumi. Edinburgh: AK Press, 1986.

———. *What is Philosophy?* Translated by Hugh Tomlinson and Graham Burchell. New York: Columbia University Press, 1994.

Deleuze, Gilles, and Claire Parnet. *Dialogues.* Translated by Hugh Tomlinson and Barbara Habberjam. London: Athlone Press, 1987.

Demus, Otto. *Byzantine Mosaic Decoration: Aspects of Monumental Art in Byzantium.* 2d ed. Boston: Boston Book and Art Shop, 1955.

Derrida, Jacques. *Speech and Phenomena: And Other Essays on Husserl's Theory of Signs.* Translated by David B. Allison. Evanston, Ill.: Northwestern University Press, 1973.

———. *Dissemination.* Translated by Barbara Johnson. London: Athlone Press, 1981.

———. "The Time of a Thesis: Punctuations." In *Philosophy in France Today,* edited by Alan Montefiore. Cambridge: Cambridge University Press, 1983.

Destombes, Marcel. *Mappemondes A.D. 1200–1500*. Catalogue préparé par la Commission des Cartes Anciennes de l'Union Géographique Internationale. Amsterdam: Israel, 1964.

*Det babyloniska skapelseeposet "enuma elis."* See *Enuma Elish*.

Detienne, Marcel, and Jean-Pierre Vernant. *Cunning Intelligence in Greek Culture and Society.* Translated by Janet Lloyd. Chicago: University of Chicago Press, 1991.

d'Harnoncourt, Anne, and Walter Hopps. "Etant donnés: 1° la chute d'eau, 2° le gaz d'éclairage: Reflections On a New Work by Marcel Duchamp." *Bulletin of Philadelphia Museum of Art* 64, nos. 299 and 300 (April-June and July-September 1969).

d'Harnoncourt, Anne, and Kynaston McShine. *Marcel Duchamp.* New York: Museum of Modern Art, 1973.

Dhorme, Edouard. *A Commentary on the Book of Job.* Translated by Harold Knight. London: Nelson, 1967.

Diamond, Cora. *The Realistic Spirit: Wittgenstein, Philosophy and the Mind.* Cambridge, Mass.: MIT Press, 1991.

Dicks, D. R. *The Geographical Fragments of Hipparchus.* London: Athlone Press, 1960.

Dilke, O. A. W. *Greek and Roman Maps.* Ithaca, N.Y.: Cornell University Press, 1985.

Dimock, George E. Jr. "The Name of Odysseus." In *Essays on the "Odyssey": Selected Modern Criticism,* edited by Charles E. Taylor. Bloomington: Indiana University Press, 1969.

Dodds, E. R. "On Misunderstanding the *Oedipus Rex.*" *Greece and Rome* 13 (1966): 37–49.

Doel, Marcus A. *Poststructuralist Geographies: The Diabolical Art of Spatial Science.* Edinburgh: Edinburgh University Press, 1999.

———. "Gunnar Olsson's Transformers: The Art and Politics of Rendering the Co-Relation of Society and Space in Monochrome or Technicolor." *Antipode* 35 (2003): 140–67.

Dooley, Mark. *The Politics of Exodus: Søren Kierkegaard's Ethics of Responsibility.* New York: Fordham University Press, 2001.

Douglas, Charlotte. *Kazimir Malevich.* London: Thames and Hudson, 1994.

Drake, H. A. *Constantine and the Bishops: The Politics of Intolerance.* Baltimore, Md.: Johns Hopkins University Press, 2000.

Drake, James. "The Naming Disease: How Jakobson's Essay of Aphasia Initiated Postmodernist Deceits." *Times Literary Supplement,* September 4, 1998: 14–15.

Drutt, Matthew. *Kazimir Malevich: Suprematism.* New York: Solomon R. Guggenheim Foundation, 2003.

DuBois, Page. *Torture and Truth.* New York: Routledge, 1991.

Duchamp, Marcel. *La Mariée mise à nu par ses célibataires, même.* (The Green Box.) Paris: Éditions Rrose Sélavy, 1934. Translated as *The Bride Stripped Bare by Her Bachelors, Even: A Typographic Version by Richard Hamilton of Marcel Duchamp's "Green Box."* New York: Wittenborn, 1960.

———. "I Propose to Strain the Laws of Physics." Interview by Francis Roberts. *Art News* 67, no. 8 (December 1968): 46–47.

———. *Duchamp du signe: Ecrits.* Réunis et présentés par Michel Sanouillet. Paris: Flammarion, 1974.

———. *Manual of Instructions for Étant donnés: 1° la chute d'eau, 2° le gas d'éclairage.* Phildelphia: Philadelphia Museum of Art, 1978.

———. *The Complete Works of Marcel Duchamp.* Edited by Arturo Schwarz. 2 vols. London: Thames and Hudson, 1997.

———. *In the Infinitive.* A typotranslation by Richard Hamilton and Ecke Bonk of Marcel Duchamp's *White Box.* European distribution, Köln: Buchhandlung Walter König, 1999.

Duffy, Bruce. *The World As I Found It.* New York: Ticknor and Fields, 1987.

Dufour, Dany-Robert. *Les mystères de la trinité.* Paris: Gallimard, 1990.

Dummett, Michael. *Frege: Philosophy of Language.* New York: Harper and Row, 1973.

———. *The Interpretation of Frege's Philosophy.* London: Duckworth, 1981.

———. *Frege: Philosophy of Mathematics.* Cambridge, Mass.: Harvard University Press, 1991.

Duro, Paul, ed. *The Rhetoric of the Frame.* Cambridge: Cambridge University Press, 1996.

Duve, Thierry de. "The Readymade and the Tube of Paint." *Artforum* 24, no. 9 (May 1986): 110–21.

———. *Pictorial Nominalism: On Marcel Duchamp's Passage from Painting to the Readymade.* Minneapolis: University of Minnesota Press, 1991.

———, ed. *The Definitively Unfinished Marcel Duchamp.* Cambridge, Mass.: MIT Press, 1991.

———. *Kant after Duchamp.* Cambridge, Mass.: MIT Press, 1996.

Düchting, Hajo. *Paul Cézanne, 1839–1906: Nature into Art.* Translated by Michael Hulse. Köln: Taschen, 1991.

Dünkelsbühler, Ulrike Oudée. *Reframing the Frame of Reason: "Trans-lation" In and Beyond Kant and Derrida.* Translated by Max Statkiewicz. Amherst, Mass.: Humanity Books, 2002.

Eco, Umberto. *Den frånvarande strukturen: Introduktion till den semiotiska forskningen.* Translated by Estrid Tenggren. Lund: Bo Cavefors, 1971.

———. *A Theory of Semiotics.* Bloomington: Indiana University Press, 1976.

———. *Semiotics and the Philosophy of Language.* London: Macmillan Press, 1984.

———. *Art and Beauty in the Middle Ages.* Translated by Hugh Bredin. New Haven, Conn.: Yale University Press, 1986.

———. *Foucault's Pendulum.* Translated by William Weaver. New York: Harcourt Brace Jovanovich, 1989.

———. *Baudolino.* Translated by William Weaver. London: Secker and Warburg, 2002.

Edgerton, Samuel Y. Jr. *The Renaissance Rediscovery of Linear Perspective.* New York: Basic Books, 1975.

———. "From Mental Matrix to Mappamundi to Christian Empire: The Heritage of Ptolemaic Cartography in the Renaissance." In *Art and Cartography: Six Historical Essays,* edited by David Woodward. Chicago: University of Chicago Press, 1987.

———. *The Heritage of Giotto's Geometry: Art and Science on the Eve of the Scientific Revolution.* Ithaca, N.Y.: Cornell University Press, 1991.

Edson, Evelyn. *Mapping Space and Time: How Medieval Mapmakers Viewed Their World.* London: British Library, 1999.

Edwards, Peter. *Torture.* New ed. Philadelphia: University of Pennsylvania Press, 1996.

Ehrman, Bert T. *Lost Scriptures: Books that Did Not Make It into the New Testament.* Oxford: Oxford University Press, 2003.

Einem, Herbert von. *Der Mainzer Kopf mit der Binde.* Arbeitsgemeinschaft für Forschung des Landes Nordrhein-Westfalen, Geisteswissenschaften, Heft 3.. Köln: Westdeutscher Verlag, 1955.

Eliade, Mircea. *The Sacred and the Profane.* Translated by Willard R. Trask. New York: Harcourt, Brace and Company, 1959.

Else, Gerald F. *The Origin and Early Form of Greek Tragedy.* Cambridge, Mass.: Harvard University Press, 1965.

———. *Aristotle's "Poetics": The Argument.* Cambridge Mass.: Harvard University Press, 1967.

*Encyclopaedia Judaica.* Jerusalem: Keter Publishing House, 1971.

Enskat, Rainer. *Kant's Theorie des geometrischen Gegenstandes: Untersuchungen über die Voraussetzungen der Entdeckbarkeit geometrischer Gegenstände bei Kant.* Berlin: Walter de Gruyter, 1978.

*Enuma Elish:*

*Det babyloniska skapelseeposet "enuma elis."* Translated by Alfred Haldar. Uppsala: Almqvist och Wiksell, 1952.

*Enuma elish: Det babyloniska skapelseeposet.* Translated with commentary by Ola Wikander. Stockholm: Wahlström och Widstrand, 2005.

*The Epic of Gilgamesh.* See *Gilgamesh.*

Erdin, Franz. *Das Wort Hypostasis: Seine bedeutungsgeschichtliche entwicklung in der altchristlichen Literatur bis zum Abschluss der trinitarischen auseinandersetzungen.* Freiburg im Breisgau: Herder und Co., 1939.

Eriksson, Gunnar. *Rudbeck 1630–1702: Liv, lärdom, dröm i barockens Sverige.* Stockholm: Atlantis, 2002.

Eshleman, Clayton. *Juniper Fuse: Upper Paleolithic Imagination and the Construction of the Underworld.* Middletown, Conn.: Wesleyan University Press, 2003.

Evans, Helen C. *Byzantium: Faith and Power (1261–1557).* New York: Metropolitan Museum of Art, 2004.

Farinelli, Franco. *I segni del mondo: Imagine cartografica e discorso geografico in età moderna.* Firenze: La Nuova Italia, 1992.

———. "Squaring the Circle, or the Nature of Political Identity." In *Limits of Representation,* edited by Franco Farinelli, Gunnar Olsson and Dagmar Reichert. München: Accedo, 1994.

———. "Did Anaximander ever Say (or Write) any Words? The Nature of Cartographical Reason." *Ethics, Place and Environment* 1 (1998): 135–44.

———. "The Geography of Zero." In *Bodies and Space: Gunnar's Travels,* edited by Marco Picone. Palermo: Università di Palermo, Laboratorio Geografico, 2002.

———. *Geografia: Un'intoduzione ai modelli del mondo.* Torino: Einaudi, 2003.

Farson, Daniel. *The Gilded Gutter Life of Francis Bacon.* London: Vintage, 1994.

Feigl, Herbert, and W. S. Sellars, eds. *Readings in Philosophical Analysis.* New York: Appleton-Century-Crofts, 1949.

Feiler, Bruce. *Abraham: In Search of the Father of Civilisation.* London: Judy Piatkus, 2002.

Ferguson, Robert. *The Short Sharp Life of T. E. Hulme.* London: Allen Lane, 2002.

Ferlin, Nils. *Barfotabarn.* Stockholm: Albert Bonnier, 1933.

Fernández-Armesto, Felipe. *So You Think You're Human? A Brief History of Humankind.* Oxford: Oxford University Press, 2005.

Ferreira, M. Jamie. *Transforming Vision: Imagination and Will in Kierkegaard's Faith.* Oxford: Clarendon Press, 1991.

Florschütz, Gottlieb. *Swedenborg and Kant: Emanuel Swedenborg's Mystical View of Humankind, and the Dual Nature of Humankind in Immanuel Kant.* West Chester, Penn.: Swedenborg Foundation, 1993.

Fogelin, Robert J. "Three Platonic Analogies." *Philosophical Review* 80 (1971): 371–82.

Forest, Jean-Daniel. *L' épopée de Gilgamesh et sa postérité: Introduction au langage symbolique.* Paris: Méditerranée, 2002.

Foster, Benjamin R., ed. *Before the Muses: An Anthology of Akkadian Literature.* 2 vols. Bethesda, Md.: CDL Press, 1996.

Foucault, Michel. *Discipline and Punish: The Birth of the Prison.* Translated by Alan Sheridan. Harmondsworth: Penguin, 1979.

———. *The History of Sexuality: An Introduction.* Translated by Robert Hurley. Harmondsworth: Penguin, 1981.

———. *This is not a Pipe.* Translated by James Harkness. Berkeley and Los Angeles: University of California Press, 1983.

Frayn, Michael. *Copenhagen.* New York: Methuen, 1998.

Freedberg, David. *The Power of Images: Studies in the History and Theory of Response.* Chicago: University of Chicago Press, 1989.

Frege, Gottlob: "Über Sinn und Bedeutung." *Zeitschrift für Philosophie und Philosophische Kritik* 100 (1892): 25–50.

———. *Translations from the Philosophical Writings of Gottlob Frege.* Edited by Peter T. Geach and Max Black. Oxford: Blackwell, 1980.

French, Marilyn. *Shakespeare's Division of Experience.* New York: Ballantine, 1981.

Friedman, John Block. *The Monstrous Races in Medieval Art and Thought.* Cambridge, Mass.: Harvard University Press, 1981.

Friedman, Michael. *Kant and the Exact Sciences.* Cambridge, Mass.: Harvard University Press, 1992.

Friis Johansen, Kirsten. *Den europæiske filosofis historie.* Vol. 1, *Antiken.* København: Nyt Nordisk Forlag Arnold Busk, 1991.

Frye, Northrop. *The Great Code: The Bible and Literature.* New York: Harcourt Brace Jovanovich, 1981.

———. *Words with Power: Being a Second Study of the Bible and Literature.* New York: Harcourt Brace Jovanovich, 1990.

Gablik, Suzi. *Magritte.* London: Thames and Hudson, 1970.

Gasché, Rodolphe. *The Tain of the Mirror: Derrida and the Philosophy of Reflection.* Cambridge, Mass.: Harvard University Press, 1986.

Gautier Dalché, Patrick. *Géographie et culture: La représentation de l'espace du VIe au XIIe siècle.* Aldershot: Ashgate, 1997.

———. *La 'Descriptio Mappe Mundi' de Hugues de Saint Victor.* Paris: Études Augustiennes, 1988.

Geoffrey, Stephen. *The Songs of Homer.* Cambridge: Cambridge University Press, 1962.

Gernet, Louis. *The Anthropology of Ancient Greece.* Baltimore, Md.: Johns Hopkins University Press, 1981.

Gerstenberger, Erhard. "Covenant and Commandment." *Journal of Biblical Literature* 84 (1965): 38–51.

———. *Wesen und Herkunft des "Apodiktischen Rechts."* Neukirchen-Vlyun: Neukirchener, 1965.

Geus, Klaus. *Eratosthenes von Kyrene: Studien zur hellenistischen Kultur- und Wissenschaftsgeschichte.* München: Beck, 2002.

Gibbons, Sarah L. *Kant's Theory of Imagination: Bridging Gaps in Judgment and Experience.* Oxford: Clarendon Press, 1994.

*Gilgamesh, Epic of:*

*The Babylonian Gilgamesh Epic: Introduction, Critical Edition and Cuneiform Texts.* 2 vols.

Edited and translated by Andrew R. George. Oxford: Oxford University Press, 2003.

*The Epic of Gilgamesh.* Translated by N. K. Sandars. Harmondsworth: Penguin, 1960.

*The Epic of Gilgamesh: The Babylonian Epic Poem and Other Texts in Akkadian and Sumerian.* Edited and translated by Andrew R. George. London: Penguin, 1999.

*The Epic of Gilgamesh: A New Translation, Analogues, Criticism.* Translated by Benjamin R. Foster. New York: Norton, 2001.

*L'épopée de Gilgamesh: Poème babylonien.* Translated by G. Contenau. Paris: L'artisan de livre, 1939.

*Gilgamesh.* Translated from the Sin-leqi-sunnini Version by John Gardner and John Maier: New York: Knopf, 1984.

*Gilgamesh: A New English Version.* Translated by Stephen Mitchell. New York: Free Press, 2004.

*Gilgamesh-eposet.* Translated by Knut Tallqvist. Stockholm: Geber, 1945.

Girard, René. *Job: The Victim of His People.* Translated by Yvonne Freccero. Stanford, Calif.: Stanford University Press, 1987.

Girard, Xavier. *La chapelle de Rosaire.* Paris: Éditions de la Réunion des Musées Nationaux, 1992.

Golding, John. *Duchamp: The Bride Stripped Bare by Her Bachelors, Even.* London: Penguin, 1973.

———. *Visions of the Modern.* London: Thames and Hudson, 1994.

———. *Paths to the Absolute: Mondrian, Malevich, Kandinsky, Pollock, Newman, Rothko and Still.* London: Thames and Hudson, 2000.

Good, Edwin M. *In Turns of Tempest: A Reading of Job with a Translation.* Stanford, Calif.: Stanford University Press, 1990.

Gordis, Robert. *The Book of Job: Commentary, New Translation, and Special Studies.* New York: Jewish Theological Seminary, 5738/1978.

Gorlé, Dinda L. *Semiotics and the Problem of Translation: With Special Reference to the Semiotics of Charles S. Peirce.* Amsterdam: Rodopi, 1994.

Gosling, J. C. B. *Plato.* London: Routledge and Kegan Paul, 1973.

Gould, Rupert T. *The Marine Chronometer: Its History and Development.* London: J. D. Potter, 1923.

———. *John Harrison and His Timekeepers.* London: National Maritime Museum, 1978.

Gow, Andrew. "Gog and Magog on Mappaemundi and Early Printed World Maps: Orientalizing Ethnography in the Apolictic Tradition." *Journal of Early Modern History* 2 (1998): 61–88.

Gow, Andrew, and Jolanta Pekacz. *A Bibliography of Scholarship on Mappaemundi and Early World Maps.* http://www.ualberta.ca/~agow/mapbib.html.

Grabar, André. *Christian Iconography: A Study of Its Origins.* Princeton, N.J.: Princeton University Press, 1968.

Grafton, Anthony. *Leon Battista Alberti: Master Builder of the Italian Renaissance.* London: Penguin, 2002.

Grant, Kenneth. *Nightside of Eden.* London: Frederick Muller, 1977.

———. *Outside the Circles of Time.* London: Frederick Muller, 1980.

Grant, Robert M., ed. *Gnosticism: A Source Book of Heretical Writings from the Early Christian Period.* London: Collins, 1961.

Green, Geoffrey. *Literary Criticism and the Structures of History: Erich Auerbach and Leo Spitzer.* Lincoln: University of Nebraska Press, 1982.

Green, Ronald M. *Kierkegaard and Kant: The Hidden Debt.* Albany: State University of New York Press, 1992.

Greenberg, Moshe. "Job." In *The Literary Guide to the Bible,* edited by Robert Alter and Frank Kermode. London: Collins, 1987.

Greimas, Algirdas Julien. *On Meaning: Selected Writings in Semiotic Theory.* Translated by Paul J. Perron and Frank H. Collins. Minneapolis: University of Minnesota Press, 1987.

Gren, Martin. *Earth Writing: Exploring Representation and Social Geography In-Between Meaning/Matter.* Gothenburg: Department of Human and Economic Geography, University of Gothenburg, 1994.

Griffin, Jasper, ed. *Sophocles Revisited: Essays Presented to Sir Hugh Lloyd-Jones.* Oxford: Clarendon Press, 1999.

Griffiths, Alan: "Euenius the Negligent Nightwatchman (Herodotus 9.92–6)." In *From Myth to Reason? Studies in the Development of Greek Thought,* edited by Richard Buxton. Oxford: Oxford University Press, 1999.

Gulyga, Arsenij. *Immanuel Kant: His Life and Thought.* Translated by M. Despalatovic. Boston: Birkhäuser, 1987.

Guyer, Paul. *Kant and the Claims of Knowledge.* Cambridge: Cambridge University Press, 1987.

———. *The Cambridge Companion to Kant.* Cambridge: Cambridge University Press, 1992.

Gärdenfors, Peter. *Hur Homo blev sapiens: Om tänkandets evolution.* Nora: Nya Doxa, 2000.

Haas, Frans A. J. de. *John Philoponus' New Definition of Prime Matter: Aspects of its Background in Neoplatonism and the Ancient Commentary Tradition.* Leiden: E. J. Brill, 1997.

Hacker, P. M. S. *Insight and Illusion: Wittgenstein on Philosophy and the Metaphysics of Experience.* Oxford: Oxford University Press, 1972.

Hahn-Woernle, Birgit. *Die Ebstorfer Weltkarte.* Ebstorf: Das Kloster, 1989.

Hanfling, Oswald. *Wittgenstein and the Human Form of Life.* London: Routledge, 2002.

Hansen-Møller, Jette. *Den skjulte diagonal: En landskabsfortelling i ord og billeder.* København: Christain Ejlers' Forlag, 1995.

Hanson, R. P. C. *The Search for the Christian Doctrine of God: The Arian Controversy 318–381.* Edinburgh: T. and T. Clark, 1988.

Harley, J. B. *The New Nature of Maps: Essays in the History of Cartography.* Edited by Paul Laxton. Baltimore, Md.: Johns Hopkins University Press, 2001.

Harris, Roy. *Language, Saussure and Wittgenstein: How to Play Games with Words.* London: Routledge, 1988.

Harvey, David. "Cosmopolitanism and the Banality of Geographical Evils." *Public Culture* 4 (2000): 529–64.

Harvey, P. D. A. *Medieval Maps.* London: British Library, 1991.

———. *Mappa Mundi: The Hereford World Map.* London: British Library, 1996.

Havelock, Eric A. *The Greek Concept of Justice: From Its Shadow in Homer to Its Substance in Plato.* Cambridge, Mass.: Harvard University Press, 1978.

Hawkes, C. F. C. *Pytheas: Europe and the Greek Explorers.* Oxford: Blackwell, 1977.

Hedenius, Ingemar. *Fyra dygder.* Stockholm: Bonnier, 1955.

Hederman, Mark Patrick. "Cinema and the Icon." *Crane Bag* 8 (1984): 90–99.

Heeg, Erik van den, and Sven Olov Wallenstein, eds. *Tankens arkipelag: Moderna kantläsningar.* Stehag: Brutus Östling, 1992.

Hegel, G. W. F. *The Phenomenology of Mind.* Translated by J. B. Baillie. New York: Harper Torchbooks, 1967.

Heidegger, Martin. *Kant and the Problem of Metaphysics.* Translated by Richard Taft. Bloomington: Indiana University Press, 1984.

Heidel, Alexander. *The Gilgamesh Epic and Old Testament Parallels.* Chicago: University of Chicago Press, 1946.

———. *The Baylonian Genesis: The Story of Creation.* 2d ed. Chicago: University of Chicago Press, 1951.

Heidel, William Arthur. *The Frame of the Ancient Maps: With a Discussion of the Discovery of the Sphericity of the Earth.* New York: American Geographical Society, 1937.

Held, George F. "Parallels between the *Gilgamesh Epic* and Plato's *Symposium.*" *Journal of Near Eastern Studies* 42 (1983): 133–41.

Henderson, Linda Dalrymple. *The Fourth Dimension and Non-Euclidean Geometry in Modern Art.* Princeton, N.J.: Princeton University Press, 1983.

Henne, Helmut, and Kristine Kaiser, eds. *Fritz Mauthner—Sprache, Literatur, Kritik: Festakt und Symposion zu seinem 150. Geburtstag.* Tübingen: Niemeyer, 2000.

Hennig, Richard. *Terrae Incognitae.* 2 vols. Leiden: E. J. Brill, 1944, 1950.

Henry, Patrick. "What Was the Iconoclastic Controversy About?" *Church History* 45 (1976): 16–31.

Herrin, Judith. *The Formation of Christendom.* Princeton, N.J.: Princeton University Press, 1989.

Herskovits, Annette. *Language and Spatial Cognition: An Interdisciplinary Study of the Prepositions in English.* Cambridge: Cambridge University Press, 1986.

Hester, Marcus B. *The Meaning of Poetic Metaphor: An Analysis in the Light of Wittgenstein's Claim that Meaning is Use.* The Hague: Mouton, 1967.

Heubeck, Alfred. *Die homerische Frage: Ein Bericht über die Forschung der letzten Jahrzehnte.* Darmstadt: Wissenschaftliche Buchgesellschaft, 1974.

Hilmy, Stephen S. *The Later Wittgenstein: The Emergence of a New Philosophical Method.* Oxford: Basil Blackwell, 1987.

Hobbes, Thomas. *Leviathan.* Edited with an introduction by C. B. Macpherson. London: Penguin, 1968.

Hoffman, Yair. *The Blemished Perfection: The Book of Job in Context.* Sheffield: Sheffield Academic Press, 1996.

*The Holy Bible.* See *Bible.*

Homer. *The Iliad.* Translated by Richmond Lattimore. Chicago: University of Chicago Press, 1962.

———. *Odyséen.* Translated by Erland Lagerlöf. Lund: Gleerups, 1957.

———. *Odissea.* Translated by Rosa Calzecchi Onesti. Torino: Einaudi, 1963.

———. *The Odyssey: The Story of Odysseus.* Translated by W. H. D. Rouse. New York: The New American Library, 1937.

———. *The Odyssey.* Translated by E. V. Rieu. Harmondsworth: Penguin, 1946.

———. *The Odyssey.* Translated by Robert Fitzgerald. Garden City, N.Y.: Doubleday, 1963.

———. *Sången om Odysseus.* Translated by Tord Bæckström. Stockholm: Forum, 1975.

Horowitz, Wayne. *Mesopotamian Cosmic Geography.* Winona Lake, Ind.: Eisenbrauns, 1998.

Horvath, Brook, and Irving Malin, eds. *Pynchon and "Mason and Dixon."* Cranbury, N.J.: Associated University Press, 2000.

Hosking, Geoffrey. *Russia and the Russians: A History.* Cambridge, Mass.: Harvard University Press, 2001.

Hughes, Christopher. *Kripke: Names, Necessity and Identity.* Oxford: Oxford University Press, 2004.

Hugh of St. Victor. *The "Didascalicon" of Hugh of St. Victor: A Medieval Guide to the Arts.* Translated and edited by Jerome Taylor. New York: Columbia University Press, 1991.

Hulme, T. E. "Romanticisim and Classicism." In *The Collected Writings of T. E. Hulme,* edited by Karen Csengeri. Oxford: Clarendon Press, 1994.

Husserl, Edmund. *Cartesian Meditations: An Introduction to Phenomenology.* Translated by Dorion Cairns. The Hague: Martinus Nijhoff, 1973.

Høffding, Harald. *A History of Philosophy.* Translated by B. E. Meyer. New York: Dover, 1955.

Høy, Pia. *Det dekonstruerede maleri: Marcel Duchamps Etant donnés.* København: Rævens Sorte Bibliotek, 2001.

Illich, Ivan. *In the Vineyard of the Text: A Commentary to Hugh's"Didascalicon."* Chicago: University of Chicago Press, 1993.

Jacob, Christian. *Géographie et ethnographie en Grèce ancienne.* Paris: Armand Colin, 1991.

———. *L'empire des cartes: Approche théorique de la cartographie à travers l'histoire.* Paris: Albin Michel, 1992.

———. "Mapping in the Mind: The Earth from Ancient Alexandria." In *Mappings,* edited by Denis Cosgrove. London: Reaktion Books, 1999.

Jacobsen, Thorkild. "*The Gilgamesh Epic:* Romantic and Tragic Vision." In *Lingering over Words: Studies in Ancient Near Eastern Literature in Honor of William L. Moran,* edited by Tzvi Abusch et al. Atlanta, Ga.: Scholars Press, 1990.

Jaeger, Werner. *Paideia: Die Formung des Griechischen Menschen.* 3 vols. Berlin: Walter de Gruyter, 1934, 1944, 1947.

Jakobson, Roman, and Morris Halle. *Fundamentals of Language.* The Hague: Mouton, 1956.

James, Carol P. "Reading Art Through Duchamp's Glass and Derrida's Glas." *SubStance* 10 (1981): 105–28.

Jammer, Max. *Concept of Space: The History of Theories of Space in Physics.* 2d ed. Cambridge, Mass.: Harvard University Press, 1969.

Janik, Allan, and Stephen Toulmin: *Wittgenstein's Vienna.* New York: Simon and Schuster, 1973.

Jaspers, Karl. *Kant.* Translated by Ralph Manheim. New York: Harcourt, Brace and World, 1962.

Jeansonne, Sharon Pace. *The Women of Genesis: From Sarah to Potiphar's Wife.* Minneapolis, Minn.: Fortress Press, 1990.

Jensen, Ole Michael. "Red River Valley: Geo-Graphical Studies in the Landscape of Language." *Environment and Planning D: Society and Space* 11 (1993): 295–301.

———. *Vægge: Kundskabens projektioner/Walls: Projections of* Knowledge. Copenhagen: Scandinavian University Press, 1995.

Jensen, Povl Johs. *Plotin.* København: Ejnar Munksgaard, 1948.

*The Jerusalem Bible.* See *Bible.*

Johannisson, Karin. *Nostalgia: En känslans historia.* Stockholm: Bonniers, 2001.

John of Damascus. *On the Divine Images: Three Apologies Against Those Who Attack the Divine Images.* Translated by David Anderson. Crestwood, N.Y.: St. Vladimir's Seminary Press, 1980.

Jonas, Hans. *The Gnostic Religion: The Message of the Alien God and the Beginnings of Christianity.* Boston: Beacon Press, 1963.

Jones, Amelia. *Postmodernism and the En-Gendering of Marcel Duchamp.* Cambridge: Cambridge University Press, 1994.

Jonsson, Inge, and Olle Hjern. *Swedenborg: Sökaren i naturens och andens världar. Hans verk och efterföljd.* Stockholm: Proporius, 1976.

Joselit, David. *Infinite Regress: Marcel Duchamp 1910–1941.* Cambridge, Mass.: MIT Press, 1998.

Joyce, James. *Ulysses.* New York: Random House, 1934.

———. *Finnegans Wake.* New York: Viking Press, 1939.

Judovitz, Dalia. "Rendezvous with Marcel Duchamp: Given." In *Marcel Duchamp: Artist of the Century,* edited by Rudolf E. Kuenzli and Francis M. Naumann. Cambridge, Mass.: MIT Press, 1989.

Jullien, François. *Un sage est sans idée: Ou l'autre de la philosophie.* Paris: Seuil, 1998.

Jung, Carl Gustav. *Answer to Job.* Translated by R. F. C. Hull. Princeton, N.J.: Princeton University Press, 1973.

Järv, Harry. "Biblioteket i Alexandria: Världens viktigaste kulturinstitution." In *Alexandria: Musernas stad,* edited by Stefan Hilding and Suzanne Inger Sörling. Stockholm: Rubicon, 1997.

Kafka, Franz. *The Trial.* Translated by Villa and Edwin Muir. New York: Schocken, 1968.

Kahn, Charles H. *Anaximander and the Origins of Greek Cosmology.* New York: Columbia University Press, 1960.

Kamerbeck, J. C. *The Plays of Sophocles.* Leiden: E. J. Brill, 1967.

Kandinsky, Wassily. *Concerning the Spiritual in Art.* Translated by M. T. H. Stadler. New York: Dover, 1977.

———. *Point and Line to Plane.* Translated by H. Dearstyne and H. Rebay. New York: Dover, 1979.

Kant, Immanuel. *Critique of Judgment.* Translated by James Creed Meredith. Oxford: Clarendon Press, 1952.

———. *Critique of Practical Reason.* Translated by Mary Gregor. Cambridge: Cambridge University Press, 1997.

———. *Critique of Pure Reason.* Translated by F. Max Müller. Garden City, N.Y.: Doubleday, 1966.

———. *Critique of Pure Reason.* Translated by Norman Kemp Smith. London: Palgrave Macmillan, 1973, reissued 2003.

———. *Géographie (Physische Geographie).* Paris: Aubier, 1999.

———. *Grounding for the Metaphysics of Morals.* Translated by James W. Ellington. Indianapolis, Ind.: Hackett, 1981.

———. *Immanuel Kant's sämmtliche Werke.* Teil 6, *Immanuel Kant's Schriften zur Physischen Geographie.* Edited by Friedrich Wilhelm Schubert. Leipzig: L. Voss, 1839.

———. *Träume eines Geistersehers—erläutert durch Träume der Metaphysik* (1766). Translated by John Manoleso in *"Dreams of a Spirit Seeker" by Immanuel Kant and Other Related Writings.* New York: Vintage Press, 1962.

Kaplan, Robert. *The Nothing That Is: A Natural History of Zero.* Oxford: Oxford University Press, 2001.

Karatani, Kojin. *Architecture as Metaphor: Language, Number, Money.* Translated by Sabu Kohso. Cambridge, Mass.: MIT Press, 1995.

Karjalainen, Pauli Tapani. "Place and Intimate Sensing." *Nordisk Geografisk Tidskrift* 27 (1998): 3–16.

Kaufmann, Walter. *Tragedy and Philosophy.* Garden City, N.Y.: Doubleday, 1968.

Kearney, Richard. *The Wake of Imagination.* London: Routledge, 1994.

———. *Poetics of Imagining: Modern to Post-modern.* Edinburgh: Edinburgh University Press, 1998.

Keller, Pierre. *Kant and the Demand of Self-Consciousness.* Cambridge: Cambridge University Press, 1998.

Kelly, Joseph F. *The Problem of Evil in the Western Tradition: From the Book of Job to Modern Genetics.* Collegeville, Minn.: Liturgical Press, 2002.

Kelly, J. N. D. *Early Christian Creeds.* 3d ed. New York: David McKay Company, 1972.

Kemp, Martin. "Science, Non-Science and Nonsense: The Interpretation of Brunelleschi's Perspective." *Art History* 1 (1978): 134–61.

———. *The Science of Art: Optical Themes in Western Art from Brunelleschi to Seurat.* New Haven, Conn.: Yale University Press, 1990.

———. *Behind the Picture: Art and Evidence in the Italian Renaissance.* New Haven, Conn.: Yale University Press, 1997.

Kemp Smith, Norman. *A Commentary on Kant's "Critique of Pure Reason."* 2d ed. London: Macmillan, 1923.

Kenny, Anthony. *Wittgenstein.* Cambridge, Mass.: Harvard University Press, 1973.

Kierkegaard, Søren. *Concluding Unscientific Postscript.* Translated by D. F. Swenson and Walter Lowrie. Princeton, N.J.: Princeton University Press, 1944.

———. *Fear and Trembling: A Dialectical Lyric.* Translated by Walter Lowrie. Princeton, N.J.: Princeton University Press, 1941.

———. *Repetition.* Translated by Walter Lowrie. Princeton, N.J.: Princeton University Press, 1962.

Kimble, George H. T. *Geography in the Middle Ages.* London: Methuen, 1938.

King, Ross. *Brunelleschi's Dome: The Story of the Great Cathedral in Florence.* London: Chatto and Windus, 2000.

Kirk, Geoffrey Stephen. *The Songs of Homer.* Cambridge: Cambridge University Press, 1962.

———. *Myth: Its Meaning and Functions in Ancient and Other Cultures.* Berkeley and Los Angeles: University of California Press, 1970.

——— et al. *The Presocratic Philosophers: A Critical History with a Selection of Texts.* 2d ed. Cambridge: Cambridge University Press, 1983.

Kirwan, L. P. "The Christian Topography and the Kingdom of Axum." *Geographical Journal* 138 (1972): 166–77.

Kitto, H. D. F. *Poiesis: Structure and Thought.* Berkeley and Los Angeles: University of California Press, 1966.

Knight, G. Wilson. *The Wheel of Fire: Essays in the Interpretation of Shakespeare's Sombre Tragedies.* London: Oxford University Press, 1930.

Knox, Bernard M. W. *The Heroic Temper: Studies in Sophoclean Tragedy.* Berkeley and Los Angeles: University of California Press, 1964.

———. *Oedipus at Thebes: Sophocles' Tragic Hero and His Time.* 2d ed. New Haven, Conn.: Yale University Press, 1966.

Koerner, Joseph Leo. *The Moment of Self-Portraiture in German Renaissance Art.* Chicago: University of Chicago Press, 1993.

Körner, Stephan. *Kant.* Harmondsworth: Penguin, 1955.

———. "Laws of Thought." In *The Encyclopedia of Philosophy*, edited by Paul Edwards. New York: Macmillan, 1967.

Kojève, Alexandre. *Introduction to the Reading of Hegel: Lectures on "The Phenomenology of Spirit."* Translated by James H. Nichols Jr. Ithaca, N.Y.: Cornell University Press, 1980.

Krauss, Rosalind E. *Bachelors.* Cambridge, Mass.: MIT Press, 1999.

Kripke, Saul A. *Naming and Necessity.* Oxford: Basil Blackwell, 1981.

Kristeva, Julia. "The Speaking Subject." In *On Signs*, edited by Marshall Blonsky. Oxford: Basil Blackwell, 1985.

Kristoferson, Kris. *Border Lord.* New York: Resaca, 1972.

Kronholm, Tryggve. *De tio orden.* Stockholm: Verbum, 1992.

Kuehn, Manfred. *Kant: A Biography.* Cambridge: Cambridge University Press, 2001.

Kuenzli, Rudolf, and Francis M. Naumann, eds. *Marcel Duchamp: Artist of the Century.* Cambridge, Mass.: MIT Press, 1989.

Kugler, Hartmut. "Die Ebstorfer Weltkarte: Ein europäisches Weltbild in deutschen Mittelater." *Zeitschrift für deutsches Altertum und deutsche Literatur* 116 (1987): 1–29.

———, ed. *Ein Weltbild vor Columbus: Die Ebstorfer Weltkarte Interdisciplinäres Colloquium 1988.* Weinheim: VCH, Acta Humaniora, 1991.

Kuh, Katharine. *The Artist's Voice.* New York: Harper and Row, 1960.

Kuryluk, Ewa. *Veronica and Her Cloth: History, Symbolism, and the Structure of a "True" Image.* Oxford: Basil Blackwell, 1991.

Kuschel, Karl-Josef. *Abraham: A Symbol of Hope for Jews, Christians and Muslims.* London: SCM Press, 1995.

Lacan, Jacques. *Écrits: A Selection.* Translated by Alan Sheridan. London: Tavistock, 1977.

———. *The Four Fundamental Concepts of Psycho-Analysis.* Edited by Jacques-Alain Miller. Translated by Alan Sheridan. New York: W. W. Norton, 1981.

Lachterman, David Rapport. *The Ethics of Geometry: A Genealogy of Modernity.* New York: Routledge, 1989.

Lagercrantz, Olof. *Dikten om livet på den andra sidan: En bok om Emanuel Swedenborg.* Stockholm: Wahlström och Widstrand, 1996.

Lakoff, George. *Women, Fire, and Dangerous Things: What Categories Reveal About the Mind.* Chicago: University of Chicago Press, 1987.

Lakoff, George, and Mark Johnson. *Metaphors We Live By.* Chicago: University of Chicago Press, 1980.

———. *Philosophy in the Flesh: The Embodied Mind and Its Challenge to Western Thought.* New York: Basic Books, 1999.

Lakoff, George, and Rafael Núñez. *Where Mathematics Comes From: How the Embodied Mind Brings Mathematics into Being.* New York: Basic Books, 2000.

Lakoff, George, and Mark Turner. *More than Cool Reason: A Field Guide to Poetic Metaphor.* Chicago: University of Chicago Press, 1989.

Lakoff, Robin. *Language and Woman's Place.* New York: Harper and Row, 1975.

Lamb, Jonathan. *The Rhetoric of Suffering: Reading the Book of Job in the Eighteenth Century.* Oxford: Clarendon Press, 1995.

Lambert, W. G. *Babylonian Wisdom Literature.* Oxford: Clarendon Press, 1960.

Landen, Annette. *Läsvägar.* Lund: Institutionen för konstvetenskap, 1979.

Lane Fox, Robin. *The Search for Alexander.* Boston: Little, Brown and Company, 1980.

Lang, Uwe Michael. *John Philoponus and the Controversies over Chalcedon in the Sixth Century: A Study and Translation of the "Arbiter."* Sterling, Va.: Peeters, 2001.

Latour, Bruno, and Peter Weibel, eds. *Iconoclash: Beyond the Image Wars in Science, Religion and Art.* Cambridge, Mass.: MIT Press, 2002.

Lawrence, James F., ed. *Testimony to the Invisible: Essays on Swedenborg.* West Chester, Penn.: Swedenborg Foundation, 1995.

Lecoq, Danielle. "La 'mappemonde' du *De arca noe mystica* de Hugues de Saint-Victor (1128–1129)." In *Géographie du monde au moyen âge et à la renaissance,* edited by Monique Pelletire. Paris: Éditions du CTHS, 1989.

Lecouteux, Claude. *Les monstres dans la pensée médiévale européenne: Essai de présentation.* Paris: Presses de l'Université de Paris-Sorbonne, 1993.

Le Doeuff, Michèle. *The Philosophical Imaginary.* Translated by Colin Gordon. London: Athlone Press, 1989.

Leithäuser, Joachim G. *Mappae Mundi: Die geistige Eroberung der Welt.* Berlin: Safari, 1958.

Lerer, Seth, ed. *Literary History and the Challenge of Philology.* Stanford, Calif.: Stanford University Press, 1996.

Lewis, M. J. T. *Surveying Instruments of Greece and Rome.* Cambridge: Cambridge University Press, 2001.

Lindberg, David C. *Roger Bacon and the Origins of Perspectiva in the Middle Ages: A Critical Edition and English Translation of Bacon's "Perspectiva."* Oxford: Clarendon Press, 1996.

Linde, Ulf. *Efter hand: Texter 1950–1985.* Stockholm: Bonniers, 1985.

———. *Marcel Duchamp.* Stockholm: Rabén och Sjögren, 1986.

———. "Perifera notiser om centralperspektivet." *Artes* 70, no. 3 (1991): 45–64.

Lindkvist, Karl Gunnar. *Studies on the Local Sense of the Prepositions 'in', 'at, 'on', and 'to' in Modern English.* Lund: Lund Studies in English, no. 20, 1968.

Linnér, Sture. *Homeros.* Stockholm: Norstedt, 1985.

Logan, Beryl, ed. *Immanuel Kant's "Prolegomena to Any Future Metaphysics" in Focus.* London: Routledge, 1996.

Lotman, Ju. M., and B. A. Uspenskij. *The Semiotics of Russian Culture.* Edited by Ann Shukman. Translated by N. F. C. Owen. Ann Arbor: Department of Slavic Languages and Literature, University of Michigan, 1984.

Lovelace, Richard. *The Poems of Richard Lovelace.* Edited by C. H. Wilkinson. Oxford: Clarendon Press, 1930.

Loyola, Ignatius. *The "Spiritual Exercises" of St. Ignatius: Based on Studies in the Language of the Autograph.* Translated by Louis J. Puhl. New York: Vintage Books, 2000.

Lugt, Pieter van der. *Rhetorical Criticism and the Book of Job.* Leiden: E. J. Brill, 1995.

Lyotard, Jean-François. *Les TRANSformateurs DUchamp.* Paris: Éditions Galilée, 1977. Translated by Ian McLeod as *Duchamp's TRANSformers.* Venice, Calif.: Lapis Press, 1990.

Machotka, Pavel. *Cézanne: Landscape into Art.* New Haven, Conn.: Yale University Press, 1996.

Macmullen, Ramsay. *Christianity and Paganism in the Fourth to Eighth Centuries.* New Haven, Conn.: Yale University Press, 1997.

Magee, Bryan. *Confessions of a Philosopher: A Journey Through Western Philosophy.* New York: Random House, 1997.

Maguire, Henry. *The Icons of Their Bodies: Saints and their Images in Byzantium.* Princeton, N.J.: Princeton University Press, 1996.

Maier, John, ed. *Gilgamesh: A Reader.* Wauconda, Ill.: Bolchazy-Carducci Publishers, 1997.

Makkreel, Rudolf A. *Imagination and Interpretation in Kant: The Hermeneutical Import of "The Critique of Judgment."* Chicago: University of Chicago Press, 1990.

Malantschuk, Gregor. *Kierkegaard's Thought.* Princeton, N.J.: Princeton University Press, 1971.

Malcolm, Norman. *Ludwig Wittgenstein: A Memoir.* Oxford: Oxford University Press, 1958.

———. *Wittgenstein: Nothing is Hidden.* Oxford: Basil Blackwell, 1986.

———. *Wittgenstein: A Religious Point of View.* London: Routledge, 1993.

Malevich, Kazimir S. *Essays on Art 1915–1933.* Edited by Troels Andersen. Translated by Xenia Glowacki-Prus and Arnold McMillin. 4 vols. København: Borgen, 1968–78.

Mallarmé, Stéphane. *Mallarmé.* Edited with an introduction and prose translation by Anthony Hartley. Harmondsworth: Penguin, 1965.

Maly, Kenneth, ed. *The Path of Archaic Thinking: Unfolding the Work of John Sallis.* Albany: State University of New York Press, 1995.

Manetti, Antonio di Tuccio. *The Life of Brunelleschi by Antonio di Tuccio.* Edited by Howard Saalman. Translated by Catherine Engass. University Park: Pennsylvania State University Press, 1970.

Marin, Louis. *Utopics: Spatial Play.* Atlantic Highlands, N.J.: Humanities Press, 1984.

Marx, Karl. *Capital.* Translated by Samuel Moore and Edward Aveling. New York: International Publishers, 1967.

Mathews, Thomas F. *The Clash of Gods: A Reinterpretation of Early Christian Art.* Rev. ed. Princeton, N.J.: Princeton University Press, 2003.

Mauthner, Fritz. *Beiträge zu einer Kritik der Sprache.* Stuttgart: J. G. Gotta, 1901–1902.

May, J. A. *Kant's Concept of Geography and Its Relation to Recent Geographical Thought.* Toronto: University of Toronto Press, 1970.

McGuinness, Brian. *Wittgenstein: A Life. Young Ludwig (1889–1921).* Berkeley and Los Angeles: University of California Press, 1988.

Megill, Allan. *Prophets of Extremity: Nietzsche, Heidegger, Foucault, Derrida.* Berkeley and Los Angeles: University of California Press, 1985.

Melnick, Arthur. *Space, Time and Thought in Kant.* Dordrecht and Boston: Kluwer, 1989.

Melville, Herman. *Moby-Dick: or, The Whale.* San Francisco: Arion Press, 1979.

Meri, Veijo. *Kvinna i spegeln.* Translated by Bertel Kihlman. Stockholm: Natur och Kultur, 1964.

Merlan, Philip. "Plotinus." In *The Encyclopedia of Philosophy,* edited by Paul Edwards. New York: Macmillan, 1967.

Merleau-Ponty, Maurice. *Phenomenology of Perception.* Translation by Colin Smith. London: Routledge and Kegan Paul, 1962.

———. *Sense and Non-Sense.* Translated by Hubert L. Dreyfus and Patricia A. Dreyfus. Evanston, Ill.: Northwestern University Press, 1964.

———. *The Visible and the Invisible.* Translated by Alphonso Lingis. Evanston, Ill.: Northwestern University Press, 1968.

Meuris, Jacques. *René Magritte, 1898–1967.* Translated by Michael Scuffil. Köln: Taschen, 1997.

Meyendorff, John. *Christ in Eastern Christian Thought.* Washington, D.C.: Corpus Books, 1969.

Meyer, Marvin, with Esther A. de Boer. *The Gospels of Mary: The Secret Tradition of Mary Magdalene, the Companion of Jesus.* New York: Harper San Francisco, 2004.

Miles, Jack. *God: A Biography.* New York: Simon and Schuster, 1995.

———. *Christ: A Crisis in the Life of God.* London: William Heinemann, 2001.

Miller, Konrad. *Mappae mundi: Die ältesten Weltkarten.* Vol. 5, *Die Ebstorfkarte.* Stuttgart: Jos. Rot'sche, 1896.

Milner, John. *Kazimir Malevich and the Art of Geometry.* New Haven, Conn.: Yale University Press, 1996.

Monk, Ray. *Ludwig Wittgenstein: The Duty of Genius.* London: Jonathan Cape, 1990.

Monmonier, Mark S. *How to Lie with Maps.* 2d ed. Chicago: University of Chicago Press, 1996.

Monod, Jacques. *Chance and Necessity: An Essay on the Natural Philosophy of Modern Biology.* Translated by Austryn Wainhouse. New York: Knopf, 1971.

Mulhall, Stephan. *On Being in the World: Wittgenstein and Heidegger on Seeing Aspects.* London: Routledge, 1990.

———. *Inheritance and Originality: Wittgenstein, Heidegger and Kierkegaard.* Oxford: Clarendon Press, 2001.

Myskja, Bjørn K. *The Sublime in Kant and Beckett: Aesthetic Judgment, Ethics and Literature.* Berlin: Walter de Gruyter, 2002.

Neef, Heinz-Dieter. *Die Prüfung Abrahams: Eine exegetisch-theologische Studie zu Gen. 22:1–19.* Stuttgart: Calwer, 1998.

Nelson, James, ed. *Wisdom: Conversations with the Elder Wise Men of Our Day.* New York: W. W. Norton, 1958.

Nemo, Philippe. *Job and the Excess of Evil.* Translated by Michael Kigel. Pittsburgh, Penn.: Duquesne University Press, 1998.

Nesbit, Molly. "The Language of Industry." In *The Definitively Unfinished Marcel Duchamp,* edited by Thierry de Duve. Cambridge, Mass.: MIT Press, 1991.

Netz, Reviel, "The Aesthetics of Mathematics: A Study." In *Visualization, Explanation and Reasoning Styles in Mathematics,* edited by Paolo Mancosu et al. Dordrecht: Springer, 2005.

*New Catholic Encyclopedia.* 2d ed. Washington, D.C.: Catholic University of America, 2003.

Newman, Barnett. *Selected Writings and Interviews.* Edited by John P. O'Neill. New York: Knopf, 1990.

Newman, John Henry. *Arians of the Fourth Century.* New ed. Westminster, Md.: Christian Classics, 1968.

Newsom, Carol A. *The Book of Job: A Contest of Moral Imaginations.* Oxford: Oxford University Press, 2003.

Newsom, Carol A., and Sharon Ringe. *Women's Bible Commentary.* Louisville, Ky.: Westminster John Knox Press, 1998.

Nietzsche, Friedrich. *The Birth of Tragedy.* Translated by Walter Kaufmann. New York: Vintage Books, 1967.

———. "On Truth and Lie in an Extra-Moral Sense." In *The Portable Nietzsche,* edited by Walter Kaufmann. Harmondsworth: Penguin, 1976.

Nikulin, Dmitri. *Matter, Imagination and Geometry: Ontology, Natural Philosophy and Mathematics in Plotinus, Proclus and Descartes.* Aldershot: Ashgate, 2002.

Noegel, Scott B. *Janus Parallelism in the Book of Job.* Sheffield: Sheffield Academic Press, 1996.

Noort, Edward, and Eibert Tigchelaar. *The Sacrifice of Isaac: The Aqeda (Genesis 22) and its Interpretations.* Leiden: E. J. Brill, 2002.

Norris, Christopher. *Derrida.* London: Fontana, 1987.

North, Robert. *A History of Biblical Map Making.* Wiesbaden: Dr. Ludwig Reichert, 1979.

Novotny, Fritz. *Cézanne und das Ende der Wissenschaftlichen Perspektive.* Wien: Verlag von Anton Schroll, 1938.

Nussbaum, Martha. *Love's Knowledge: Essays on Philosophy and Literature.* Oxford: Oxford University Press, 1992.

Ogden, C. G., and I. A. Richards. *The Meaning of Meaning: A Study of The Influence of Language upon Thought and of the Science of Symbolism.* 4th ed. London: Kegan Paul, Trench, Trubner and Company, 1936.

Ohly, Friedrich. "Deus Geometra: Skizzen zur Geschichte einer vorstellung von Gott." In *Tradition als historische Kraft: Interdisziplinäre Forschungen zur Geschichte des früheren Mittelalters,* edited by Norbert Kamp and Joachim Wollasch. Berlin: Walter de Gruyter, 1982.

Oliva, Achille Bonito, ed. *Figurabile: Francis Bacon.* Milano: Electa, 1993.

Ollman, Bertell. *Alienation: Marx's Conception of Man in Capitalist Society.* Cambridge: Cambridge University Press, 1971.

Olsson, Anders. *Läsningar av INTET.* Stockholm: Albert Bonniers, 2000.

Olsson, Gunnar. "Complementary Models: A Study of Colonization Maps." *Geografiska Annaler* 50 B (1968): 115–32.

———. "Inference Problems in Locational Analysis." In *Behavioral Problems in Geography: A Symposium,* edited by R. G. Cox and R. G. Golledge. Evanston, Ill.: Northwestern University, Studies in Geography, no. 17, 1969.

———. *Birds in Egg/Eggs in Bird.* London: Pion, 1980.

———. "On Yearning for Home: An Epistemological View of Ontological Transfomations." In *Humanistic Geography and Literature,* edited by D. C. D. Pocock. London: Croom Helm, 1981.

———. *Antipasti.* Göteborg: Korpen, 1990.

———. *Lines of Power/Limits of Language.* Minneapolis: University of Minnesota Press, 1991.

———. "Chiasm of Thought-and-Action." *Environment and Planning D: Society and Space* 11 (1993): 279–94.

———. "Job and the Case of the Herbarium." *Environment and Planning D: Society and Space* 12 (1994): 221–25.

———. "Die Projektion des Begehrens/Das Begehren der Projektion." In *Räumliches Denken,* edited by Dagmar Reichert. Zürich: vdf Hochschulverlag AG an der ETH Zürich, 1996.

———, ed. *Chimärerna: Porträtt från en forskarutbildning.* Stockholm: Nordplan, 1996.

———. "Stadier på livets väg." In *Att famna en ton,* edited by Gunnar Olsson. Uppsala: Acta Universitatis Upsaliensis, 1998.

———. "Towards a Critique of Cartographical Reason." *Ethics, Place and Environment* 1 (1998): 145–55.

———. "Horror vacui." In *Svensk kulturgeografi: En exkursion inför 2000-talet,* edited by Martin Gren and P.O. Hallin. Lund: Studentlitteratur, 1998.

———. "Constellations," *Sistema Terra* 8, issue 1–3 (1999): 139–44.

———. "Landscape—Border Station Between Stonescape and Mindscape." In *La con-*

*struzione del Paesaggio Siciliano: Geografie scrittori a confronto,* edited by Girolamo Cusimano. Palermo: Università degli Studi di Palermo, 1999.

———. "Detta är detta," *Nordisk Estetisk Tidskrift* 20–21 (2000): 39–54.

———. "Skattkammarön." In *Kulturens plats/Maktens rum,* edited by Martin Gren et al. Stehag: Brutus Östling, 2000.

———. "Washed in a Washing Machine™." In *Postmodern Geography: Theory and Praxis,* edited by Claudio Minca. Oxford: Blackwell, 2001.

———. "Glimpses." In *Geographical Voices: Fourteen Autobiographical Essays,* edited by Peter Gould and Forrest R. Pitts. Syracuse, N.Y.: Syracuse University Press, 2002.

Olvera, Karen M. *Marcel Duchamp's "The Large Glass" as "Negation of Women."* Ann Arbor, Mich.: UMI, 1986.

Onasch, Konrad. *Die Ikonenmalerei: Grundzüge einer systematischen Darstellung.* Leipzig: Koehler und Armelang, 1968.

Ophir, Adi. *Plato's Invisible Cities: Discourse and Power in the "Republic."* London: Routledge, 1991.

Oppenheim, A. Leo. *The Interpretations of Dreams in the Ancient Near East: With a Translation of an Assyrian Dreambook.* Philadelphia: American Philosophical Society, 1956.

Ortega y Gasset, José. *Man and People.* Translated by Willard R. Trask. New York: Norton, 1957.

Orwell, George. *Animal Farm.* New York: Harcourt Brace and Company, 1946.

Ouspensky, Leonide. *Theology of the Icon.* Crestwood, N.Y.: St. Vladimir's Seminary Press, 1978.

*Oxford English Dictionary.* The Compact Edition. Oxford: Oxford University Press, 1971.

Padovan, Richard. *Proportion: Science, Philosophy, Architecture.* London: E. and F. N. Spon, 1999.

Panofsky, Erwin. *Perspective as Symbolic Form.* Translated by Christopher S. Wood. New York: Zone Books, 1991.

Papineau, David. *Thinking about Consciousness.* Oxford: Oxford University Press, 2002.

Parfitt, Tudor. *The Lost Tribes of Israel: The History of a Myth.* London: Weidenfeld and Nicholson, 2002.

Parpola, Simo. *Assyrian Prophesies.* Helsinki: Helsinki University Press, 1997.

Parry, Kenneth. *Depicting the World: Byzantine Iconophile Thought of the Eighth and Ninth Centuries.* Leiden: E. J. Brill, 1996.

Parry, Milman. *The Making of Homeric Verse: The Collected Papers of Milman Parry.* Edited by Adam Parry. Oxford: Clarendon Press, 1971.

Paz, Octavio. *Marcel Duchamp: Appearance Stripped Bare.* Translated by Rachel Phillips and Donald Gardiner. New York: Viking Press, 1978.

Pears, David. *The False Prison: A Study of the Development of Wittgenstein's Philosophy.* 2 vols. Oxford: Clarendon Press, 1987, 1988.

Pelikan, Jaroslav. *The Christian Tradition: A History of the Development of Doctrine.* Vol. 1, *The Emergence of the Catholic Tradition (100–600).* Chicago: University of Chicago Press, 1971.

———. *The Christian Tradition: A History of the Development of Doctrine.* Vol. 2, *The Spirit of Eastern Christendom (600–1700).* Chicago: University of Chicago Press, 1974.

———. *Imago Dei: The Byzantine Apologia for Icons.* Princeton, N.J.: Princeton University Press, 1990.

———. *What Has Athens to Do with Jerusalem? Timaeus and Genesis in Counterpoint.* Ann Arbor: University of Michigan Press, 1997.

Peppiatt, Michael. *Francis Bacon: Anatomy of an Enigma*. London: Weidenfeld and Nicholson, 1996.

Perdue, Leo G., and W. Clark Gilpin, eds. *The Voice from the Whirlwind: Interpreting the Book of Job*. Nashville, Tenn.: Abingdon Press, 1992.

Perelman, Chaïm, and L. Olbrechts-Tyteca. *The New Rhetoric: A Treatise on Argumentation*. Translated by John Wilkinson and Purcell Weaver. Notre Dame, Ind.: University of Notre Dame Press, 1971.

Pérez Paoli, Ubaldo Ramón. *Der plotonische begriff von ῾ΨΠΟΣΤΑΣΙΣ und die augustinische Bestimmung Gottes als Subiectum*. Würzburg: Augustinus Verlag, 1990.

Perkins, Robert L., ed. *Kierkegaard's "Fear and Trembling": Critical Appraisals*. Tuscaloosa: University of Alabama Press, 1981.

Perloff, Marjorie. *Wittgenstein's Ladder: Poetic Language and the Strangeness of the Ordinary*. Chicago: University of Chicago Press, 1996.

Petitot-Cocorda, Jean. *Morphogenèse du sens: Pour un schématisme de la structure*. Paris: Presses Universitaires de France, 1985.

———. *Physique du sens: De la théorie des singularités aux sémio-narrative*. Paris: CNRS, 1992.

Petrova, Evgenija, et al. *Malevich: Artist and Theoretician*. Translated by Sharon McKee. Paris: Flammarion, 1990.

Philo, Chris. "Reflections on Gunnar Olsson's Discourse in Contemporary Human Geography." *Environment and Planning D: Society and Space* 2 (1984): 217–40.

———. "Escaping Flatland: A Book Review Essay Inspired by Gunnar Olsson's *Lines of Power/Limits of Language*." *Environment and Planning D: Society and Space* 12 (1994): 229–52.

Pickles, John. *A History of Spaces: Geographic Reason, Mapping and the Geo-Coded World*. London: Routledge, 2004.

Picone, Marco. "Anaximander versus Anaximander." In *Bodies and Space: Gunnar's Travels*, edited by Marco Picone. Palermo: Università di Palermo, Laboratorio Geografico, 2002.

Piltz, Anders. "Ansikte, mask, person: Om den europeiska människobilden." In *Libens merito: Festskrift till Stig Strömholm*, edited by Olle Matsson et al. Uppsala: Kungliga Vetenskapsamhällets i Uppsala Handlingar, vol. 21, 2001.

Pitkin, Hanna Fenichel. *Wittgenstein and Justice: On the Significance of Ludwig Wittgenstein for Social and Political Thought*. Berkeley and Los Angeles: University of California Press, 1972.

Plato. *Collected Dialogues of Plato*. Edited by Edith Hamilton and Huntington Cairns. Bollingen Series LXXI. Princeton, N.J.: Princeton University Press, 1961.

———. *The Republic of Plato*. Translated by Allan Bloom. New York: Basic Books, 1991.

———. *The Republic of Plato*. Translated by Francis MacDonald Cornford. Oxford: Oxford University Press, 1945.

———. *The Republic*. Translated by Tom Griffith and Giovanni R. F. Ferrari. Cambridge: Cambridge University Press, 2000.

———. *Republic*. Translated by G. M. A. Grube. Indianapolis, Ind.: Hackett, 1974.

———. *The Republic*. Translated by Desmond Lee. London: Penguin, 2003.

———. *The Republic*. Translated by Paul Shorey. Cambridge, Mass.: Harvard University Press, 1956.

Platon. *Skrifter: Gästabudet*. Translated by Jan Stolpe. Stockholm: Atlantis, 2000.

———. *Staten.* Translated by Jan Stolpe. Stockholm: Atlantis, 2003.

Plotinus. *Enneads.* Translated by A. H. Armstrong. Cambridge, Mass.: Harvard University Press, 1985.

Pritchard, James B., ed. *Ancient Near Eastern Texts Relating to the Old Testament.* Princeton, N.J.: Princeton University Press, 1955.

Ptolemy. *Ptolemy's "Almagest."* Translated and annotated by Gerald James Toomer. London: Duckworth, 1984.

Pucci, Pietro. *Odysseus Polutropos: Intertextual Readings in the "Odyssey" and the "Iliad."* Ithaca, N.Y.: Cornell University Press, 1987.

Puttfarken, Thomas. *The Discovery of Pictorial Composition: Theories of Visual Order in Painting 1400–1800.* New Haven, Conn.: Yale University Press, 2000.

Pynchon, Thomas. *Mason & Dixon.* New York: Henry Holt and Company, 1997.

Radermacher, Hans. *Kant, Swedenborg, Borges: Mit Paraphrasen von Jorge Luis Borges.* Bern: Peter Lang, 1986.

Rajchman, John. *The Deleuze Connections.* Cambridge, Mass.: MIT Press, 2000.

Ramírez, José L. *Skapande mening—en begreppsgenealogisk studie om rationalitet, vetenskap och planering.* Stockholm: Nordplan, 1995.

Ramírez, Juan Antonio. *Duchamp: Love and Death, Even.* Translated by Alexander R. Tulloch. London: Reaktion Books, 1998.

Rapaport, I. *The Babylonian Poem "Enuma Elish" and Genesis Chapter One: A New Theory on the Relationship Between the Ancient Cuneiform Composition and the Hebrew Scriptures.* Melbourne: Hawthorn Press, 1979.

Reck, Erich H., ed. *From Frege to Wittgenstein: Perspectives on Early Analytic Philosophy.* Oxford: Oxford University Press, 2002.

Reeve, C. D. C. *Philosopher-Kings: The Argument of Plato's "Republic."* Princeton, N.J.: Princeton University Press, 1988.

———. *Substantial Knowledge: Aristotle's Metaphysics.* Indianapolis, Ind.: Hackett, 2000.

Reichert, Dagmar. "Obituary." *Ethics, Place and Environment* 1 (1998): 157–64.

Reinhardt, Karl. *Sophocles.* Translated by H. and D. Harvey. Oxford: Basil Blackwell, 1979.

Rendsburg, Gary A. *The Redaction of Genesis.* Winona Lake, Ind.: Eisenbrauns, 1986.

Rescher, Nicholas. *Many-valued Logic.* New York: McGraw-Hill, 1969.

Reutersward, Carl Fredrik. *Titta, jag är osynlig!* Stockholm: Gedins Förlag, 1989.

Reventlow, Henning Graf. *Gebot und Predigt im Dekalog.* Gütersloh: G. Mohn, 1962.

Rewald, John. *Cézanne: A Biography.* New York: Harry N. Abrams, 1990.

———. *The Paintings of Paul Cézanne: A Catalogue Raisonné.* Vol. 1, *The Texts;* vol. 2, *The Plates.* New York: Harry N. Abrams, 1996.

Rhees, Rush, ed. *Ludwig Wittgenstein: Personal Recollections.* Oxford: Basil Blackwell, 1981.

Richardson, John T. E. *Mental Imagery and Human Memory.* London: Macmillan, 1980.

Ricoeur, Paul. *Memory, History, Forgetting.* Translated by Kathleen Blamey and David Pellauer. Chicago: University of Chicago Press, 2004.

Rider, Sharon P. *Avoiding the Subject: A Critical Inquiry into Contemporary Theories of Subjectivity.* Stockholm: Thales, 1998.

Rist, J. M. *Plotinus: The Road to Reality.* Cambridge: Cambridge University Press, 1967.

Robinson, Richard. *Plato's Earlier Dialectic.* 2d ed. Oxford: Clarendon Press, 1953.

Romm, James S. *The Edges of the Earth in Ancient Thought: Geography, Exploration, and Fiction.* Princeton, N.J.: Princeton University Press, 1992.

Roseman, Christina Horst. "Pytheas of Massalia: A Critical Examination of the Texts." Ph.D. diss., University of Washington. Ann Arbor, Mich.: University Micofilms International, 1983.

———. *Pytheas of Massalia: On the Ocean.* Chicago: Ares Publishers, 1994.

Rosengren, Mats. *Psychagōgia—konsten att leda själar: Om konflikten mellan retorik och filosofi hos Platon och Chaïm Perelman.* Göteborg: Filosofiska institutionen, Göteborgs universitet, 1997.

Ross, David John Athole. *Alexander Historiatus: A Guide to Medieval Illustrated Alexander Literature.* London: Warburg Institute, 1963.

Rothko, Mark. *The Artist's Reality: Philosophies of Art.* Edited and with an introduction by Christopher Rothko. New Haven, Conn.: Yale University Press, 2004.

Rotman, Brian. *Signifying Nothing: The Semiotics of Zero.* London: Macmillan, 1987.

Royce, Josiah. *The World and the Individual.* New York: Dover, 1959.

Rubenstein, Richard E. *When Jesus Became God: The Epic Fight over Christ's Divinity in the Last Days of Rome.* New York: Harcourt Brace and Company, 1999.

Ruberg, Uwe. "Mappae mundi des Mittelaters in Zusammenwirkung von Text und Bild." In *Text und Bild: Aspekte des Zusammenwirkens zweier Künste in Mittelalter und früher Neuzeit,* edited by Christel Meier and Uwe Ruberg. Wiesbaden: Dr. Ludwig Reichert Verlag, 1980.

Russell, Bertrand. "Meinong's Theory of Complexes and Assumptions." *Mind* 13 (1904): 204–19.

———. *Introduction to Mathematical Philosophy.* London: George Allen and Unwin, 1920.

Russell, Jeffrey Burton. *Inventing the Flat Earth: Columbus and Modern Historians.* New York: Praeger, 1991.

Russo, Joseph et al., eds. *A Commentary on Homer's "Odyssey."* 3 vols. Oxford: Clarendon Press, 1992.

Ruthrof, Horst. *The Body in Language.* London: Cassell, 2000.

Said, Edward W. *The World, the Text, and the Critic.* Cambridge, Mass.: Harvard University Press, 1983.

Sallis, John. *Being and Logos: The Way of Platonic Dialogue.* Atlantic Higlands, N.J.: Humanities Press, 1986.

———. *Spacings—Of Reason and Imagination in Texts of Kant, Fichte, Hegel.* Chicago: University of Chicago Press, 1987.

———. *Shades—Of Painting at the Limit.* Bloomington: Indiana University Press, 1998.

———. *Chorology: On Beginning in Plato's "Timaeus."* Bloomington: Indiana University Press, 1999.

———. *Force of Imagination: The Sense of the Elemental.* Bloomington: Indiana University Press, 2000.

Sandars, Nancy K. *Poems of Heaven and Hell from Ancient Mesopotamia.* Harmondsworth: Penguin, 1971.

Sanouillet, Michel, and Elmer Peterson, eds. *The Writings of Marcel Duchamp.* New York: Da Capo Press, 1973.

Sarton, George. *A History of Science.* Vol. 2, *Hellenistic Science and Culture in the Last Three Centuries B.C.* Cambridge, Mass.: Harvard University Press, 1959.

Sartre, Jean-Paul. *Imagination.* Translated by W. Williams. Ann Arbor: University of Michigan Press, 1962.

Saussure, Ferdinand de. *Course in General Linguistics.* Translated by Roy Harris. London: Duckworth, 1983.

Scafi, Alesssandro. "Mapping Eden: Cartographies of the Earthly Paradise." In *Mappings,* edited by Denis Cosgrove. London: Reaktion Books, 1999.

Scarry, Elaine. *The Body in Pain: The Making and Unmaking of the World.* Oxford: Oxford University Press, 1985.

Schapiro, Meyer. *Paul Cézanne.* New York: Abrams, 1990.

———. *Words, Script, and Pictures: Semiotics of Visual Language.* New York: George Braziller, 1996.

Scheindlin, Raymond P. *The Book of Job.* New York: W. W. Norton, 1998.

Schiappa, Edward. *The Beginnings of Rhetorical Theory in Classical Greece.* New Haven, Conn.: Yale University Press, 1999.

Schildt, Göran. *Cézanne.* Stockholm: Wahlström och Widstrand, 1946.

Schlögel, Karl. *Im Raume lesen wir die Zeit: Über Ziviliationsgeschihte und Geopolitk.* München: Carl Hanser Verlag, 2003.

Schmitt, Evmarie. *Cézanne in Provence.* New York: Prestel, 1995.

Schopenhauer, Arthur. *The World as Will and Representation.* Translated by E. F. J. Payne. New York: Dover, 1966.

Schroeder, Frederic M. *Form and Transformation in the Philosophy of Plotinus.* Montreal: McGill-Queen's University Press, 1992.

Schönborn, Christoph von. *L'icône de Christ.* Fribourg: Éditions Universitaires Fribourg Suisse, 1976.

Scott, George Ryley. *The History of Torture Throughout the Ages.* 1940. Reprint, London: Kegan Paul, 2004.

Searle, John R. *The Construction of Social Reality.* London: Penguin, 1996.

Seel, Gerhard, ed. *Ammonius and the Seabattle: Texts, Commentary and Essays.* Berlin: Walter de Gruyter, 2001.

Segal, Charles. *Oedipus Tyrannus: Tragic Heroism and the Limits of Knowledge.* New York: Twayne Publishers, 1993.

Seife, Charles. *Zero: The Biography of a Dangerous Idea.* New York: Penguin, 2000.

Seigel, Jerrold. *The Private Worlds of Marcel Duchamp: Desire, Liberation, and the Self in Modern Culture.* Berkeley and Los Angeles: University of California Press, 1995.

Seldes, Lee. *The Legacy of Mark Rothko.* New York: Holt, Rinehart and Winston, 1978.

Seligman, Paul. *The Apeiron of Anaximander: A Study in the Origin and Function of Metaphysical Ideas.* London: Athlone Press, 1962.

Serres, Michel. *Le système de Leibniz et ses modèles mathémathiques.* Paris: Presses Universitaires de France, 1968.

———. *Les origines de la géométrie.* Paris: Flammarion, 1993.

———. *La légende des anges.* Paris: Flammarion, 1993.

Shakespeare William. *The Complete Works of William Shakespeare.* Rev. ed. Edited by Hardin Craig and David Benington. Glenview, Ill.: Scott, Foresman and Company, 1973.

Shell, Susan Meld. *The Embodiment of Reason: Kant on Spirit, Generation, and Community.* Chicago: University of Chicago Press, 1996.

Sigvardson, Joakim. *Immanence and Transcendence in Thomas Pynchon's "Mason and Dixon." A Phenomenological Study.* Stockholm: Almqvist och Wiksell, 2002.

Simmons, W. Sherwin. *Kasimir Malevich's "Black Square" and the Genesis of Suprematism 1907–1915.* New York: Garland, 1981.

Singer, Peter. *Writings on an Ethical Life.* London: Fourth Estate, 2001.

Sklar, Lawrence. *Theory and Truth: Philosophical Critique within Foundational Science.* Oxford: Oxford University Press, 2000.

Smith, Jonathan Z. *To Take Place: Toward Theory in Ritual.* Chicago: University of Chicago Press, 1987.

Smith, Mark S. *The Early History of God: Yahweh and the Other Deities in Ancient Israel.* New York: Harper and Row, 1990.

Snaith, Norman H. *The Book of Job: Its Origin and Purpose.* London: SCM Press, 1968.

Sobel, Dava. *Longitude: The Story of a Lone Genius Who Solved the Great Scientific Problem of His Time.* New York: Walker, 1995.

Sofokles. *Kung Oidipus.* Translated by Björn Collinder. Sockholm: Forum, 1972.

Sollers, Philippe. *Les passions de Francis Bacon.* Paris: Gallimard, 1996.

Sommerbrodt, Ernst. *Die Ebstorfer Weltkarte: Text und Tafelband.* Hannover: Historischen Vereins für Niedersachsen, 1891.

Sophocles. *The Oedipus Cycle.* Translated by Dudley Fitts and Robert Fitzgerald. New York: Harcourt, Brace and World, Harvest Edition, 1949.

———. *Oedipus the King.* Translated by Stephen Berg and Diskin Clay. Oxford: Oxford University Press, 1978.

———. *Oedipus the King.* In *The Three Theban Plays.* Translated by Robert Fagles. Harmondsworth: Penguin, 1984.

Sorabji, Richard, ed. *Philoponus and the Rejection of Aristotelian Science.* London: Duckworth, 1987.

Sparke, Matthew. "Escaping the Herbarium: A Critique of Gunnar Olsson's 'Chiasm of Thought-and-Action.'" *Environment and Planning D: Society and Space* 12 (1994): 207–20.

Spiegel, Shalom. *The Last Trial: On the Legends and Lore of the Command to Abraham to Offer Isaac as a Sacrifice: The Akedah.* Philadelphia: Jewish Publication Society of America, 1967.

Spuches, Giulia de, ed. *Atlante virtuali.* Palermo: Università di Palermo, Laboratorio Geografico, 2002.

Stahl, Nanette. *Law and Liminality in the Bible.* Sheffield: Sheffield Academic Press, 1995.

Stamm, Johan Jacob, with Maurice Edward Andrew. *The Ten Commandments in Recent Research.* London: SCM Press, 1967.

Stanford, W. B. *The Ulysses Theme: A Study in the Adaptability of a Traditional Hero.* 2d ed. Ann Arbor: University of Michigan Press, 1978.

Statten, Henry. *Wittgenstein and Derrida.* Oxford: Basil Blackwell, 1985.

Stead, Christopher. *Divine Substance.* Oxford: Clarendon Press, 1977.

———. *Doctrine and Philosophy in Early Christianity: Arius, Athanasius, Augustine.* Aldershot: Variorum, 2000.

Steadman, Philip. *Vermeer's Camera: Uncovering the Truth Behind the Masterpieces.* Oxford: Oxford University Press, 2001.

Steefel, Lawrence D., Jr. *The Position of Duchamp's Glass in the Development of His Art.* New York: Garland, 1977.

Steiner, George. *Antigones.* Oxford: Clarendon Press, 1984.

Stenius, Erik. *Tankens gryning: En studie över den västerländska filosofins ursprungsskede.* 2d ed., rev. Helsingfors: Söderström, 1975.

———. *Wittgenstein's "Tractatus": A Critical Exposition of its Main Lines of Thought.* Ithaca, N.Y.: Cornell University Press, 1964.

Stevens, Wallace. *Collected Poems.* New York: Vintage, 1990.

Strawson, B. F. *The Bounds of Sense: An Essay on Kant's "Critique of Pure Reason."* London: Methuen, 1966.

Suncliffe, Barry. *The Extraordinary Voyage of Pytheas the Greek, the Man Who Discovered Britain.* London: Allen Lane, 2001.

Svare, Helge. *In situ corpori: Kant's Theory of Experience Interpreted in the Context of His Theory of the Embodied Mind.* Trondheim: Norwegian University of Science and Technology, 2003.

Sylvester, David. *The Brutality of Fact: Interviews with Francis Bacon.* 3d and enlarged ed. London: Thames and Hudson, 1986.

———. *Looking Back at Francis Bacon.* London: Thames and Hudson, 2000.

——— et al. *René Magritte: Catalogue raisonné.* Supplement 5. Antwerp: Fonds Mercator, 1997.

Tarasov, Oleg. *Icon and Devotion: Sacred Spaces in Imperial Russia.* Translated and edited by Robin Milner-Gulland. London: Reaktion Books, 2002.

Taylor, Mark C. *Altarity.* Chicago: University of Chicago Press, 1987.

Thesleff, Holger. *Platon.* Lund: Pegas, 1990.

Thomason, Bill. *God on Trial: The Book of Job and Human Suffering.* Collegeville, Minn.: Liturgical Press, 1997.

Thomson, J. Oliver. *History of Ancient Geography.* Cambridge: Cambridge University Press, 1948.

Tigay, Jeffrey H.: *The Evolution of the Gilgamesh Epic.* Philadelphia: University of Pennsylvania Press, 1982.

Tomasello, Michael. *The Cultural Origins of Human Cognition.* Cambridge, Mass.: Harvard University Press, 1999.

Tomkins, Calvin. *Duchamp: A Biography.* London: Pimlico, 1996.

Toomer, Gerald James. "Ptolemy." *Dictionary of Scientific Biography.* New York: Scribner, 1970–.

Toulmin, Stephen. *The Uses of Argument.* Cambridge: Cambridge University Press, 1964.

Tranströmer, Tomas. *Den stora gåtan.* Stockholm: Bonnier, 2004.

Trías, Eugenio. *La filosofía y su sombra.* Barcelona: Editorial Seix Barral, 1969. Translated by Kenneth Krabbenhoft as *Philosophy and its Shadow.* New York: Columbia University Press, 1983.

———. *Lógica del límite.* Barcelona: Destino, 1991.

———. *Pensar la religión.* Barcelona: Destino, 1997. Translated by Anna-Lena Renqvist as *Att tänka det religiösa.* Stockholm: Thales, 2003.

———. *La razón fronteriza.* Barcelona: Destino, 1999.

———. *Los límites del mundo.* Barcelona: Destino, 2000.

Trible, Phyllis. *Texts of Terror: Literary-Feminist Readings of Biblical Narratives.* Philadelphia: Fortress Press, 1984.

Tufte, Edward R. *Envisioning Information.* Cheshire, Conn.: Graphics Press, 1990.

———. *Visual Explanations: Images and Quantities, Evidence and Narrative.* Cheshire, Conn.: Graphics Press, 1997.

———. *The Visual Display of Quantitative Information.* 2d ed. Cheshire, Conn.. Graphics Press, 2001.

Vaihinger, Hans. *The Philosophy of "As-if": A System of the Theoretical, Practical and Religious Fictions of Mankind.* Translated by C. K. Ogden. London: Routledge and Kegan Paul, 1924.

Vawter, Bruce. *On Genesis: A New Reading.* London: Chapman, 1977.

Vellacott, Philip. *Sophocles and Oedipus.* Ann Arbor: University of Michigan Press, 1971.

Verene, Donald Phillip. *Knowledge of Things Human and Divine: Vico's "Science" and "Finnegans Wake."* New Haven Conn.: Yale University Press, 2003.

Vernant, Jean-Pierre, and Pierre Vidal-Naquet. *Myth and Tragedy in Ancient Greece.* Translated by Janet Lloyd. New York: Zone Books, 1988.

Vico, Giambattista. *The New Science of Giambattista Vico.* Translated by Thomas Goddard Bergin and Max Harold Fisch. Ithaca, N.Y.: Cornell University Press, 1984.

Vitruvius. *The Ten Books on Architecture.* Translated by M. H. Morgan. New York: Dover, 1960.

Wakeman, Mary K. *God's Battle with the Monster: A Study in Biblical Imagery.* Leiden: E. J. Brill, 1973.

Waldman, Diane. *Mark Rothko, 1903–1970.* London: Thames and Hudson, 2001.

Warring, Lennart, and Taina Kantola. *Gilgamesheposet: 'Han som såg Djupet.'* Stockholm: Natur och Kultur. 2001.

Watson, James D. *The Double Helix.* New York: New American Library, 1969.

Weiss, Jeffrey, ed. *Mark Rothko.* New Haven, Conn.: Yale University Press, 1998.

Wénin, A., ed. *Studies in the Book of Genesis: Literature, Redaction and History.* Leuven: Leuven University Press, 2001.

Westenholz, Ulla, and Aage: *Gilgamesh og Enuma Elish: Guder og mennesker i oldtidens Babylon.* København: Spektrum, 1997.

Westrem, Scott D. *The Hereford Map: A Transcription and Translation of the Legends with Commentary.* Turhout: Brepols, 2001.

Wheatley, Paul. *The Pivot of the Four Quarters: A Preliminary Enquiry into the Origins and Character of the Ancient Chinese City.* Chicago: Aldine, 1971.

———. *The Places Where Men Pray Together: Cities in Islamic Lands, Seventh through the Tenth Centuries.* Chicago: University of Chicago Press, 2001.

Wheelock, Arthur K., Jr. *Vermeer and the Art of Painting.* New Haven, Conn.: Yale University Press, 1995.

White, John. *The Birth and Rebirth of Pictorial Space.* 2d ed. London: Faber, 1967.

Whitehead, Alfred North, and Bertrand Russell. *Principia Mathematica.* 2d ed. London: Cambridge University Press, 1925.

Whitman, Cedric. *Homer and the Heroic Tradition.* Cambridge, Mass.: Harvard University Press, 1958.

Whitman, Walt. *Leaves of Grass.* "Death-Bed Edition" of 1881. New York: New American Library, Signet Classic, 1955.

Whybray, Norman. *Job.* Sheffield: Sheffield Academic Press, 1998.

Wildberg, Christian. *John Philoponus' Criticism of Aristotle's Theory of Aether.* Berlin: Walter de Gruyer, 1988.

Wilford, John Noble. *The Mapmakers: The Story of the Great Pioneers in Cartography—From Antiquity to the Space Age.* London: Pimlico, 2002.

Wilke, Jürgen. *Die Ebstorfer Weltkarte.* Bielefeld: Verlag für Regionalgeschichte, 2001.

Wilkinson, Lynn R. *The Dream of an Absolute Language: Emanuel Swedenborg and French Literary Culture.* Albany, N.Y.: State University of New York Press, 1996.

Wirz, Paul. *Kataragama: The Holiest Place in Ceylon.* Colombo: Lakehouse Investments, 1972.

Wittgenstein, Ludwig. *Tractatus Logico-Philosophicus.* Translated by D. F. Pears and B. F. McGuiness. London: Routledge and Kegan Paul, 1961.

———. *Philosophical Investigations.* Translated by G. E. M. Anscombe. Oxford: Basil Blackwell, 1953.

———. *Philosophical Grammar.* Translated by Anthony Kenny. Oxford: Oxford University Press, 1974.

———. *Culture and Value.* Edited by G. H. von Wright in collaboration with Heikki Nyman. Translated by Peter Winch. Oxford: Basil Blackwell, 1980.

———. *Zettel.* 2d ed. Edited by G. E. M. Anscombe and G. H. von Wright. Translated by G. E. M. Anscombe. Oxford: Basil Blackwell, 1981.

Wolde, Ellen, ed. *Job 28: Cognition in Context.* Leiden: E. J. Brill, 2003.

Wolf, Armin. "News on the Ebstorf World Map: Date, Origin, Authorship." In *Géographie du monde au moyen âge et à la renaissance,* edited by Monique Pelletire. Paris: Éditions du CTHS, 1989.

Wolf, Bryan Jay. *Vermeer and the Invention of Seeing.* Chicago: University of Chicago Press, 2001.

Wolf, Hope Nash. *A Study in the Narrative Structure of Three Epic Poems: Gilgamesh, The Odyssey, Beowulf.* New York: Garland, 1987.

Wolfers, David. *Deep Things out of Darkness: The Book of Job. Essays and a New English Translation.* Grand Rapids, Mich: Eerdmans, 1995.

Wolska, Wanda. *La "Topographie chrétienne" de Cosmas Indicopleustès: Theologie et science au VIe siecle.* Paris: Presses Universitaires de France, 1962.

Wood, Beatrice. "Marcel." In *Marcel Duchamp: Artist of the Century,* edited by Rudolf Kuenzli and Francis N. Naumann. Cambridge, Mass.: MIT Press, 1989.

Woodward, David: "Reality, Symbolism, Time, and Space in Medieval World Maps." *Annals of the American Association of Geographers* 75 (1985): 510–21.

———. "Medieval Mappaemundi." In *The History of Cartography,* vol. 1: *Cartography in Prehistoric, Ancient, and Medieval Europe and the Mediterranean,* edited by J. B. Harley and David Woodward. Chicago: University of Chicago Press, 1987.

Wright, Georg Henrik von. *An Essay in Deontic Logic and the General Theory of Action.* Acta Philosophica Fennica, Fasc. XXI. Amsterdam: North-Holland, 1968.

Wright, John Kirtland. *The Geographical Lore of the Time of the Crusades: A Study in the History of Medieval Science and Tradition in Western Europe.* New York: American Geographical Society, 1925.

Yates, Frances A. *The Art of Memory.* Chicago: University of Chicago Press, 1966.

———. *Giardano Bruno and the Hermetic Tradition.* Chicago: University of Chicago Press, 1978.

Zern, Leif. *Älskaren och mördaren.* Stockholm: Alba, 1984.

Zeruneith, Keld. *Træhesten: Fra Odysseus til Sokrates. En bevidsthedshistorie.* København: Gyldendal, 2002.

Zhadova, Larissa A. *Malevich: Suprematism and Revolution in Russian Art 1910–1930.* London: Thames and Hudson, 1982.

Zinn, Grover A., Jr. "Hugh of St.. Victor and the Art of Memory." *Viator* 5 (1974): 211–34.

———. "Hugh of St Victor, Isaiah's Vision, and *De arca Noe.*" In *The Church and the Arts,* edited by Diana Woods. Oxford: Blackwell, 1992.

Zornberg, Avivah Gottlieb. *The Beginning of Desire: Reflections on Genesis.* New York: Doubleday, 1995.

# PROPER NAMES

Naming is the name of the game, fix-points in the mapping of what it means to be human.

*Note:* Page numbers in boldface indicate key references in the text.

# (IN)DEFINITE DESCRIPTIONS

Lines of power thrown onto *mappae* of imagination, crumbled, stained and with an odor distinctly their own. Ironic meanings (in)definitely described as a power which never sits still, language itself a swarm of pathways without beginnings and without ends. Merely an impression that counts, yet another frequency distribution vainly searching for its interpretation. Multiple stories groping the way.

*Note:* Page numbers in boldface indicate key passages in the text.

# ACKNOWLEDGMENTS

Now, towards the end of this long endeavor, it is a great pleasure to acknowledge the help and encouragement from a large number of individuals and institutions. I am especially grateful to Riksbankens Jubileumsfond for generous help in the final production process; T. David Brent and Elizabeth Branch Dyson, the two editors without whose understanding everything would be different; Tommy Westberg and Christina Wernström for help with the illustrations; Richard Allen for his sensitive approach to copyediting; Jan and Torkel Öhman for technical assistance with the computer; and the many librarians (none mentioned and none forgotten) at the Ekonomikum in Uppsala and at the Center for Advanced Study in the Behavioral Sciences, Stanford.

In addition, I wish to thank the following establishments for the use of copyrighted and other material, Alte Pinakothek, München; Bridgeman Art Library, London; Éditions Flammarion, Paris; Galerie Christine et Isy Brachot, Bruxelles; Galleria degli Uffici, Firenze; Henry Holt and Company, New York; Kunsthistorisches Museum, Wien; Los Angeles County Museum, Los Angeles; Metropolitan Museum of Art, New York; Moderna museet, Stockholm; Musée National Message Biblique Marc Chagall, Nice; Musei Vaticani, Città del Vaticano; Museum Gustavianum, Uppsala; Museum of Modern Art, New York; Nationalmuseum, Stockholm; Philadelphia Museum of Art, Philadelphia; Princeton University Press, Princeton; The Solomon R. Guggenheim Museum, New York; Staatliche Kunstsammlung, Weimar; The State Russian Museum, St. Petersburg; The State Tretyakov Gallery, Moscow; Uppsala University Library, Uppsala; Vintage and Anchor Books, New York.